AF412269

INTERNATIONAL UNION OF CRYSTALLOGRAPHY
CRYSTALLOGRAPHIC SYMPOSIA

Molecular Structure:
Chemical Reactivity and
Biological Activity

Edited by

John J. Stezowski
Universitat Stuttgart

Co-edited by

Jin-Ling Huang
Fuzhou University

and

Mei-Cheng Shao
Beijing University

INTERNATIONAL UNION OF CRYSTALLOGRAPHY
OXFORD UNIVERSITY PRESS
1988

Oxford University Press, Walton Street, Oxford OX2 6DP
Oxford New York Toronto
Delhi Bombay Calcutta Madras Karachi
Petaling Jaya Singapore Hong Kong Tokyo
Nairobi Dar es Salaam Cape Town
Melbourne Auckland
and associated companies in
Berlin Ibadan

Oxford is a trade mark of Oxford University Press

Published in the United States
by Oxford University Press, New York

© *The various contributors listed on pp. xiii–xxvi, 1988*

British Library Cataloguing in Publication Data
Molecular structure
1. Molecules. Structure
I. Stezowski, John J.   II. Huang, Jin-
Ling   II. Shoa, Mei-Cheng
IV. International Union of Crystallography
V. Series
539'.12
ISBN 0–19–855279–3

Library of Congress Cataloging in Publication Data
Molecular structure: chemical reactivity and biological activity
edited by John J. Stezowski, Jin-Ling Huang, and Mei-Cheng Shao.
p.   cm.—(International Union of Crystallography
crystallographic symposia; 2)
Includes bibliographies and index.
1. Structure-activity relationships (Biochemistry)—Congresses.
2. Molecular structure—Congresses. 3. Reactivity (Chemistry)—
Congresses. 4. Biomolecules—Reactivity—Congresses.
I. Stezowski, John J. II. Huang, Jin-Ling. III. Shao, Mei-Cheng.
IV. International Union of Crystallography. V. Series.
QP517.S85M65 1988   574.8'8—dc19   87–35007

ISBN 0 19 855279 3

Printed in Great Britain
at the University Printing House, Oxford
by David Stanford
Printer to the University

PREFACE

Most physical scientists would agree that a major goal in science today is to gain an understanding of physical, chemical and biological processes at atomic and molecular levels, that is, in terms of structure. Structure can be characterized in many ways, from a very simple structural formula to complex wave functions that can be used to provide mathematical models for atomic and molecular structure.

A major interest in chemistry, physics and biology is to correlate molecular structure with chemical reactivity and biological activity. A variety of approaches is taken to achieve this end. Equally diverse are the systems studied. Nonetheless investigators in the field share many common interests and stand to profit greatly from information exchange.

The international symposium on "Molecular Structure: Chemical Reactivity and Biological Activity", held in Beijing, People's Republic of China from 15 to 21 September, 1986, brought together numerous experts active in this field. Many subdisciplines were represented: mineralogy, chemistry, biology and physics, to name a few. Different approaches, both experimental and theoretical, were described and new results were reported.

This volume reproduces the character of the symposium. It contains papers reflecting different philosophies for establishing an understanding of chemical reactivity and biological activity in terms of "molecular structure". Careful consideration of the results achieved by investigators using different methods or studying different systems should provide readers with valuable stimulation. In addition, the book provides the scientists with an overview of the status of research in several areas of current interest.

John J. Stezowski
Jin-Ling Huang
Mei-Cheng Shao

CONTENTS

xi

CONTRIBUTORS

Israel Agranat
Department of Organic
    Chemistry
The Hebrew University of
    Jerusalem
Jerusalem 91904
Israel

David Ajò
Istituto di Chimica e
    Technologia dei
    Radioelementi del CNR
Padova
Italy

D. Altermatt
Institute for Materials
    Research
McMaster University
1280 Main Street West
Hamilton
Ontario L8S 4M1
Canada

K. Angermund
Max-Planck-Institut für
    Kohlenforschung
Kaiser-Wilhelm Platz 1
D-4330 Mülheim a.d. Ruhr
Federal Republic of
    Germany

Ettore Appella
National Cancer Institute
National Institutes of
    Health
Bethesda
Maryland 20205
USA

A. Ballio
Dipartimento Biochimico
    dell'Università La
    Sapeinza
Rome
Italy

J. Ellis Bell
Department of Biochemistry
University of Rochester
Medical Center
Rochester
New York 14642
USA

Jan-Eric Berg
Department of Structural
    Chemistry
Arrhenius Laboratory
University of Stockholm
S-106 91 Stockholm
Sweden

Renzo Bertoncello
Dipartimento di Chimica
    Inorganica
    Metallorganica ed
    Analitica
Università
Padova
Italy

Clive E. Briant
School of Chemistry and
    Chemical Technology
University of Bradford
Bradford BD7 1DP
England

S.M. Bradley
Institute for Materials
Research
McMaster University
1280 Main Street West
Hamilton
Ontario L8S 4M1
Canada

I.D. Brown
Institute for Materials
    Research
McMaster University
1280 Main Street West
Hamilton
Ontario L8S 4M1
Canada

Vilma Busetti
Dipartimento di Chimica
    Organica
Università
Padova
Italy

Zhao Cai
Department of Chemistry
Peking University
Beijing
China

Maurizio Casarin
Istituto di Chimica e
    Technologia dei
    Radioelementi del CNR
Padova
Italy

S. Castellano
Dipartimento Chimico
    del'Università
Parma
Italy

S. Cerrini
Istituto di Strutturistica
    Chimica
`G. Giacomello'-C.N.R.
C.P. 10
00016 Monterotondo
Stazione
(Rome)
Italy

P. Chinna Chenchaiah
Department of Chemistry
Brock University
Saint Catharines
Ontario L2S 3A1
Canada

Connie G. Chidester
Physical and Analytical
    Chemistry
The Upjohn Company
Kalamazoo
Michigan 49001
USA

David W. Christianson
Gibbs Chemical
    Laboratories
Department of Chemistry
Harvard University
Cambridge
Massachusetts 02138
USA

Vivian Cody
Medical Foundation of
    Buffalo Inc.
73 High Street
Buffalo
New York 14203
USA

Shmuel Cohen
Department of Inorganic
   and Analytical
   Chemistry
The Hebrew University of
   Jerusalem
Jerusalem 91904
Israel

Ingeborg Csöregh
Department of Structural
   Chemistry
Arrhenius Laboratory
University of Stockholm
S-106 91 Stockholm
Sweden

Claude C.J. Culvenor
Division of Animal Health
CSIRO
Parkville
Victoria
Australia 3052

Mátyás Czugler
Central Research Institute
   for Chemistry of the
   Hungarian Academy of
   Sciences
P.O. Box 17
1525 Budapest
Hungary

C.J. De Ranter
Laboratorium voor
   Analytische Chemie en
   Medicinale
   Fysicochemie
Katholieke Universiteit
   Leuven
Instituut voor
   Farmaceutische
   Wetenschappen
Van Evenstraat 4
B-3000 Leuven
Belgium

Jürgen Deuter
Organisch-Chemisches
   Institut der
   Universität Heidelberg
Im Neuenheimer Feld 270
D-6900 Heidelberg
Federal Republic of
   Germany

W.L. Duax
Medical Foundation of
   Buffalo Inc.
73 High Street
Buffalo
New York 14203
USA

David J. Duchamp
Physical and Analytical
   Chemistry
The Upjohn Company
Kalamazoo
Michigan 49001
USA

Jack D. Dunitz
Organic Chemistry
   Laboratory
Swiss Federal Institute of
   Technology
ETH-Zentrum
CH-8092 Zürich
Switzerland

G.P. Elliott
School of Chemistry
University of Bristol
Cantock's Close
Bristol
England BS8 1TS

A. Evidente
Dipartimento di Chimica
   Organica e Biologia
   dell'Università
Napoli
Italy

Hai-fu Fan
Institute of Physics
Academia Sinica
Beijing
China

Felix Frolow
Department of Structural
    Chemistry
The Weizmann Institute of
    Science
Rehovot 76100
Israel

Heng Fu
Institute of Chemistry
Academia Sinica
Beijing
China

Takaji Fujiwara
Faculty of Science
Shimane University
Nishikawazucho
Matsue
Shimane 690
Japan

Elena Gaggelli
Dipartimento di Chimica
Università di Siena
Pian dei Mantellini
44-53100 Siena
Italy

A. Gavezzotti
Dipartimento di Chimica
    Fisica ed
      Elettrochimica e
        Centro CNR
Università di Milano
21033 Milano
Italy

Rodney J. Geue
Research School of
    Chemistry
The Australian National
    University
Canberra
ACT 2601
Australia

Jenny P. Glusker
Institute for Cancer
    Research
Fox Chase Cancer Center
7701 Burholme Avenue
Philadelphia
Pennsylvania 19111
USA

R. Goddard
Max-Planck-Institut für
    Kohlenforschung
Kaiser-Wilhelm Platz 1
D-4330 Mülheim a.d. Ruhr
Federal Republic of
    Germany

Barry M. Goldstein
Department of Biophysics
University of Rochester
    Medical Center
Rochester
New York 14642
USA

Carlo N. Gramaccioli
Department of Earth
    Sciences
University of Milan
Via Botticelli 23
I-20133 Milano
Italy

Gaetano Granozzi
Dipartimento di Chimica
    Inorganica
    Metallorganica ed
    Analitica
Università
Padova
Italy

Arthur Greenberg
Chemistry Division
New Jersey Institute of
    Technology
Newark
New Jersey 07102
USA

J.F. Griffin
Medical Foundation of
    Buffalo Inc.
73 High Street
Buffalo
New York 14203
USA

Yue Guan
Department of Chemistry
Peking University
Beijing
China

E. Hadjoudis
Chemistry Department
Nuclear Research Center
    "Demokritos"
Aghia Paraskevi
Attiki
Greece

Herbert Hauptman
Medical Foundation of
    Buffalo Inc.
73 High Street
Buffalo
New York 14203
USA

Frank C. Hawthorne
Department of Geological
    Sciences
The University of Manitoba
Winnipeg
Manitoba
Canada R3T 2N2

Rita Grønbæk Hazell
Department of Chemistry
Aarhus University
DK-8000 Aarhus C
Denmark

Miriam Hirschberg
Department of Chemical
    Physics
The Weizmann Institute of
    Science
Rehovot 76100
Israel

Dorothy Hodgkin
Crab Mill
Ilmington
Shipston-on-Stour
Warwickshire CV36 4LE
England

Herbert L. Holland
Department of Chemistry
Brock University
Saint Catharines
Ontario L2S 3A1
Canada

Håkon Hope
Department of Structural
 Chemistry
The Weizmann Institute of
 Science
Rehovot 76100
Israel
and
Department of Chemistry
University of California
Davis
California 95616
USA

Judith A.K. Howard
School of Chemistry
University of Bristol
Cantock's Close
Bristol
England BS8 1TS

J.L. Huang
Fuzhou University
Fuzhou
China

J.Q. Huang
Fuzhou University
Fuzhou
China

M.D. Huang
Fujian Institute of
 Research on the
 Structure of Matter
Academia Sinica
Fuzhou
China

G. Huttner
Anorganisch-Chemisches
 Institut
Universität Heidelberg
Im Neuenheimer Feld 270
D-6900 Heidelberg
Federal Republic of
 Germany

Hermann Irngartinger
Organisch-Chemisches
 Institut der
 Universität Heidelberg
Im Neuenheimer Feld 270
D-6900 Heidelberg
Federal Republic of
 Germany

Reiner Jahn
Organisch-Chemisches
 Institut der
 Universität Heidelberg
Im Neuenheimer Feld 270
D-6900 Heidelberg
Federal Republic of
 Germany

Palle Jørgensen
Department of Chemistry
Aarhus University
DK-8000 Aarhus C
Denmark

Derry W. Jones
School of Chemistry and
 Chemical Technology
University of Bradford
Bradford BD7 1DP
England

Leemor Joshua-Tor
Department of Structural
 Chemistry
The Weizmann Institute of
 Science
Rehovot 76100
Israel

Neville R. Kallenbach
Department of Biology
University of Pennsylvania
Philadelphia
Pennsylvania 19104
USA

Dietmar Kallfass
Organisch-Chemisches
    Institut der
    Universität Heidelberg
Im Neuenheimer Feld 270
D-6900 Heidelberg
Federal Republic of
    Germany

Alajos Kálmán
Central Research Institute
    for Chemistry of the
    Hungarian Academy of
    Sciences
P.O. Box 17
1525 Budapest
Hungary

Isabella L. Karle
Laboratory for the
    Structure of Matter
Naval Research Laboratory
Code 6030
Washington
D. C. 20375-5000
USA

C. Krüger
Max-Planck-Institut für
    Kohlenforschung
Kaiser-Wilhelm Platz 1
D-4330 Mülheim a.d. Ruhr
Federal Republic of
Germany

Robert Langridge
Department of
    Pharmaceutical
    Chemistry
Computer Graphics
    Laboratory
University of California
San Francisco
California 94143-0446
USA

Finn Krebs Larsen
Department of Chemistry
Aarhus University
DK-8000 Aarhus C
Denmark

G. Le Bas
Laboratoire de Physique
E.R. 180
Centre Pharmaceutique
92290 Chatenay Malabry
France

Bente Lebech
Department of Physics
Research Establishment
    Risø
DK-4000 Roskilde
Denmark

Renli Li
Department of Medicinal
    Chemistry
Beijing Medical University
Beijing
China

X.T. Lin
Fujian Institute of
    Research on the
    Structure of Matter
Academia Sinica
Fuzhou
China

Y.H. Lin
Fujian Institute of
    Research on the
    Structure of Matter
Academia Sinica
Fuzhou
China

Joel F. Liebman
Department of Chemistry
University of Maryland
Baltimore County
Catonsville
Maryland 21228
USA

Janusz S. Lipkowski
Institute of Physical
    Chemistry
Polish Academy of Science
Kasprzaka 44/52
01 224 Warszawa
Poland

William N. Lipscomb
Gibbs Chemical
    Laboratories
Department of Chemistry
Harvard University
Cambridge
Massachusetts 02138
USA

Weigin Liu
Department of Medicinal
    Chemistry
Beijing Medical University
Beijing
China

Karen L. Lobb
Department of Chemistry
University of Louisville
Louisville
Kentucky 40292
USA

J.X. Lu
Fujian Institute of
    Research on the
    Structure of Matter
Academia Sinica
Fuzhou
China

S.F. Lu
Fujian Institute of
    Research on the
    Structure of Matter
Academia Sinica
Fuzhou
China

Maureen F. Mackay
Department of Chemistry
La Trobe University
Bundoora
Victoria
Australia 3083

Thomas C.W. Mak
Department of Chemistry
The Chinese University of
    Hong Kong
Shatin
New Territories
Hong Kong

Vincenzo Malatesta
Istituto Guido Donegani
Centro Ricerche Novara
via G. Fauser 4
28100 Novara
Italy

Lj. Manojlović-Muir
Department of Chemistry
University of Glasgow
Glasgow G12 8QQ
Great Britain

Victor E. Marquez
Medicinal Chemistry
    Section
Laboratory of Pharmacology
    and Experimental
    Therapeutics
National Cancer Institute
National Institutes of
    Health
Bethesda
Maryland 20205
USA

L. Massa
Department of Physics and
    Astronomy
Hunter College
The City University of
    New York
695 Park Avenue
New York
New York 10021
USA

Josè A. Mayoral
Departamento de Quimica
    Organica
Universidad
Zaragoza
Spain

Edgar F. Meyer, Jr.
Biographics Laboratory
Department of Biochemistry
    and Biophysics
Texas A&M University
College Station
Texas 77843-2128
USA

I. Moustakali-Mavridis
Chemistry Department
Nuclear Research Center
    "Demokritos"
Aghia Paraskevi
Attiki
Greece

Benito Munoz
Department of Chemistry
Brock University
Saint Catharines
Ontario L2S 3A1
Canada

Ven L. Narayanan
National Cancer Institute
Landow Building
Room 5C-18
Bethesda
Maryland 20892
USA

Luigi R. Nassimbeni
Department of Physical
    Chemistry
University of Cape Town
Rondebosch 7700
South Africa

Margaret L. Niven
Department of Physical
    Chemistry
University of Cape Town
Rondebosch 7700
South Africa

C.M. Nunn
School of Chemistry
University of Bristol
Cantock's Close
Bristol
England BS8 1TS

Yuji Ohashi
Department of Chemistry
Ochanomizu University
Otsuka
Bunkyo-ku
Tokyo 112
Japan

Henricus C. Ottenheijm
Department of Organic
    Chemistry
University of Nijmegen
The Netherlands

Loraine M. Pschigoda
Physical and Analytical
    Chemistry
The Upjohn Company
301 Henrietta Street
Kalamazoo
Michigan 49007
USA

Minxie Qian
Institute of Chemistry
Academia Sinica
Beijing
China

R. Radhakrishnan
Biographics Laboratory
Department of Biochemistry
    and Biophysics
Texas A&M University
College Station
Texas 77843-2128
USA

G. Randazzo
Dipartimento di Chimica
    Organica e Biologia
    dell'Università
Napoli
Italy

Graziella Ranghino
Istituto Guido Donegani
Centro Ricerche Novara
via G. Fauser 4
28100 Novara
Italy

Wolfgang Reimann
Organisch-Chemisches
    Institut der
Universität Heidelberg
Im Neuenheimer Feld 270
D-6900 Heidelberg
Federal Republic of
    Germany

Mary Frances Richardson
Department of Chemistry
Brock University
Saint Catharines
Ontario L2S 3A1
Canada

Roland K. Robins
Nucleic Acid Research
    Institute
Costa Mesa
California 92626
USA

Miriam Rossi
Department of Chemistry
Vassar College
Box 484
Poughkeepsie
New York 12601
USA

N. Rysanek
Laboratoire de Physique
E.R. 180
Centre Pharmaceutique
92290 Chatenay Malabry
France

Nereo Sacchi
Istituto Guido Donegani
Centro Ricerche Novara
via G. Fauser 4
28100 Novara
Italy

Mark A. Saper
Department of Structural
    Chemistry
The Weizmann Institute of
    Science
Rehovot 76100
Israel

Alan M. Sargeson
Research School of
    Chemistry
The Australian National
    University
Canberra
ACT 2601
Australia

Yoshio Sasada
Department of Life
    Sciences
Tokyo Institute of
    Technology
Nagatsuta
Midori-ku
Yokohama 227
Japan

Nadrian C. Seeman
Department of Biological
    Sciences
State University of New
    York at Albany
Albany
New York 12222
USA

A Segre
Istituto di Strutturistica
    Chimica
`G. Giacomello'-C.N.R.
C.P. 10
00016 Monterotondo
Stazione
(Rome)
Italy

Wolfgang Seidel
EP-FE-EA
Gebäude 431
Beyer AG
Pharmaceutical Research
    Center
D-5600 Wuppertal
Federal Republic of
    Germany

M.Y. Shang
Fujian Institute of
    Research on the
    Structure of Matter
Academia Sinica
Fuzhou
China

Julian D. Shaw
School of Chemistry and
    Chemical Technology
University of Bradford
Bradford BD7 1DP
England

Gil Shoham
Department of Inorganic
    and Analytical
    Chemistry
The Hebrew University of
    Jerusalem
Jerusalem 91904
Israel

M. Simonetta
Dipartimento di Chimica
    Fisica ed
    Elettrochimica e
    Centro CNR
Università di Milano
21033 Milano
Italy

Harkishan Singh
Department of
    Pharmaceutical
    Sciences
Panjab University
Chandigarh 160014
India

G.D. Smith
Medical Foundation of
    Buffalo Inc.
73 High Street
Buffalo
New York 14203
USA

F.G.A. Stone
School of Chemistry
University of Bristol
Cantock's Close
Bristol
England BS8 1TS

Yu.T. Struchkov
A.N. Nesmeyanov Institute
    of Organoelement
    Compounds
Academy of Sciences of the
    USSR
28 Vavilov St.
Moscow 117813
USSR

Rachel Michal Suissa
Department of Organic
    Chemistry
The Hebrew University of
    Jerusalem
Jerusalem 91904
Israel

Joel L. Sussman
Department of Structural
    Chemistry
The Weizmann Institute of
    Science
Rehovot 76100
Israel

Paul A. Sutton
Medical Foundation of
    Buffalo Inc
73 High Street
Buffalo
New York 14203
USA

Lori Takahashi
Biographics Laboratory
Department of Biochemistry
    and Biophysics
Texas A&M University
College Station
Texas 77843-2128
USA

Youqi Tang
Department of Chemistry
Peking University
Beijing
China

Robert Thomas
Department of Chemistry
Brookhaven National
    Laboratories
New York
New York 11973
USA

Enzo Tiezzi
Dipartimento di Chimica
Università di Siena
Pian dei Mantellini
44-53100 Siena
Italy

Ken-ichi Tomita
Faculty of Pharmaceutical
    Sciences
Osaka University
Yamadaoka
Suita
Osaka 565
Japan

Jan P. Tollenaere
Department of Theoretical
    Medicinal Chemistry
Janssen Pharmaceutica
Research
    Laboratories
B-2340 Beerse
Belgium

Camillo Tosi
Istituto Guido Donegani
Centro Ricerche Novara
via G. Fauser 4
28100 Novara
Italy

G. Tsoucaris
Laboratoire de Physique
E.R. 180
Centre Pharmaceutique
92290 Chatenay Malabry
France

A. Uchida
Department of Life Science
Faculty of Science
Tokyo Institute of
    Technology
Nagatsuta
Midori-ku
Yokohama 227
Japan

Gianni Valensin
Dipartimento di Chimica
Università di Siena
Pian dei Mantellini
44-53100 Siena
Italy

Albert Van Donkelaar
Division of Protein
    Chemistry
CSIRO
Parkville
Victoria
Australia 3052

F. Villain
Laboratoire de Physique
E.R. 180
Centre Pharmaceutique
92290 Chatenay Malabry
France

Andrea Vittedini
Istituto di Chimica e
    Technologia dei
    Radioelementi del CNR
Padova
Italy

Inger Wahlberg
Research Department
Swedish Tobacco Company
P.O. Box 17007
S-104 62 Stockholm
Sweden

Shuyu Wang
Department of Medicinal
    Chemistry
Beijing Medical University
Beijing
China

Edwin Weber
Institut für Organische
    Chemie und Biochemie
    der Universität Bonn
Gerhard-Domagk-Strasse 1
D-5300 Bonn 1
Federal Republic of
    Germany

William J. Welsh
Department of Chemistry
University of Saint Louis
Saint Louis
Missouri 63121
USA

Donald E. Williams
Department of Chemistry
University of Louisville
Louisville
Kentucky 40292
USA

H.G. Wittmann
Max-Planck-Institut for
    Molecular Genetics
D-1000 Berlin
Federal Republic of
    Germany

Xiaojie Xu
Department of Chemistry
Peking University
Beijing
China

A. Yamano
Department of Life Science
Faculty of Science
Tokyo Institute of
    Technology
Nagatsuta
Midori-ku
Yokohama 227
Japan

A. Yonath
Department of Structural
    Chemistry
Weizmann Institute of
    Science
76100-Rehovot
Israel
and
Max-Planck Research Unit
D-2000 Hamburg
Federal Republic of
    Germany

Qi-Tai Zheng
Institute of Materia
    Medica
Chinese Academy of Medical
    Sciences
Beijing
China

Yao Zhengui
Department of Chemistry
Peking University
Beijing
China

H.H. Zhuang
Fujian Institute of
    Research on the
    Structure of Matter
Academia Sinica
Fuzhou
China

1. Crystallographic studies of the flexibility of hormones, drugs and antibiotics and their biological activity

W. L. Duax, J. F. Griffin and G. D. Smith

1. INTRODUCTION

An X-ray crystallographic determination provides a highly accurate picture of molecular geometry in a specific solid state environment. Studying the same compound in different crystal forms provides additional information on the molecular flexibility of a compound (Duax et al, 1975; Griffin et al, 1984). For most uncharged organic molecules such as steroids, a structure observed in the solid state is at or very near a local minimum energy conformation. If the energy of the global minimum is 2-3 kcal/mol lower than that of any metastable state, it is highly probable that a crystal incorporating the minimum energy conformation will be formed preferentially. If two or more conformations of a flexible molecule are of nearly equal energy they will often form crystals incorporating both conformers either as crystallographically independent molecules with dissimilar environments (Kilbourn et al, 1970) or as disordered molecules with partial occupancy in a single site (Rohrer et al, 1980). Because the active forms of drugs and hormones may not necessarily be low energy forms it is important to have reliable information on the conformation and relative energies of molecules that may not be readily crystallizable. Such information can be derived from solution and gas phase

measurements and theoretical calculations. X-ray
crystallographic studies provide a useful guide to the
interpretation of data from these and other sources (Duax et
al, 1981).

2.   ESTROGEN STRUCTURE AND RECEPTOR BINDING

On the basis of an examination of models of estradiol and
the potent synthetic estrogen diethylstilbestrol (DES) (Fig.
1a and b), Keasling and Schueler (1950) proposed that the
structural requirements for estrogenicity were a specific
distance (14.5 Å) between two hydroxyl groups separated by a
flat hydrophobic region. Subsequent X-ray crystallographic
studies have revealed that the distances between the terminal
oxygen groups in estradiol (Busetta et al, 1971) and
diethylstilbestrol (Busetta et al, 1973) are 10.9 and 12.1 Å
(Fig. 2a and b), respectively, and that the molecules
(particularly DES) are not as flat as their conventional
chemical drawings might suggest. The small but significant
difference between the oxygen-oxygen distances in these
structures cannot be overcome by molecular flexibility. The
fairly rigid fused ring system of estradiol and the central
double bond in DES prevent the oxygen from getting more than
0.1Å closer to each other. If a specific distance between two
hydroxyl groups is essential for estrogenic activity the 1.2 Å
disparity between this distance in estradiol and DES could
indicate that a water molecule plays a significant role in
acting as a link between estradiol and the receptor (Duax et
al, 1980). The distance between O(3) and a hydrogen-bonded
oxygen in estradiol hydrate is 12.1 Å, identical to the
distance in DES (Fig. 3).

Although the oxygen-oxygen distance in DES is nearly
fixed, conformational variation in the overall shape of the
molecule is possible. Five distinct crystallographic
observations of DES have resulted from studies of an anhydrous
(Weeks et al, 1970) and three solvated crystal forms (Busetta

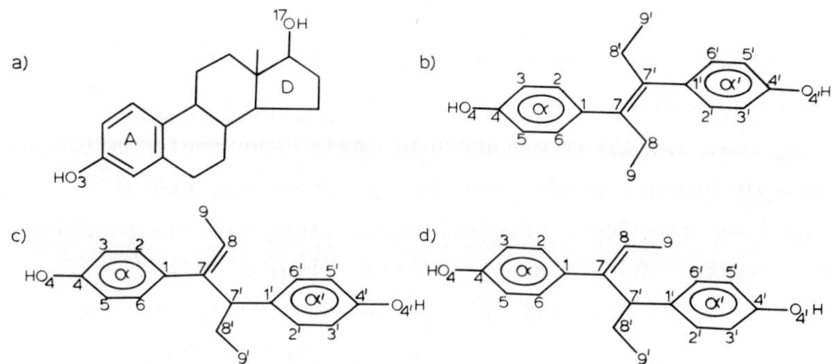

Figure 1. Chemical diagrams, atomic numbering and ring
identification for (a) estradiol, (b) diethylstilbestrol, DES,
(c) Z-pseudo diethylstilbestrol (ZPD) and (d) E-pseudo
diethylstilbestrol (EPD).

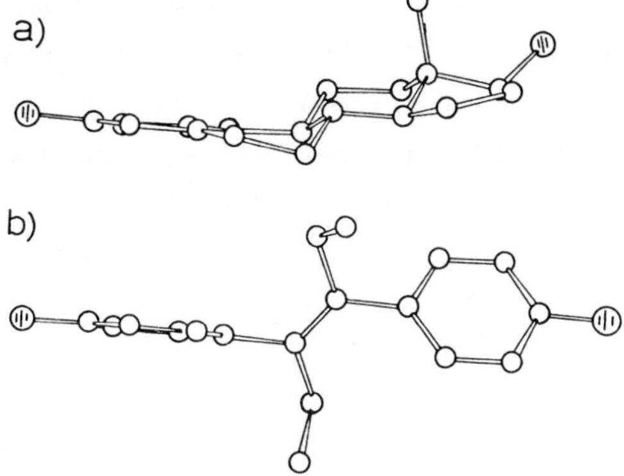

Figure 2. Comparison of the overall conformation of (a)
estradiol and (b) DES (solvated), viewed parallel to the plane
of the A-ring in estradiol and one of the phenyl rings in DES.

et al, 1973). The molecules in the solvated crystal forms are
significantly different from the anhydrous conformer, and
nearly identical to one another. When the molecule from the
anhydrous crystal is compared with one of the molecules from
the solvated crystals (Fig. 4), half of the molecule is seen
to be nearly identical and the other half differs by a
rotation of the ring. These conformations are not
particularly flat. The dihedral angles between the phenyl

rings are 0° in the anhydrous form and 60° to 70° in the
solvated molecules. The molecular mechanics program MM2p
(Allinger et al, 1983) was used to refine the conformational
isomers of DES afforded by the X-ray determinations of
solvated and anhydrous crystal forms. The relative energies
calculated for the anhydrous and the solvated molecules are
3.1 kcal/mol and 3.4 kcal/mol, and the calculated dipole
moments are 0.0 and 2.2, respectively. The dipole moment of
the molecule may well have a significant influence on
complementarity of hormone-receptor interaction. The
calculated dipole for estradiol in its solid state
conformation is 2.3. The solid-state observation and the
energy calculations of the two forms argues in favor of their
being of comparable (near minimum) energy. Molecular
environment will determine which of the two forms
predominates. It is unlikely that the two conformations
compete equally for the receptor site or are equally effective
at eliciting hormonal response. The receptor may select one
of the conformers. Hospital et al. (1975) have concluded on
the basis of hydrogen-bonding patterns and overall
conformational features that it is the conformation observed
in the solvated crystals that is responsible for both receptor
oxygen groups in estradiol (Busetta et al, 1971) and
diethylstilbestrol (Busetta et al, 1973) are 10.9 and 12.1 Å

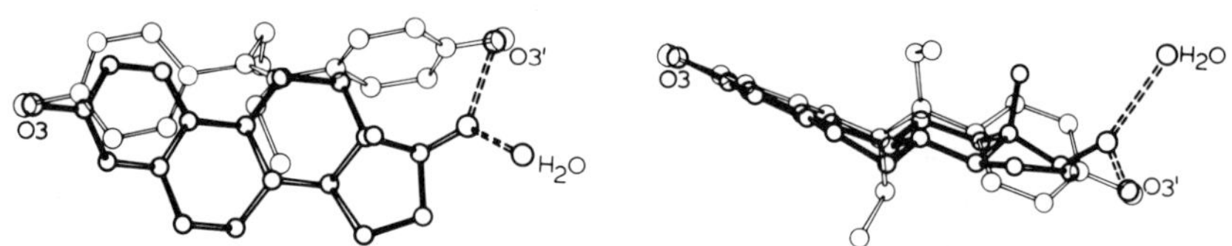

Figure 3. Superposition drawing of solvated estradiol and
DES, maximizing relative positioning of hydrophilic groups and
hydrophobic bulk. 03' is the hydroxyl group of an adjacent
molecule in crystals of estradiol. Hydrogen bonds are
indicated as broken bonds.

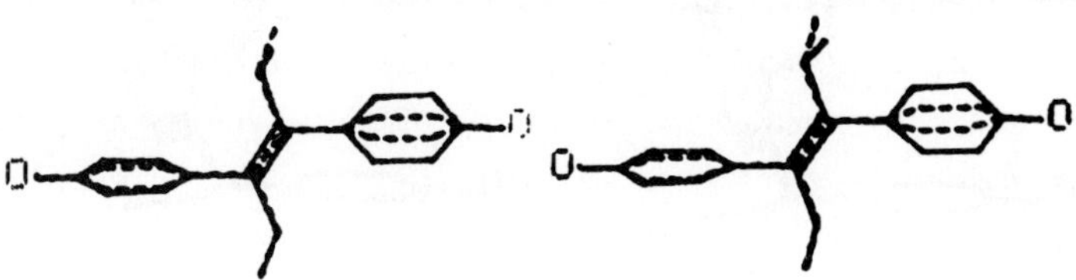

Figure 4. Stereo comparison of the conformation of DES in the
solvated (solid) and anhydrous (dashed) forms.

(Fig. 2a and b), respectively, and that the molecules
(particularly DES) are not as flat as their conventional
moments of estradiol and the DES conformer seen in the
solvated crystals also supports the contention that it is this
DES conformer that mimics estradiol in its binding to the
estrogen receptor.

A diethylstilbestrol analog, called pseudo-DES, in which
the double bond is between C(7) and C(8) rather than C(7) and
C(7') can exist in Z and E configurations (Fig. 1c and d).
Biological testing revealed that while both isomers retained
appreciable but different affinity for the estrogen receptor,
the Z isomer (ZPD) has twice the uterotropic activity of the E
(EPD) isomer (Korach et al, 1983). Since a significant
conformational difference between the two might account for
the observed activity difference, the X-ray crystal structure
analyses of the two forms and the 1-OH derivative of ZPD* were
undertaken. The results revealed striking similarities in the
observed conformation despite differences in composition,
connectivity, hydrogen bonding and crystal packing. Four
torsion angles fully define the overall conformation of these
molecules. The range in variation of three of these four

*1-OH-ZPD is according to the numbering of Figure 1, 9-OH-ZPD.

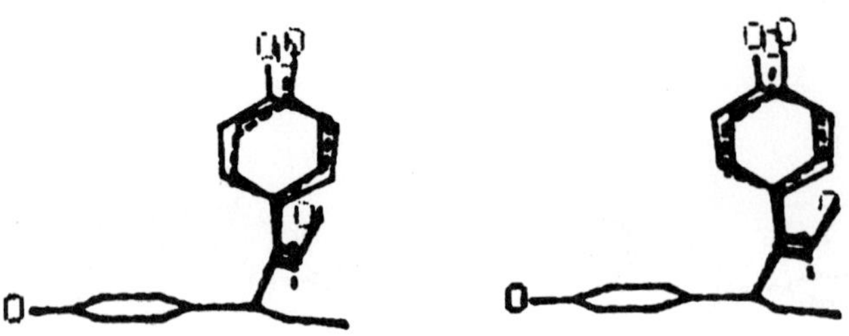

Figure 5.   Stereo view of the superposition of ZPD, EPD, and
1-OH-ZPD, illustrating the similarity in their
crystallographically observed conformations.

torsion angles is 5° or less. The only significant variation
is in the C(2)-C(1)-C(7)=C(8) torsion angle which differs by
14.4° between EPD and ZPD. The difference is a direct
consequence of the stereochemical differences between the E
and Z configurations. All three structures are observed to
have a bent conformation in which the phenyl rings are
oriented at right angles to one another (Fig. 5). This
conformation is completely unlike the conformations of
estradiol and DES.

When the crystallographically observed conformations of
EPD, ZPD and 1-OH-ZPD were subjected to energy minimization
the structures retained their overall conformation, providing
persuasive evidence that the molecule is unperturbed by the Z
to E variation or the addition of a hydroxyl group and the
associated change in crystal packing and intermolecular
contacts.

The consistency of the crystallographic results suggests
that the bent form is the energy minimum conformation for ZPD,
EPD and 1-OH-ZPD. However, previous studies indicate that
this conformation is unlikely to be compatible with any
significant degree of estrogenic biological activity. For
this reason it became important to determine whether an
extended DES-like conformation might constitute a metastable
state capable of inducing estrogenic biological response. The

crystallographically observed structures of ZPD and EPD were
transformed to the DES-like extended conformation and
subjected to energy minimization. As a result of extremely
close contacts between the hydrogens on C(9) and the α ring in
EPD (Fig. 6a), the extended conformation is stereochemically
untenable and the molecule refined to a bent conformation,
resembling in overall shape the crystallographically observed
structure. In contrast to this, the extended conformation of
the more active Z isomer does not incorporate intolerable
nonbonded interactions and refinement indicates that a
metastable state exists which resembles the DES conformation
(Fig. 6b).

The relative energies of the metastable extended form of
the active ZPD isomer and the minimum energy form observed in
the crystal structure are -3.7 kcal/mol and -5.6 kcal/mol
respectively. These relative energies are consistent with
observed differences in activity; ZPD is ~ 10% as active as
DES (Korach et al, 1983) the latter is constrained to the
active extended form 100% of the time. These results are also
consistent with the A-ring binding/D ring acting model (Duax,

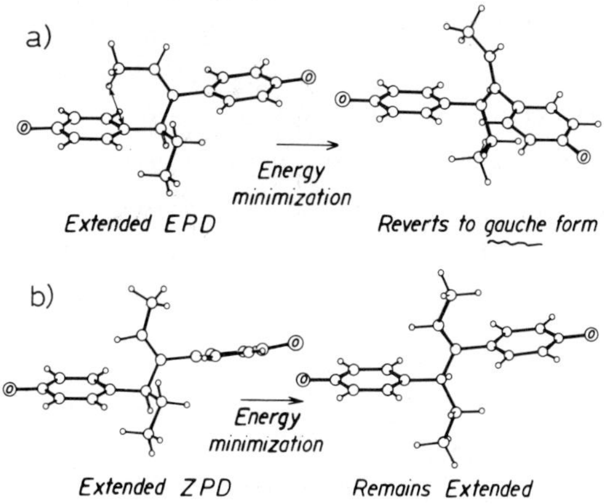

Figure 6. When theoretical DES-like extended conformers of
(a) EPD and (b) ZPD were used as the starting point for energy
minimization, EPD reverted to a bent conformation but ZPD
retained an extended conformation.

et al, 1984) and are a good example of the combined use of X-ray structure determination and molecular mechanics calculations in understanding estrogen receptor binding and activity.

3.   **METHADONE CONFORMATION, RECEPTOR BINDING AND ACTIVITY**

The activity of diphenylpropylamine analgetics is highly dependent upon the chirality at atomic positions C(5) and C(6) (Fig. 7). Although reports in the literature vary in their estimates of relative potency [15-18], (6R)-methadone, (5S)-isomethadone, and (5R,6R)-erythro-5-methylmethadone are found to have potencies comparable to morphine while (5S,6S)-erythro-5-methylmethadone is from 10 to 70 times greater. In contrast, (6S)-methadone and (5R)-isomethadone [19] have less than one-tenth the activity of their respective isomers. Surprisingly the racemate of threo-5-methylmethadone is totally devoid of activity despite the presence of the combined chiralities of the most active methadone and isomethadone molecules in its (5S,6R)-isomer. Because these molecules are potentially flexible, considerable effort has been directed toward determining the conformation or conformations responsible for opioid receptor binding. Thus far, solution spectral studies, (Henkel et al, 1974; Henkel et al, 1976) quantum mechanical (Loew et al, 1976) and force field empirical energy calculations, (Froimowitz, 1982) and X-ray crystallographic investigations (Shefter, 1974; Hanson et al, 1958; Bye, 1976; Bye, 1974) have been conducted in an effort to provide a consistent picture of the conformational requirements for activity.

Molecular mechanics calculations suggested that the diphenylpropylamines have a great deal of conformational heterogeneity (Froimowitz, 1982). For each of the common methadone analogues five or more grossly different conformations were calculated to be within 5 kcal/mol energy of one another. No clear picture emerged of a single conformation that might be responsible for analgetic action.

remarkable similarity is seen in the relative positions of the
vicinal hydrogens. This similarity, in spite of the
interchange of the Me and amine substituents on C(6), suggests
that the near eclipsing of the C(6) hydrogen and the phenyl
substituted C(4) may be by far the most energetically
acceptable arrangement for 5-methylmethadone isomers. The
conformation is also compatible with the observed proton
coupling constant (Henkel et al, 1976) and the lowest energy
molecular mechanics calculation for (5S, 6R)-methylmethadone
(Froimowitz, 1982).

Comparison of the solid state and solution data with
molecular mechanics calculations suggests that while a
conformation observed in the solid state may be at or very
near the global minimum energy conformation, it may not be the
active conformation, particularly in the case of relatively
flexible molecules of moderate activity. The data suggest

TABLE 1

Torsion angles of crystallographically observed conformations
of methadone and its congeners, $\tau_1=C(17)-C(16)-C(4)-C(5)$,
$\tau_2=C(15)-C(10)-C(4)-C(5)$, $\tau_3=C(2)-C(3)-C(4)-C(5)$, $\tau_4=C(3)-$
$C(4)-C(5)-C(6)$, $\tau_5=C(1)-C(2)-C(3)-C(4)$, $\tau_6=C(4)-C(5)-C(6)-N$,
$\tau_7=C(5)-C(6)-N-C(8)$, $\tau_{7'}=C(5)-C(6)-N-C(9)$

Compound	τ_1	τ_2	τ_3	τ_4	τ_5	τ_6	τ_7	$\tau_{7'}$
Normethadone[27]	-20°	98°	175°	74°	168°	-165°	71°	-53°
	-22	100	172	75	177	-167	73	-53
(6R)-methadone[25]	-34	96	-174	76	157	-146	75	-53
(5S)-isomethadone[24]	-25	86	-167	66	176	-152	81	-154
(5S,6R)-threo-5-methylmethadone[28]	-38	93	-166	71	-123	-138	78	-52
(5S,6S)-erythro-5-methylmethadone[17]	-40	100	175	64	175	97	92	-145

Three families of conformers were suggested as possible
candidates for the active form and entirely different
conformations were proposed to be responsible for the action
of different isomers.

The X-ray crystallographically determined conformations
of salts of normethadone, (6R)-methadone, (5S)-isomethadone,
(5S,6R)-threo-5-methylmethadone, and (5S,6S)-erythro-5-
methylmethadone defined by the eight torsion angles of Fig. 7
are compared in Table 1. The conformations of the molecules
are remarkably consistent despite variability in their anions,
crystal habit, and intermolecular association. There is a
minor difference observed in the position of the C(1) Me in

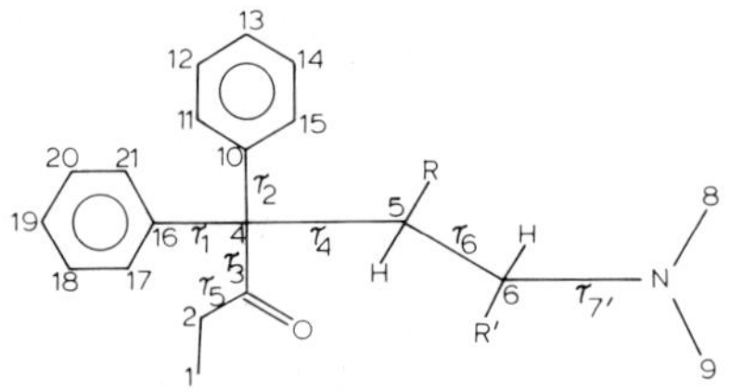

Figure 7. Atomic numbering and torsion angles defining the
molecular conformations of normethadone (R=R'=H), methadone
(R=H, R'=CH$_3$), isomethadone (R=CH$_3$, R'=H) and 5-
methylmethadone (R=R'=CH$_3$).

the threo compound and variability is observed in the nitrogen
methyl positions. The only significant difference is the
(+)clinal conformation of τ_6 observed in the most active
analogue, (5S,6S)-erythro-5-methylmethadone.

The consistency of the conformations of these structures
suggests that there may be less flexibility in the torsion
angles involving the quaternary C(4) than is suggested by
either the quantum chemical (Loew et al, 1976) or molecular
mechanics calculation (Froimowitz, 1982).

When the threo and erythro structures are viewed in
Newman projections along the C(5)-C(6) bond (Fig. 8), a

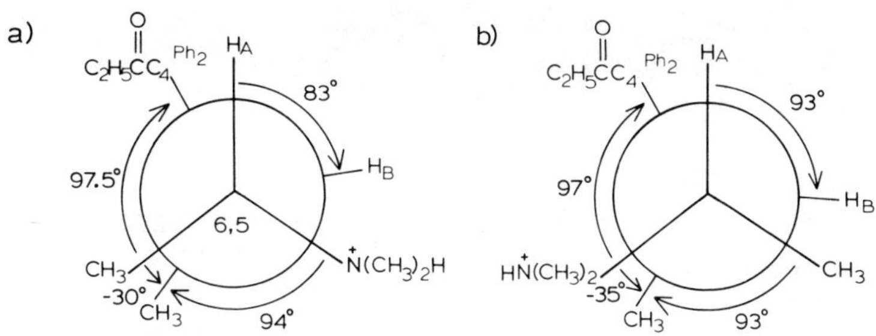

Figure 8.   Newman projection [C(6)→C(5)] of (a) (5S,6R)-threo
and (b) (5S,6S)-erythro-5-methylmethadone illustrating
orthogonality of the hydrogen atoms in the nearly eclipsed
observed conformations.

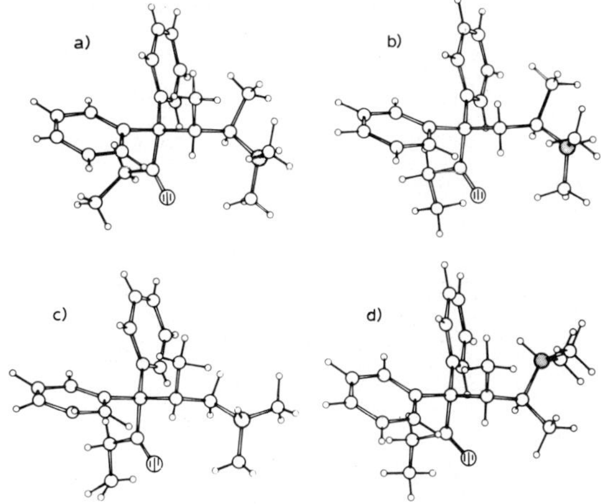

Figure 9.   Observed conformation of (a) (5S,6R)-threo-5-
methylmethadone, (b) (6R)-methadone, (c) (5S)-isomethadone,
and (d) (5S,6S)-erythro-5-methylmethadone (N is stippled).

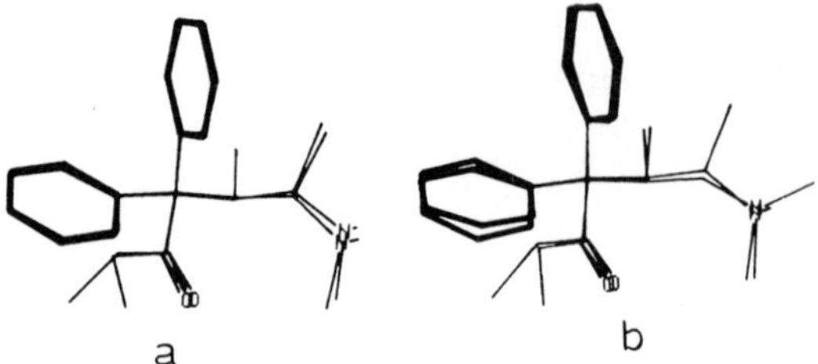

Figure 10. A comparison of the crystallographically observed
inactive (5S,6R)-threo-5-methylmethadone and (a) (6R)-
methadone and (b) (5S)-isomethadone.

that for protonated (6R)-methadone and (5S)-isomethadone the
lowest energy conformation is one in which nitrogen is in the
extended conformation but that in solution some fraction of
the population of molecules is in the (+)clinal conformation.

TABLE 2

Relative energies (kcal/mol) of the extended and (+)clinal
conformations of methadone (M) analogues. The structures are
tabulated in order of decreasing activity.

| | Conformational Energy | |
	Extended	(+)Clinal
erythro-5-methyl-M	35	23
M	15	18
iso-M	13	23
threo-5-methyl-M	23	35

The crystallographically observed conformations of (6R)-
methadone, (5S)-isomethadone and (5S,6S)-erythro-5-

methylmethadone are illustrated in Fig. 9. The (5S,6R) isomer
of the threo structure has a conformation nearly identical to
that of (6R)-methadone and differs from that of (5S)-
isomethadone only in the orientation of the amine Me
substituents (Fig. 10). This demonstration of the stability
of extended conformers of the active compounds, (6R)-methadone
and (5S)-isomethadone, and the totally inactive (5S,6R)-threo-
5-methylmethadone may indicate that opioid receptors are
capable of adjusting to the presence of a 5S or 6R substituted
Me, but not to both in the same molecule. Alternatively, it
may be that extended conformers are not suitable substrates
for the opioid receptor and that alternate conformers of (6R)-
methadone and (5S)-isomethadone are responsible for their
receptor affinity. Several plausible explanations have been
proposed (Duax et al, 1984) for the dramatic difference in the
potency of the erythro and threo isomers.

The fact that (5S,6S)-erythro-5-methylmethadone is
observed crystallographically as a unique conformation
relative to the other methadones (Fig. 9) has special
significance and might account for its greatly enhanced
potency. The combination of chiral centers may be responsible
for promoting a conformation which has greater affinity for
opioid receptors than other methadone compounds.

The molecular mechanics program MM2p was used to refine
the conformations of the (+)clinal and extended isomers of
(6R)-methadone, (5S)-isomethadone, (5S,6R)-threo-5-
methylmethadone and (5S,6S)-erythro-5-methylmethadone. The
results of these calculations are presented in Table 2. The
calculations indicate that in the case of the more potent
erythro structure the crystallographically observed (+)clinal
conformer is greatly preferred. In the case of the inactive
threo structure the crystallographically observed extended
conformation is preferred and the (+)clinal may even be
forbidden. The calculations suggest that isomethadone should
also favor the extended (crystallographically observed)

conformation but that both conformers of methadone are of
nearly equal energy.

The X-ray data, energy calculations and relative
biological activities can be interpreted as showing the
(+)clinal conformation is optimal for opioid receptor binding.
The very potent _erythro_ isomer appears to be constrained to
the active (+)clinal conformation while the inactive _threo_
isomer is constrained to the inactive extended conformation.
Although the minimum energy conformations of methadone and
isomethadone correspond to the inactive conformation, both are
capable of taking up the active (+)clinal shape. The weaker
activity of methadone and isomethadone relative to _erythro_-5-
methylmethadone could be due to the fact that a proportionally
smaller fraction of M and IM molecules in any ensemble are in
the active conformer.

4.   IONOPHORES, FLEXIBILITY AND ION SELECTIVITY

Ionophores are compounds that induce or facilitate ion
transport across membranes (Dobler, 1981). The resultant
changes in ion gradients and transmembrane potentials
influence a wide spectrum of biological activities (Pressman
_et al_, 1982). Their usefulness depends upon their ion
selectivity and efficiency of transport. The ion selectivity
must be largely influenced by the relative binding energies of
the different cations. Among the factors influencing this

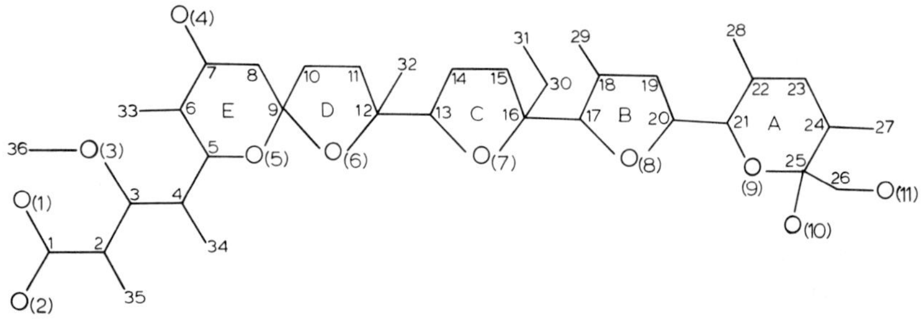

Figure 11.   Chemical structure, numbering scheme, and ring
identification of monensin A.

would be cation size and charge, number, type and distribution
of ligands, and the flexibility of the ionophore (Pressman et
al, 1982; Eisenman et al, 1975; Gresh et al, 1981). However,
the relative importance of these parameters is disputed. In
order to attempt to relate structural factors to the
selectivity of ionophores, the structures of the free acids
should be compared with those of at least two cations with
different selectivities.

The antibiotic monensin A (Fig. 11) is a biologically
active compound produced by a strain of Streptomyces
cinnamonensis. A member of the family of monocarboxylic acid,
polycyclic, polyether antibiotics, it induces monovalent
cation permeability, exhibits selectivity for Na^+ over K^+ and
has a greater stability constant with Na^+ than K^+.

Monensin-A is the ionophore for which the most
crystallographic data is available. The structures of the
free acid (Lutz et al, 1971) hydrated and anhydrous forms of
the Na^+ complex (Duax, et al, 1980) the NaBr complex (Ward et
al, 1978) the hydrated K^+ complex (Duax et al, 1986) and the
Ag^+ complex (Pinkerton et al, 1970) have been determined.

Superficially, all the cation complexes of monensin A are
similar. In the K^+ complex, as in other complexes, two
intramolecular hydrogen bonds involving the carboxyl oxygens
and two hydroxy groups at the opposite end of the molecule are
found. These head-to-tail hydrogen bonds (O(1)-O(11), 2.51Å
and O(2)-O(10), 2.62Å), common in this class antibiotics,
produce a pseudocyclic conformation which serves to form the
cavity in which the cation is coordinated. This
intramolecular hydrogen bonding and cation coordination is
illustrated in Figure 12.

Figure 13 illustrates the magnitudes of the deviations of
the individual ring torsion angles of the complexes from those
of the free acid. The conformations of all the complexes
except the K^+ complex differ from the free acid in a very
similar fashion. The K^+ complex, however, exhibits

significant differences from the other complexes in ring D.
These torsion angle changes in ring D correspond to a
pseudorotation from one envelope form to another. This
pseudorotation acts as a hinge to expand the coordination
sphere relative to that found in the Na$^+$ complex to
accommodate the larger K$^+$ ion. This expansion is illustrated
in Figure 14.

Molecular mechanics calculations [13] were performed on a
36 atom fragment from C(4) to C(13) (Fig. 11) which includes
the spiro-fused ring and adjacent atoms. The calculation
indicates that the conformations observed in the free acid and
all except the K$^+$ complex are essentially indistinguishable
from the minimum energy conformation. Fixing the value of the
C(9)-(10)-C(11)-C(12) torsion angle to its observed value for
results are qualitative, but they are consistent with the both
the Na$^+$ and K$^+$ complex and again minimizing the energy of
these fragments yields an energy for the Na$^+$-like fragment
2.3kcal/mol lower than that for the K$^+$ ion form. These
conclusion that the introduction of the larger K$^+$ ion into the
coordination site of monensin A necessitates the distortion of

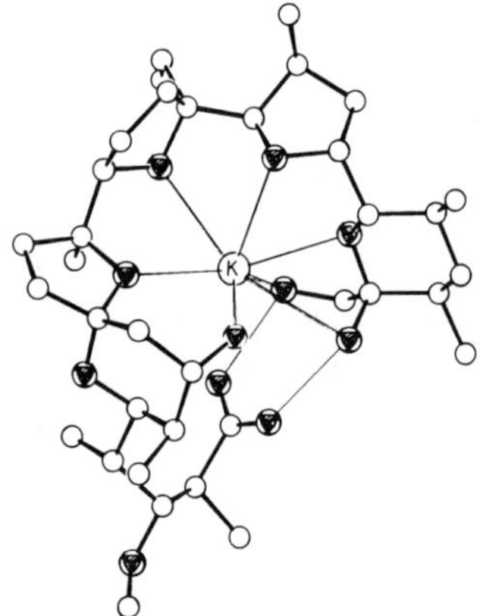

Figure 12. An illustration of the intramolecular hydrogen
bonding and the cation coordination in the potassium monensin
A complex. Oxygens are depicted by   . Narrower lines
indicate K$^+$ coordination and intramolecular H bonds.

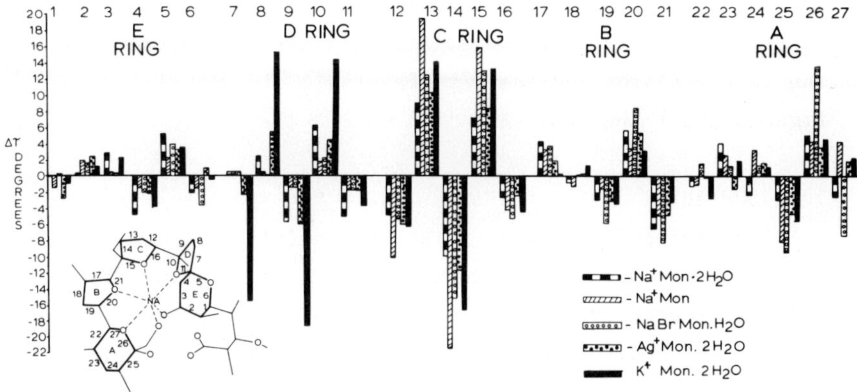

Figure 13.   The differences in the intraring torsion angles of the metal complexes of monensin A from those of the free acid.

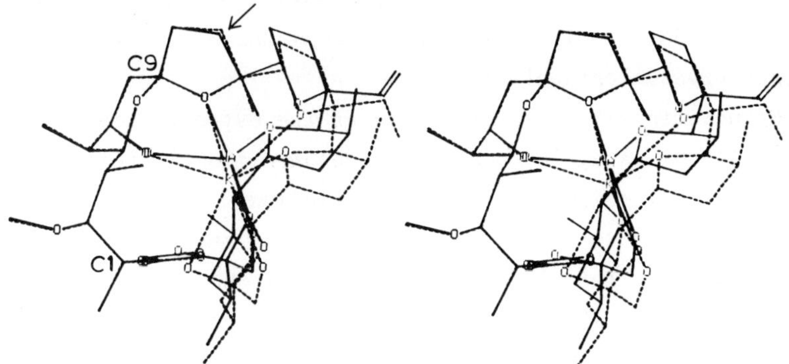

Figure 14.   Stereo diagram illustrating the expansion of the coordination sphere in the K⁺ complex (dashed) compared to the Na⁺ complex (solid).   This expansion is achieved by a pseudorotation in the D-ring (arrow).

a constrained portion of the molecule from its low energy conformation.

The molecular constraints do not permit a uniform distribution of coordinating ligands around the Na⁺ or K⁺ ions.   However, these distortions are more severe in the case of the K⁺ ion (Duax et al, 1986).

We conclude that the selectivity exhibited by monensin A

for Na^+ over K^+ is due to the energetic cost of distorting a constrained portion of the molecule in order to accommodate the larger K^+ ion and the less favorable distribution of ligands about the ion.

5.   SUMMARY

The three-dimensional structural details of hormones, drugs and antibiotics determined by X-ray crystallographic techniques provide reliable information on the global minimum energy conformations of these molecules or local minimum energy conformations that are within a few kcal/mol of the global minimum. In favorable cases, state of the art molecular mechanics calculations provide quantitative agreement with X-ray results and information on the relative energy of other local minimum energy conformations not observed crystallographically.

Structures with high affinity for the estrogen receptor almost without exception contain a phenolic ring that appears to be essential to initiate receptor binding. Crystallographic determination and molecular mechanics calculations for the synthetic estrogen DES and its metabolites ZPD and EPD indicate that DES is constrained to an extended conformation essential for estrogen potency, that EPD is constrained to a bent conformation that can bind to the estrogen receptor but fails to illicit strong hormonal response and that while the low energy bent conformer is capable of binding to the receptor, the higher energy metastable extended conformer of ZPD is probably responsible for its estrogenic activity.

The analgesic activity of methadone and some of its derivatives is attributed to its direct binding to opioid receptors. Crystallographic and spectral analysis of methadone, isomethadone, _erythro_-5-methylmethadone and _threo_-5-methylmethadone suggest that the active conformation is one in which the nitrogen atom is rotated back toward the phenyl ring. Energy calculations suggest that this conformation is favored in the potent _erythro_ enantiomer of 5-methylmethadone, and is prohibited in the totally inactive _threo_ enantiomer, and that methadone and isomethadone have the flexibility to readily take up active as well as inactive conformations.

X-ray crystallographic structure determinations of complexed and uncomplexed ionophores can offer insight into the molecular details of the mechanism of ion capture and release. A thorough analysis of the differences in coordination number, geometric arrangement of the coordinating oxygens, bond distances and bond strengths to the ions in the selectivity sequence of an ionophore can identify the structural basis for control of activity. Comparison of the crystal structures of Na^+ and K^+ complexes of monensin revealed that suitable coordination of K^+ involved the distortion of a five membered ring from the conformation observed in all other X-ray studies of monensin. Molecular mechanics calculations indicated that this distortion was energetically unfavorable and must be a contributing factor to the preference of monensin for Na^+.

ACKNOWLEDGEMENTS: Research supported in part by NIAMDD Grant No. AM26546, GMS Grant No. GM32812, DRR Grant No. RR05716 and a grant from the Margaret L. Wendt Foundation. The organization and analysis of the data base associated with this investigation and several of the illustrations were carried out by use of the PROPHET system, a unique national computer resource sponsored by the NIH.

REFERENCES

ALLINGER, N.L. and YUH, Y. (1979). MM1/MMP1. A program for general molecular mechanics calculations with the 1973 force field. QCPE No. 400. MM2. Program with more recent force field. QCPE No. 395. MM2p is a version of MM2 containing the π treatment from MMP1 amended by D.C. Rohrer (1983).

BUSETTA, B. and HOSPITAL, M. (1971). Acta Cryst. **B28**, 560–567.

BUSETTA, B., COURSEILLE, C. and HOSPITAL, M.(1973). Acta Cryst. **B29**, 2456–2462.

BYE, E.(1976). Acta Chem. Scand. Sect. **B30**, 323.

BYE, E.(1974). Ibid. **28**, 5.

DOBLER, M. (1981). "Ionophores and Their Structures". John Wiley and Sons, New York.

DUAX, W.L. and NORTON, D.A. (1975). "Atlas of Steroid Structure", Vol. 1. Plenum Press, New York.

DUAX, W.L. and WEEKS, C.M.(1980). "Estrogens in the Environment", pp.11–31. Elsevier/North Holland, New York.

DUAX, W.L., SMITH, G.D. and STRONG, P.D.(1980). J. Amer. Chem. Soc. **102**, 6725.

DUAX, W.L., GRIFFIN, J.F. and ROHRER, D.C. (1981). J. Am. Chem. Soc. **103**, 6705–6712.

DUAX, W.L., GRIFFIN, J.F., ROHRER, D.C., WEEKS, C.M. and EBRIGHT, R.H. (1984). "Biochemical Actions of Hormones", Vol. XI. Academic Press, New York.

DUAX, W.L., SMITH, G.D., GRIFFIN, J.F. and STRONG, P.D. (1984). Tetrahedron **40**, 467–471.

DUAX, W.L., PANGBORN, W.A. and LANGS, D.A. (1986). American Crystallographic Association Meeting, Hamilton, Ontario, Canada, Abstract No. PB56.

EDDY, N.B. and MAY, E.L.(1952). J. Org. Chem. **17**, 321.

FROIMOWITZ, M. (1982). Ibid **25**, 689.

GRESH, N., ETCHEBEST, C. DELALUZ ROJAS, O. and PULLMAN, A. (1981). Inter. J. Quantum Chemistry: Quantum Biology Sym. 8, pp.109–116. John Wiley and Sons, Inc.

GRIFFIN, J.F., DUAX, W.L. and WEEKS, C.M.(1984). "Atlas of Steroid Structure", Vol. 2. Plenum Press, New York.

HANSON, A.W. and AHMED, F.R.(1958). Acta Cryst. **11**, 724.

HENKEL, J.G., BELL, K.H. and PORTOGHESE, P.S.(1974). J. Med. Chem. **47**, 124.

HENKEL, J.G., BERG, E.D. and PORTOGHESE, P.S. (1976). Ibid. 19, 1308.

HOSPITAL, M., BUSETTA, B., CORSEILLE, C. and PRECIGOUX, G. (1975). J. Steroid Biochem. 6, 221–225.

KILBOURN, B.T. and OWSTON, P.G. (1970). J. Chem. Soc. (B), 1–5.

KORACH, K.S. and DUAX, W.L.(1983). 65th Annual Meeting of the Endocrine Society, San Antonio, Texas, Abstract No. 1047.

LOEW, G.H., BERKOWITZ, D.S. and NEWTH, R.C. (1976). Ibid. 19, 863.

LUTZ, W.K., WINKLER, F.K. and DUNITZ, J.D. (1971). Helv. Chim. Acta. 54, 1103.

MAY, E.L. and EDDY, N.B. (1952). Ibid. 17, 1210.

PERT, C.B. and SNYDER, S.H. (1973). Proc. Natl. Acad. Sci. U.S.A. 70, 2243.

PINKERTON, M. and STEINRAUF, L.K. (1970) J. Mol. Biol. 49, 533.

PORTOGHESE, P.S. (1965). J. Pharm. Sci. 55, 865.

PORTOGHESE, P.S., PAUPAERT, J.H., LARSON, D.L., GROUTAS, W.C., MEITZNER, G.D., SWENSON, D.C., SMITH, G.D. and DUAX, W.L. (1982). J. Med. Chem. 25, 684.

PRESSMAN, B.C. and FAHIM, M. (1982). Ann. Rev. Pharmacol. Toxicol. 22, 465–490.

ROHRER, D.C. and FULLERTON, D.S. (1980). Acta Cryst. B36, 1565–1568.

SHEFTER, E. (1974). Ibid. 17, 1037.

WARD, D.L., WEI, K.T., HOOGERIDGE, J.C. and POPOV, A.I. (1978). Acta Cryst. B34, 110.

WEEKS, C.M., COOPER, A. and NORTON, D.A. (1970). Acta Cryst. B26, 429–433.

2. Steroidal neuromuscular blocking agents: discovery of Chandonium

Harkishan Singh

Neuromuscular blocking agents interrupt transmission of nerve impulse at the skeletal neuromuscular junction. The quest for an ideal neuromuscular blocking agent continues. The literature on neuromuscular blocking agents has been reviewed recently (Singh et al., 1984). One category of such agents are azasteroidal chemically. In this presentation is described the work on azasteroidal neuromuscular blocking agents carried out by us leading to the discovery of clinically effective agent chandonium iodide (Singh, 1985).

We initiated a programme on the synthesis of bisonium steroids as potential neuromuscular blocking agents, with one or both of the cationic systems present as part of the steroid ring skeleton at different interonium distances (Singh et al., 1973). The first active compound synthesized was chemically novel 4,17a-dimethyl-4,17a-diaza-D-homo-5α-androstane dimethiodide (HS-342) (1). In the anaesthetized cat, HS-342 exhibited nondepolarizing neuromuscular blocking activity approximately equal to that of tubocurarine (Marshall et al., 1973a, 1973b). It had duration of action one-third that of tubocurarine and had rapid onset of action. It also possessed ganglion blocking activity. The rapid onset and short duration of action proved to be an important lead.

A related compound 4-methyl-17β-dimethylamino-4-aza-5α-androstane dimethiodide (HS-467) (2) (Singh and Paul, 1974)

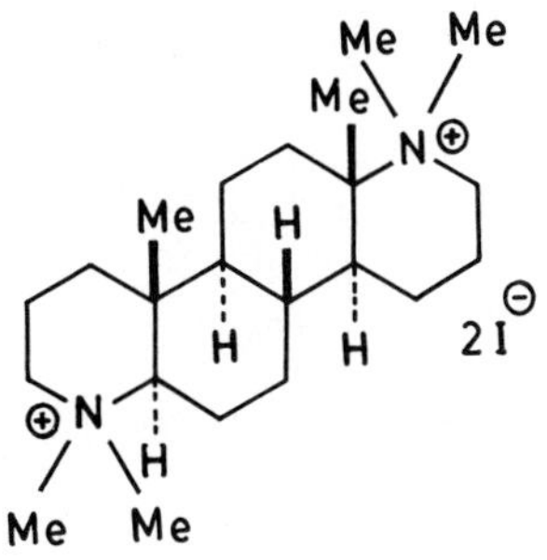

which has somewhat extended interonium distance was approximately equipotent to tubocurarine, but exhibited extremely weak ganglion blocking activity (Gandiha et al., 1974); the duration of action was again short and the onset rapid.

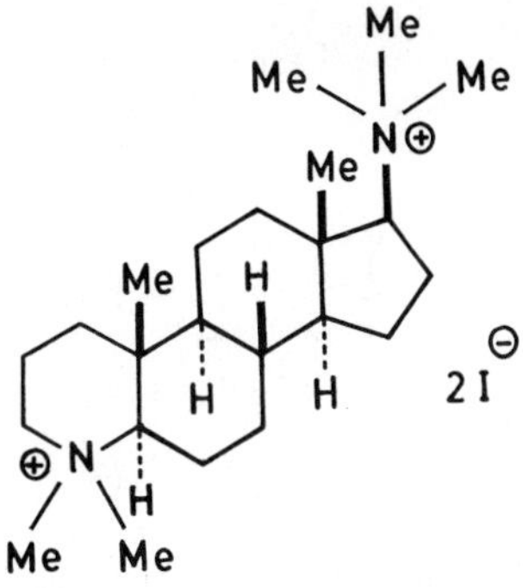

In the series the compound 17a-methyl-3β-pyrrolidino-17a-aza-D-homo-5-androstene dimethiodide (HS-310) (3) (Singh and Paul, 1974), christened as chandonium iodide after the name of Chandigarh city, proved to be of particular interest. It possesses a powerful nondepolarizing neuromuscular blocking activity of short duration and rapid onset, being only slightly less active than pancuronium (Gandiha et al., 1974, 1975). It has little or no ganglion blocking activity. The saturated congener 17a-methyl-3β-pyrrolidino-17a-aza-D-homo-5α-androstane dimethiodide (dihydrochandonium iodide) (HS-692) (4) and analogues possessing bulkier cationic heads were synthesized (Singh et al., 1979a) but these showed

lesser potency (Teerapong *et al.*, 1979). The chandonium
analogues containing acetylcholine and choline moieties,

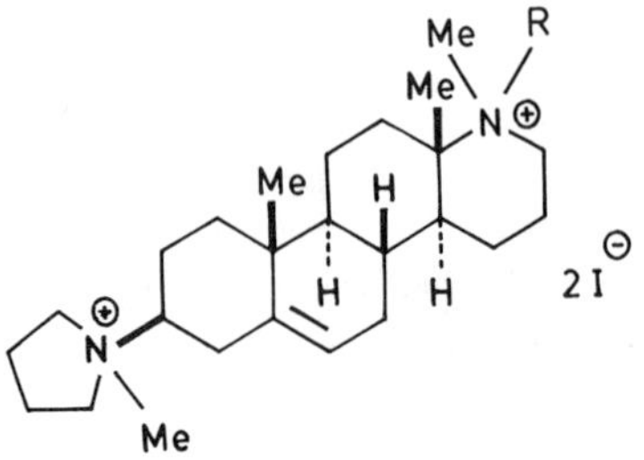

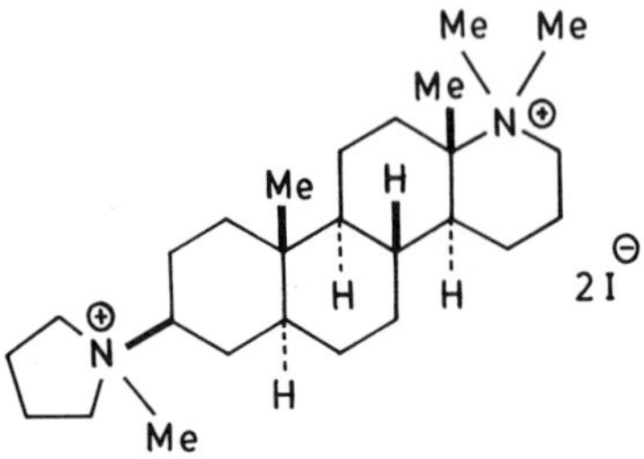

3 (R = Me)

5 (R = CH$_2$CH$_2$OAc)

6 (R = CH$_2$CH$_2$OH)

4

HS-627 (5) and HS-626 (6) (Singh *et al.*, 1979b) were approximately equipotent with chandonium iodide as neuromuscular
blocking agents in anaesthetized cat (Marshall *et al.*, 1981).

Out of the azasteroids synthesized by us chandonium
iodide (3) remains the most promising compound. The
toxicity studies carried out at the Central Drug Research
Institute, Lucknow, India, on rats and monkeys have not
revealed any adverse effects. The Institute is monitoring
the clinical studies, which are at phase II at the moment.
The results are encouraging.

The molecular structures of bisquaternary steroids
HS-342 (1), HS-467 (2) and chandonium iodide (3) have been
determined crystallographically (Palmer *et al.*, 1980). Ring B

in chandonium iodide is a half-chair and ring D in HS-467
is a distorted enevelope. All other rings in the steroid
skeletons of the three molecules are in the chair conformation, and all rings are <u>trans</u>-connected. Chandonium iodide
appears to be the least rigid of the three molecules; there
is a limited rotation about C(3)-N$^+$ bond. In HS-342 (1) the
interonium distance is 0.817 nm and the combination of neuromuscular and ganglionic blocking actions is probably due to
a shorter interonium distance. The ganglionic blocking
activity in HS-467 (2) (interonium distance 0.902 nm) is
extremely weak, and in chandonium iodide (3) (interonium
distance 1.029 nm) there is little or no ganglionic blocking
activity.

It is of interest to note that on the basis of molecular orbital calculations on structural models corresponding to
chandonium iodide (3), dihydrochandonium iodide (4), HS-467
(2) and certain other azasteroidal neuromuscular blockers
there has been shown a high degree of delocalisation of
positive charge (Chaudhary <u>et al</u>., 1985). May be the drug-
receptor complex may not involve binding of quaternary
nitrogen (onium centre) with the anionic site of the receptor.
Instead, a more general cationic head comprising onium
nitrogen and adjacent atoms may be taking part in electrostatic interactions.

It may be mentioned that crystal and molecular
structure studies have been carried out of 17a-methyl-3β-
pyrrolidino-17a-aza-D-homo-5-androstene (HS-309) (Mazid <u>et al</u>.,
1977), the tertiary amine corresponding to chandonium iodide
(3); 17a-methyl-3β-pyrrolidino-17a-aza-D-homo-5α-androstane
(HS-691) (Husain <u>et al</u>., 1982), the tertiary amine corresponding to dihydrochandonium iodide (4); and 17a-(2-hydroxyethyl)-
3β-pyrrolidino-17a-aza-D-homo-5-androstene (HS-625) (El-Shora
<u>et al</u>., 1984) and its dimethiodide (HS-626) (6) (El-Shora

et al., 1982).

The crystallographic studies have given important information about the stereochemical and other features which have been of help in interpretting and correlating the biological activity. In addition, it has been possible to unequivocally assign the structures and configurations to the intermediates which have hitherto been postulated on the basis of reaction mechanistic and other chemical evidences.

REFERENCES

CHAUDHARY, A.K., GOMBAR, V.K. and SINGH, H. (1985). Indian Journal of Chemistry 24B, 2.

El-SHORA, A.I., PALMER, R.A., SINGH, H., BHARDWAJ, T.R. and PAUL, D. (1982). Journal of Crystallographic and Spectroscopic Research 12, 255.

El-SHORA, A.I., PALMER, R.A., SINGH, H., BHARDWAJ, T.R. and PAUL, D. (1984). Journal of Crystallographic and Spectroscopic Research 14, 89.

GANDIHA, A., MARSHALL, I.G., PAUL, D., RODGER, I.W., SCOT, W. and SINGH, H. (1975). Clinical and Experimental Pharmacology and Physiology 2, 159.

GANDIHA, A., MARSHALL, I.G., PAUL, D. and SINGH, H. (1974). Journal of Pharmacy and Pharmacology 26, 871.

HUSAIN, J., TICKLE, I.J., PALMER, R.A., SINGH, H., BHARDWAJ, T.R. and PAUL, D. (1982). Acta Crystallographica B38, 130.

MARSHALL, I.G., HARVEY, A.L., SINGH, H., BHARDWAJ, T.R. and PAUL, D. (1981). Journal of Pharmacy and Pharmacology 33, 451.

MARSHALL, I.G., PAUL, D. and SINGH, H. (1973a). Journal of Pharmacy and Pharmacology 25, 441.

MARSHALL, I.G., PAUL, D. and SINGH, H. (1973b). European Journal of Pharmacology 22, 129.

MAZID, M.A., PALMER, R.A., SINGH, H. and PAUL, D. (1977). Acta Crystallographica B33, 3641.

PALMER, R.A., KALAM, M.A., SINGH, H. and PAUL, D. (1980). Journal of Crystal and Molecular Structure 10, 31.

SINGH, H. (1985). Indian Journal of Pharmaceutical Sciences 47, 29.

SINGH, H., BHARDWAJ, T.R., AHUJA, N.K. and PAUL, D. (1979a). Journal of Chemical Society, Perkin Transactions I 305.

SINGH, H., BHARDWAJ, T.R. and PAUL, D. (1979b). Journal of Chemical Society, Perkin Transactions I 2451.

SINGH, H., CHAUDHARY, A.K., BHARDWAJ, T.R. and PAUL, D. (1984). Journal of Scientific and Industrial Research 43, 306.

SINGH, H. and PAUL, D. (1974). Journal of Chemical Society, Perkin Transactions I 1475.

SINGH, H., PAUL, D. and PARASHAR, V.V. (1973). Journal of Chemical Society, Perkin Transactions I 1204.

TEERAPONG, P., MARSHALL, I.G., HARVEY, A.L., SINGH, H., PAUL, D. and BHARDWAJ, T.R. (1979). Journal of Pharmacy and Pharmacology 31, 521.

3. NMR analysis of conformation-activity relationships in a homologous series of synthetic muscle relaxing agents

Elena Gaggelli, Enzo Ticzzi and Gianni Valensin

1. INTRODUCTION

p(o-alkyloxybenzamido)benzoate diethyl (2-hydroxyethyl) methyl

ammonium halides are synthesized drugs that exhibit a smooth

muscle relaxing property (Ghelardoni et al, 1973). Pharmacolo-

gic activity strongly depends on length and position of the

alkyloxy side chain: ortho-substituted compounds are by far

the most effective with a sharp maximum of activity shown by

the octyloxy (n=7) derivative.

Conformation and dynamics of the first term of the series

(n=2) were previously delineated (Valensin et al, 1984; Sega

et al, 1984) by measuring proton NMR selective and nonselecti-

---

§Contribution from the Department of Chemistry, University of

Siena, Pian dei Mantellini 44, 53100 Siena, Italy

ve spin-lattice relaxation rates, proton-proton intramolecular NOEs and carbon-13 NMR relaxation rates. When extending proton relaxation investigations to other terms of the series, it was found (Sega et al, 1985) that the spatial arrangement of the two aromatic moieties substantially changes with increasing length of the alkyloxy chain. The most active octyloxy derivative was shown to assume a unique configuration where the angles between each aromatic ring and the plane of the amide unit were almost equal ($\pm 55°$ and $\pm 53°$).

In this paper we present the whole picture of NMR data for the four compounds of the series, wherefrom a complete delineation of motional and conformational dynamics is reached. Besides the aromatic backbone, attention will be also focussed on the hydrocarbon side chains in order to accomplish a thorough understanding of structural parameters relevant for biological activity.

3. RESULTS and DISCUSSION

It is known (Allerhand et al, 1971) that carbon-13 spin-lattice relaxation rates (R_1's) are almost exclusively determined by dipolar interaction with directly bonded or nearby protons, thus allowing delineation of molecular dynamics.

If relaxation rates of protonated backbone carbons are first considered, it is noticed that the relaxation rate of C-5 is always faster than those of the other aromatic carbons, suggesting that C-2--C-5 is always the main rotation axis for ring A despite of increasing length and weight of the alkyloxy chain in ortho. This motion, as well as librational motions of ring A, slow down in other terms of the series which is very

Table 1

^{13}C relaxation rates for solutions 0.2 M in DMSO-d$_6$

Carbon	n=2	n=3	n=7	n=9
C-3	3.00	3.37	4.37	5.18
C-4	2.77	3.73	4.06	4.83
C-5	4.93	6.71	6.02	7.19
C-6	3.11	3.86	4.93	4.97
C-9,9'	2.86	4.11	3.66	3.85
C-10,10'	2.72	3.51	3.68	3.94
C-13	5.92	5.71	6.41	6.41
C-14	4.63	4.59	5.46	4.93

likely to be due to effects of the alkyloxy chain. It is interesting to note that, as a consequence of the relative minimum
reached in the relaxation rate of C-5 for n=7, the octyloxy
derivative exhibits the lowest difference between the rates of
the main rotational motion and librational motion.

The C-8--C-11 axis is naturally the main rotational axis
for ring B. The relaxation rates of C-9,9' and C-10,10' stea-
dily increase along the series, but the overall increment (1.0
s^{-1}) is much less pronounced than that observed for ring A
(2.0 s^{-1}). Clearly, motions of ring B are less affected by the
alkyloxy chain. Librational motions of ring B therefore become
faster than those of ring A: for n=2, n=3 and to a lesser extent n=7, the two librational motions accur at the same or at

very similar velocities; whereas dephasing is accomplished for n=9. The most important feature is the matching of librational motions with the main rotational motion for n=7, while librational motions of the two rings are still governed by approximately the same correlation time.

Motional delineation of side chains was then attempted in the light of a rather surprising intramolecular dipolar connectivity between protons of the terminal methyl of the octyloxy chain and protons of the methyls of the ethyl groups bonded to the quaternary nitrogen (Sega et al, 1985).

Quantitative evaluation of motional features within the four compounds was done in terms of the model-free approach of Lipari and Szabo (1982 a,b) that yields information on internal motions in terms of a "generalized" order parameter, S^2, and of an "effective" correlation time $_e$. Such parameters are calculated from carbon-13 spin-lattice relaxation rates and NOEs at one frequency or, alternatively, from carbon-13 spin-lattice relaxation rates at two frequencies.

C-6 was chosen as the carbon atom of the backbone to which refer motional features of carbons within the side chains. It was in fact noticed that, while all the aromatic protonated carbons of ring A but C-5 exhibit the same relaxation behavior (Table 1), the NOE was similar for C-3 and C-4 but much lower for C-6. This fact agrees with a strong dipolar interactions between H-6 proton and the first methylene of the alkyloxy chain which yields a certain slow-motion contribution to the observed NOE, strongly suggesting that C-6 is by far the most hindered carbon atom.

The measured S^2's are reported in Table 2.

Table 2

Measured order parameters in solutions 0.2 M in $DMSO_6$

Carbon	n=2	n=3	n=7	n=9
C-16	0.730	0.711	0.815	0.489
C-17	0.075	0.052	0.114	0.002
C-18	0.984	0.720	0.880	0.826
C-19	0.048	0.057	0.051	0.002
C-21	---	0.054	0.026	0.004
C-22	---	---	0.316	0.099

The order parameter represents a model-independent measurement of the degree of spatial restriction underwent by the motion. the fact that high values of S^2 are always found for C-18 is a further proof of restricted motion within the first methylene of the alkyloxy chain in agreement with the dipolar interaction with C-6. A high value of S^2 yields in fact evidence of C-H vectors reorienting at the same rate as the C-6--H-6 vector. A very interesting feature is the large value of S^2 found for C-22 in the octyloxy derivative (but not in the decyloxy one). If it is considered that C-22 has a middle chain position relatively far from the almost anchored C-18, it may be stated that the conformational freedom of the octyloxy chain is somehow reduced with respect to the decyloxy chain. This is further ratified by the fact that for the least restricted carbon atom for n=7 (C-21) S^2 is more than one order of magnitude greater than for the least restricted carbon atoms for n=9 (C-19).

By considering the whole S^2 values for the two side chains it is apparent that the most restricted motions belong to the n=7 compound where, in particular, the alkyl substituents undergo very hindered reorientations. The conclusion is that the most active octyloxy derivative displays the lowest conformational freedom, most probably due to the existence of a "closed" structure that does actually exist at the solid state (Dapporto and Sega, 1986).

REFERENCES

DAPPORTO,P. and SEGA,A. (1986) Acta Cryst. Ser.C in press.

GHELARDONI,M.,PESTELLINI,V.,PISANTI,N. and VOLTERRA,G. (1973) J.Med.Chem. **16,** 1063.

LIPARI,G. and SZABO,A. (1982a) J.Am.Chem.Soc. **104,** 4546.

LIPARI,G. and SZABO,A. (1982b) J.Am.Chem.Soc. **104,** 4559.

SEGA,A.,GHELARDONI,M.,PESTELLINI,V.,POGLIANI,L. and VALENSIN, G. (1984) Org.Magn.Reson. **22,** 649.

SEGA,A.,GAGGELLI,E. and VALENSIN,G. (1985) Magn.Reson.Chem. **23,** 649.

VALENSIN,G.,POGLIANI,L.,GHELARDONI,M.,PESTELLINI,V. and SEGA, A. (1984) Can.J.Chem. **62,** 2131.

4. Molecular mechanics, crystallography, and drug research

David J. Duchamp, Loraine M. Pschigoda and Connie G. Chidester

ABSTRACT

The CONFOS molecular mechanics program and its associated set of force field parameters has been under development at Upjohn for many years. This system has the ability to calculate almost all types of organic molecules found in drugs, including complicated heterocyclic molecules, and has a number of features designed for use in drug design research. Recent work has emphasized the molecular mechanics of organosulfur and organophosphorus compounds The set of experimental structures used in parameterization has grown to over 300 experimental structures. Some of the features of CONFOS are discussed, and force field parameterization and verification statistics are presented.

1. INTRODUCTION

The use of molecular mechanics calculations as a tool in the interpretation and extension of molecular conformations from crystal structure analysis is now well established. The CONFOS molecular mechanics program and its associated force field have been in use and under periodic development in our laboratory for many years. The primary goal of our system is to produce accurate three-dimensional structural results, and accurate energy differences between the various possible conformations of a given molecule. In recent years we have been extending the program and the force field to cover all the functional groups usually found in drug molecules. This effort is still underway, so this paper is a status report of the project. Recent work includes extensions to handle more functional groups involving nitrogen, sulfur, and phosphorus.

2.   ENERGY CALCULATION

The strain energy of the molecule is calculated according to the following expression:

$$E_{strain} = E_{bond} + E_{angle} + E_{torsion} + E_{out\text{-}of\text{-}plane} + \tag{1}$$
$$E_{electrostatic} + E_{non\text{-}bonded} + E_{H\text{-}bond}$$

where the individual energy terms are given by the following equations:

$$E_{bond} = \sum_{bonds} f_c \, (d - d_o)^2 \tag{2}$$

where d is the bond distance, and f_c and d_o are parameters in the force field characteristic of the particular type of covalent bond.

$$E_{angle} = \sum_{bond\ angles} f_c \, [\,(\theta - \theta_o)^2 - g\,(\theta - \theta_o)^3\,] \tag{3}$$

where θ is the bond angle, f_c and θ_o are parameters in the force field characteristic of the particular type of covalent bond angle, and g is a constant.

$$E_{torsion} = \sum_{torsion\ angles} f_\omega \, [\,1 - \cos(n\,\omega)\,] \tag{4}$$

where ω is one of the torsion angles about a particular bond, f_ω is a force field parameter characteristic of the particular type of torsion angle, and n is an integer selected to give the correct rotational dependency.

$$E_{out\text{-}of\text{-}plane} = \sum_{sp^2\ atoms} f_c \, (d_i)^2 \tag{5}$$

where d_i is the distance of the planar sp^2 atom from the plane defined by the three atoms bonded to it, and f_c is a force field parameter characteristic of the particular type of sp^2 atom.

$$E_{electrostatic} = \sum (q_i \, q_j / r_{ij})\,exp\,(-0.33\,r_{ij}) \tag{6}$$

where q_i and q_j are the charges on atoms i and j, r_{ij} is the distance between atoms i and j. The summation is over all atom pairs where 1) the atoms are not bonded to each other or to the same atom, and 2) the atoms do not form an $H \cdots Acceptor$ pair of a hydrogen bond. The exponential may be regarded as a factor which accounts for shielding by intervening atoms at larger distances.

$$E_{non\text{-}bonded} = \sum f\,exp\,(-g\,r_{ij}) - e\,r_{ij}^{-6} \tag{7}$$

where f, g, and e are force field paramters characteristic of the kinds of atoms in the atom pair, and r_{ij} and the summation range are as in (6).

$$E_{\text{H-bond}} \;=\; \sum_{\substack{\text{H}\cdots\text{Acceptor}\\ \text{pairs}}} f_{hb}\, exp\,(\text{-}\, g_{hb}\, r_{ij}) \text{ - e } r_{ij}^{-6} \qquad (8)$$

where parameters are as in (7), except the non-bonded f and g are replaced by f_{hb} and g_{hb} which are characteristic of the particular type of hydrogen bond.

The actual energy minimized by the program is E_{total}, given by

$$E_{total} \;=\; E_{strain} \;+\; \sum E_{extra} \qquad (9)$$

where the E_{extra} terms are "extra potentials", sometimes called "forcing potentials". These "extra potentials" are a method of applying one or more constraints on the molecule to allow study of conformations with other than the minimum energy. A typical question which might be asked using this method is: *"How much energy will it cost to ... ?"*

3.   ELEMENTS   SUPPORTED

In CONFOS calculations, all atoms are identified by their chemical kind, *e.g.* nitrogen, then by a subtype, usually indicative of a functional group, *e.g.* amide. The program currently recognizes the 18 elements shown below:

H	**B**	**C**	**N**	**P**	**O**	**S**	**F**	**Cl**
Br	**Li**	**Na**	**K**	**Mg**	**Ca**	**Si**	**Fe**	**Zn**

Subtypes and force field parameters have been developed for C, H, N, O, S, P, F, Cl, Br, Na, K, and Ca. Na, K, and Ca are supported as cations only.

4.   ATOM SUBTYPES

Atoms in molecules are further subdivided into "subtypes", which are automatically assigned by the program from the bonding and from allowed valences of the elements. New subtypes have been created when necessary to differentiate structural features of certain functional groups. The subtypes usually have obvious chemical meanings, but this is not strictly so.

The subtypes currently supported for carbon atoms are shown below:

Primary   Secondary   Tertiary   Quaternary   <u>Hydrocarbon</u>
Cyclopropane   Cyclobutane
SP (Linear)   Unsaturated   C-Unsaturated
Aromatic   Aromatic Bridgehead

Carbonyl   Carboxyl   Carboxylate   Carbonate   <u>Oxygenated</u>
Gamma-onate (like Acetylacetonate)

Di-halogenated   Tri-Halogenated   Di-hetero   <u>Halogenated</u>

Amide   Isoamide Urea Isourea Carbamate   <u>Nitrogen</u>
Amidine Guanidine Amidinum   Guanidinium   <u>w/Oxygen</u>

Anion   3-Ring Anion   SP^2 Anion   <u>Anionic</u>

Each of these subtypes are precisely defined, usually in the way expected from the name, *e.g.* **Primary** means an sp^3 tetrahedral carbon with 3 hydrogens.

5.   ADDITIONAL CONFOS FEATURES

The CONFOS program provides for **automatic initialization from atomic coordinates.** The program first determines **connectivity** (which atoms are covalently bonded), then automatically assigns **bond types** (single, double, triple or aromatic), atom **subtypes**, and **partial atomic charges. Potential function parameters** are automatically assigned from the force field. **Hydrogen bonds** are detected by distance and atom kind and subtype criteria. The only exceptions to automatic initialization are that the user is required to: 1) specify the nitrogen bearing a positive charge in some complicated heterocycles, *e.g.* puromycin, 2) identify anionic carbon atoms, and 3) override the program on whether a nitrogen is planar or pyramidal in some complicated molecules.

The CONFOS program provides for **in-crystal minimization**, a method for simulating the environment seen by a molecule in a crystal. Since the conformation of a molecule in a crystal is sometimes greatly influenced by its neighbors in the crystal, such a method is essential for reproducing crystal structure results in molecular mechanics calculations. The simulated crystal environment is set up by automatic generation of all symmetry-related atoms within a specific distance, usually about 7 Å. Space group symmetry, handled automatically by the program, and unit-cell parameters are held fixed during minimization. The positions of all atoms in the asymmetric unit are varied, with symmetry-related counterparts following along. This method does not allow variation of unit cell parameters, nor does it calculate an accurate lattice energy; for such calculation, a true lattice energy calculating program such as Busing's WMIN (Busing, 1970) should be used.

CONFOS provides for **multiple molecule minimization** (Duchamp, 1977). Up to ten molecules can be interspersed and simultaneously minimized, while connected at certain user selected points by "extra potentials". This method is very useful in conformational matching of drug molecules which have different chemical structures, but are known to bind at the same receptor site.

6. CHARGE ASSIGNMENT

Charges are automatically assigned to atoms by an empirical scheme designed to be very fast and more accurate than simple quantum mechanical schemes. The program first assigns any formal charges, using a table which currently contains 26 formal charges indexed by element and subtype. Next charges are adjusted between neighboring bonded atoms using bond moments. The program first looks for a bond moment in a table of special bond moments (currently 48 in length), indexed by elements, subtypes, and bond type; then in a general bond moment table (currently 54 long), indexed only by elements and bond type. After this initial assignment, several steps are performed to allow for charge sharing around positive nitrogen, induction effects, carbon-carbon charge sharing, and aromatic ring "meta" effects.

Charge assignments were checked by comparing 189 calculated dipole moments with experimental dipole moments, selected primarily from gas phase determinations. The current average difference between the observed and calculated dipole moments is 0.21 Debyes; the average bias is -0.01 Debyes.

7. FORCE FIELD VERIFICATION RESULTS

We have developed a large test set of experimental structures, selected to give accurate experimental examples of the functional groups being parameterized. Some fairly large and fairly complicated molecules are included, since experience has shown that it is not sufficient to use only small molecules as test structures. Our current set of test structures consists of 339 experimental data sets: 106 determined by microwave spectroscopy, 169 by crystal structure analysis (42 neutron diffraction and 127 X-ray diffraction), 59 by electron diffraction, 3 by raman spectroscopy, and 2 by infrared spectroscopy.

During each cycle of force field parameter development, each of these structures is minimized, and the calculated results are compared with the experimental. A special computer program is used to correlate the massive amounts of data produced. The current agreement statistics are given in Table 1. For each structural feature, the number of different types (potential parameter sets), the total number of observations, and the average differences between experimental and calculated values is given. Since the number of observations varies greatly with type, two averages are given: for the "type" average, the observations for each type were averaged together, then an average was taken over the types; the "obs" average was a straight average over all observations. The "non-bonded contacts" listed are very close, usually repulsive, contacts.

Table 1. Overall   Agreement (September 1986)

	Number		Average Difference	
	Types	Obs	Type	Obs.
Bond Angles	386	7467	1.3°	1.3°
Bond Distances	273	5348	0.011 Å	0.013 Å
Torsion Angles	279	9063	3.3°	2.7°
Out-of-Plane	20	819	0.012 Å	0.011 Å
Hydrogen Bonds	77	857	0.087 Å	0.081 Å
Non-Bonded Contacts	44	5430	0.105 Å	0.06 Å

The agreement is quite good, considering the variety of functional groups covered. Analysis on an element-by-element basis gives comparable results.

For molecules not in the test set, reasonable expectations of accuracy for calculated structural features are about 1.5 times the values reported in the agreement table. That is: about 1.5° to 2.0° for bond angles, about 0.015 Å for bond distances and out-of-plane distances, about 5° for torsion angles, and about 0.1 Å to 0.15 Å for hydrogen bonds and non-bonded contacts.

REFERENCES

BUSING, W. R. (1970). *Trans. Am. Crystallogr. Assoc.* **6**, 57.

DUCHAMP, D.J. (1977). In *Algorithms for Chemical Computations* (ed. R. E. Christoffersen) pp. 98-121. American Chemical Society, Washington.

5. Conformational analysis of fentanyl and its derivatives

Jan P. Tollenaere

1. INTRODUCTION

Analgesic narcotics continue to draw the attention of pharma-
cologists, medicinal chemists, crystallographers and theoreticians. A wealth of structural information is available
which allows one to classify the various classes of analgesic
narcotics on the basis of precise X-ray structural data
(Tollenaere et al, 1979).

In this contribution it will be shown how the combined use
of X-ray crystallographic data and quantum chemical calculations may add to our knowledge of the conformational properties of fentanyl and some of its derivatives. More specifically, using computer graphics techniques it will be illustrated that comparing the conformational domains by superposition of the conformational energy maps may lead to the
identification of a conformational subspace common to fentanyl
and some of its closely related congeners.

2. METHODS

The wealth of information regarding the many conformations of
structurally flexible molecules calls for appropriate handling
techniques. In this respect the introduction of molecular
computer graphics facilities in various research laboratories
(Hopfinger, 1984; Vinter, 1985) comes to the rescue of the

medicinal chemist in his need to visualize the results of, for example, a conformational analysis.

The system developed in our laboratory with the acronym AIDA (Aid in Interactive Drug Analysis) is a highly modular system designed for modeling of small drug molecules composed of a maximum of about 200 atoms. AIDA is implemented on an IBM 3090 mainframe computer and connected to the monochrome Tektronix 4054 graphics microcomputer. AIDA is interfaced to a number of well-known Quantum Chemistry Program Exchange routines. A more detailed description of AIDA has recently been published elsewhere (Tollenaere et al, 1986a).

As can be seen from Fig. 1 the conformations of the 4-anilidopiperidine fragment of fentanyl and its congeners are fairly similar. The question as to whether this conformational similarity is dictated by crystal packing forces or constitutes an intrinsic property of this fragment can only be answered by theoretical conformational analysis. However, when topographically comparing molecules it is wise to ascertain that their physical bulk properties are somewhat of the same order of magnitude. This being the case for the present series of compounds (Tollenaere et al, 1986b) one may attempt

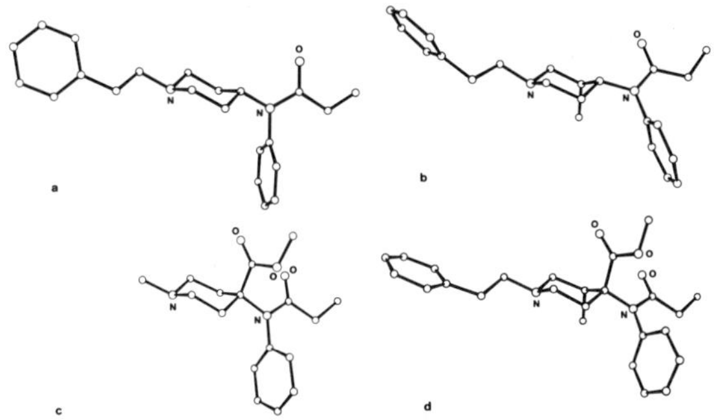

FIG. 1. Crystal structure conformations of a: fentanyl (Peeters et al, 1979); b: R 26 800 (Humblet); c: R 32 395 is the N-Me derivative of carfentanil (Durant); d: lofentanil (De Ranter).

to correlate the biological activity differences for this
series of compounds with the possible observed or calculated
conformational differences.

3. RESULTS

The conformational energy maps $E = f(\varphi_1, \varphi_2)$ based on
PCILO calculations are combined in a Venn diagram shown in
Fig. 2. The diagram displays the 3 Kcal/mole contours (solid
lines) of fentanyl, which encompasses all the low energy regions of its congeners. More interesting to note is the nar-
rowly defined conformational region represented by the shaded
area, which is common to all compounds studied. This common

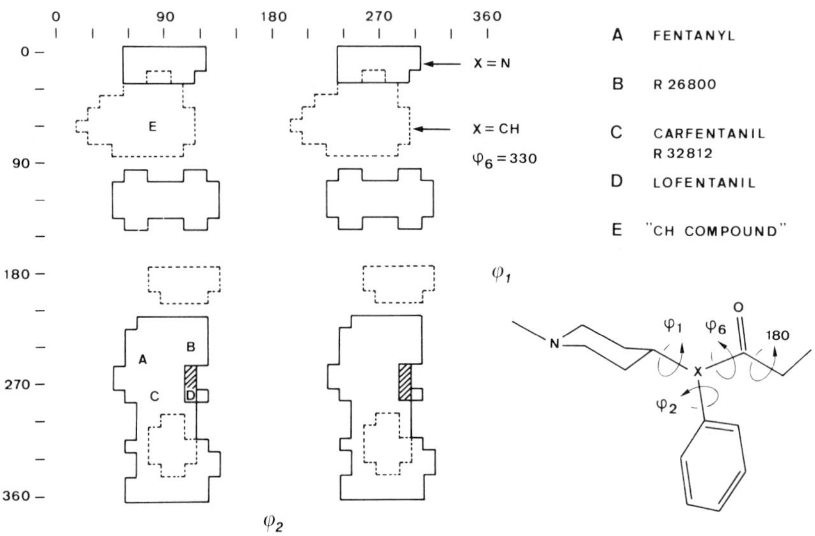

FIG. 2. Venn diagram showing the 3 Kcal/mole contours of
fentanyl (solid line) and the "CH-compound" (dashed line).
The calculated PCILO minima are indicated by A, B, C, D and
E. The shaded area represents the conformational space common
to fentanyl, R 26 800, carfentanil, lofentanil and R 32 812
(3-eq Me isomer of lofentanil).

conformation characterized by (φ_1 = 270° $\pm$ 10, $\varphi_2 \approx$ 110°) thus defines the probably biologically active or recepter-preferred conformation of the 4-anilidopiperidine moiety.

The calculation of the solvent-accessible surface area of fentanyl and its congeners led to the observation that the anilido N atom is completely shielded by its surrounding atoms and, therefore, unable to interact directly with the receptor atoms. In this sense the anilido N atom may be thought of as a structural spacer holding the anilidophenyl ring and the amidic bond in an appropriate mutual orientation. However, knowing that the replacement of the anilido N atom in fentanyl by a CH group entails a virtually complete loss of morphino-mimetic activity (Lobbezoo $\underline{et\ al}$, 1981), the prediction was made that the conformational minimum energy region of the "CH-compound" should be located outside the common region of fentanyl and its congeners. In fact, the dashed line 3 Kcal/mole contour of the "CH-compound" shown in Fig. 2 shows no commonality with the shaded area. Further calculations show that the "CH-compound" can mimic to a reasonable extent the receptor-preferrred conformation only at an energy expenditure of the order of 4 to 5 Kcal/mole.

Inspection of the separate conformational $E(\varphi_1, \varphi_2)$ maps (not shown due to space limitations) indicates that appropriate substitution strongly influences the conformationally flexibility of a molecule. The conformational flexibility or entropy can be calculated using eq (3):

$$F = \sum_{i=0}^{n} e^{-\Delta E_i / RT}$$

$$RT^2 \left(\frac{d \ln F}{dT} \right)_V = \frac{1}{F} \sum_{i=0}^{n} e^{-\Delta E_i / RT} \Delta E_i$$

$$S = R \left[\ln F + T \left(\frac{d \ln F}{dT} \right)_V \right] \qquad \text{eq. 3}$$

where F is the partition function summed over the N = 36^2 states with conformational energy ΔE_i and R and T the gas

constant and the absolute temperature respectively. The results listed in Table 1 show that the conformational entropy
values S offer a measure for conformational flexibility.

TABLE 1

<u>Conformational entropy</u>

	S (e.u.)
Fentanyl	10.13
R 26 800	6.81
Carfentanil	7.05
Lofentanil	4.44
3-eq-Me-lofentanil	6.53
CH-fentanyl	8.17

4. SUMMARY

It has been shown that the combination of X-ray data and
quantum chemical calculations can be used to address the problem of locating the most probable receptor-preferred conformation of a series of structurally related but conformationally flexible molecules. The statistical mechanical
expression for entropy provides an unbiased way to quantitate
conformational flexibility. The use of interactive molecular
computer graphics is indispensable.

ACKNOWLEDGEMENTS: The author would like to thank H. Moereels,
M. Van Loon and L. Geentjens for support during all stages of
preparation of this manuscript.

REFERENCES
DE RANTER, C.J. Private communication.
DURANT, F. Private communication.
HOPFINGER, A.J. (1984). <u>Pharmacy Int.</u> 224

HUMBLET, C. Private communication.

LOBBEZOO, M.W., SOUDIJN, W. and VAN WIJNGAARDEN, I. (1981). J. Med. Chem. 24, 777.

PEETERS, O.M., BLATON, N.M., DE RANTER, C.J., VAN HERK, A.M. and GOUBITZ, K. (1979). J. Cryst. Mol. Struct. 9, 153.

TOLLENAERE, J.P., MOEREELS, H. and RAYMAEKERS, L.A. (1979). Atlas of the Three-Dimensional Structure of Drugs. Elsevier/North-Holland Biomedical Press, Amsterdam, New York.

TOLLENAERE, J.P., MOEREELS, H. and VAN LOON, M. (1986a). Topics in Molecular Pharmacology, A.S.V. Burgen, G.C.K. Roberts and M.S. Tute, eds., Vol. 3, pp. 195-213. Elsevier Science Publishers, Amsterdam, New York.

TOLLENAERE, J.P., MOEREELS, H. and VAN LOON, M. (1986b). Progress in Drug Research, E. Jucker, ed., Vol. 30, pp. 91-126. Birkhäuser Verlag, Basel.

VINTER, J.G. (1985). Chemistry in Britain, 32.

6. Characteristics of peptide conformations

Isabella Karle

ABSTRACT

Although the sequence of residues in a peptide can be readily
determined and the bond lengths and bond angles are nearly
the same in all peptides, the conformation of a peptide is
very difficult to predict. These types of molecules are very
flexible, and, although the values of the torsion angles in
the backbone generally fall within the limits calculated by
Ramachandran, many different conformations are possible. Crystal structure analyses have shown that some peptides exist
as multiple conformers, side-by-side, in a unit cell. Examples
are cyclic hexaglycyl with four quite different conformations
for the backbone and Leu-enkephalin that forms an infinite,
anti-parallel beta-sheet with four different conformers side-
by-side. On the other hand, larger peptides have remarkably
constant conformations in polymorphic crystals despite different solvents, different packing, different intermolecular
hydrogen bonds and even limited side group substitutions. A
unique conformation is maintained by the cyclic decapeptide
antamanide in five different polymorphs. Another example
of constant conformation is afforded by a linear decapeptide
and a hexadecapeptide with the same sequence for residues
1 to 10. Residues 1-9 have an identical helical conformation
in both. Contrary to observations in structures containing

helical segments in other peptides and proteins and contrary
to theoretical calculations on the association of alpha-
helices, the alpha-helices in the present structures do not
pack in an antiparallel fashion. Since both the decapeptide
and hexadecapeptide crystallize in Pl cells, and since there
is only one peptide molecule per unit cell in each case, the
alpha-helices must associate in a parallel fashion.

1.   INTRODUCTION

Peptides found in animal, plant and fungal sources, as well
as in micro-organisms, perform many diverse biological functions. They are regulators for bodily processes such as
growth, appetite and blood pressure. They act as opiates,
antitoxins, neurotoxins, antibiotics and ionophores. In order
to begin to understand the diversity of peptide functions,
one of the fundamental pieces of information is a knowledge
of the geometry and folding of peptide molecules. It is
with this purpose in mind that x-ray diffraction analyses
of single crystals of peptides are performed.

     The sequence of residues in a peptide is usually known
before a structure analysis begins.   Bond-lengths and bond
angles differ relatively little in all peptides (Benedetti,
1977). The torsional angle about the $C'-N$ bond, often
called the peptide or amide bond, is generally close to 180°
($\pm$5°). The ranges of rotation about the remaining two
types of bonds in a peptide backbone, $N-C^\alpha$ and $C^\alpha-C'$, generally fall within the limitations determined by calculation
using hardsphere models (Ramachandran and Sasisekharan, 1968).
Without even considering the rotational flexibility of the
side-chains, within the limitations of the rotational freedom
about the $N-C^\alpha$ and $C^\alpha-C'$ bonds, there are a very large
number of conformational possibilities for a peptide backbone.
The following questions arise: Do peptides exist in many conformations? Is there a preferred or stable conformation for

a particular sequence? How do these flexible molecules fold?
X-ray structure analyses to date have provided some answers.

2.   MULTIPLE CONFORMATIONS

Cyclic hexaglycyl crystallizes in the triclinic system with
four quite different conformations of the backbone occurring
side-by-side in the unit cell (Karle and Karle, 1963). The
four conformers occur in the ratio of 4:2:1:1. The most
prevalent conformer has two internal hydrogen bonds in beta-
bends of Type I, Fig. 1, whereas the remaining conformers
participate only in intermolecular hydrogen bonding. The
ease of forming several conformers is probably facilitated
by the absence of any side-chains in this peptide which
permits a greater latitude in the rotational ranges about
the bonds in the backbone.

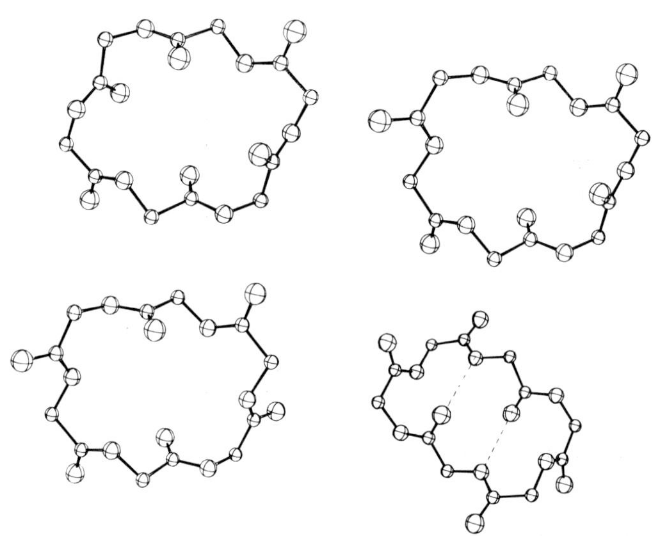

FIG. 1. Four conformers of cyclic hexaglycyl that occur in
the same triclinic cell.

Leu-enkephalin, Tyr-Gly-Gly-Phe-Leu, a natural opiate found in the brain and gut, is a linear peptide that crystallizes with four different conformers side-by-side in the crystal (Karle et al, 1983). In Fig. 2 it can be seen that the backbone is fairly similar in all four conformers. However, the side-chains, depicted by dark atoms, assume a number of entirely different conformations, particularly Tyr and Leu. The schematic diagram on the left in Fig. 3 shows how the backbones of the four confomers are directed to form an infinite anti-parallel beta-sheet. The right diagram is oriented in the same manner and shows the arrangement of the backbone atoms forming the beta-sheet with every NH and CO moiety utilized in the hydrogen bonding. The side-chains are directed above or below the beta-sheet into a thick layer of cocrystallized solvent consisting of water and dimethylform-amide.

Are there always multiple conformations? The answer is negative, especially for larger peptides. The larger peptides appear to have a unique folding pattern probably because

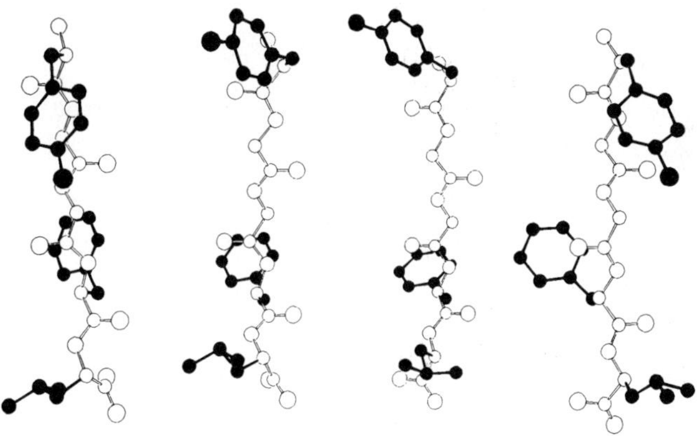

FIG. 2. Four conformers of Leu-enkephalin that crystallize side-by-side in a unit cell of the crystal.

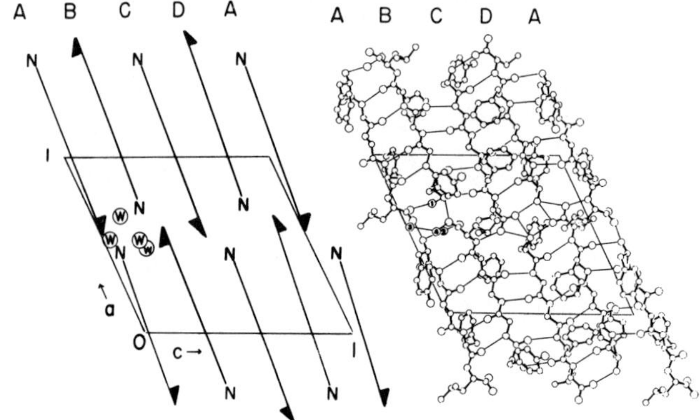

FIG. 3. The infinite anti-parallel beta-sheet formed by the
four different conformers of Leu-enkephalin. The schematic
diagram on the left shows the orientations of the backbones
in the four conformers. Four water molecules are denoted by W.

there are many more intramolecular interactions, such as
internal hydrogen bonds, and hydrophobic attractions, as compared to lattice forces, such as intermolecular hydrogen
bonds.

3.   UNIQUE CONFORMATIONS IN CYCLIC PEPTIDES

How can the invariability of conformation be tested? Fortu-
nately many peptides crystallize from a variety of solvents
and form polymorphs. That is, a particular peptide may crystallize in different crystal forms with different space groups
and different arrangements of the molecules with respect to
their neighbors. In this manner, the influence of the presence and polarity of solvent, or the attraction between peptide molecules or lack of it, can be examined.

Cycloamanide A, cyclic(Pro-Val-Phe-Phe-Ala-Gly), isolated from Amanita phalloides, has been found in two polymorphic forms (Karle and Chiang, 1984). One form has four H_2O

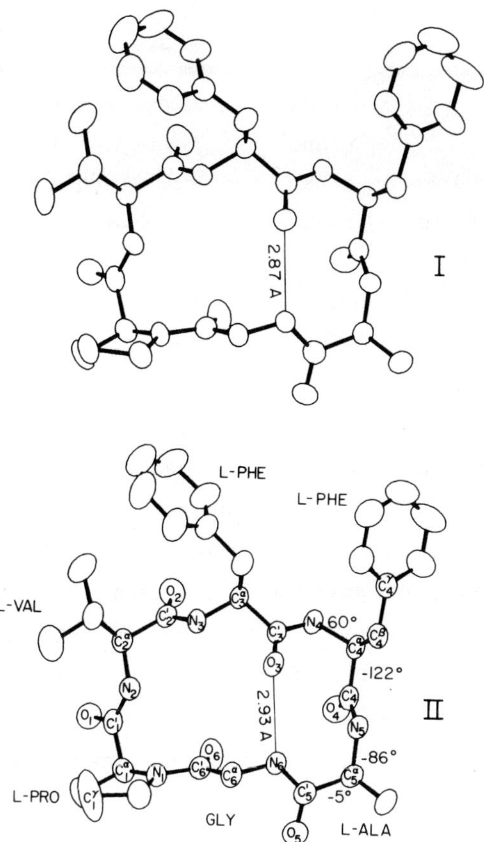

FIG. 4.   Comparison of the conformations of cycloamanide in
two different polymorphs.   Form I cocrystallizes with 4 H_2O;
Form II cocrystallizes with 1 H_2O and 3 C_2H_5OH.

molecules cocrystallized per asymmetric unit, the other has
one H_2O and three C_2H_5OH molecules cocrystallized.   In Fig. 4
is shown a comparison of the peptide molecule in the two crystal forms.   Not only is the conformation almost identical, but
an aberrant beta-bend occurs in both polymorphs.   Since the
corner residues are L-Phe and L-Ala, a Type I bond is expected, but a type II' is formed instead.   Type II' beta-bends

occur for D,L residues and are energetically unfavorable for L,L residues (Ramachandran et al, 1968). In the Type II' beta-bend in cycloamanide A, the $C^\beta(Phe)\cdots N(Ala)$ separation is only 2.84 Å despite an enlargement of 2-4° in the values of the $C^\beta_4 C^\alpha_4 C'_4$ and $C^\alpha_4 C'_4 N_5$ angles from the average of observed values in other peptides. In the two polymorphs, the cyclic peptide molecules stack over each other in the same manner, by a two-fold screw operation, into infinite columns. The differences in packing occur in the solvent channels between the peptide columns.

Antamanide, a cyclic decapeptide isolated from _Amanita phalloides_, and a biologically active synthetic analog (both from the laboratory of Prof. Theodor Wieland, Heidelberg) crystallize from completely nonpolar solvents, as well as polar solvents, into a number of different polymorphs. X-ray diffraction analyses have been completed on the polymorphs listed in Table 1. The contacts between neighboring peptide molecules in polymorph I are mostly by hydrogen bonds via intervening water molecules. In polymorphs II and III, adjacent peptide molecules are held together by hydrophobic contacts, where the rings of Pro[3] and Phe[4] stack over each other and the stack is continued by the stacked rings of Pro[8] and Phe[9] from a neighboring molecule. The process is repeated to form infinite stacks of alternating Pro and Phe rings along either side of each molecule. In polymorph IV, the packing forces are largely hydrophobic with a confluence of valyl, prolyl and phenylalanyl side chains at oblique angles with respect to each other. Despite different polarities of solvents, different cocrystallized solvent, different amounts of hydration, different associations between neighboring peptide molecules, different extrinsic hydrogen bonding (if any), and even different side chains on residues 4 and 6, the conformation of the peptide in all the polymorphs remains unchanged. The conformations of natural anta-

Table 1

Polymorphs of antamanide (AA) and synthetic symmetric [Phe⁴Val⁶]
antamanide (SAA)

	I	II	III		IV
Polymorph	$AA \cdot 8H_2O \cdot X^*$	$SAA \cdot 3H_2O \cdot Y^*$	$SAA \cdot 12H_2O$		$SAA \cdot 5H_2O$
Solvent	CH_3CN/H_2O	MeAc/n-Hex	Dioxane/H_2O^+		Acetone/H_2O
			Acetone/CH_3CN^+		or EtOH/H_2O
					or DMSO/H_2O
Space Group	$P2_12_12_1$	$P2_12_12$	$P2_12_12$		$P2_12_12_1$
a (Å)	15.88	14.85	15.10	15.08	20.19
b (Å)	34.88	21.77	22.01	21.92	21.12
c (Å)	13.61	12.06	11.02	11.01	16.13
Vol/Z ($Å^3$)	1884	1949	1831	1819	1719
Reference	Karle et al 1979	Karle 1977a	Karle 1986	Karle 1977b	Karle et al 1986

*X or Y: Additional cocrystallized solvent, disordered and un-
dentified.
+Same structure for peptide molecule; probably same structure
 for water molecules. Later paper has more complete refinement
 of 16 partially or fully occupied water sites.

manide in polymorph I and the active synthetic analog in
polymorph II are shown in Fig. 5

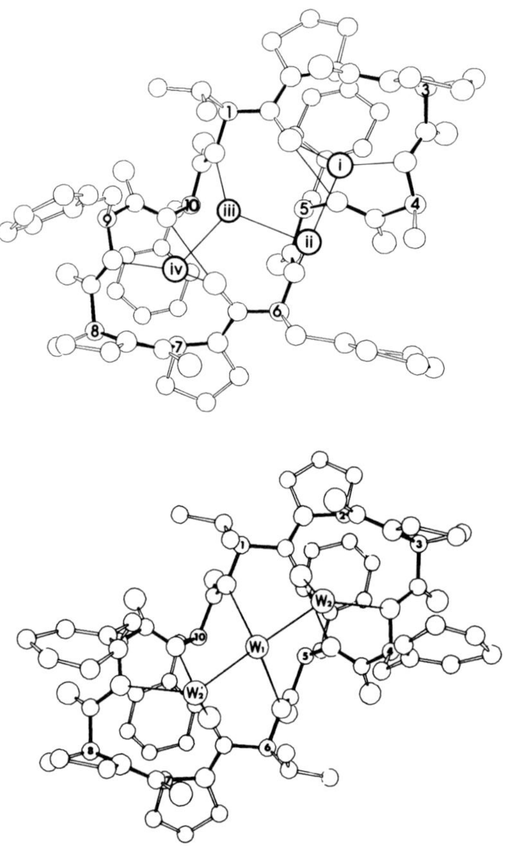

FIG. 5. Comparison of the conformations of antamanide with four intrinsic water molecules (upper diagram) and [Phe^{4}Val^6]antamanide with three intrinsic water molecules (lower diagram).

The one condition that does affect conformation of cyclic peptides is complexation with ions. Uncomplexed peptides generally do not have a suitable cavity already preformed, or a suitable polar region, to accept a metal ion directly. A large amount of folding and twisting of the backbone occurs during complexation (Karle, 1985).

4.   UNIQUE CONFORMATION IN LINEAR PEPTIDES

Structures of polymorphs of linear peptides have not been determined yet. The example that will be discussed here is an apolar hexadecapeptide and a decapeptide having a common 1-10 sequence. The syntheses and crystallizations were performed by Prof. P. Balaram and M. Sukumar, Bangalore. The hexadecapeptide, designed to be an apolar analog of the naturally-occurring membrane-modifying peptide zervamicin II, has the following sequence: Boc-Trp-Ile-Ala-Aib-Ile^5-Val-Aib-Leu-Aib-Pro^{10}-Ala-Aib-Pro-Aib-Pro^{15}-Phe-OMe. Both the decapeptide and the hexadecapeptide crystallize in space group P1 with one molecule of peptide per cell and very few water molecules.

The conformations of the decapeptide and hexadecapeptide are compared in Figs. 6 and 7, respectively (Karle et al, 1986; Karle et al, 1987). The helical peptides form infinite columns by head-to-tail hydrogen bonding. In each peptide, the N of the indole group of Trp^1 donates a proton to form hydrogen bonds directly with carbonyl oxygens of the molecule in the cell above. In the decapeptide, there is the bifurcated H-bond between N(Trp) and O(8) and O(9), while in the hexadecapeptide there is the H-bond between N(Trp) and O(14). There are no direct head-to-tail H-bonds between backbone atoms. The intervention of a small number of water molecules acting as bridges makes possible head-to-tail linkages such as N(1)···W···O(7) in the decapeptide and N(2)···W···W···O(14) in the hexadecapeptide.

Both peptides form an almost identical helix for residues 1-9, beginning with one or two 3_{10}-type helical segments and continuing with five or four alpha-type helical segments. There is very little difference in the torsional angles of the backbone in the two peptides for residues 1-9. The side-chain orientations are also almost the same except for the isoleucyl

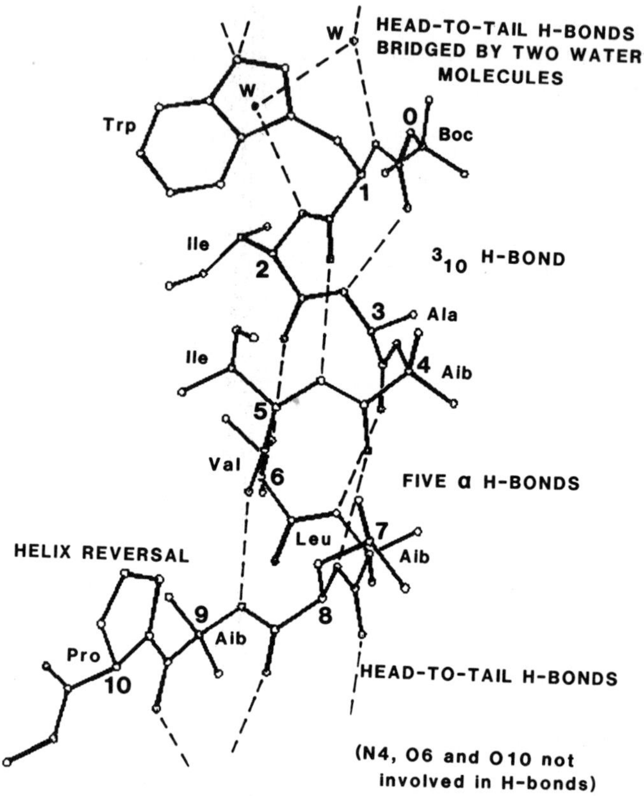

FIG. 6. Membrane modifying decapeptide. View along the helix. One molecule per triclinic cell. Hydrogen bonds are indicated by dashed lines. Water molecules labelled W.

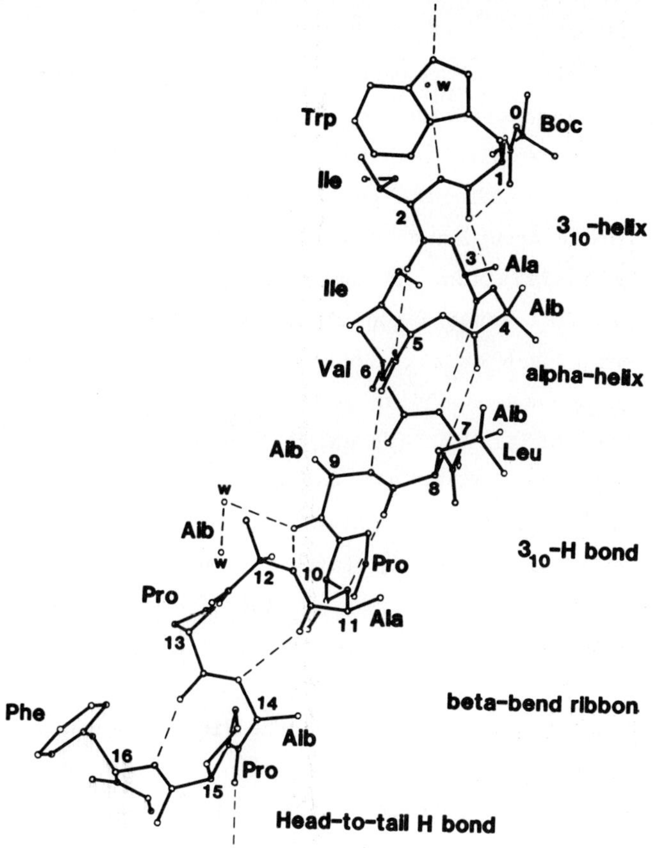

FIG. 7. Membrane modifying hexadecapeptide with the same 1-10 sequence as the decapeptide in Fig. 6. View along the helix.

side chain in residue 2. In the hexadecapeptide the remaining seven residues, among which are three prolyl residues, form a ribbon of three beta-bends. This ribbon also coils into a spiral. Thus, the 16-residue peptide has a sequence of three different kinds of helices: 3_{10}-helix, alpha-helix and beta-ribbon.

A most curious occurrence in crystals of both the decapeptide and hexadecapeptide is the parallel packing of the alpha-helices. Parallel packing is dictated by the fact that there is only one molecule per unit cell for each peptide (triclinic space group) and that all repetitions of molecules are only by translation. A survey of protein structures shows antiparallel association of alpha-helices. A nonapeptide fragment (Bosch et al, 1985b) of the C-terminal of alamethicin and an undecapeptide model (Bosch et al, 1982) of the N-terminal of alamethicin as well as alamethicin itself (Fox and Richards, 1982), all exhibit antiparallel packing of the alpha-helices in the crystal. Furthermore, theoretical calculations of minimum energy favor antiparallel association of alpha-helices (Hol and de Maeyer, 1984). Thus, the parallel packing of the peptides shown in Figs. 6 and 7 is quite unusual.

Almost all the structures described in this presentation have been solved by the symbolic addition procedure for phase determination (Karle and Karle, 1966). This research was supported in part by National Institute of Health Grant GM30902.

REFERENCES

BENEDETTI, E. (1977). _Peptides_. Proceedings of the Fifth American Peptide Symposium, M. Goodman and J. Meienhofer, eds., Wiley, N.Y. pp. 257-273.
BOSCH, R., JUNG, G., SCHMITT, H. and WINTER, W. (1985a). _Biopolymers_ 24 961-978.

BOSCH, R., JUNG, G., SCHMITT, H. and WINTER, W. (1985b). Biopolymers 24 979-999.

FOX, R. O. and RICHARDS, F. M. (1982). Nature 300, 325-330.

HOL, W. G. J. and de MAEYER, M. C. H. (1984). Biopolymers 23, 809-817.

KARLE, I. L. (1977a). J. Am. Chem. Soc. 99, 5152-5157.

KARLE, I. L. (1977b). Proc. Natl. Acad. Sci. USA 74, 2602-2606.

KARLE, I. L. (1985). Biomolecular Stereodynamics III, Proc. of the Fourth Conversation, R. H. Sarma and M. H. Sarma, eds., Adenine Press, Guilderland, N.Y., pp. 197-215.

KARLE, I. L. (1986). Int. J. Peptide Protein Res. 28, 6-14.

KARLE, I. L. and KARLE, J. (1963). Acta Cryst. 16, 969-975.

KARLE, I. L., WIELAND, T., SCHERMER, D. and OTTENHEYM, H.C.J. (1979). Proc. Natl. Acad. Sci. USA 76, 1532-1536.

KARLE, I. L. and KARLE, J., CAMERMAN, A. and CAMERMAN, N. (1983). Acta Cryst. B39, 625-637.

KARLE, I. L. and CHIANG, C. C. (1984). Acta Cryst. C40, 1381-1386.

KARLE, I. L. and WIELAND, T. (1986). Int. J. Peptide Protein Res. (In Press).

KARLE, I. L., SUKUMAR, M. and BALARAM, P. (1986). Proc. Natl. Acad. Sci. USA, In Press.

KARLE, I. L., FLIPPEN-ANDERSON, J., SUKUMAR, M. and BALARAM, P. (1987). To be published.

KARLE, J. and KARLE, I. L. (1966). Acta Cryst. 21, 849-859.

RAMACHANDRAN, G. N. and SASISEKHARAN, V. (1968). Adv. Protein Chem. 23, 283-438.

7. X-ray structure of phomopsin A, a hexapeptide mycotoxin

Maureen F. Mackay, Albert Van Donkelaar and C. C. J. Culvenor

1.   INTRODUCTION

Phomopsin A is the principal mycotoxin isolated from cultures of the fungus *Phomopsis leptostromiformis* (Kuhn) Bubak *ex* Lind. (Culvenor, et al., 1977; Lanigan et al., 1979) and is the causative agent of lupinosis. Lupinosis, a mycotoxicosis of sheep, cattle, horses and other domestic livestock grazing lupins (Lupinus spp.) or post-harvest lupin roughage (Van Warmelo et al., 1970; Gardiner and Petterson, 1972), is characterized by severe liver damage. The disease is of considerable importance in Australia and other countries (Gardiner, 1967; Marasas, 1978), where lupins, being legumes, are cultivated as a high protein seed crop. The lupin kernel is estimated to contain 40% crude protein (Coffey, 1986). Phomopsin A is produced on lupin stubbles infected with *Phomopsis leptostromiformis*, the levels usually being below 20 mg/kg, but occasionally reaching 40 mg/kg (Hancock, 1986). The fungus also infects lupin seed (Petterson and Wood, 1986), so that livestock could be at risk if fed lupin seed severely infected with the fungus.

Phomopsin A induces mitotic abnormalities shown to be due to a colchicine – like arrest of mitosis (Tönsing et al., 1984), the main target organ being the liver. Sheep acutely affected show a massive accumulation of lipid in their liver cells.   Phomopsin A is extremely toxic, in sheep, the LD_{50} being about 20–30 μg/kg body-weight given subcutaneously (Jago et al., 1982) and about 1 mg/kg intraruminally (Peterson,1986). The hazard due to phomopsin A poisoning is enhanced by the cumulative nature of the toxicity.

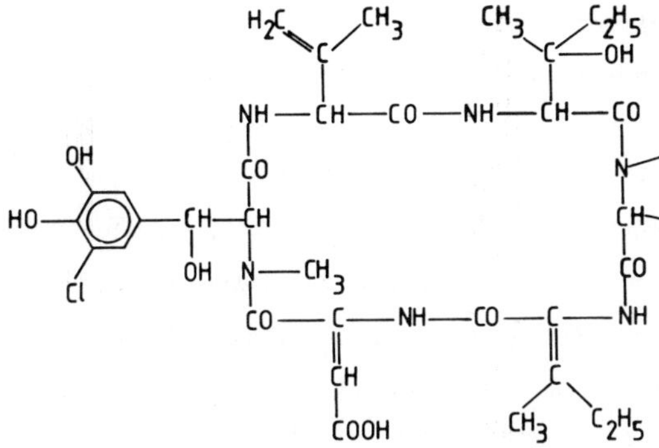

(1)

Phomopsin A is a hexapeptide made up of 2,3- and 3,4-didehydro and 3-hydroxyamino-acid residues *viz*. N-methyl-3-(3-chloro-4,5-dihydroxyphenyl)-3-hydroxyalanine, 3,4-didehydro-valine, 3-hydroxyisoleucine, 3,4-didehydroproline, E-2,3-didehydroisoleucine and E-2,3-didehydroaspartic acid, and was initially reported to be a cyclic peptide 1 (Culvenor et al., 1983).   This was subsequently amended to the linear structure 2 on the basis of FAB mass spectral data and further NMR studies (Edgar et al., 1986).   Structure 2, however, was not wholly

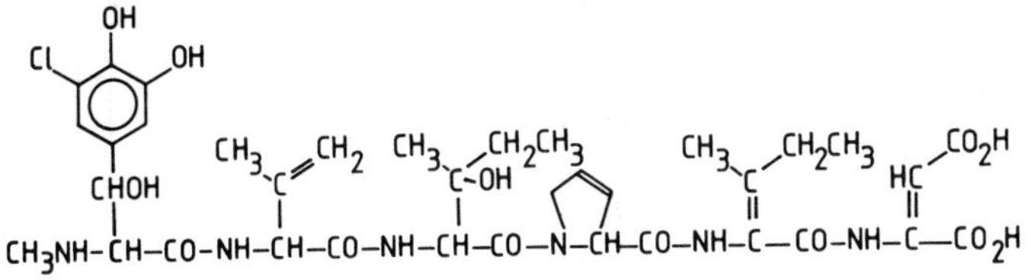

(2)

satisfactory in explaining certain properties of phomopsin A, for example a lack of borate complexing consistent with one rather than two phenolic hydroxyl groups (Edgar _et al._, 1985).   In an attempt to complete definition of the molecular structure of the mycotoxin, an X-ray analysis was undertaken.

2.   **CRYSTALLOGRAPHY**

Hydrated orthorhombic crystals of phomopsin A, $2(C_{36}H_{45}C\ell N_6O_{12})$. $5H_2O$, $M = 1666.6$, belong to the space group $P2_12_12_1$ with $a = 18.756(3)$, $b = 22.321(3)$, $c = 23.940(6)$ Å, $U = 10023(5)$ Å^3, $F(000) = 3528$, $D_C = 1.10$ g cm^{-3}, $Z = 4$ and $\mu(Cu K\alpha) = 11.0$ cm^{-1}.   The structure was solved by direct methods with MULTAN 78 (Main _et al._, 1978) with diffractometer data measured with Cu Kα radiation ($\bar{\lambda} = 1.5418$ Å) at 288(1) K. Refinement with 3633 observed 1.1 Å data ($I > \sigma I$) with anisotropic temperature factors given to the $C\ell$ atoms and isotropic temperature factors given to C,N and O, converged at R = 0.13.   The parameters were refined in two blocks with unit weights.

(3)

The molecular structure of phomopsin A is given as **3**.   The molecule has a linear configuration with a 13-membered macrocyclic ring which incorporates a _meta_-fused benzene ring. Phomopsin A resembles the cyclopeptide group of alkaloids (Tschesche and Kaussman, 1975), in that the latter are tetra- or pentapeptides with a modified (decarboxylated and dehydrated) 3-(*p*- or *m*-hydroxyphenyl)-3-hydroxylalanine unit linked through an ether bridge to a 3-hydroxyamino acid, to form a 13- or 14-membered ring. However, in the alkaloids, the peptide chain is connected through the amino group of the phenylalanine unit rather than through the carboxyl group as in phomopsin A.

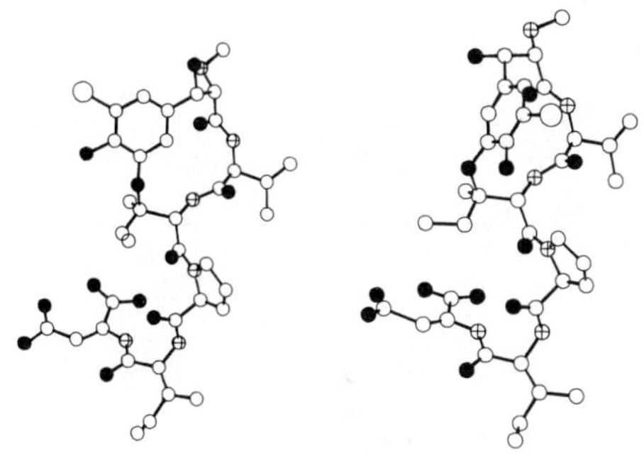

Fig. 1.   Perspective views of the two independent molecules in the crystal.

The two molecules in the asymmetric unit adopt different conformations – see Fig.1. The major difference is the relative orientation of the 3-chloro-4,5-dihydroxyphenyl moiety in the macrocyclic ring. The face of the ring, which is towards the observer in the molecule at the left in the diagram, has turned in the other molecule to be face-on to the C(6)-C(7) segment of the macrocyclic ring. The aromatic ring apparently is able to flip over by rotation about an axis joining O(2) and C(11). Phomopsin A is present in the crystal as a zwitterion with ionization at the didehydroaspartic acid residue and protonation at the N-methyl nitrogen of the substituted phenylalanine moiety.

3.   **SUMMARY**

Phomopsin A is a potent microtubule inhibitor and thus could represent a new class of compounds with importance in antitumour or anthelmintic chemotherapy (Dustin, 1984). Microtubules constitute a major component of the cytoskeleton of living cells, and are formed by the polymerization of the globular protein tubulin ($M \approx 110,000$). The tubulin dimer is made up of two almost identical subunits, $\alpha-$ and $\beta-$tubulin. A diagrammatical representation of the structure of a microtubule is given by Snyder and McIntosh (1976). The plant toxins, colchicine **4**, vinblastine **5** (Moncrief and Lipscomb, 1966) and maytansine **6** (Bryan et al., 1973) also inhibit microtubule formation. Phomopsin A, vinblastine and maytansine bind at

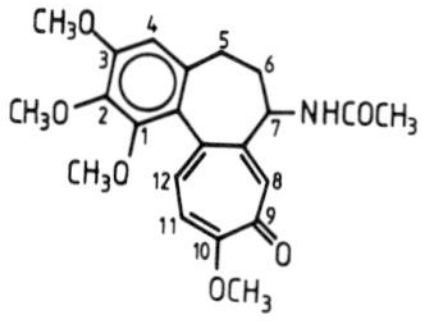

(4 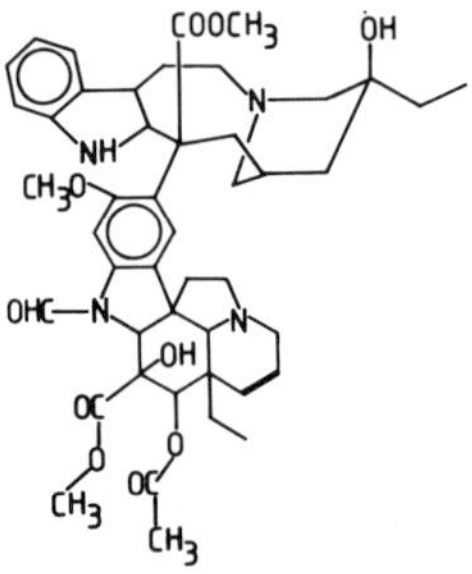)   (5 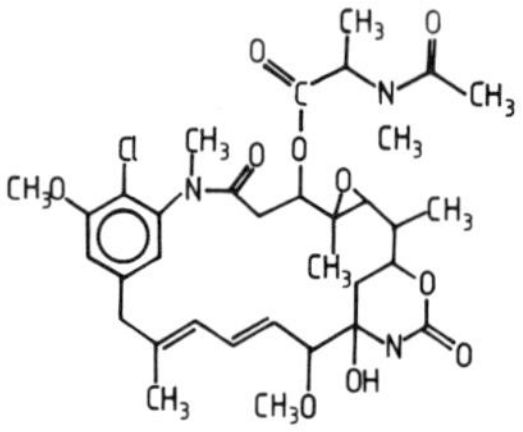)   (6)

or near the same tubulin site which is different from the colchicine binding site (Lacey et al., 1986). The three plant toxins have been shown to exhibit useful antitumour activity (Fitzgerald, 1976; Kupchan et al., 1978; Zavala et al., 1978; Quinn and Beisler, 1981), vinblastine and maytansine being powerful anti-leukaemic agents (Johnson et al., 1960; Kupchan et al., 1972).

Phomopsin B **7**, also isolated from *Phomopsis leptostromiformis*, and the two chemical derivatives of phomopsin A, phomopsinamine A **8** and octohydrophomopsin A **9** also are potent microtubule inhibitors (Lacey et al., 1986). The phomopsins and vinblastine are the most potent inhibitors being 30-fold more potent than colchicine and 5-fold more potent than maytansine. When the phomopsins, vinblastine or maytansine are bound to tubulin, the binding of colchicine to the protein is enhanced quite dramatically (Lacey et al., 1986).

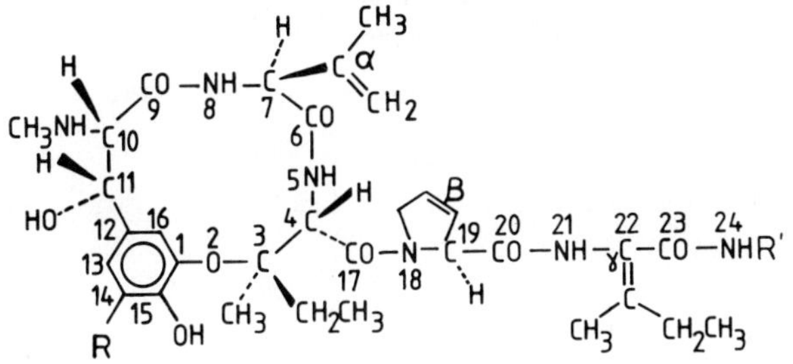

(**7**) R = H, R′ = C(COOH)≡CH(COOH)

(**8**) R = Cℓ, R′ = H

(**9**) R = Cℓ, R′ = CH(COOH)—CH(COOH)

(reduction at sites α, β and γ)

REFERENCES

BRYAN, R.F., GILMORE, C.J. and HALTIWANGER, R.C. (1973). J. Chem. Soc. Perkin Trans.II, 897-902.

COFFEY, R.S. (1986). J. Agric. West. Aust. **27**, 55-56.

CULVENOR, C.C.J., BECK, A.B., CLARKE, M., COCKRUM, P.A., EDGAR, J.A., FRAHN, J.L., JAGO, M.V. LANIGAN, G.W., PAYNE, A.L., PETERSON, J.E., PETTERSON, D.S., SMITH, L.W. and WHITE, R.R. (1977). Aust. J. Biol. Sci. **30**, 269-277.

CULVENOR, C.C.J., COCKRUM, P.A., EDGAR, J.A. FRAHN, J.L.,

GORST-ALLMAN, C.P., JONES, A.J., MARASAS, W.F.O., MURRAY, K.E., SMITH, L.W., STEYN, P.S., VLEGGAAR, R. and WESSELS, P.L. (1983). J. Chem. Soc. Chem. Commun., 1259-1262.

DUSTIN, P. (1984). Microtubules, 2nd rev. ed., Springer-Verlag, Berlin, p. 423.

EDGAR, J.A., CULVENOR, C.C.J., FRAHN, J.L., JONES, A.J., GORST-EDGAR, J.A., FRAHN, J.L., COCKRUM, P.A. and CULVENOR, C.C.J. (1986): in Micotoxins and Phycotoxins (eds. P.S. Steyn and R. Vleggaar) Elsevier, Amsterdam, pp. 169-184.

ALLMAN, C.P., MARASAS, W.F.O., STEYN, P.S., VLEGGAAR, R. and WESSELS, P.L. (1985) in: Trichothecenes and Other Mycotoxins (ed. J. Lacey) Wiley, Chichester, pp. 317-324.

FITZGERALD, T.J. (1976). Biochem. Pharmacol. 25, 1383-1387.

GARDINER, M.R. (1967). Adv. Vet. Sci. 11, 85-138.

GARDINER, M.R. and PETTERSON, D.S. (1972). J. Comp. Pathol. 82, 5-13.

HANCOCK, G.R. (1986). Personal communication.

JAGO, M.V., PETERSON, J.E., PAYNE, A.L. and CAMPBELL, D.G. (1982). Aust. J. Exp. Biol. Med. Sci. 60, 239-251.

JOHNSON, I.S. WRIGHT, H.F., SVOBODA, G.H. and VLANTIS, J. (1960). Cancer Res. 20, 1016-1022.

KUPCHAN, S.M., KOMODA, Y., COURT, W.A., THOMAS, G.J., SMITH, R.M., KARIM, A., GILMORE, C.J., HALTIWANGER, R.C., and BRYAN, R.F. (1972). J. Amer. Chem. Soc. 94, 1354-1356.

KUPCHAN, S.M., SNEDEN, A.T., BRANFMAN, A.R., HOWIE, G.A., REBHUN, L.I., McIVOR, W.E., WANG, R.W. and SCHNAITMAN, T.C. (1978). J. Med. Chem. 21, 31-37.

LACEY, E., EDGAR, J.A. and CULVENOR, C.C.J. (1986). Biochem. Pharmacol. (in press).

LANIGAN, G.W., PAYNE, A.L., SMITH, L.W., WOOD, P. McR. and PETTERSON, D.S. (1979). Appl. Environ. Microbiol. 37, 289-292.

MAIN, P., HULL, S.E., LESSINGER, L. GERMAIN, G., DECLERCQ, J.P. and WOOLFSON, M.M. (1978). MULTAN 78. A system of computer programs for the automatic solution of crystal structures from X-ray diffraction data.   Univ. of York England, and Louvain Belgium.

MARASAS, W.F.O. (1978) in: Mycotoxic Fungi, Mycotoxins, Mycotoxicoses (eds. T.D. Wyllie and L.G. Morehouse) Vol. 2, Marcel Dekker., New York, pp. 161, 186, 213 and 275, and references cited therein.

MONCRIEF, J.W. and LIPSCOMB, W.N. (1966). Acta Cryst. 21, 322-331.

PETERSON, J.E. (1986).   Proceedings of the 4th International Lupin Conference, August, Geraldton, Australia (in press).

PETTERSON, D.S. and WOOD, P.McR. (1986). J. Agric. West. Aust. 27, 53-54.

QUINN, F.R. and BEISLER, J.A. (1981). J. Med. Chem. 24, 251- 256.

SNYDER, J.A. and McINTOSH, J.R. (1976). Annu. Rev. Biochem. 45, 699-720.

TÖNSING, E.M., STEYN, P.S., OSBORN, M. and WEBER, K. (1984). Eur. J. Cell. Biol. 35, 156-164.

TSCHESCHE, R. and KAUSSMANN, E.U. (1975) in: The Alkaloids (ed. R.H.F. Manske) Vol. 15, Academic Press, New York, p. 165.

Van WARMELO, K.T., MARASAS, W.F.O., ADELAAR, T.F., KELLERMAN, T.S., VAN RENSBURG, I.B.J. and MINNE, J.A. (1970). J. S. Afr. Vet. Med. Assoc. 41, 235-247.

ZAVALA, F. GUENARD, D. and POTIER, P. (1978). Experienta, 34, 1497-1499.

8. 'Uncommon' aminoacids and their derivatives: conformation in the crystal and in the vapour phase

David Ajò, Maurizio Casarin, Vilma Busetti, Gaetano Granozzi, Josè A. Mayoral and Henricus C. Ottenheijm

1. DEHYDROAMINOACID DERIVATIVES

In many naturally occurring peptides, α,β-dehydroaminoacid residues (Fig.1) are present (Ajò et al, 1980b).

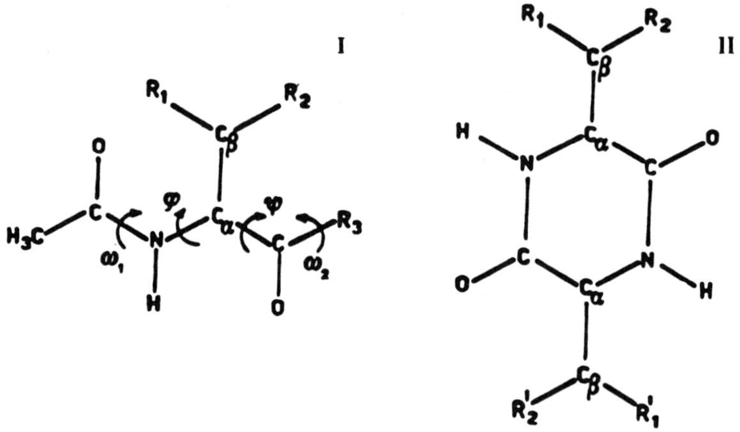

Fig. 1.: Acylic (I) and cyclic (II) α,β-dehydroaminoacid derivatives.

Conformational and electronic studies (Ajò et al, 1984a, b) of these molecules are of interest, in order to investigate the mechanism of their asymmetric hydrogenation.

1.1 Crystal and molecular structure

The conformational behavior that can be inferred from theoretical studies [see in Fig.2 the conformational energy map of the Z-dehydrophenylalanine (Δ^Z-Phe) residue] is quite different from that reported for saturated peptides. The low-energy region enclosing minima I and II appears to be peculiar to dehydro residues, due to conjugation effects. On the other hand, the amplitude of the regions below a given conformational energy level, taken as gauge of conformational flexibility, is similar to that for saturated residues.

In general, X-ray data support our conformational predictions, since experimental conformations correspond to low calculated energy regions, and several predicted minimum energy regions are populated. The presence of non-populated minima in the maps should suggest to extend our studies to systems containing dehydroaminoacid residues in different environments and bearing a larger series of side chains: actually, the most recent results concern N-acyldehydroamino acid N-alkylamides (Aubry <u>et al</u>, 1984; Aubry <u>et al</u>, 1985). However, none of the reported conformations populate the absolute minimum. This point is still open to discussion.

1.2 UV photoelectron studies

The conformational behaviour of dehydroamino acid derivatives is related to the properties of the unsaturated amino group. Mixing of doubly occupied ethylene π_{CC} with imidic π_N gives rise to π^- and π^+ combinations. Separation ($\Delta\pi$) between π^- and π^+ in the photoelectron (PE) spectra (Ajò <u>et al</u>, 1980a) of N-acetyldehydroalanine (which is planar in the crystal) is similar to the $\Delta\pi$ value reported for simple planar systems. Therefore, we feel confident that a planar conformation is maintained in the gas phase also, in agreement with theoretical conformational predicitions (Ajò <u>et al</u>, 1979).

On the contrary, β-substituted derivatives exhibit nonplanar conformations in the crystal state. Moreover, theoretical conformational analysis predicts the existence of four energy minima, all of them non-planar. Actually, in the PE spectra of N-acetyldehydrophenylalanine, the Δπ value is significantly lower than the corresponding one of N-acetyldehydroalanine. This is related to the smaller conjugation extent.

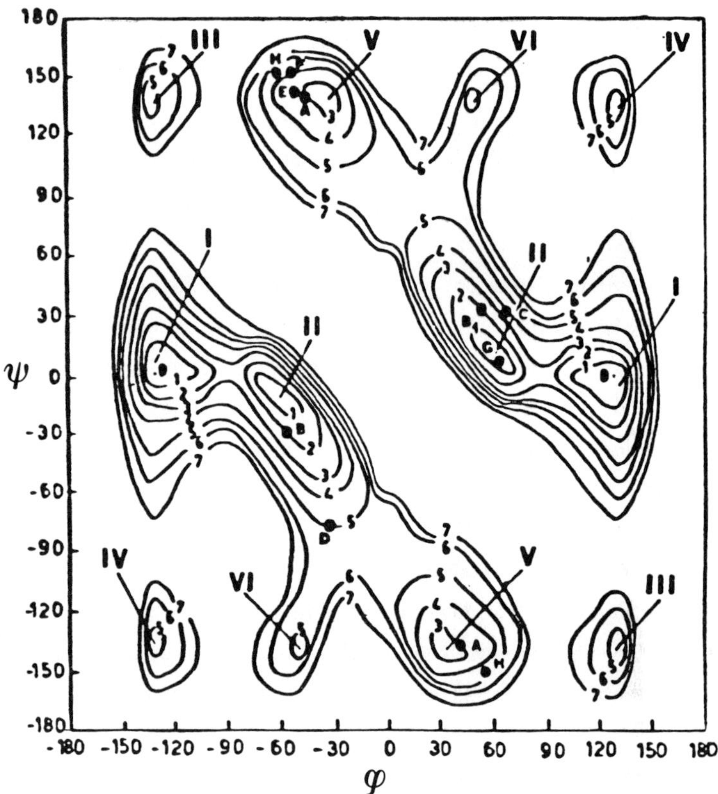

Fig. 2. The 0° value of the torsion angles corresponds to cis conformations: A,B: N-acetyl-(Δ^Z-Phe)$_2$-Gly (Pieroni et al, 1975), C,D: N-acetyl-(Δ^Z-Phe)$_2$-Ala (Pieroni et al, 1976, 1977), E: N-acetyl-Δ^Z-Phe-Pro (Ajò et al, 1982), F: N-acetyl-Δ^Z-Phe-Ala (Busetti et al, 1984), G: N-pivaloyl-Pro-Δ^Z-Phe-NH-i-Pr (Aubry et al, 1985), H: N-acetyl-Δ^Z- Phe-NH-Me (Aubry et al, 1985)

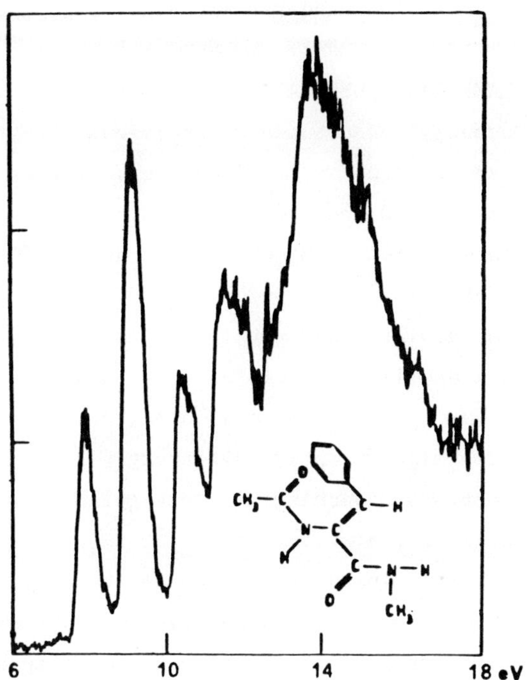

Fig. 3.: HeI PE spectrum of N-acetyl-Δ^{Z}-Phe-NH-Me

On the basis of PE results, non-planar conformations were proposed for all of β-substituted systems, which is in agreement
with theoretical predictions (Ajò et al, 1984b).

We have recently extended our studies to N-acyldehydro-
aminoacid N-alkylamides (see paragraph 1.1). They are models
of peptide units. We succeeded in obtaining good PE spectra
(Fig. 3) that can be interpreted by comparison with simple
related amides, even though deeper discussion will require
results of a larger series of compounds.

In conclusion, information obtained by PE spectroscopy,
in connection with more usual techniques, provides a degree of
knowledge of the electronic structure and conformational

behaviour of dehydroaminoacid derivatives, which is now comparable with that of the saturated amino acid derivatives.

2. N-HYDROXYAMINOACID DERIVATIVES

We have recently undertaken an analogous structural investigation of another class of "uncommon" systems, namely N-hydroxy-aminoacids and their derivatives, chemically and biologically related to α,β-dehydroaminoacids (Ottenheijm and Herscheid, 1986).

The molecular conformation of two N-acetyl-N-hydroxy-aminoacids (I and II, that differ only in a β-methyl group) is similar as far as the ϕ torsion angle is concerned but the γ values are significantly different (Fig. 4). On the other hand their crystal packing, in particular hydrogen-bond network is surprisingly similar!

These preliminary results point to an interesting conformational behaviour of these systems, and encourage us to pursue the investigation of N-hydroxyaminoacid derivatives bearing different side chains and terminal groups. PE and theoretical conformational investigations are in progress.

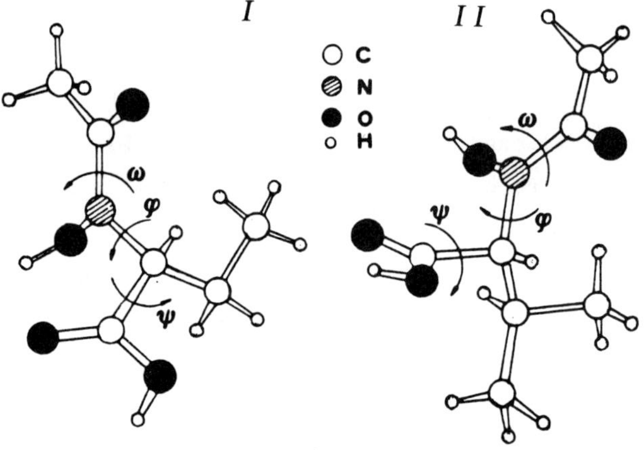

Fig. 4.: Crystal state conformation of N-acetyl-N-hydroxy-valine (I) and N-acetyl-N-hydroxy-α-aminobutyric acid (II).

ACKNOWLEDGMENTS

We thank Mr. Franco De Zuane for invaluable technical assistance.

REFERENCES

AJO`, D.,BUSETTI, V. and GRANOZZI, G. (1982). Tetrahedron 38, 3329.

AJO`, D.,BUSETTI, V.,OTTENHEIJM, H.C.J. and PLATE, R. (1984). Acta Crystallographica C40, 324, and references therein.

AJO`, D.,CASARIN, M.,GRANOZZI, G.,OTTENHEIJM, H.C.J. and PLATE, R. (1984). Reclueil des Travaux Chimiques Pays-Bas 103, 365, and references therein.

AJO`, D.,GRANOZZI, G.,CILIBERTO,E. and FRAGALA,I. (1980), Journal of the Chemical Society,Perkin Trans.II 483.

AJO`, D.,GRANOZZI, TONDELLO, E. and DEL PRA, A. (1980). Biopolymers 19, 469, and references therein.

AUBRY, A., ALLIER, F., BOUSSARD, G. and MARRAUD, M. (1985). Biopolymers 24, 639.

BUSETTI, V., AJO', D. and CASARIN, M. (1984),Acta Crystallographica C40, 1245.

OTTENHEIJM, H.C.J. and HERSCHEID, J.D.M. (1986), Chemical Reviews 86, 697.

PIERONI, O., FISSI, A., MERLINO, S. and CIARDELLI, F.(1976-77) Israel Journal of Chemistry 15, 22.

PIERONI, O., MONTAGNOLI, G., FISSI, A., MERLONI, S. and CIARDELLI, F. (1975), Journal of the American Chemical Society 97, 6820.

9. Conformational studies on some tobacco cembranoids

Jan-Eric Berg and Inger Wahlberg

ABSTRACT

Results from X-ray diffraction determinations have, in absentia, given an insight into the conformations about the 5,6 bond in (1S,2E,4S,6R,7E,11S)-2,7,12(20)-cembratriene-4,6,11-triol, [1], (1S,2E,4R,6R,7E,11S,12S)-6-acetoxy-11,12-epoxy-2,7-cembradiene-4,-ol, [2], (1S,2E,4S,6R,7S,8S,11E)-7,8-epoxy-2,11-cembradiene-4,6,-diol, [3], (1S,2E,4S,6R,7R,8R,-11E)-7,8-epoxy-2,11-cembradiene-4,6,-diol, [4]. An analysis of the ^{1}H and ^{13}C NMR spectra suggests that these conformations are essentially retained in solution and that other cembranoid alcohols or acetates are comformationally reminiscent of [1-4]. We have now carried out molecular mechanics (MM2) calculations using X-ray diffraction data For [1-4] as inputs. These show that, for each compound, the geometry at the local energy minimum is close to that existing in the crystalline state.

1. INTRODUCTION

Some sixty diterpenoids of the cembrane class have so far been isolated from tobacco. The (1S,2E,4R,6R,7E,11E)- and (1S,2E,-4S,6R,7E,11E)-2,7,11-cembratriene-4,6-diols, [5, 6], are the

major components.   Their   gross   structures were reported by
Roberts and Rowland (1962), but it was not until 1975 that the
relative stereochemistry of [5] was determined by X-ray analysis (Springer et al, 1975).   Ozonolysis was used to settle the
absolute configuration of C-1 to S (Aasen et al, 1975).

The stereochemistry of [6] was resolved by chemical correlation in 1982 (Wahlberg et al).   The long time required to determine the stereochemistries of these two, as well as other
tobacco cembranoids, can be ascribed to the fact that, since
so little is known about their conformational properties, NMR
cannot normally be used for unambiguous stereochemical assignments.   As a remedy, we decided to initiate a conformational
study of tobacco cembranoids. Another impetus was the interesting discovery from X-ray analyses that substantial conformational differences with respect to the C-4 to C-8 portion of
the molecule do exist within this group of compounds (Wahlberg
et al, 1984).   In this paper we will present some recent
results obtained by using molecular mechanics calculations and
NMR methods.   Four compounds [1-4], whose structures had been
determined previously (Wahlberg et al, 1982; Behr, et al,
1980; Wahlberg et al, in press), will be discussed.

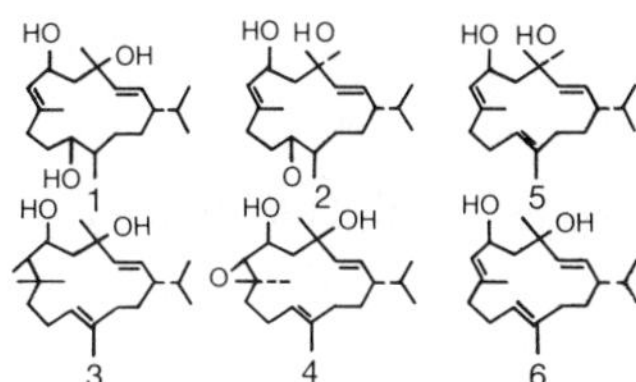

Fig. 1. Compounds [1-6].

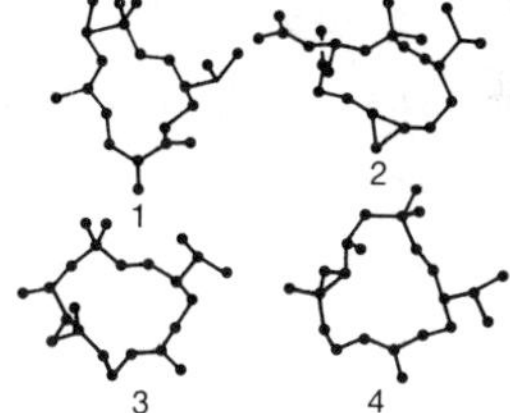

2. DISCUSSION

The energies of these four X-ray structures were initially
minimized using Allinger's MM2 program (QCPE No. 400). Our
results for the 11,12-epoxide, [2], illustrated in Fig. 2,
show that the structure representing the local energy minimum

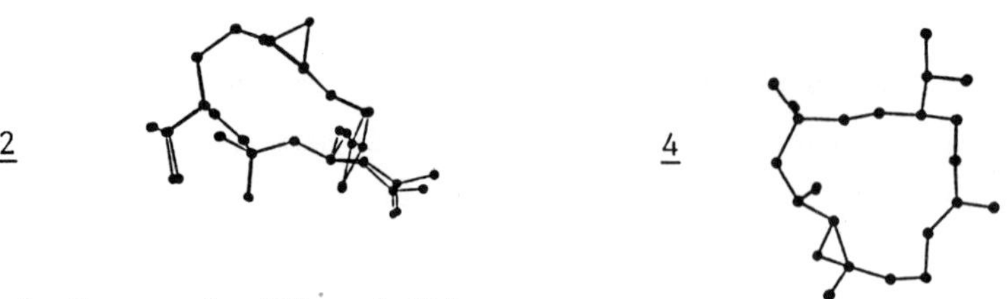

Fig. 2. Compounds [2] and [4].

is close to that found in the crystalline state.

A similar resemblance between the conformation in the crystalline state and that representing the local energy minimum was found for [4], a compound conformationally different from [2] with respect to the C-4 to C-6 portion of the molecule (Fig. 2).

The mean deviations with respect to bond lengths, angles and dihedral angles observed between the X-ray structures of [1-4] and the corresponding structures representing local energy minima are summarized:

Compound	[1]	[2]	[3]	[4]
Bonds	0.029Å	0.016	0.033	0.022
Bond Angles	3.6°	1.2	3.0	1.7
Dihedral Angles	15.2°	7.6	6.9	5.6

The major discrepancies are found around the hydroxyl and epoxide groups, a result that may be accounted for by crystal packing effects. Intermolecular hydrogen bonding is most likely also a factor of importance. It has been shown by the X-ray studies that the acetoxyl group at C-6 in one molecule of the 11,12-epoxide [2] participates in hydrogen bonding with the hydroxyl group at C-4 in another creating long chains. In the case of the 7R,8R-epoxide [4], the epoxide oxygen and the hydroxyl hydrogens at C-6 are the interacting species.

It was also of interest to explore whether the local energy minimum found is also the global one. Compound [4] was

selected for this study. All possible conformers of this compound were generated by using a computer program, RNGCFM from
QCPE (No. 510), which opens the ring at a given bond and then
rotates all the dihedral angles in the ring with a given increment. The possible conformers are separated from the physically impossible ones by the program, which takes bond lengths
and van der Waals contacts into account during the generation
of new conformers and rejects those having abnormal values.
Bond closure constraints are also applied to reject structures
that do not lead to the formation of a closed ring system. Of
the 5 billion structures generated, 46 were considered possible and unique. These were subjected to molecular mechanics
minimization.   Three groups of conformers were distinguished.
In the first, Fig. 3a, both the hydroxyl group at C-6 and the
epoxide group are located on the reverse side of the ring
framework as compared with the X-ray structure. In the second, Fig. 3b, the hydroxyl group at C-6 has a reverse alloca-
tion as compared with that in the X-ray structure. In the
third, Fig. 3c, the conformation is virtually superimposable
on that of the refined X-ray structure, the maximum deviations
being: bond length 0.004Å, bond angle 2.9° and dihedral angle
11.0°.   This conformer is also the one having the lowest energy, indicating that it represents the global minimum energy.

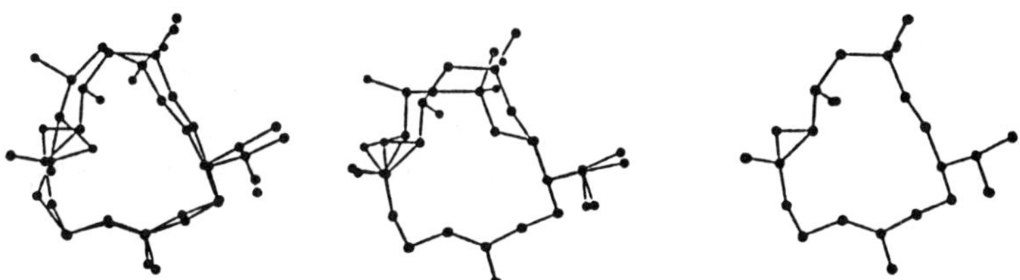

Fig. 3. Results from the conformational energy search for [4].
Depicted are examples from groups 1 to 3 as images a) to c)
respectively.

It was also of interest to explore whether the conformation of [4] in the crystalline state is retained in solution. A detailed analysis of its ^{1}H NMR spectrum by using ^{1}H - ^{1}H shift correlation spectroscopy and spin simulation studies was carried out (low temperature NMR work has not yet been performed). The coupling constants obtained were in harmony with the pertinent dihedral angles found in the crystalline state, indicating that no major conformational change takes place on going from the crystalline to the solution state (Wahlberg et al, in press).

3.CONCLUDING COMMENTS

Future work will include the effects of inter- and intramolecular hydrogen bonds in the minimizations. We are particularly interested in calculations comprising cembranoids connected by hydrogen bonds to a chain. We will also continue calculations of global minima but with a higher resolution.

REFERENCES

AASEN, A.J., JUNKER, N., ENZELL, C.R., BERG J.-E. and PILOTTI, A.-M. (1975). Tetrahedron Letters, 2607.

BEHR, D., WAHLBERG, I., NISHIDA, T., ENZELL, C.R., BERG, J.-E. and PILOTTI, A.-M. (1980). Acta Chemica Scandinavia B34, 195.

ROBERTS, D.L. and ROWLAND, R.L. (1962). Journal of Organic Chemistry 27, 3989.

SPRINGER, J.P., CLARDY, J., COX, R.H., CUTLER, H.G. and COLE, R.J., (1975). Tetrahedron Letters, 2737.

WAHLBERG, I., WALLIN, I., NARBONNE, C., NISHIDA, T., ENZELL, C.R. and BERG, J.-E. (1982). Acta Chemica Scandinavia B36, 147.

WAHLBERG, I., ARNDT, R., WALLIN, I., VOGT, C., NISHIDA, T., and ENZELL, C.R. (1984). Acta Chemica Scandinavia B38, 21.

WAHLBERG, I., EKLUND A.-M., VOGT, C., ENZELL, C.R. and BERG, J.-E. Acta Chemica Scandinavia in press.

10. Structure–activity relationships in Fusicoccin: ^{1}H NMR study of Fusicoccin and of some of its derivatives and analogs

A. Ballio, S. Castellano, S. Cerrini, A. Segre, A. Evidente and
G. Randazzo

ABSTRACT.

An ^{1}H N.M.R. study was performed on fusicoccin (FC) and on some
of its derivatives and analogs. All spectra were completely
analyzed in term of chemical shifts and coupling constants. All
possible conformations, leading to ring closure and compatible
with ^{3}J(H-H) values were generated by a computer program. Only
one conformation of FC and its derivatives is obtained compatible with the above constrain. Results in solution show that
all the active derivatives exhibit a similar conformation.

INTRODUCTION.

Information on the structural requirements necessary for biological activity of fusicoccin were obtained by comparing effects _in vitro_ and _in vivo_ of some derivatives and analogs with
those obtained on fusicoccin itself (Ballio et al.,1981 a,b).
It was suggested, on the basis of the crystal structures
(Cerrini et al.,1986), that the conformation of the tricarbocyclic system of the aglycone may be an important factor for

biological activity. With the purpose of evaluating the role of the conformation of the aglycone and of the whole molecule, a [1]H N.M.R. study was performed on fusicoccin and on some of its derivatives and analogs.

RESULTS AND DISCUSSION.

[1]H N.M.R. study at 500 MHz, in CDC13, was performed on: fusicoccin (FC) (1), its aglycone (FCA) (2), 7(S)-9(S)-diaster-eoisomer of FCA (3), FC-perhydrogenated derivative (4), cotylenol (5), FC-epimer in 3 (6), and a tetracyclic derivative of FCA (7). For each product, [1]H N.M.R. spectra were complete-ly analyzed with a least-square program; in all cases the r.m.s. deviation is less than 0.4 Hz. Chemical shifts and coupling constants for all derivatives are reported in TABLE 1 and 2 . Rather precise values of torsion angles can be obtained from ^{3}J(HC-CH,sp3) values (Haasnoot et al., 1980). Even though condensed rings may give anomalous values of ^{3}J, by using Haasnoot equation, a set of possible ranges for torsion angles was derived. The conformational study was performed as follows: each ring was "opened" and every possible conformation, obtai-nable by allowing torsion angles to rotate by suitable steps in all possible ranges, was calculated. Only those conformations leading to ring closure (within 0.1 A) were considered. The obtained conformations were compared with the N.M.R. data; all those grossly incompatible were discarded. Finally all torsion angles were refined by allowing smaller steps of variability

TABLE 1.

Chemical shifts (p.p.m. from TMS).

	1	2	3	4	5	6	7
1	5.325	5.362	5.364	1.496	5.517	5.080	⎡1.563
2							⎣2.166
3	2.788	2.780	2.760	2.556		2.530	2.168
4	1.684	1.696	1.758	2.020	1.405	1.408	1.755
4'	1.553	1.563	1.620	1.261	1.982	1.848	1.610
5	1.942	1.955	1.673	2.354	1.269	1.370	1.856
5'	1.604	1.602	1.580	2.240	2.009	2.004	1.457
6	2.778	2.827	2.766		2.931	2.909	2.124
7	2.020	1.990	1.903	2.493	1.938	1.948	2.096
8	3.958	3.838	3.907	3.923	3.934	3.988	3.660
9	3.914	3.963	4.807	3.716	4.066	3.846	4.041
12	3.772	3.843	3.744	4.008	1.850	3.806	3.873
12'					1.687		
13	2.374	2.478	2.610	2.384	2.142	2.590	2.404
13'	2.198	2.147	2.547	2.260	2.097	2.168	2.328
15	3.380	3.310	2.901	3.222	3.266	3.188	3.300
16	3.420	3.385	3.637	1.075	3.358	3.805	3.300
16'	3.362	3.402	3.442		3.076	3.708	3.211
17	0.865	0.880	0.974	1.092	0.805	0.874	1.053
18	1.100	1.190	1.320	1.120	1.214	1.180	1.153
19	4.304	3.393	3.561	4.380	0.957	1.002	3.300
19'	3.794	3.578	3.380	3.743			3.561
20	1.110	0.950	1.043	1.067	1.034	1.050	0.960
22	2.068			2.100			
23	3.361	3.343	3.372		3.403	3.360	3.323
24	5.090			5.041		4.981	
25	3.699			3.606		3.615	
26	4.979			5.040		3.836	
27	3.589			3.502		3.492	
28	3.682			3.799		3.690	
29	3.541			3.527		3.512	
29'	3.476			3.499		3.371	
31	2.145			2.170		-	
33	5.802			⎤1.508 / 1.457		5.768	
34_b	5.125			⎤0.841		5.150	
34_c	5.123			⎦		5.142	
35	1.270			1.150		1.272	
36	1.260			1.160		1.264	

TABLE 2.

$\underline{J}_{ij}$ coupling constants.

i-j	1	2	3	4	5	6	7
1 - 1							-13.3
1 - 3	2.0	2.0	1.7	-	-	2.0	
1 - 6	2.0	2.0	1.7	-	2.6	2.0	
3 - 4	7.8	7.7	2.5	9.0		4.8	6.8
3 - 4'	6.4	6.6	9.0	7.7		7.4	1.1
4 - 4'	-12.8	-12.9	-12.9		-13.0	-12.3	-12.5
4 - 5	7.5	7.3	8.2	9.9	10.9	8.4	7.0
4 - 5'	5.7	5.6	7.3	9.1	7.5	5.6	12.8
4'- 5'	7.2	7.2	11.8	7.8	7.1	6.5	6.7
4'- 5	1.1	1.5	6.6	1.8	3.4	0.9	1.2
5 - 5'	-12.9	-13.1	-12.9	-12.8	-13.4	-13.0	-13.0
5 - 6	8.5	8.5	1.4		8.7	8.1	8.3
5'- 6	5.0	4.9	7.1		8.9	5.9	10.3
6 - 7	0	0	11.3		0	0	2.0
7 - 8	4.2	4.7	3.7	2.8	4.3	5.0	5.0
8 - 9	10.1	10.3	3.7	8.5	10.1	10.0	10.0
12 - 12'					-12.0		
12 - 13	6.0	5.7	5.1	7.7	7.2	6.0	3.0
12'- 13					1.7		
12 - 13'	4.8	5.0	0.8	9.5	8.4	3.9	0
12'- 13'					10.1		
13 - 13'	-15.5	-15.8	-16.6		-15.8	-15.7	-16.8
3 - 16	6.0	6.7	5.2			4.2	4 ~
3 - 16'	2.9	0.8	4.7			3.5	8.6
16 - 16'	-9.0	-9.3	-9.1		-9.5	-9.2	-9.2
4 - 16					1.7		
15 - 19	10.1	10.8	10.0	10.2			10 ~
15 - 19'	4.2	4.2	4.8	4.2			5 ~
19 - 19'	-10.2	-9.3	-10.0	-10.0			-10 ~
24-25	4.0			4.0		3.6	
25-26	9.5			9.5		9.9	
26-27	9.5			9.5		9.8	
27-28	9.5			9.8		9.8	
28-29	4.4			5.0		4.5	
28-29'	4.0			4.8		7.0	

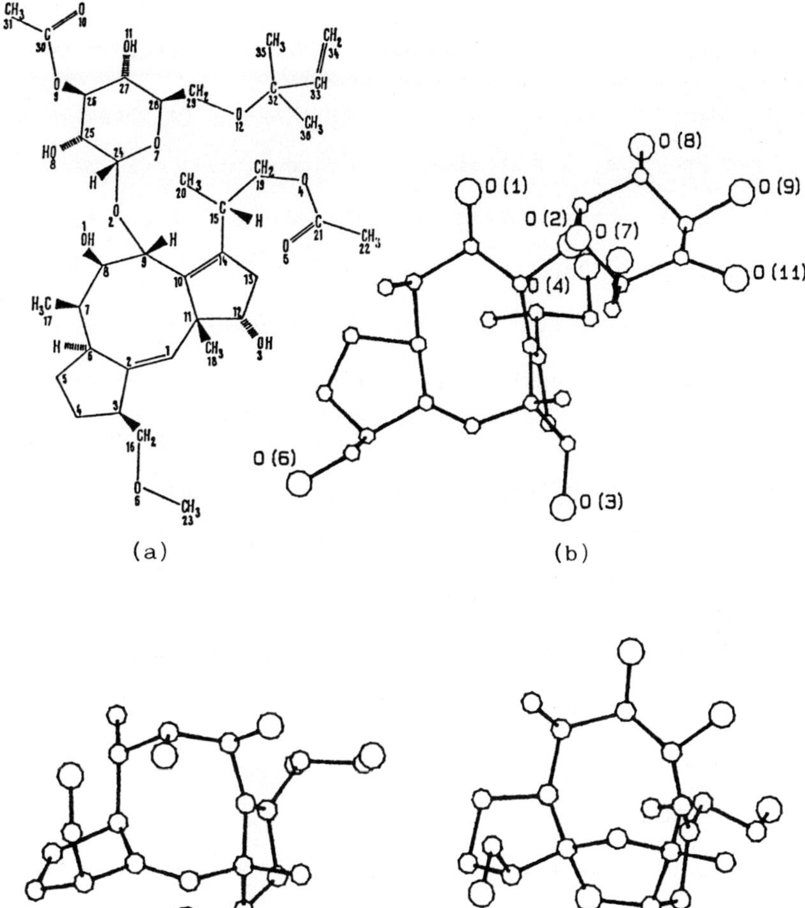

FIGURE: a) Fusicoccin FC 1.   b) Molecular structure of 1, showing the conformation of the carbon skeleton; the side chains on O(4),O(6),O(9) and O(12) are not shown due to lack of N.M.R. data; only C and O atoms are drawn for sake of clarity. c) Molecular structure of 3. d) Molecular structure of 7.

within the possible sets of solutions. In this way it is possible to get separately the conformation of the carbon skeleton of the aglycone and of the sugar. The relative orientation of the two moyeties was obtained by bidimensional NOE measurements. The structures obtained in this way for FC and for some of its derivatives are shown in the FIGURE.

The main result of this analysis is that all active derivatives ($\underline{1},\underline{2}$ and $\underline{5}$) exhibit in solution a similar conformation. Among the inactive compounds, $\underline{6}$ shows the same conformation, while $\underline{3}$ and $\underline{7}$ have a completely different one. Thus the inactivity of $\underline{6}$ must be attributed only to the chirality of C(3). The solid state conformation of $\underline{1},\underline{3}$ and $\underline{7}$ is known and corresponds to that found in solution. N.M.R. data support the role of the conformation of the tricarbocyclic system as a topological key in determinig the biological activity.

REFERENCES

BALLIO, A., DE MICHELIS, M. I., LADO, P. and RANDAZZO, G. (1981a). Physiol. Plant. 52,471, and ref. quoted therein.

BALLIO, A., FEDERICO, R. and SCALORBI D. (1981b). Physiol. Plant. 52,476, and ref. quoted therein.

CERRINI, S., FEDELI, W., GAVUZZO, E. and MAZZA, F. (1986). Abstracts, Int. Symp. on Molecular Structure, Beijing, p.91, and ref. quoted therein.

HAASNOOT, C. A. G., DE LEEUW, F. A. A. M. and ALTONA, C. (1980). Tetrah. 36,2783,and ref. quoted therein.

11. Crystal structures, conformational and molecular shape analysis and quantitative structure–anticonvulsant activity relationships of cinnamamides

Xiaojie Xu, Yue Guan, Zhao Cai, Yao Zhengui, Youqi Tang, Heng Fu, Minxie Qian, Shuyu Wang, Renli Li and Weigin Liu

1. INTRODUCTION

Since 1974 Beijing Medical University has studied folk medicines, among them white pepper and radish, which have been used as antiepileptic drugs in northern China for many years. Piperine, **1** was found to be the active constituent responsible for their antiepileptic activity.

$$CH=CH-CH=CH-C(=O)-N$$

1

In an effort to open up a new source for analogous drugs and to simplify the synthetic problem, an ethylene linkage was deleted from the chemical structure of **1**; N-(3,4-methylenedioxy-cinnamoyl)-piperdine, **2** was synthesized.

$$A \quad CH=CH-C(=O) \quad B \quad N \quad C$$

2

After systematic pharmacological experiments, chronic and acute toxicological tests (Pei Yin-quan, et al., 1978), **2** (Antiepilepsirine) was tested clinically for the treatment of epilepsy and found to be an effective broad spectrum antiepileptic drug with relatively few unfavorable side effects.

To improve the anticonvulsant activity of **2**, structural modifications were carried out for three moieties of the molecule, A, B, & C, in the illustrated structural formula.

Modication of moiety A: Substitution in the benzene ring with various groups leads to obvious changes in anticonvulsant activity (Zhang Xiao-hui, et al., 1980).

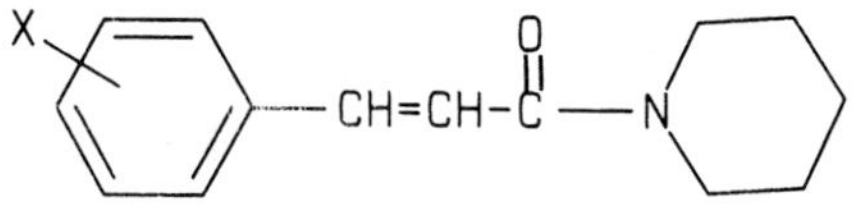

$X = 3\text{-}Cl,\ 3\text{-}F,\ 4\text{-}F,\ 4\text{-}Br,\ 4\text{-}Cl,\ 3\text{-}I,\ 3\text{-}CF_3$

Modification of moiety C:

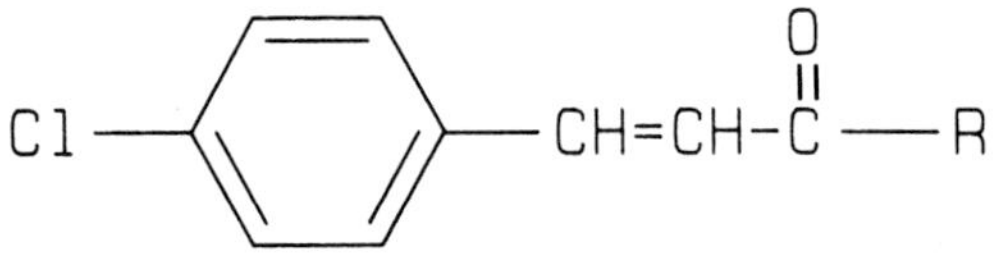

$R = NH_2,\ NH(CH_2)_3CH_3,\ NHCH(CH_3)_2$
$NH(CH_2)_4CH_3,$

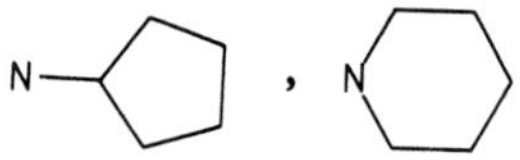

Among this series, cinnamoyl isopropylamine, cinnamoyl _sec_-butylamine and cinnamoyl piperidine are the most active.

Modification of moiety B:

$$R-\text{benzene ring}-X-\overset{\overset{O}{\|}}{C}-N\text{(piperidine)}$$

$$R-\text{benzene ring}-X-\overset{\overset{O}{\|}}{C}-NHCH(CH_3)_2$$

$$X = -CH=CH-, \; -CH_2-CH_2-, \; -CH_2-$$
$$R = H, \; 3,4\text{-}OCH_2O\text{-}, \; 4\text{-}Cl, \; 4\text{-}NO_2$$

Shortening the B portion to one carbon atom decreases anticonvulsant activity, whereas substitution in the para position of the benzene ring with chlorine may enhance anticonvulsant activity.

Thirty eight substituted cinnamamides were synthesized and subsequently subjected to Hansch analysis for the study of structure-activity relationships. With consideration of the accumulated data for these relationships, a series of cinnamamides, 12 compounds with varying activities were subjected to crystal structure determinations, and 28 compounds to quantum-mechanical and molecular-mechanical calculations. Quantitive correlation of structure and activity based upon molecular shape analysis was carried out. These studies are believed to be relevant to understanding the mechanism of the anticonvulsant activity.

2. STRUCTURAL FEATURES

2.1 X-Ray Crystallographic Data

X-ray data were collected with a Nicolet R3 diffractometer for crystals of the 12 compounds (compound 2 and those pre-

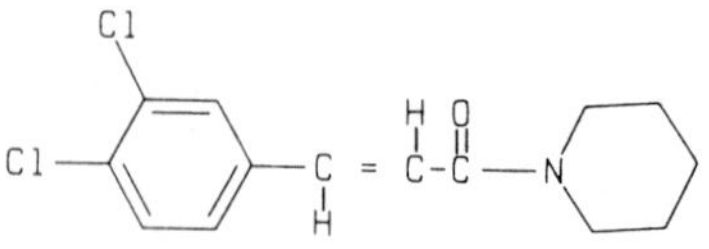

CINNAMAMIDES

3

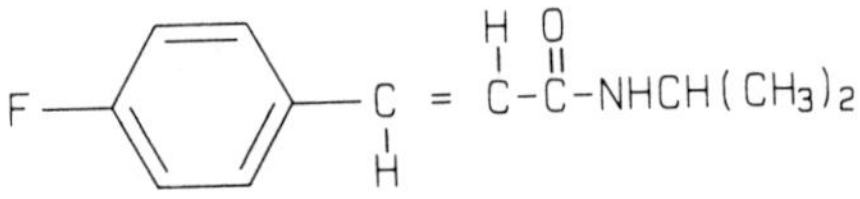

9

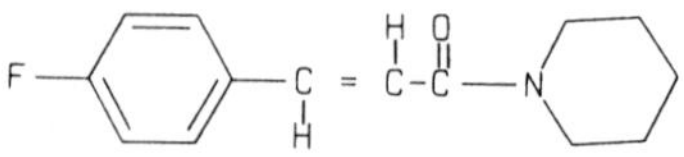

4

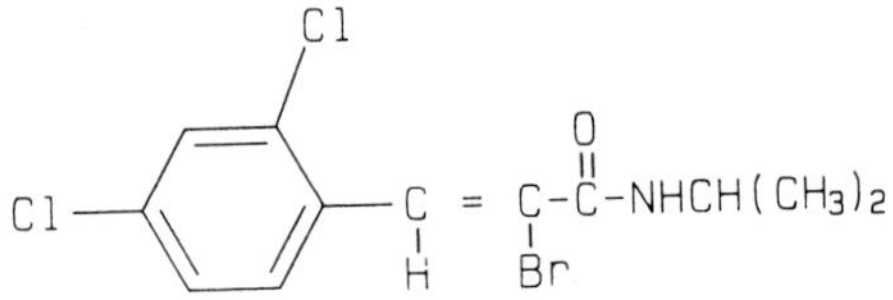

10

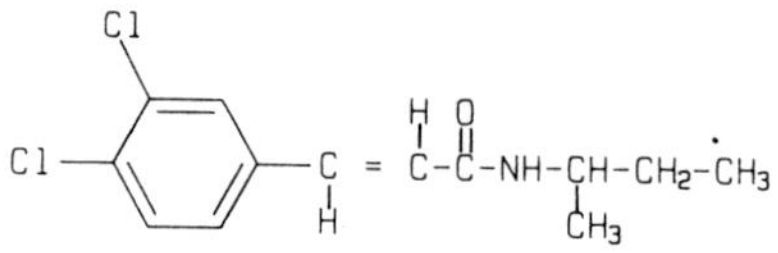

5

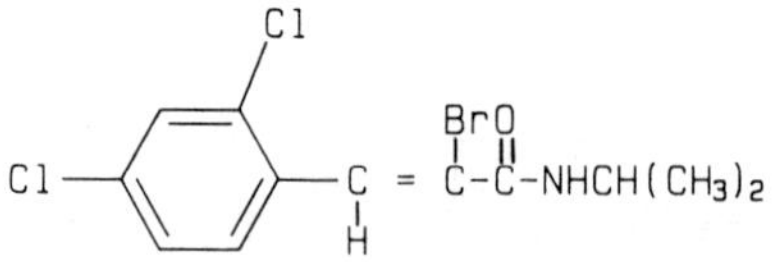

11

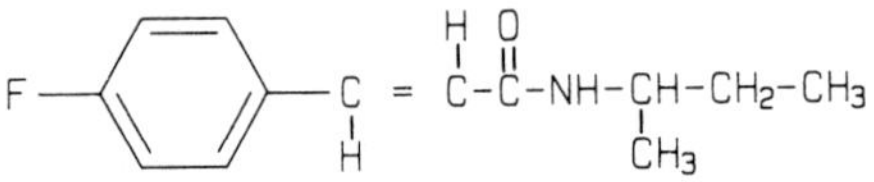

6

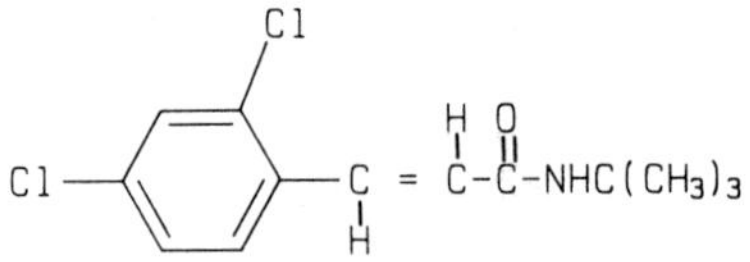

12

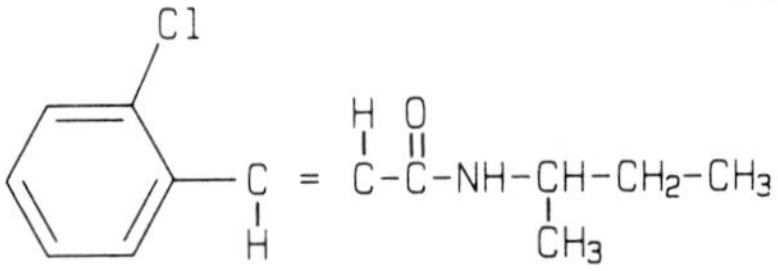

7

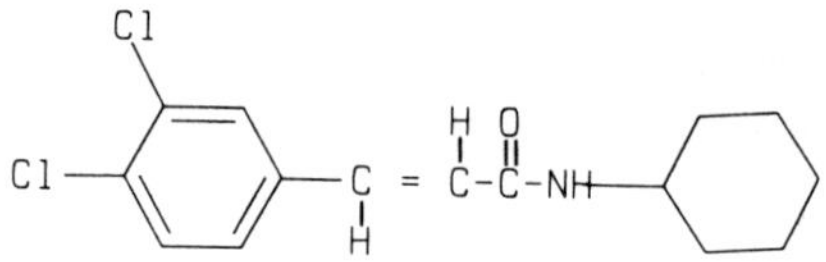

13

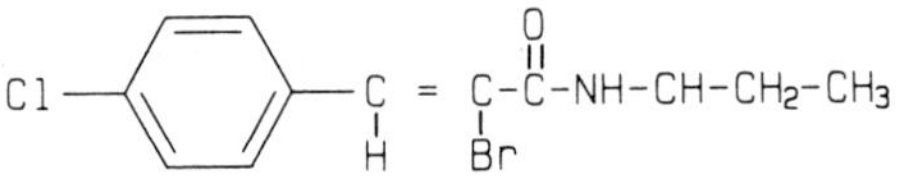

8

Fig. 1. The chemical structures of compounds for which crystal stuctures were determined. A crystal structure was also determined for **2**.

sented in Fig. 1 and analyzed using the Nicolet SHELXTL program package with an Eclipse S/230 computer. Selected molecular plots (ball and stick and space-filling models) are presented in Fig. 2.

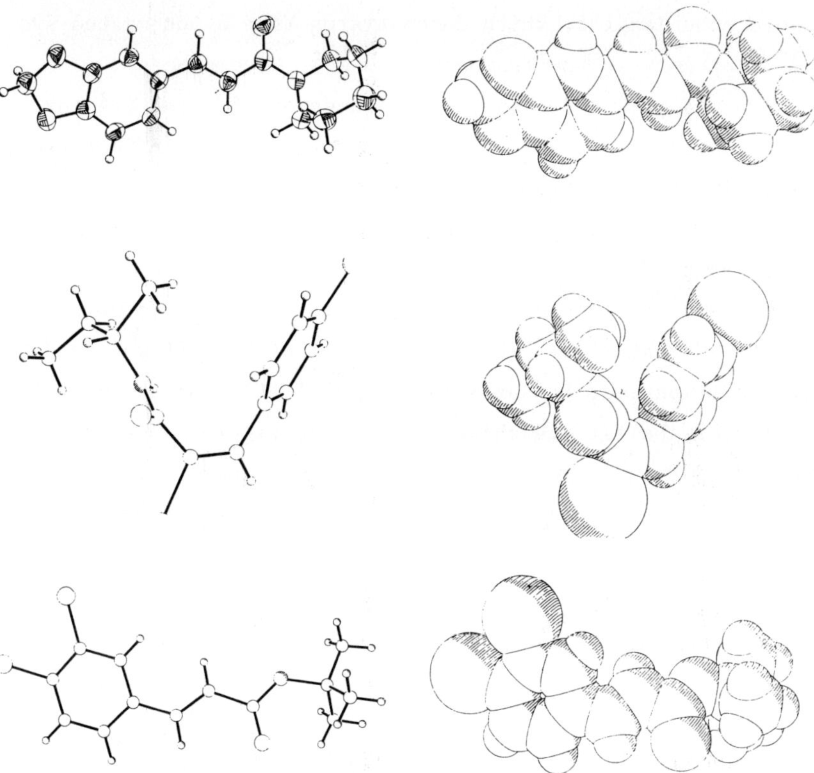

Fig 2. Selected representation for molecular entities resulting from crystal structure determinations. Depicted are ball
and stick and space filling models.

Although the chemical compositions for the compounds are
quite different, they all have a commmon framework.

The results of the crystal structures indicate that substitution, especially at the α- and/or β-positions, affects the structural features of this framework. All bond lengths in this moiety are intermediate between standard single and double bond lengths, which demonstrates that a conjugated system is present in the framework for all 12 compounds.

The phenyl-O and phenyl-N distances are 5.6 and 6 Å for the E-configuration and 4.9 and 4.2 Å for the Z-configuration, respectively. The framework consists of three small conjugated planes: the phenyl group, the double bond region and the amide group.

We used the molecular superposition technique to compare structural features . The XFIT program of SHELXTL was used to fit two molecules, for example 2 and 8, Fig. 3; the distributions of the fit are shown in the same figure. The entries D(RMS) and V in Table 1 are the root-mean-square deviation for all atoms and the common overlap volume of the two molecules described in MSA-QSAR.

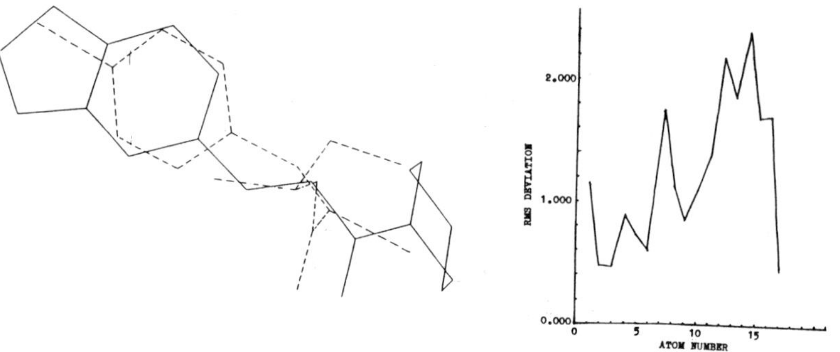

Figure 3. Molecules 2 and 7 superimposed and the RMS deviations of individual atoms.

Table 1. Selected structural parameters for compounds 2-13

Compound No.	Phenyl-O Å	Phenyl-N Å	V Å^3	D(rms) Å	Config- uration
2	5.69	6.26	148.4	0	E
3	5.64	6.27	125.5	0.795	E
4	5.64	6.26	97.6	0.905	E
5	5.35	5.75	96.1	0.884	E
6	5.33	5.75	121.7	0.134	E
7	5.70	6.10	99.0	0.763	E
8	4.85	4.10	81.3	1.195	E
9	5.72	6.14	97.0	0.797	E
10	5.61	6.32	112.7	0.178	E
11	4.98	4.27	88.7	1.144	Z
12	5.69	6.14	96.5	0.792	E
13	5.74	6.16	96.9	0.900	E

2.2 Conformational Analysis

The conformation of a drug molecule in the organism may
have great influence on its anticonvulsant activity. We can
determine conformations by X-ray crystallography. However,
the conformation in the crystal may be different from their
active conformation. It is often believed that the stable
conformer corresponding to the minimum of the potential can be
correlated with the biological activity. We performed conformational analysis for compounds 2 - 13 using the DPCILO
program to determine the stable conformer in the isolated
state, the so called dominant conformer (A. Masson, et al.,
1970).

We defined the rotation axes and the torsion angles as follows:

Take for example the conformational analysis of **2**. The conformational energy diagram for this compound (Fig. 4) was obtained by rotation around the axis defined by C(5)-C(8). The conformational energy map in Fig. 4 was derived by the simultaneous rotations around axes C(9)-C(10) and C(10)-N. The toriosn angles for the dominant conformers are shown in Fig. 5. The packing energy of the crystal conformer and the dominant conformer in the isolated state were calculated using the OPEC program.

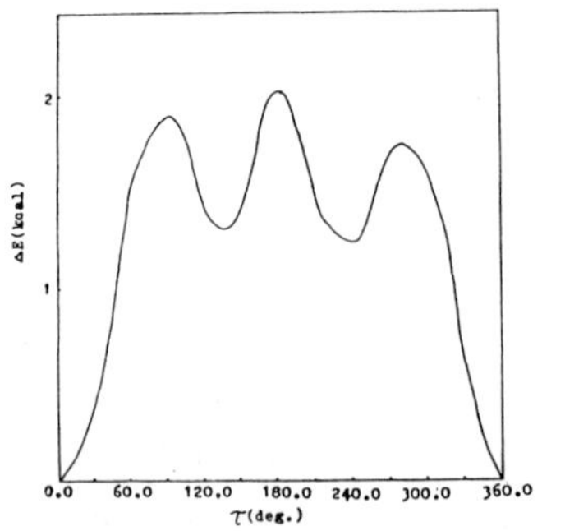
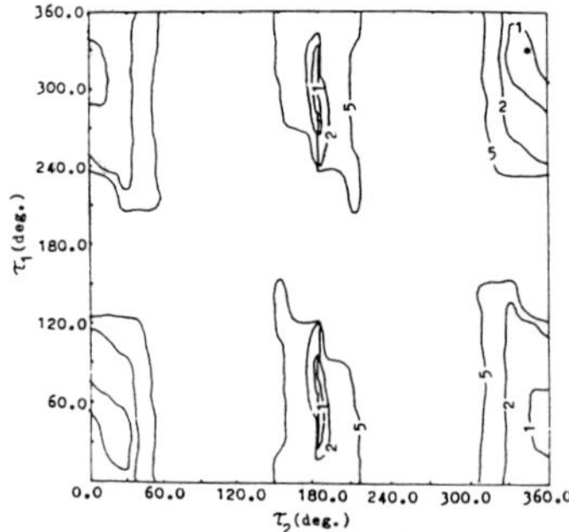

Fig 4. The conformational energy diagram (a) and conformational energy map (b) for **2**.

2.3. Molecular electronic properties and electrostatic potential fields

The drug action of a compound is expected to depend on their physical, electronic and steric properties.

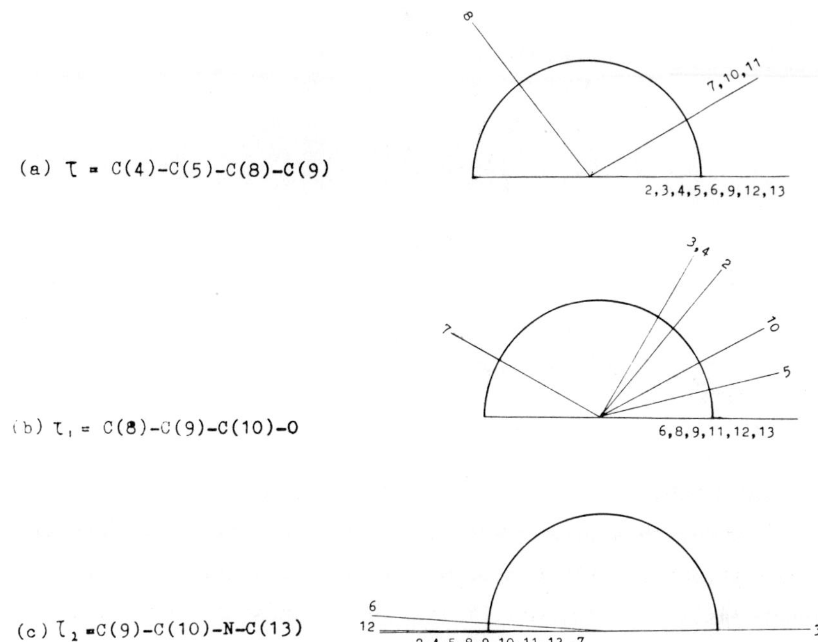

Fig. 5. Torsion angles for the dominant conformers derived from DPCILO calculations.

E(LUMO), E(HOMO), E(LUMO)-E(HOMO) and Q (atomic net charge) were determined by CNDO/2 quantum chemical calculations for 28 compounds. CNDO/2 was used to calculate the MO wave functions which were used to calculate the electrostatic potential at every point in the space. For example, the electrostatic potential map on the X-Y plane for 2 is shown in Fig. 6. In general, there is a large negative electrostatic potential field around the phenyl group substituted by one or two halogen atoms; there is another negative electrostatic potential region around the carbonyl oxygen atom.

There is a positive electrostatic region around the N atom. The volume of the substituent attached to the N atom affects the distribution of this positive electrostatic field.

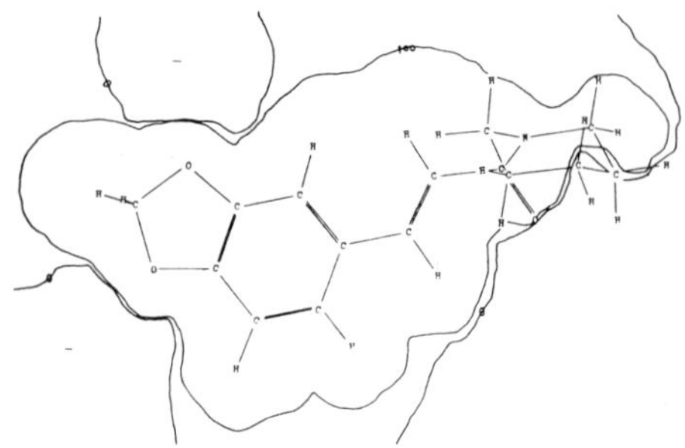

Fig. 6. Electrostatic potential in the X-Y plane for $\mathbf{2}$

4. QSAR STUDY

The Hansch approach was employed in an effort to develop correlations for the compounds in Table 2. The following equations were obtained. (N = 35, compounds $\mathbf{2}$, $\mathbf{19}$, and $\mathbf{39}$ were not included):

(1) $\log(1/C) = 0.246(\pm 0.15)\log P - 0.293(\pm 0.51)$

             $R = 0.513\ S = 0.238\ F = 11.77$

(2) $\log(1/C) = -0.287(\pm 0.19)(\log P)^2 +$
             $2.153(\pm 1.25)\log P - 0.362(\pm 2.05)$

             $R = 0.660\ S = 0.211\ F = 12.34$

(3) $\log(1/C) = -0.297(\pm 0.15)(\log P)^2 +$
         $2.173(\pm 1.02)\log P + 0.437(\pm 0.22)\Sigma\sigma - 3.435(\pm 1.68)$

             $R = 0.796\ S = 0.173\ F = 17.83$

(4) $\log(1/C) = -0.155(\pm 0.17)(\log P)^2 +$
             $1.350(\pm 1.09)\log P + 0.257(\pm 0.23)\Sigma\sigma -$
             $0.264(\pm 0.19)MR_{2,3,4} - 2.032(\pm 1.81)$

              $R = 0.844\ S = 0.156\ F = 18.64$

N = 38

(5) $\log(1/C) = -0.219(\pm 0.29)(\log P)^2 +$
             $1.647(\pm 1.40)\log P + 0.275(\pm 0.29)\Sigma\sigma -$
             $0.078(\pm 0.22)MR_{2,3,4} - 2.442(\pm 2.33)$

             $R = 0.689\ S = 0.205\ F = 7.48.$

Table 2. Parameters used in the derivation of QSAR equations
1-4 and 8-11

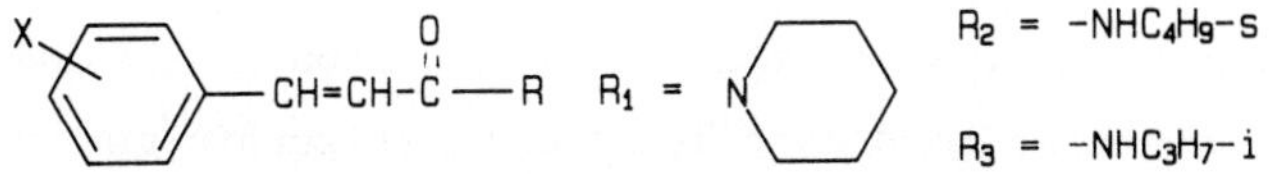

$$X\text{—}C_6H_3\text{—CH=CH—C(=O)—R} \qquad R_1 = \text{N-piperidyl} \qquad R_2 = -NHC_4H_9\text{-}s \qquad R_3 = -NHC_3H_7\text{-}i$$

log(1/c)

No.	X	R	obsd.	calcd	Δ	logP	π	Σσ	MR[a]
14	3-Cl	R_1	0.788	0.658	0.130	3.43	0.71	0.37	0.80
15	3-F	"	0.578	0.572	0.006	2.86	0.14	0.34	0.29
4	4-F	"	0.458	0.499	0.041	2.86	0.14	0.06	0.29
16	4-Br	"	0.314	0.582	0.268	3.57	0.86	0.23	1.09
17	$2,4\text{-}Cl_2$	"	0.664	0.673	0.009	4.14	1.42	0.46	1.30
3	$3,4\text{-}Cl_2$	"	0.550	0.709	0.159	4.14	1.42	0.60	1.30
18	4-Cl	"	0.606	0.621	0.015	3.43	0.71	0.23	0.80
2	$3,4\text{-}OCH_2O\text{-}$	"	0.564	0.116	0.448	2.66	-0.06	-0.32	1.00
19	$3,4,5\text{-}(OCH_3)_3$	"	0.793	0.037	0.756	2.66	-0.06	0.07	1.68
20	$4\text{-}NO_2$	"	0.268	0.292	0.024	2.44	-0.28	0.78	0.94
21	$3\text{-}NO_2$	"	0.324	0.274	0.050	2.44	-0.28	0.71	0.94
22	$3\text{-}CF_3$	"	0.921	0.744	0.177	3.60	0.88	0.43	0.70
33	$2\text{-}CF_3$	"	0.723	0.772	0.049	3.60	0.88	0.54	0.70
24	$4\text{-}CF_3$	"	0.921	0.772	0.149	3.60	0.88	0.54	0.70
25	$3\text{-}OH,4\text{-}OCH_3$	"	-0.272	-0.263	0.009	2.05	-0.69	-0.15	1.17
26	$4\text{-}OCH_3$	"	0.218	0.152	0.066	2.70	-0.02	-0.27	0.99
27	3-I	"	0.320	0.525	0.205	3.84	1.12	0.32	1.60
28	$4\text{-}OC_2H_5$	"	0.500	0.252	0.248	3.19	0.47	-0.24	1.45
29	$4\text{-}OC_3H_7\text{-}n$	"	0.290	0.285	0.005	3.77	1.05	-0.25	1.91
30	$4\text{-}OC_4H_9\text{-}n$	"	0.180	0.196	0.016	4.27	1.55	-0.32	2.37
31	3-Cl	R_2	0.410	0.717	0.307	3.67	0.71	0.37	0.80
32	3-F	"	0.495	0.674	0.179	3.10	0.14	0.34	0.29
6	4-F	"	0.495	0.602	0.107	3.10	0.14	0.06	0.29
33	4-Br	"	0.540	0.633	0.093	3.82	0.86	0.23	1.09
34	$2,4\text{-}Cl_2$	"	0.735	0.680	0.055	4.38	1.42	0.46	1.30
5	$3,4\text{-}Cl_2$	"	0.977	0.716	0.211	4.38	1.42	0.60	1.30
35	4-Cl	"	0.714	0.681	0.033	3.67	0.71	0.23	0.80
36	$4\text{-}CF_3$	"	0.772	0.819	0.047	3.84	0.88	0.54	0.70
37	$3\text{-}CF_3$	"	0.989	0.790	0.199	3.84	0.88	0.43	0.70
38	3-Cl	R_3	0.620	0.628	0.008	3.33	0.71	0.37	0.80
39	3-F	"	0.301	0.674	0.373	3.10	0.14	0.34	0.29
9	4-F	"	0.288	0.452	0.164	2.76	0.14	0.06	0.29
40	4-Br	"	0.580	0.559	0.021	3.48	0.86	0.23	1.09
11	$2,4\text{-}Cl_2$	"	0.600	0.665	0.065	4.04	1.42	0.46	1.30
41	4-Cl	"	0.801	0.592	0.209	3.33	0.71	0.23	0.80
12	$3,4\text{-}Cl_2$	"	0.498	0.701	0.203	4.04	1.42	0.60	1.30
42	$4\text{-}CF_3$	"	0.899	0.747	0.152	3.50	0.88	0.54	0.70
43	$3\text{-}CF_3$	"	0.924	0.719	0.205	3.50	0.88	0.43	0.70

[a]$MR = MR_{2,3,4}$

N is the number of data points, R is the correlation coefficient, S is the standard deviation, F is the significance test of regression analysis. In equations (1) to (5), logP is the lipid/water partition coefficient (calculated as the sum of Hammett substituent constants for the substituents on the benzene ring), MR is the molar refractivity (scaled by 0.1) of the substituents (to represent steric parameters), subscripts are used to indicate the positions of substituents.

Equation (4) gives the optimum correlation equation. Equations (1) to (3) are the stepwise development of (4). From (4) it is apparent that the anticonvulsant activities of this series of compounds are parabolically correlated with their lipid/water partition coefficients; the optimum value for logP is 4.35. The electronic and steric parameters for the benzene ring subsituents make significant contributions to the anticonvulsant activities.

Lipid/water partition coefficioents are important factors influencing physiological activtiy for the drugs affecting the central nervous systems. In general, the optimum partition coefficient, using octanol/water as the solvent system, is about 100, or logP = 2. In this series of compounds the optimum value of logP is 4.35 (the compound with logP = 2 shows very low anticonvulsant activity, e.g. compound 25, Table 2). The logP value for this group of compounds may reflect two properties, the transport of the drug and its hydrophobic binding strength with the receptor.

If only the hydrophobic constant of the substituent (π_x) is used in the correlation equation (6) is obtained. We interpret the fact that the correlation is not particularly good to indicate that hydrophobic binding between the substituents and the receptor is also an important factor.

$$(6): \quad \log(1/C) = -\,0.370(\pm0.19)(\pi_x)^2 + 0.653(\pm0.25)(\pi_x) + 0.394(\pm0.11)$$

$$N = 35 \quad R = 0.699 \quad S = 0.201$$

5. MSA-QSAR STUDIES

One theory of molecular shape analysis is outlined by A. J. Hopfinger and coworkers (1980). The biological activity and specificity of compounds are expected to depend on their physical, electronic and steric properties. Thus MSA-QSAR may be superior to QSAR using only physical-chemical and chemical structural features.

The OPEC program was used to calculate the molecular volume, surface area and free surface for each atom. SHELXTL-XFIT was used to fit two molecles. An inactive molecule, compound $\underline{48}$ in Table 3 was used as the reference. Shape similarities and differences can be described quantatively in terms of common overlap volume, V, between pairs of molecules. The common overlap steric volume, V_o, was calculated by the method described by Hopfinger (1980).

Drug molecules approach the receptor, activate it and then effect the activity. This procedure very likely involves atomic net charge, atomic free surface, distribution of electrostatic potential and molecular shape, etc. The correspondence between molecular shape of an anticonvulsant and the receptor plays an important role in affecting the activity. Consequently, we incorporated the parameters mentioned above in the QSAR parameters of the Hansch analysis. We chose **48**, Table 3 as the reference and calculated the overlap shape volume V_o and S_o ($V_o^{2/3}$) for the 26 compounds listed. MSA-QSAR analysis was performed with the parameters for the 26 compounds; the following correlation equations were obtained ($N = 24$, **5** and **47** were not included):

$$(7) \qquad \log(1/C) = -0.007(\pm0.024)S_o + 0.699(\pm0.121)$$
$$R = 0.125 \quad S = 0.285 \quad F = 0.35$$

$$(8) \qquad \log(1/C) = -0.004(\pm0.003)S_o^2 +$$
$$0.241(\pm0.174)S_o - 3.291(\pm0.104)$$
$$R = 0.555 \quad S = 0.244 \quad F = 4.679$$

Table 3. Selected parameters used in MSA-QSAR regression
equations.

$R_4 = -NHC_3H_7\text{-}n$

$R_5 = -NHC_4H_9\text{-}t$

$R_6 = -NH\text{-}$cyclohexyl

Compd.	X[a]	R	$S_o.$	logP	T_1	B_1	B_2	log(1/C)
14			29.6	3.43	-8.0	1.244	1.352	0.788
18			28.6	3.43	2.6	1.208	1.379	0.606
15			32.41	2.86	-8.0	1.244	1.352	0.578
4			25.62	2.86	-8.0	1.244	1.352	0.458
16			21.2	3.57	2.6	1.208	1.379	0.314
3			29.0	4.14	2.6	1.208	1.379	0.550
17			28.76	4.14	2.6	1.208	1.379	0.664
2			28.78	2.66	-24.71	1.234	1.365	0.564
25			26.16	2.05	-27.7	1.234	1.365	-0.272
23			30.99	3.60	-8.0	1.244	1.352	0.723
24			36.36	3.60	-8.0	1.244	1.352	0.921
31			36.67	3.67	-14.0	1.215	1.375	0.410
32			33.88	3.10	-14.0	1.215	1.375	0.495
6			35.27	3.10	14.6	1.244	1.329	0.495
34			26.81	4.38	-14.0	1.215	1.375	0.735
5			38.98	4.38	-14.0	1.215	1.375	0.977
35			27.84	3.67	-14.0	1.215	1.375	0.714
44	2-Cl	R_2	28.49	3.67	17.1	1.229	1.318	0.226
38			27.66	3.33	15.6	1.234	1.345	0.620
39			25.53	3.10	15.6	1.234	1.345	0.301
9			27.23	2.76	15.6	1.234	1.345	0.288
41			28.12	3.33	15.6	1.234	1.345	0.801
45	3,4-OCH$_2$O-	"	33.90	2.56	-27.7	1.234	1.365	0.576
46	3,4-OCH$_2$O-	R_4	34.7	2.58	-27.7	1.234	1.365	0.277
47	3,4-Cl$_2$	R_5	38.5	4.31	-8.3	1.234	1.317	0.081
48	3,4-Cl$_2$	R_6	45.79	4.84	12.5	1.284	1.305	-0.128

[a]Substituents are given only for compounds not previously defined.
R_2 and R_3 are defined in Table 2.

(9) $\log(1/C) = 0.085(\pm 0.19)\log P + 0.203(\pm 0.119)$

$R = 0.194 \quad S = 0.282 \quad F = 0.86$

(10) $\log(1/C) = -0.35(\pm 0.16)(\log P)^2 +$

$2.487(\pm 1.119)\log P - 3.78(\pm 0.09)$

$R = 0.714 \quad S = 0.206 \quad F = 10.91$

(11) $\log(1/C) = -0.005(\pm 0.002)S_o^2 +$

$0.327(\pm 0.156)S_o + 0.239(\pm 0.156)\log P - 5.35(\pm 0.09)$

$R = 0.736 \quad S = 0.204 \quad F = 7.898$

(12) $\log(1/C) = -0.359(\pm 0.176)(\log P)^2 +$

$2.541(\pm 1.203)\log P + 0.003(\pm 0.02)S_o - 3.95(\pm 0.09)$

$R = 0.716 \quad S = 0.21 \quad F = 6.993$

(13) $\log(1/C) = -0.003(\pm 0.002)S_o^2 +$

$0.226(\pm 0.159)S_o - 0.224(\pm 0.178)(\log P)^2 +$

$1.719(\pm 1.18)\log P - 6.221(\pm 0.076)$

$R = 0.816 \quad S = 0.179 \quad F = 9.438$

(14) $\log(1/C) = -0.004(\pm 0.003)S_o^2 +$

$0.259(\pm 0.17)S_o - 0.226(\pm 0.177)(\log P)^2 +$

$1.761(\pm 1.177)\log P + 3.466(\pm 6.329)B_1 - 11.035(\pm 0.076)$

$R = 0.83 \quad S = 0.177 \quad F = 7.943$

(15) $\log(1/C) = -0.003(\pm 0.003)S_o^2 +$

$0.201(\pm 0.169)S_o - 0.237(\pm 0.181)(\log P)^2 +$

$1.796(\pm 1.2)\log P + 2.083(\pm 4.612)B_2 - 8.882(\pm 0.077)$

$R = 0.825 \quad S = 0.179 \quad F = 7.69$

(16) $\log(1/C) = -0.004(\pm 0.002)S_o^2 +$

$0.246(\pm 0.121)S_o - 0.297(\pm 0.131)(\log P)^2 +$

$2.312(\pm 0.884)\log P + 15.444(\pm 7.414)B_1 +$

$10.898(\pm 5.343)B_2 - 41.283(\pm 0.054)$

$R = 0.922 \quad S = 0.126 \quad F = 16.139$

Equation (16) gives the optimum correlation. Equations (7) to (15) show its stepwise development. From equation (16) we obtained the optimum value of $S_o = 30.75$ and $\log P = 3.89$ (see **31**, Table 3). The torsion angle between the plane of the phenyl group and that of the double bond of the reference compound, **48**, a derivative of low activity, is 35.1°, (a high degree of noncoplanarity), further its S_o exceeds the optimum value. In contrast, in **31** this torsion angle is -9.3°. For **2**, the torsion angle is smaller, but the overlap volume compared with **48** is lower than the optimum value. In equation (16) bond lengths C=O (B_1) and C-N (B_2) are linearly correlated to the activity. Thus it appears that there is a rigid matching relationship in configuration and conformation between the drug molecule and the receptor.

We obtain the following equations with better correlation coefficients:

$$(17) \quad \log(1/C) = -0.003(\pm 0.003)S_o^2 +$$
$$0.176(\pm 0.165)S_o - 0.288(\pm 0.189)(\log P)^2 +$$
$$2.185(\pm 1.277)\log P - 0.005(\pm 0.006)T_1 - 6.254(\pm 0.073)$$
$$R = 0.843 \quad S = 0.171 \quad F = 8.809$$

$$(18) \quad \log(1/C) = -0.003(\pm 0.002)S_o^2 +$$
$$0.175(\pm 0.151)S_o + 0.381(\pm 0.149)\log P -$$
$$439.163(\pm 245.446)B_2^2 + 1188.611(\pm 663.692)B_2 -$$
$$807.704(\pm 0.069)$$
$$R = 0.864 \quad S = 0.16 \quad F = 10.59$$

Equation (17) indicates that the smaller the torsion angle between the planes of the C=C bond and the C=O group (T_1) is, the stronger the expected anticonvulsant activity.

Equation (18) indicates that the activity is quadratically correlated with the C-N bond length (optimum value 1.35 Å).

6: CONCLUSION

From the QSAR analysis we conclude that the lipid/water partition coefficient is an important factor affecting the pharmacological effect of this family of anticonvulsants. Their optimum partition coefficient is 4.35, which is higher than the normal value of 2. This indicates that the influence of logP on anticonvulsant activity not only involves trans-

port phenomena but also the strength of the hydrophobic binding between the substituted benzene ring and the receptor.
The electrostatic potential field surrounding the benzene ring
has a direct bearing on activity. But the volume of the substituent group and its influence on the pharmacological effect
must not be overlooked. Substituents on the benzene ring must
comply with the steric requirements of the receptor.

Our study reveals that the charge density on N and O atoms, the extent of the electrostatic field surrounding the N
atoms, the free surface of the N and O atoms and the C-N and
C=O bond lengths are all important to the pharmacological
effect.

QSAR studies show that the distance between the benzene
ring and the N and O atoms have a bearing on the pharmacological properties. The shorter the distance, the poorer the
pharmacological effect. This observation can explain why the
trans isomer of cinnamamides (with phenyl-O distances of 5.560
and phenyl-N distances of about 6.0 Å) are more potent than
the cis isomers (respective distances, 4.8-5.0 and 4.1-4.3
Å). It may also explain why the elimination of one carbon
atom in the B portion of the cinnamamides causes the pharmacological effect to be reduced markedly.

The better the coplanarity, the more active the drug. Further, the optimum length of the C-N framework bond (1.35 Å) indicates this bond should participate fully in conjugation.

From the above results, we conclude that the benzene ring
and N, and O atoms may be the three important sites of the
drug molecule. The distances between the three sites and the
configuration involved should have certain steric compatibil-
ity with the receptor.

The MSA-QSAR study makes it clear that S_o is an important factor, exceeded only by the lipid/water partition coefficient, in determining the activity. This study also indicates
that molecular shape significantly influences the pharmaco-

logical activity. However, if we consider substitution in the
A and C moieties the correlation coefficient increases dramatically. Finally, we conclude that the volume factor of the substituent group may have a dramatic influence on the pharmacological effect.

REFERENCES

PEI, Y.-Q, LI, J.-S., CAI, Z.-J., ZHANG, B.-H., TAO, C. and
KU, B.-C. (1978) National Medical Journal of China 58, 216.
ZHANG X.-H., Li, R.L., Cai, M.-S., (1980). Journal of Beijing
Medical College 12, 83.
MADDON, A., LEVY B. and MALRIEU, J.P. (1970). Theoretica
Chimica Acta 18, 197.
HOPFINGER, A.J. (1980). Journal of the American Chemical
Society 102, 7196.

12. Quantitative structure–activity relationships for a series of nitroheterocyclic drugs

C. J. De Ranter

1. INTRODUCTION

Nitroheterocyclic products were introduced in human medicine during World War II by the discovery of the antibacterial properties of nitrofurazone. It was, and even nowadays it is still used for the topical treatment of burns and wounds.

As a result of that finding research for novel nitroheterocyclic compounds with biological activity was started. This led to the synthesis of thousands of nitrofurans, nitroimidazoles and nitrothiazoles. Among the nitrofurans tested for their biological properties, only a few are used in human medicine, *e.g.* nitrofurazone, nitrofurantoin, nifurtimox. A similar development occurred for the nitrothiazoles of which niridazole is the most frequently used. The discovery of the antitrichomonal activity of azomycin (Nakamura and Umezawa, 1955), a 2-nitroimidazole antibiotic isolated from cultures of *Nocardia mesenterica*, was the starting-point for an extensive survey of the antiprotozoal activity of nitroimidazoles. Soon two compounds came to the fore : metronidazole, the less toxic, was selected for the treatment of human trichomoniasis; dimetridazole proved to be of great value in the treatment and prevention of blackhead in turkeys and swine dysentery.

In addition to their use as antiparasitic and bactericidal

agents, the nitroheterocyclic drugs are becoming of current interest in cancer therapy for their radiosensitizing and chemosensitizing properties (Stratford, 1982; Adams, 1986). Remarkably to stress is that their activities occur under anaerobiosis or conditions of low oxygen tension only (*e.g.* anaerobic protozoa, hypoxic tumour cells). The activities are strongly enhanced in micro-organisms which contain ferredoxin (Yarlet *et al.*, 1985). This selectivity is related to the prerequisite for their biological activity, i.e. reduction of the nitrogroup (Edwards *et al.*, 1973; Lindmark and Muller, 1976). During activation transient intermediates are formed which react with various cellular targets of which DNA is generally assumed to be the lethal target (Müller, 1983).

Many *in vitro* experiments have been focused on the drug-target interaction, in which nitroheterocycles were reduced either chemically or electrochemically in the presence of DNA or nucleotides. Although there is still no conformity concerning the base-specificity of this interaction, most studies reveal that primarily the guanine residue is susceptible (LaRusso *et al.*, 1978; Varghese and Whitmore, 1983; Declerck *et al.*, 1983, 1987). In the latter experiments degradation of DNA was never observed. So far, the nature of the formed DNA-adduct and of the reactive nitroheterocyclic derivatives remains unknown. For this reason and to contribute to the understanding of the chemical descriptors governing the pathways of action, a thorough study was made of the *in vitro* activity of 74 nitroheterocyclic compounds in correlation with particular physicochemical and structural parameters.

2.　　EXPERIMENTAL

The *in vitro* activity (37 °C, BHI-H agar, pH 7.4, anaerobic conditions : 5 % CO_2, 10 % H_2, 85 % N_2) was determined against *Bacteroïdes fragilis* (NCTC 9343) and *Clostridium perfringens*

(ATCC 19574) using the twofold dilution method. Reduction potentials, $E_{1/2}$ (differential pulse polarography, Britton-Robinson buffer pH 7.4), dipole moments (in 1,4-dioxan and after extrapolation to infinite dilution) and log P-values (1-octanol-water system, borate buffer pH 7.4, ionic strength in the water phase 0.15) were measured under nitrogen atmosphere.

3.   RESULTS AND DISCUSSION

To specify the structural parameters influencing the biological activity a Free-Wilson analysis (1964) was performed. The restrictions imposed by this technique reduced our set of 74 compounds to a subset of 18 products. From that analysis resulted that 92 % of the differences in activity against *Bacteroïdes* and 91 % of the anti-*Clostridium* action are due to substituent variation : the 5-nitroheterocycles being definitely more active than the 4-nitro-analogues. An alkyl substituent at position 1 distinctly strengthens the activity, especially if it is $-CH_2CH_2OH$. Finally, a methylgroup at position 5 combined with a nitro at 4 decreases the activity causing the 4-nitroimidazoles to become completely inactive.

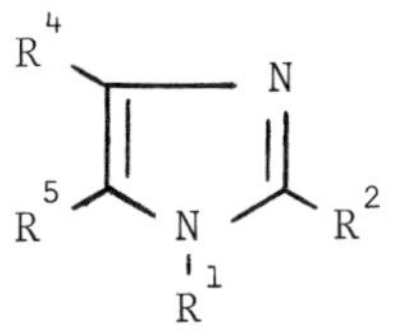

A multiple regression analysis on all 74 compounds, however, learned that only 21 % of the activity against *Bacteroïdes* and 27 % against *Clostridium* could be explained by the three selected descriptors. A plot of the *in vitro* activity against the measured dipole moments showed that the set of compounds could be subdivised in two subsets : the inactive 4-nitro and the active 2- and 5-nitroheterocycles. As many compounds were not soluble enough in the solvent used, 1,4-dioxan, the dipole moment of only 53 of the 74 products could be determined experimentally. To include the remaining

products in the correlation analysis an indicator variable IMU
was introduced: it equals 1 if the nitrogroup is at position 2 or
5, and it is zero for the 4-nitro compounds. With this variable
included the prediction of the activity increases to 54 % for
Bacteroïdes and to 45 % for *Clostridium*.

The reduction of the nitrogroup is, as mentioned, an
essential point for bioactivity. By plotting the anti-
Bacteroïdes activity as a function of the reduction potential
the nitrofurans showed up in a separate cluster (less negative,
although not more active). As nitrofurans, nitrothiazoles and
other nitro-products (pyrroles, benzofurans, thiophenes) were
present in only a very limited number, these compounds were
left out and only nitroimidazole derivatives were taken in the
subsequent analysis.

By looking very carefully through the *in vitro* data it
became clear that compounds possessing a vinyl- or styryl
substituent showed a markedly higher biological activity;thus
chance for conjugation between the substituent and the
imidazole nucleus seemed an important factor too. To take into
account these structural features two new indicator variables
were added : (1) a parameter CONJ being 1 if conjugation is
possible, (2) the dummy variable VINYL being 1 if a vinyl- or
styryl group is present. Then to find the best combination of
physicochemical and structural parameters for explaining the
biological activity of the nitroimidazole series, a factor
analysis was performed. The different parameters were ordered
in four groups, *factors*, strongly independent of each other.
In factor 1 we find the biological activity, the reduction
potential E and indicator variable IMU; in factor 2 the VINYL
parameter and the UV-absorption maximum; factor 3 contains the
log P-value and factor 4 CONJ and IX, an indicator for the
presence of $-CH_2CH_2XR$ with X=N, O or S (the latter two are
inversely correlated). The best description of the inhibitory

Bacteroïdes activity is given by the combination of the reduction potential E, VINYL and CONJ.

$$-\log \text{MIC} = 11.1\,\text{E} + 1.9\,\text{VINYL} - 1.3\,\text{CONJ} + 7.8 \quad \text{(for } \textit{Bacteroïdes}\text{)}$$
$$\quad\quad (0.3) \quad\ (0.3) \quad\quad\quad\quad\quad (0.5)$$
$$n = 60 \quad R^2 = 0.80 \quad s = 0.5 \quad F = 74$$

For *Clostridium* the best prediction is given by IMU, IX and VINYL, although the combination with E, VINYL and CONJ is nearly equally significant ($R^2 = 0.62$). The relation is now †

$$-\log \text{MBC} = 1.3\,\text{IMU} + 0.9\,\text{VINYL} + 0.6\,\text{IX} - 0.3 \quad \text{(for } \textit{Clostridium}\text{)}$$
$$\quad\quad (0.2) \quad\quad (0.3) \quad\quad\ (0.2) \quad (0.2)$$
$$n = 60 \quad R^2 = 0.65 \quad s = 0.5 \quad F = 35$$

A further examination of the 60 nitroimidazoles revealed that the whole set could be further subdivided in : (1) the 2-methyl-4- and -5-nitroimidazoles, for which 80 % of the activity against *Bacteroïdes* can be explained by changes in the reduction potential alone; and (2) the 1-methyl-4- and -5-nitroimidazoles, where the dummy variable CONJ discriminates the products in the 2-alkyl analogs (CONJ=0) and the derivatives with a substituent on C^2, capable to conjugate with the imidazole ring (CONJ=1). In all equations the coefficient of CONJ was found to be negative, and this raises the question why the phenylanalogs are less active than the alkylanalogs.

A semi-empirical conformation analysis showed that the phenylring is rotated by 60 degrees with regard to the imidazole plane preventing the phenyl ring from being coplanar with the imidazole nucleus. Vinyl- and styrylgroups, on the contrary, can turn away from the N^1 substituent permitting conjugation with the imidazole. That could explain the ten to hundredfold increase in activity when one of the latter is present. That hypothesis was confirmed by an X-ray structure analysis of niridazole (a thiazole), where both heterocyclic

---

† MBC = minimum bactericide concentration required.

rings are nearly coplanar (Peeters *et al.*, 1984).

Beside the phenyl, vinyl and styryl groups, CONJ=1 denoted also a series of nitroimidazoles containing a substituent directly connected to C^2 by means of a heteroatom. The influence of these substituents, and especially the role of the lone pair on the *in vitro* activity stays obscure.

Finally, it was found that the lipophilicity of the compounds has no significant influence on the *in vitro* activity of the microorganisms studied here.

REFERENCES

ADAMS, G.E. (1986), *Biochem. Pharmacol.* 35, 71.

DECLERCK, P.J., DE RANTER, C.J. and VOLCKAERT, G. (1983), *FEBS Lett. 164*, 145.

DECLERCK, P.J. and DE RANTER, C.J. (1987), *J. Chem. Soc. Faraday Trans. 1*, 83, in press.

EDWARDS, D.I., DYE, M. and CARNE, H. (1973), *J. Gen. Microbiol. 76*, 135.

FREE, S.M. and WILSON, J.W. (1964), *J. Med. Chem. 7*, 395.

LARUSSO, N.F., TOMASZ, M., KAPLAN, D. and MULLER, M. (1978), *Antimicrob. Agents Chemother. 13*, 19.

LINDMARK, D.G. and MULLER, M. (1976), *Antimicrob. Agents Chemother. 10*, 476.

MULLER, M. (1983), *Surgery 93*, 165.

NAKAMURA, S. and UMEZAWA, H. (1955), *J. Antibiot.* 9A, 66.

PEETERS, O.M., BLATON, N.M. and DE RANTER, C.J. (1984), *Acta Cryst.* C40, 1748.

STRATFORD, I.J. (1982), *Int. J. Radiation Oncology Biol. Phys.* 8, 391.

VARGHESE, A.J. and WHITMORE, G.F. (1983), *Cancer Res. 43*, 78.

YARLETT, N. GORRELL, T.E., MARCZAK, R. and MULLER, M. (1985), *Mol. Biochem. Parasitol. 14*, 29.

13. QSAR and molecular modelling of 1,4-dihydropyridines

Wolfgang Seidel

ABSTRACT

Conformationally restricted dihydropyridinelactones have been synthesized and used as tools to study conformational requirements for these molecules at the receptor site. The flexibility of these restricted lactone compounds has been determined by conformational rearch techniques. Using these methods, it is possible to explain the activity of compounds, the X-ray structures of which deviate considerably from the postulated ideal conformation at the receptor.

In a second approach, structure activity relationships of calcium antagonists have been compared with a QSAR analysis of calcium agonists. Agonists and antagonists showed rather similar structure activity relationships that indicate a common receptor site for both types of molecules. Satisfactory prediction of positive inotropic activity was achieved from computed increments.

To rationalize the different mode of action of agonists, steric differences between agonists and antagonists have been analysed.   The agonists show significant deviations in bond lengths, molecular volume and ring geometry. Meanwhile, antag-

onists and agonistic behaviour has been described for the
enantiomeric dihydropyridines. This indicates that conformational studies may not be the right tool to distinguish
agonists from antagonists.

1. PROBING THE RECEPTOR SITE WITH RIGID ANALOGS

The molecular geometry of 1,4-dihydropyridines has been
studied extensively by X-ray analysis during the last few
years (Triggle 1980, Fossheim 1982, Langs 1985). General conformational features of this class of compounds include a flat
boat conformation of the heterocyclic ring, a parallel orientation of the ester groups due to resonance and a more or less
pronounced perpendicular orientation of the aromatic ring with
respect to the heterocycle. This may be illustrated by the
X-ray structure of Nifedipine (Fig. 1).

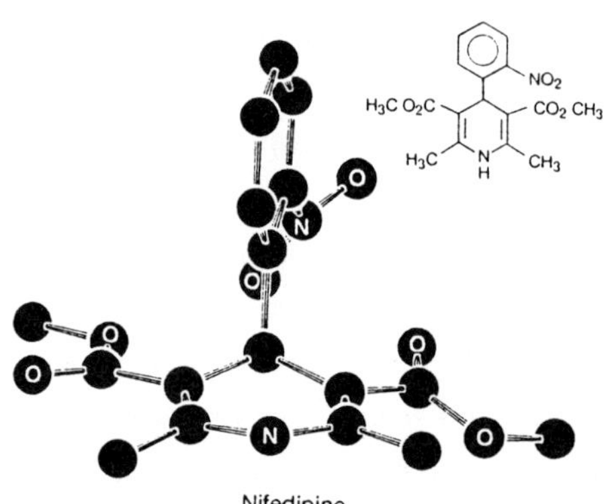

Fig. 1. The chemical structure of nifedipine and its conformation from the crystal structure determination.

The observed influence of steric bulk in ortho substituted
dihydropyridines on the biological activity has led to a hypothesis that the aromatic ring might be forced into a perpendicular conformation which would be preferred by the receptor
(Loev, 1974). To prove this hypothesis, we synthesized confor-

mationally restricted analogs (Fig. 2) within a broad range of
possible conformations to exclude those which were inactive.
The rotation around the connecting bond between two rings was
restricted by a lactone bridge.

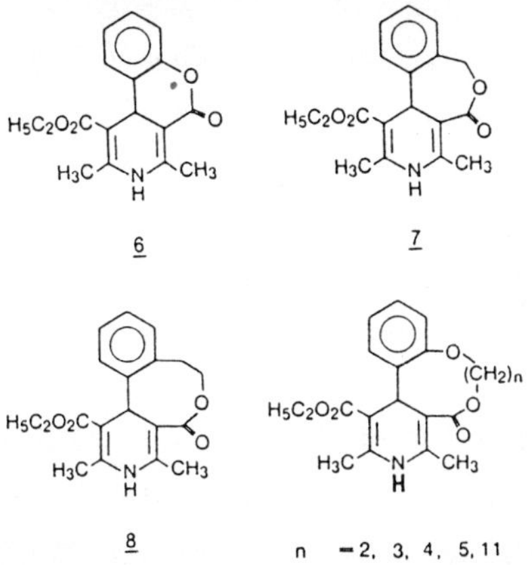

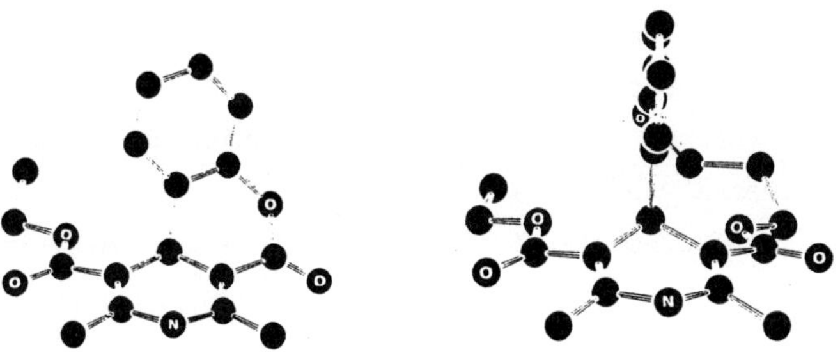

Fig. 2. Chemical structures of conformationally restricted
analogs.

The X-ray analysis of the six membered ring lactones (Fig.
3) shows that the lactone bridge forces the plane of the aromatic ring into a conformation in which it is no longer bisecting the dihydropyridine ring.

In contrast, the largest twelve-membered ring lactone, 12,
(Fig. 4) almost ideally bisects the dihydropyridine structure.   In the other compounds synthesized, the orientation of
the aromatic ring lies between the two conformations shown.

Fig. 3. Conformation for 6.   Fig. 4. Conformation for 12.

In radioligand competition experiments, the six membered ring, characterized by the largest deviation of the aromatic ring from the bisecting plane, is not active at all. The activity of the larger rings increases as the aromatic ring approaches the bisecting plane with the exception of the eleven-membered lactone ring compound. Considering these results, it is assumed that the dihydropyridine receptor does not accept a substrate with a conformation of the aromatic ring far away from the bisecting plane.

The irregularity in the eleven-membered lactone , which is very active at the receptor but shows a considerable deviation from the perpendicular orientation may be explained by the inherent flexibility of this compound. Conformational searches and vector maps clearly show that there is really enough flexibility for this ring to occupy the bisecting plane and fulfill the requirements for the receptor. It seems very promising to use the static picture derived from X-ray crystallography in combination with conformational search techniques to get a dynamic picture of molecules.

2.DIFFERENTIATION BETWEEN AGONIST AND ANTAGONIST ACTIVITY

The most obvious chemical feature of calcium agonistic dihydropyridines is the presence of a 3-nitro or lactone function replacing one of the ester groups in the antagonists. A quantitative structrue-activity relationship analysis (QSAR) of the Free Wilson type carried out with about 1220 structures revealed that their activity could be nicely described by adding up increments for the different substitutions of the two ring systems. The calculated activities are compared with the experimental results in Fig. 5.

As the increments for the agonists are very similar to those known from the antagonists, it may be concluded that both agonist and antagonist bind to the same receptor subunit. Also, the three-dimensional comparison of these two

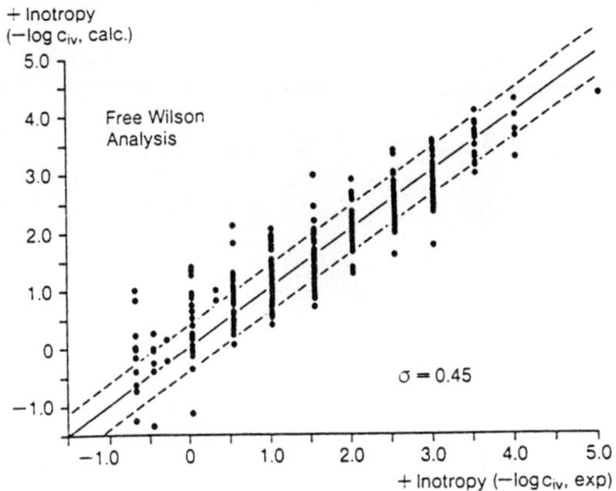

Fig. 5. Quantative structure activity relationships.

classes of calcium modulators reveals only minor differences in bond lengths and angles. The altered steric and electonic properties of the dihydropyridine ring caused by the persence of a lactone - or nitro group - result in less puckering of the dihydropyridine ring. There are also differences in charge distribution, dipole moment and hydrogen bonding between agonists and antagonists. However it has been shown in the meantime that compounds that appeared to be agonists as a racemic mixture, show antagonistic behavior as the (S)-form, whereas the (R)-form is an agonist. (Franckowiack, 1985; Hof, 1985, Fig. 6).

Facing this situation, one must confess that molecular modelling techniques and computational methads are of very limited value in differentiating between agonistic and antagonistic properties of enantiomeric compounds.

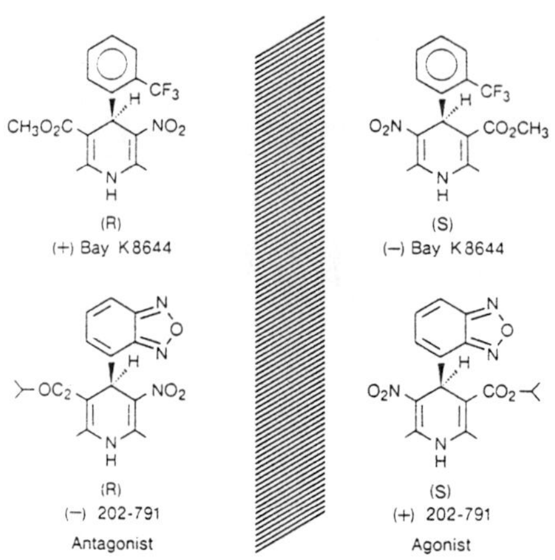

Fig. 6. Enantiomeric antagonist - agonist pairs.

REFERENCES

FOSSHEIM, R., SVARTENG, K., MOSTAD, A., ROMMING, C., SHEFTER, E., TRIGGLE, D., (1982). Journal of Medicinal Chemistry 25, 126.

FRANCKOWIACK G., BECHEM, M., SCHRAMM, M., THOMAS, G. (1985). European Journal for Pharmacology 114, 223.

HOF., R.P., RUEGG, U.T., HOR. A., VOGEL., A. (1985). Journal of Cardiovascular Pharmacology 7, 689.

LOEV, B., GOODMAN, M., TESESHI, K., MACKO. R. (1974). Journal of Medicinal Chemistry 17, 956.

TRIGGLE, A., SHEFTER, E., TRIGGLE, D. (1980). Journal of Medicinal Chemistry 23, 1442.

14. Crystal structure–function studies on some biologically active small molecules

Qi-tai Zheng

China is a country possessing abundant resources of medicinal plants which have been used for combating various kinds of human diseases for thousands of years. In the 16th century, Shi-zeng Li, a great Chinese scholar, left his classical Chinese Pharmacopeia, which is still a valued book for the evaluation of thousands of Chinese drugs.

During the past years, Chinese crystallographic groups, including ours, have determined crystal structures for about sixty compounds derived from Chinese folk medicines. Among them are preparations claimed to be treatments for tapeworm, to be antimalarials, anti-inflammatories, antirheumatics, contra-ceptives, antitumor agents, analgesics and compounds that lower enzyme activity. These compounds are tabulated here; among them are: alkaloids (**1-14**), terpenes (**15-36**), a nucleotide (**37**), amino acids (**38-40**), a steroid (**41**), esters (**42-46**), coumarins (**47-48**), glycosides (**49-50**), phenols (**51-52**), a quinone (**53**), flavonoids (**54-55**) and lignans (**56-60**). Chemical structures are illustrated schematically in Scheme 1.

Alkaloids

Brunoniune, **1**, is a new C_{20} diterpenoid alkaloid isolated from the whole <u>Delphinium brunonianum Royle</u> plant which grows in southwestern China. It is used in Tibet as a folk medicine

for treatment of influenza, itchy-rash and snake bite. The
crystal structure, which was carried out with colorless crystals kindly supplied by W.-1. Sung, demonstrated that the side
chains are: C_4-βCH, C_5-βH, C_7-αCH, C_9-βH, C_{15}-βOH, C_{16}-βCH$_2$.
There are six 6-membered rings (A-F). Rings A, B, and F dis-
play chair conformations while C, D and E are boat conformers.
The B/F rings are cis-fused while A/B and B/C are trans-fused.
The C_{17}-N and C_{16}-C_{20} bonds display double bond character. As
expected, the molecule is nonplanar with rings D, E and F out
of the mean plane of the A, B and C rings. Intermolecular contacts indicate weak hydrogen bonds $\{(C_{15}$-OH$)\cdots(C_7$-OH$)\}$ and
$\{(N)\cdots(C_7$-OH$)\}$.

Guanfu base G, **2**, (isolated from the root of <u>Aconitum</u>
<u>coreanum (Levl) Raipaics</u> is used to treat "expel wind evil", to
dissipate phlegm and to alleviate pain; it is a new C_{20}-diter-
penoid alkaloid.

Bonvalotine A, **3**, a new C_{19} diterpenoid, and its analogs
D and E (**4** and **5**) have been isolated from the roots of <u>Del-</u>
<u>phinium bonvalotii Franch</u>, a medicinal plant grown in China
that is claimed to be analgesic, anti-inflammatory and an-
tirheumatic. Delbotine, **4**, is also a new C_{19}-alkaloid; Delbox-
ine, **5**, is a C_{18}-diterpenoid alkaloid.

Isoaconitine, **6**, is a C_{19}-diterpenoid alkaloid isolated
from zicaowu, a plant grown in Yunnan province and belonging to
<u>Aconitum Ranunculaceae</u>. Zicaowu is used as an analgesic in
Chinese herbal medicine.

Verticilatine, **7**, is isolated from the water soluble
fraction of <u>Rauvolfia verticilata (Lour) Baill. f. rubrocarpa</u>
H. T. Chang. mss. and is used to treat high blood pressure.

Koumine, **8**, was isolated from <u>Gesemium elegems benth</u>
(used to treat malignant boils and cramps of limbs) which is
grown in Fujian, Guandun and Schichan provinces. The absolute
configuration was also determined to be as illustrated in
Scheme 1.

Yuehchukene, **9,** is a new bis-indole alkaloid isolated from the root of <u>Murraya paniculate (L) Jack</u>. The crystal structure was determined for the mono-N-acetyl derivative.

Tuyanhusu, **10,** is a new alkaloid isolated from <u>Tuyanhu</u> that is used as an analgesic in Chinese folk medicine. The six membered ring containing the N-atom is in a boat conformation.

Dehydrocrydaline, **11,** isolated from <u>Corydalis ambigua</u>, is used to treat cardio-vascular diseases.

Monocrotaline, **13,** is an alkaloid, isolated from <u>Crota-laria Retusa L.</u>, having cancer inhibiting action.

Oxofanchirine, **14,** was isolated from the root of <u>Stephania Tetrandra S. moore</u>, which is used as an analgesic and to treat effects of "expel wind evil" and for fever due to cold.

TERPENES

Monoterpenes

<u>Melasma arvense (Benth.) H. M. (Scrophulariaceae)</u> is a Chinese folk medicine in Yunnan province. Melasmoside, **15,** is a new monoterpenoid glucoside isolated from the aqueous soluble fraction of its roots.

Cantharidin, **16,** was isolated from Cantharides using acetone as the solvent. It has antitumor activity.

Sesquiterpenes

Qinghaosu, **17,** an active principle of <u>Artemisia annua L.</u>, gave colorless needle-like crystals. The absolute configuration was determined by the anomalous scattering of C and O atoms. The six membered C ring includes an oxygen bridge and a peroxy bridge. Perhaps these make up the active center of this antimalarial.

Table I. Compounds for which crystal structures have been carried out.

ALKALOIDS

1.	Brunonine (Nangjujian)	$C_{22}H_{33}NO_3$	
2.	Guanfu Base G (Guanfusu)	$C_{26}H_{33}NO_7 \cdot CH_3I$	(Chen et al, 1984)
3.	Bonvalotine A (Chuanxinlianjian A)	$C_{26}H_{39}NO_8$	
4.	Delbotine (Chuanxinlian D)	$C_{26}H_{43}NO_7$	
5.	Delboxine (Chuanxinlian E)	$C_{24}H_{37}NO_7$	
6.	Isoaconitine (Zicaowujian)	$C_{37}H_{47}NO_{11} \cdot HI$	(Fan and Zheng, 1978, 1980)
7.	Verticilatine (Hounoluofumujianjin)	$C_{20}H_{24}N_2O_3 \cdot HCl$	
8.	Koumine (Gouwensu)	$C_{20}H_{22}N_2O \cdot HBr$	(Rao et al, 1982)
9.	Yuehchukene	$C_{28}H_{28}N_2O$	(Kong et al, 1985)
10.	Tuyaunhusu	$C_{20}H_{19}NO_6$	
11.	Dehydrocrydaline (Tuoqingyznhusuo)	$C_{22}H_{24}NO_4 \cdot HCl$	(Rao et al, 1984)
12.	Fissistiging C (Guafumusu)	$C_{20}H_{23}NO_4$	(Xu et al, 1985)
13.	Monocrotaline (Yebaihejin)	$C_{16}H_{23}NO_6$	(Wang, 1978)
14.	Oxofanchirine (Yanhuafanjijin)	$C_{37}H_{34}N_2O_7$	

TERPENES

Monoterpenes

15.	Melasmoside (Heisuo)	$C_{19}H_{32}O_9 \cdot CH_3OH$	(Yang et al, 1983)
16.	Cantharidin (Banmaosu)	$C_{10}H_{12}O_4$	(Wang et al, 1984)

Sesquiterpenes

17.	Qinghaosu	$C_{15}H_{22}O_5$	(Qinghousu group, 1977, 1980)
18.	Bromoqinghaosu	$C_{17}H_{27}O_5Br$	(Zhang et al, 1981)
19.	Armillarine (Mihuanjun)	$C_{24}H_{38}O_6$	(He et al, 1982)
20.	Suberogorgia Sp. acid	$C_{15}H_{20}O_3$	(Niu et al, 1985)

21.	9(12)-Capnellen E-8α,10β-diol	$C_{15}H_{24}O_2$	(Me et al, 1985)
22.	Baimuxiangic acid	$C_{15}H_{24}O_3$	(He et al, 1982)

Diterpenes

23.	Diacetylrobdocalin B	$C_{24}H_{35}O_4$	
24.	Rabdocalin A (Xiangchacai A)	$C_{24}H_{30}O_9$	(Chen and Chen, 1985a)
25.	Rabdophylin G (Xiangchacai G)	$C_{22}H_{30}O_7$	(Chen et al, 1984)
26.	Robdophillin H	$C_{22}H_{30}O_7$	(Chen and Chen, 1985b)
27.	Zujiunsu	$C_{20}H_{28}O_3$	
28.	Praelolide D (Changxiaoyuenezhi)	$C_{28}H_{35}O_{12}Cl$	(Dai et al, 1985)
29.	Pseudolaric acid A (Tujinpijiasuan)	$C_{22}H_{28}O_6$	(Yao and Lin, 1984)
30.	Torreyagrandate (Xiangfeizhi)		

Triterpenes

31.	Isochuanliansu (Yichuanliansu)	$C_{30}H_{38}O_{11}$	
32.	Chuanliansu		(Gua et al, 1984)
33.	Anemosupogenin (Baitouweng)	$C_{30}H_{48}O_4$	(Liu and Huan, 1984)
34.	Picfeltarragenin A (Kuxuanshen A)	$C_{30}H_{44}O_6$	(Huan et al, 1981)
35.	Acetyltylophorin B (Waerteng)	$C_{32}H_{52}O_2$	(Xu et al, 1985)

Tetraterpene

36.	Methylsartortuoate A	$C_{41}H_{62}O_8 \cdot C_2H_5OH$	

NUCLEOTIDE

37.	c-UMP (Uridine 3'5'-cyclicphosphate)	$C_9H_{11}N_2O_8P$	

AMINO ACIDS

38. Cucurbitine (Nanansuan) $C_5H_{10}N_2O_2 \cdot HClO_4$ (Fan and Lin, 1965)
39. Quisqualic acid salt $C_5H_6N_3O_5^-$ K+
40. L-Citrulline dihydrate $C_6H_{13}N_3O_3 \cdot 2H_2O$
     and α-hydroxymethylserine $C_4H_9NO_4$ (Jin et al, 1985)

STEROID

41. Tenacissigennin (Tongguangsu) $C_{21}H_{32}O_5$ (Tungguangsu group, 1980;
                                                                                         Zheng et al, 1981)

ESTERS

42. Armillarisin A $C_{12}H_{10}O_5$
43. Arteanomalactone (Qihaoneizhi) $C_{19}H_{24}O_6$ (Lin and Zhang, 1985)
44. Glucopyranosyl-deoxyandrographolide $C_{26}H_{40}O_9$ (Ma and Chen, 1983)
45. Riligustilide (Shuanggaobenzhi) $C_{24}H_{28}O_4$ (Mun et al, 1983)
46. Chufeineizhi (Hainanchufeineizhi) $C_{19}H_{18}O_4$ (Bao et al, 1978)

COUMARINS

47. Moellendorffiline (Zoumaqingneizhi) $C_{26}H_{24}O_{10}$
48. Heliclactone (Shanzhimazhi) $C_{15}H_{16}O_4$

GLYCOSIDES

49. Pentaacetylsarmentosin (Cuipengcaodai) $C_{21}H_{27}NO_{12}$ (Gu et al, 1981)
50. Isosarmentosin (Yicuipengcaodai) $C_{11}H_{17}NO_7$ (Zheng et al, 1981)

PHENOLS

51. Gossypol (Mianfen) $C_{30}H_{30}O_8 \cdot C_2H_4O_2$ (Li and Tuan, 1982; Xu et al, 1982)
52. Agrimophol (Hecaofen) $C_{26}H_{34}O_8$ (Tuan et al, 1982)

QUINONE

53. Hypocrellin B (Zhuhuangjunsu B) $C_{30}H_{26}O_{10}$

FLOVONOIDS

54. Argutone (Madinsu) $C_8H_{10}O_5$
55. Liuflovensu $C_{17}H_{14}O_8$

LIGNANS

56. Glaberide A (Tumaodongqingjiasu)   (Zhang and Lin, 1984)
57. Schisanhenol (Wuweizifen) $C_{23}H_{30}O_6$ (Xu et al, 1982)
58. Rubschisantherin A (Wuweizichunjia) $C_{30}H_{31}O_9Br$ (He et al, 1977)
59. Wuweizichun B $C_{23}H_{27}O_7Br$ (Xu et al, 1981)
60. Angeloylgomisin (Nanwuweizizhiyi) $C_{27}H_{30}O_8$

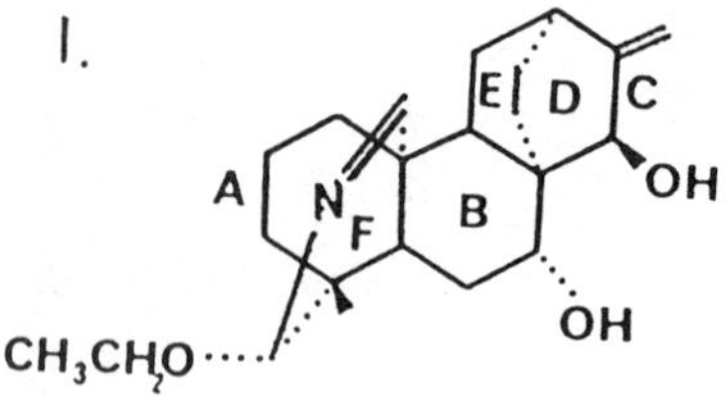

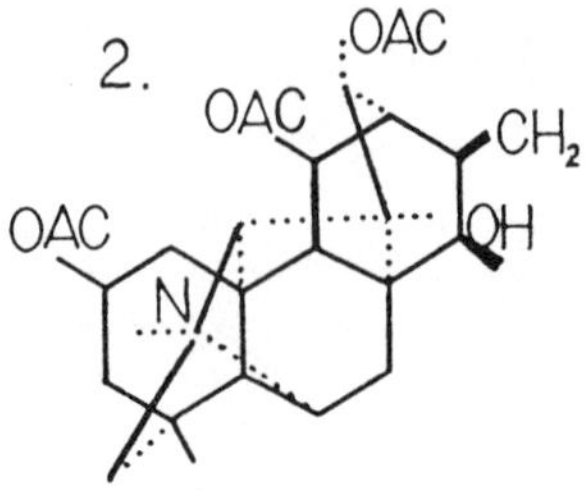

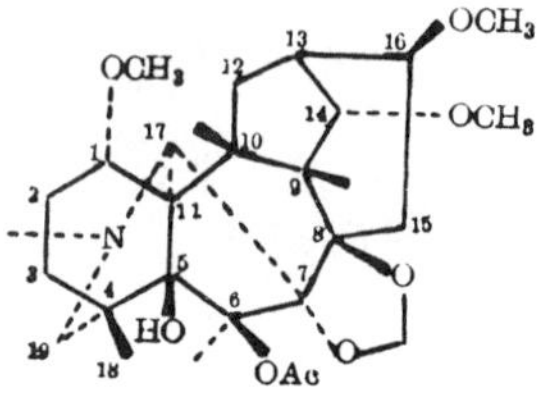

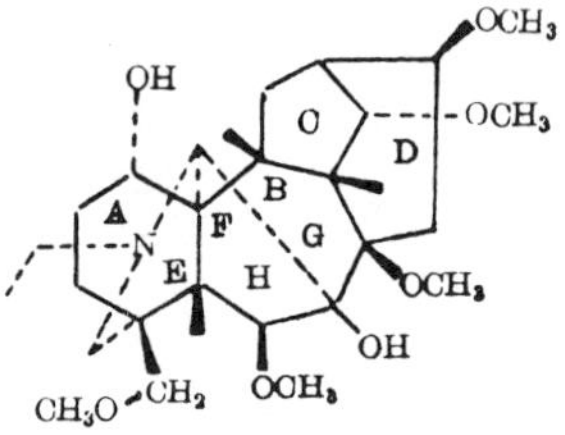

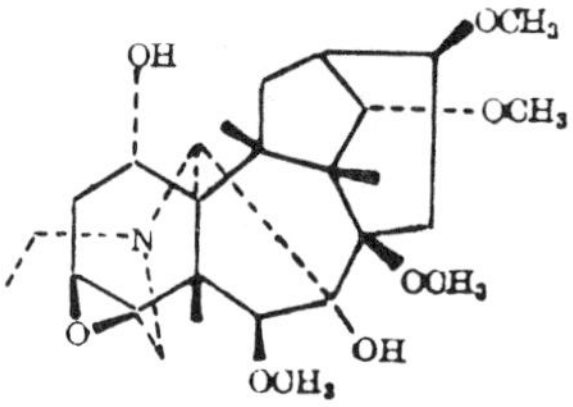

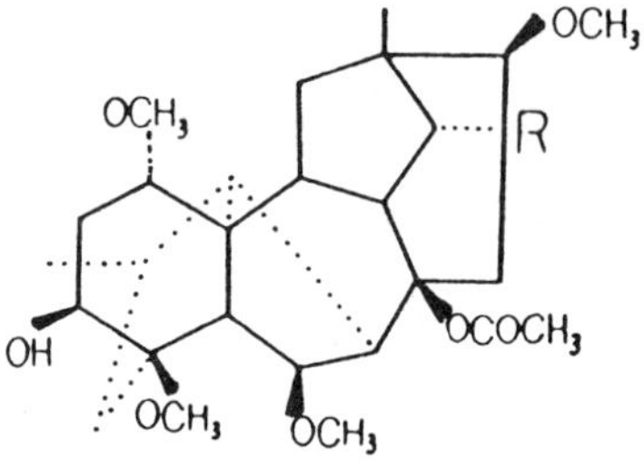

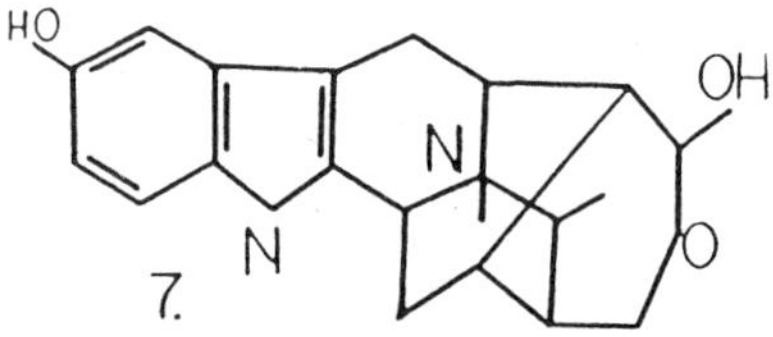

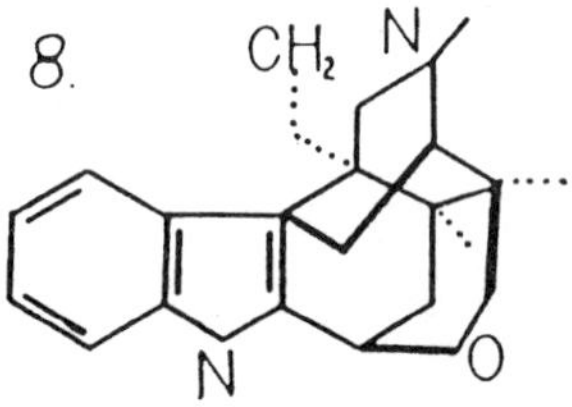

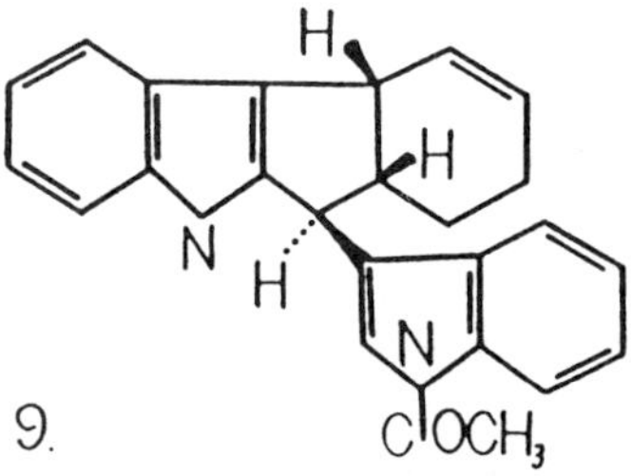

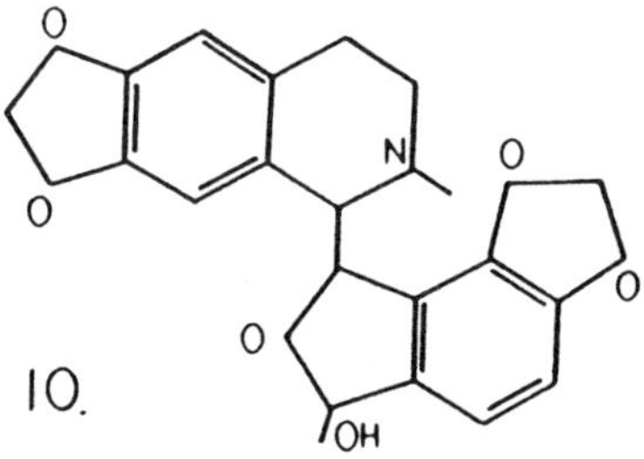

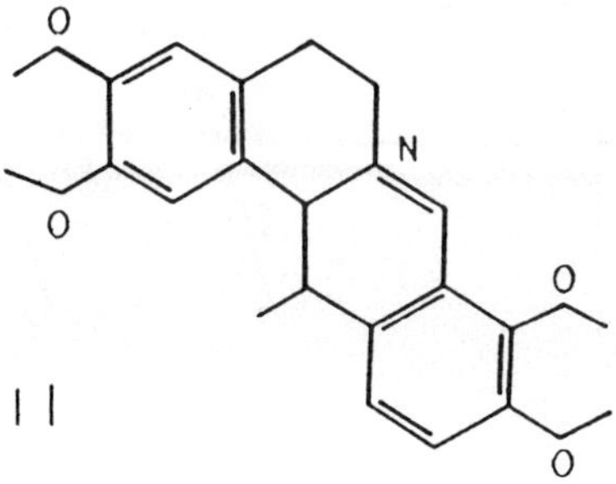

11 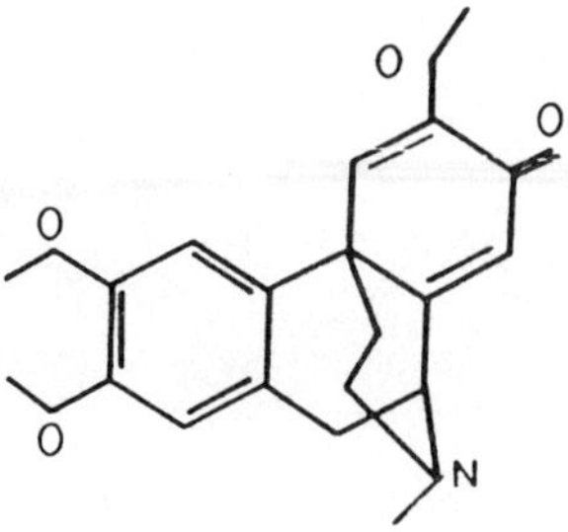
12
14 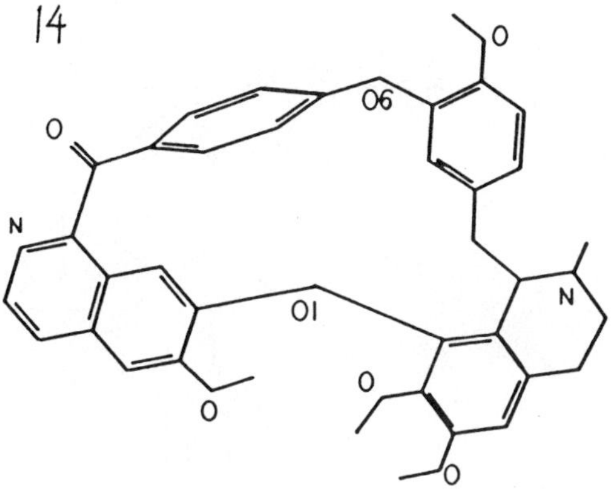
13 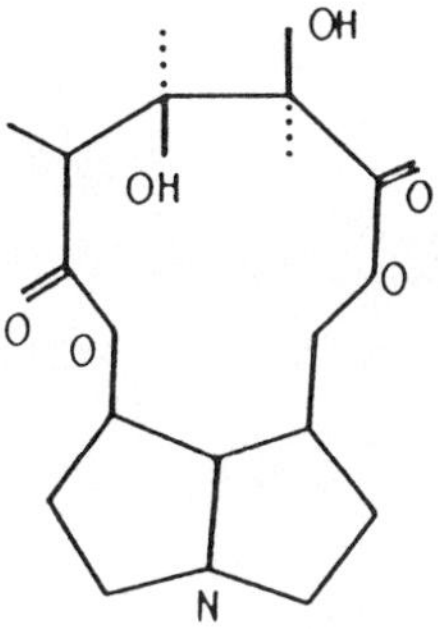 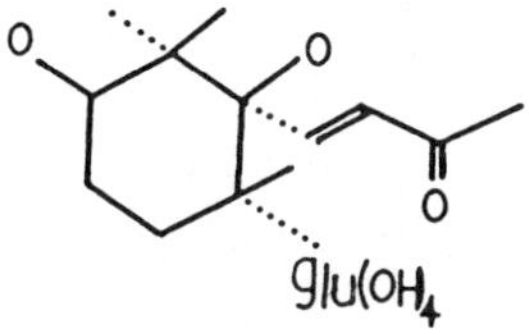
O6
O1
OH
OH
N
glu(OH_4
15. 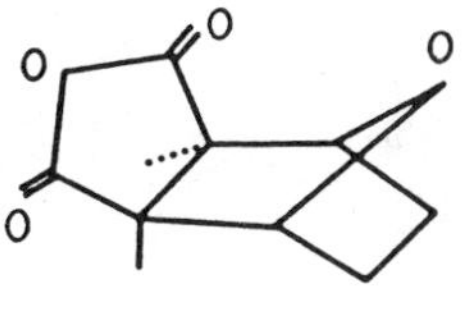
16. 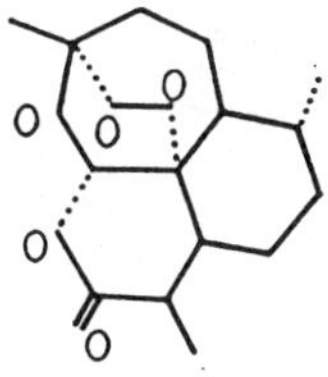
17.
18.
Br

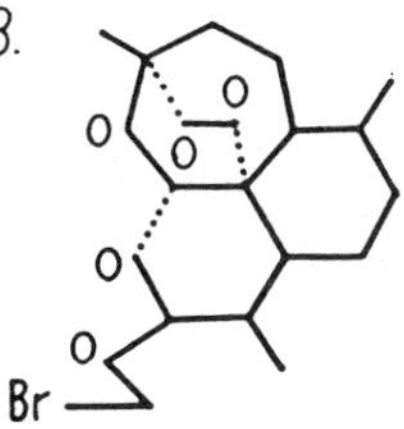

19.

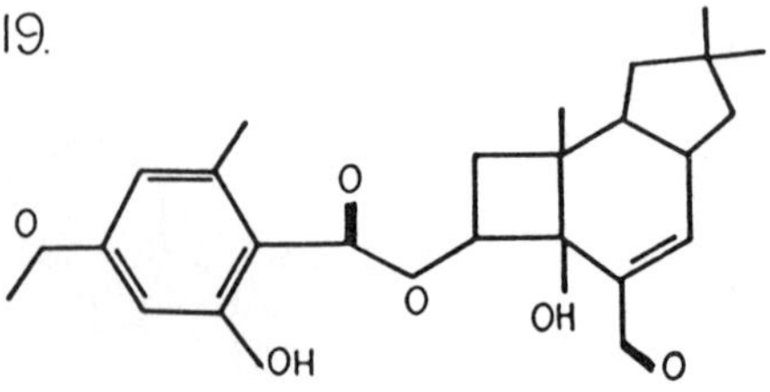

20

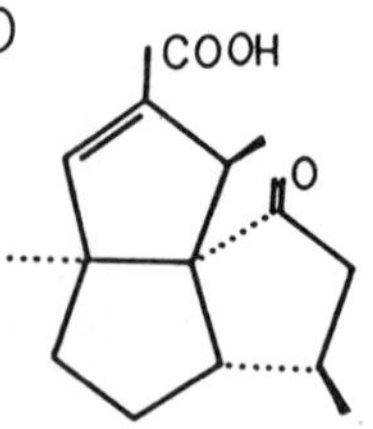

21

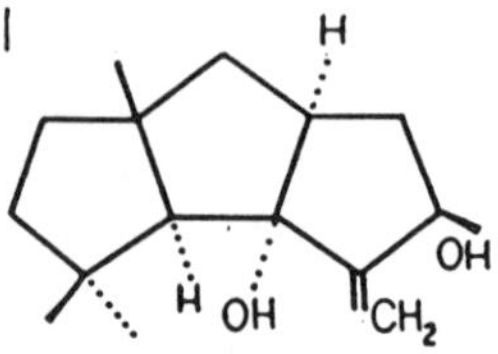

22

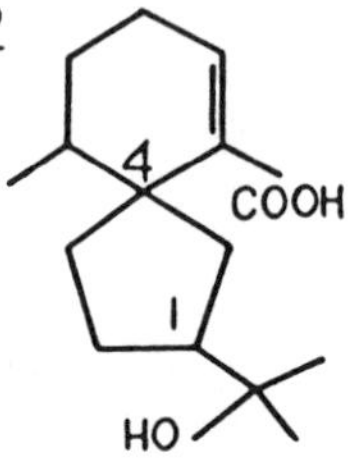

23

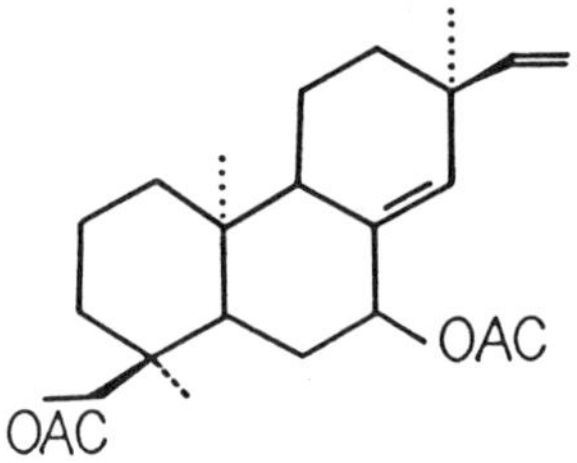

24

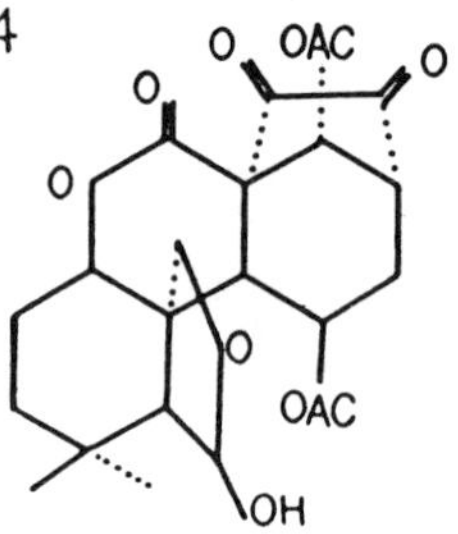

25

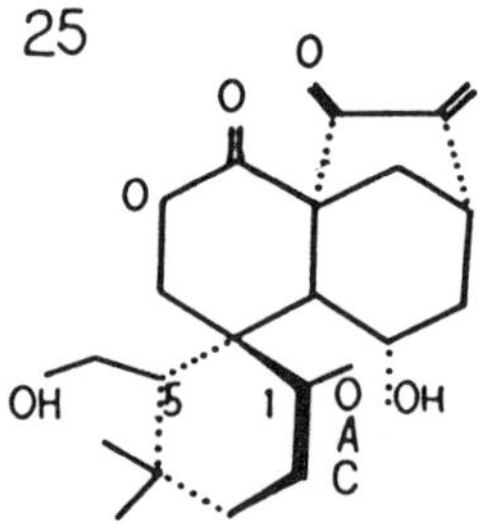

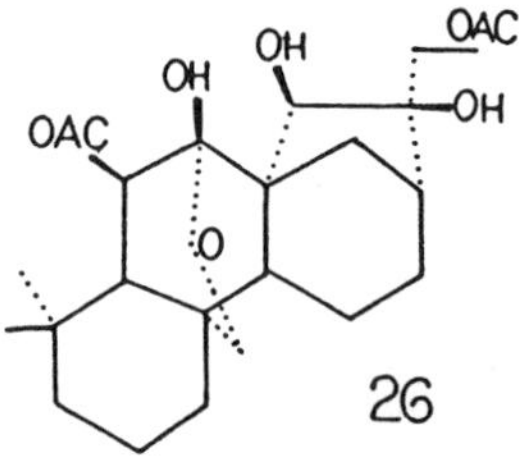

26

27

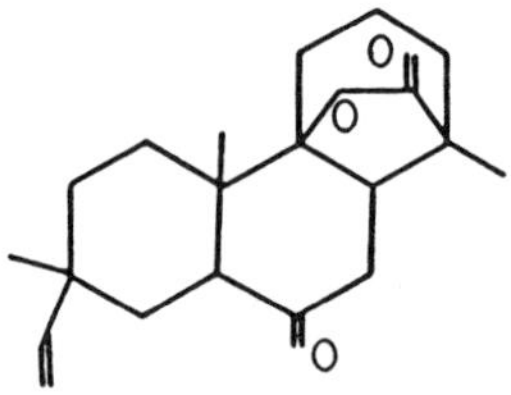

28

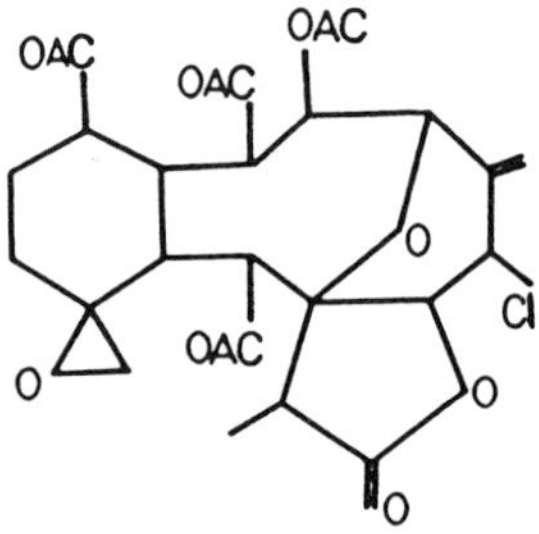

29 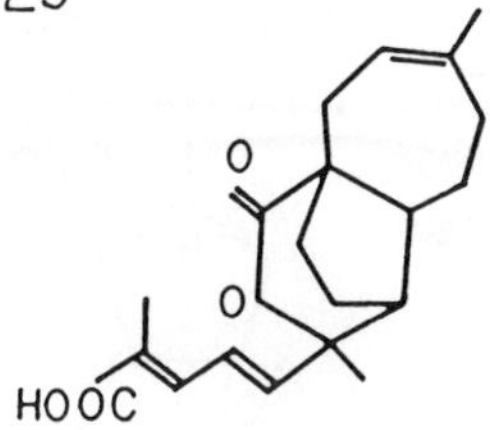
30 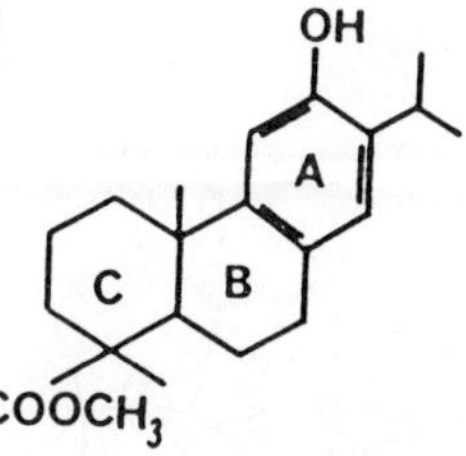
OH
A
C
B
COOCH₃
31 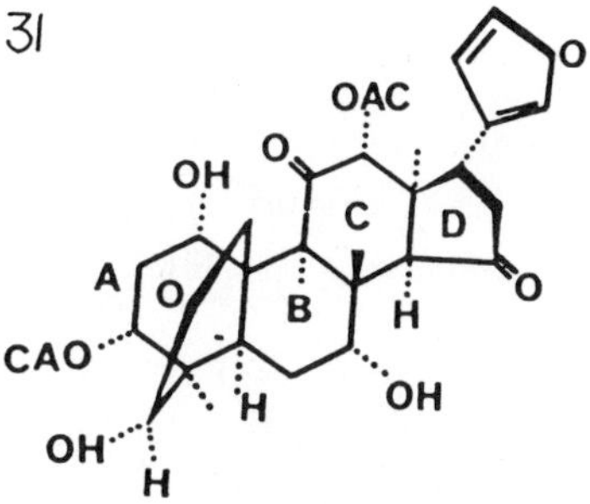
OAC
OH
A
O
B
C
D
H
CAO
OH
H
OH
H
O
32 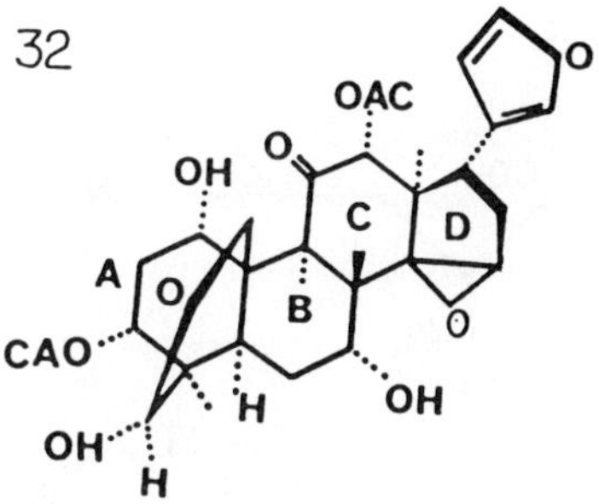
OAC
OH
A
O
B
C
D
H
CAO
OH
H
OH
H
O
O
33 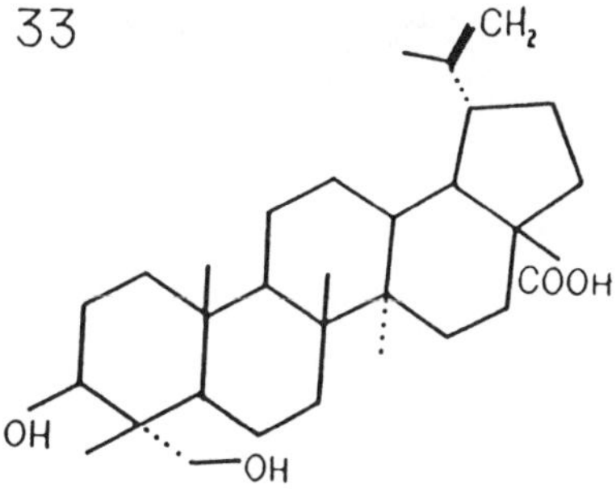
CH₂
COOH
OH
OH
34 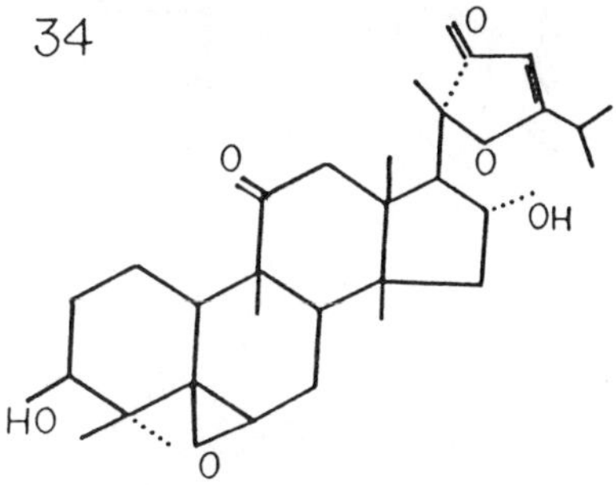 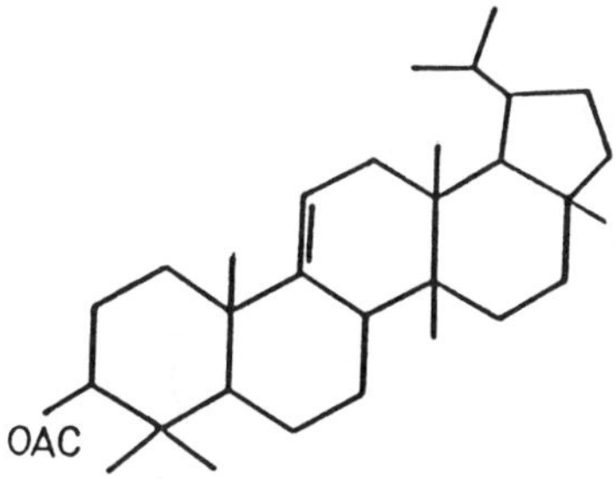
O
O
OH
HO
O
35
OAC
36
OH
OH
O
COOCH₃
O
O
O

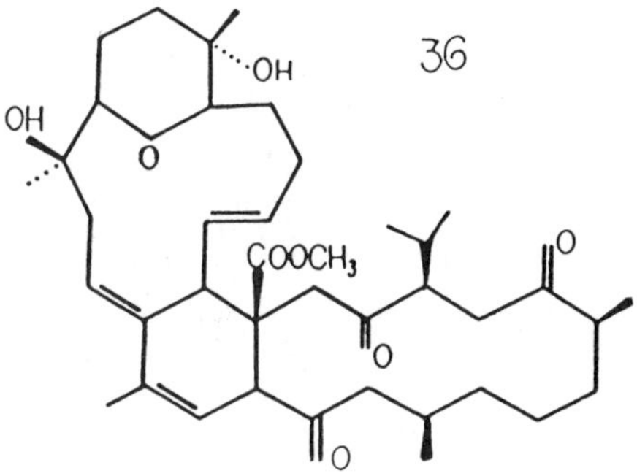

37 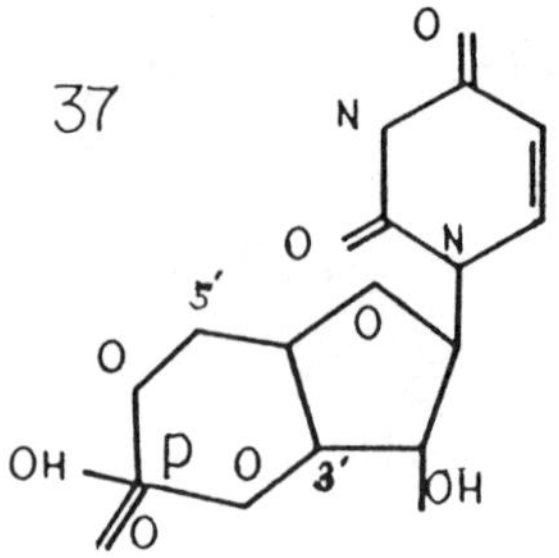
38 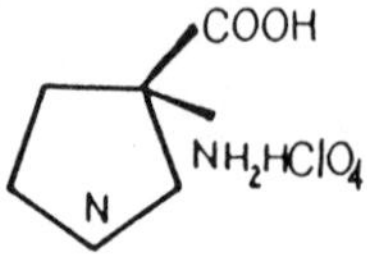
39 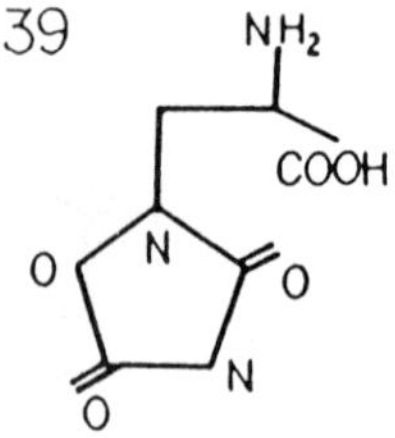
40 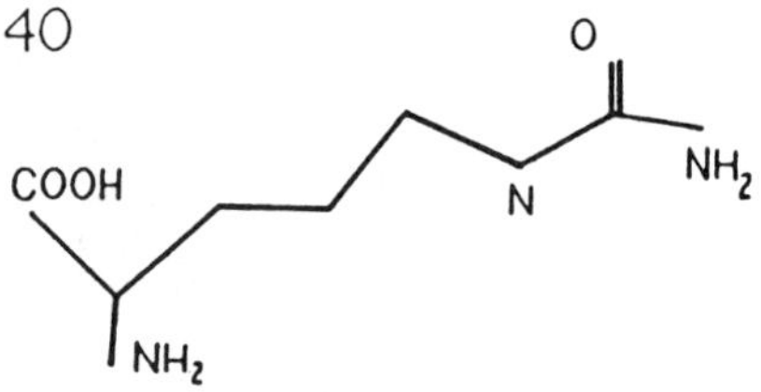 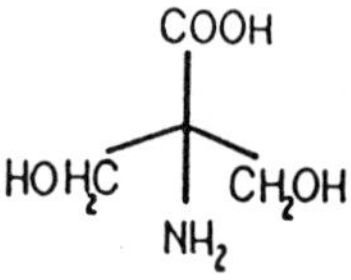
41 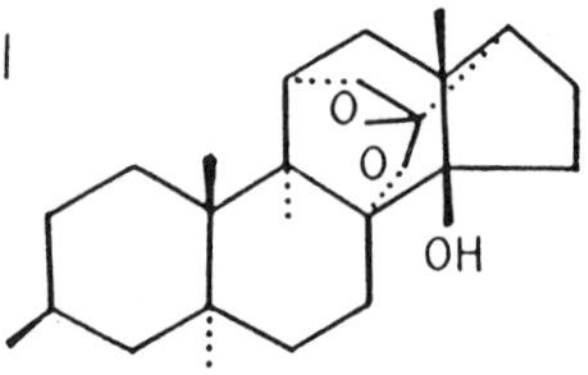
42 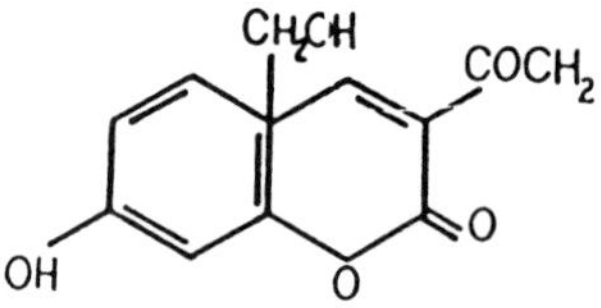
43 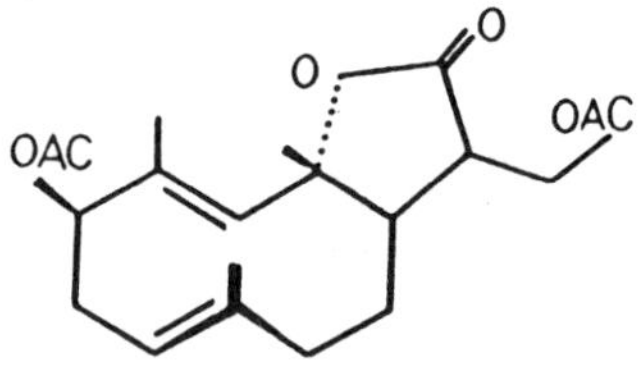
44 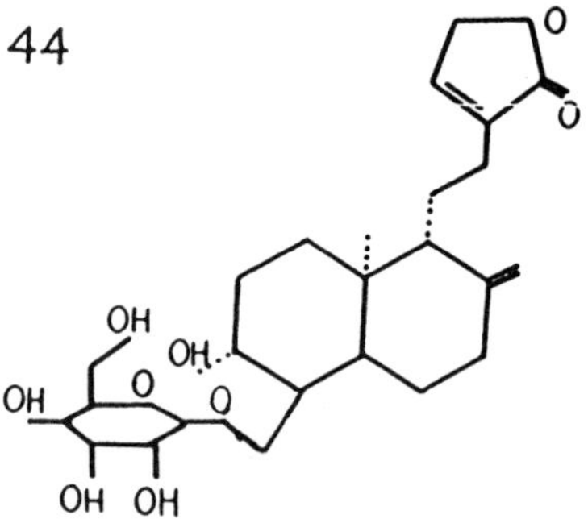 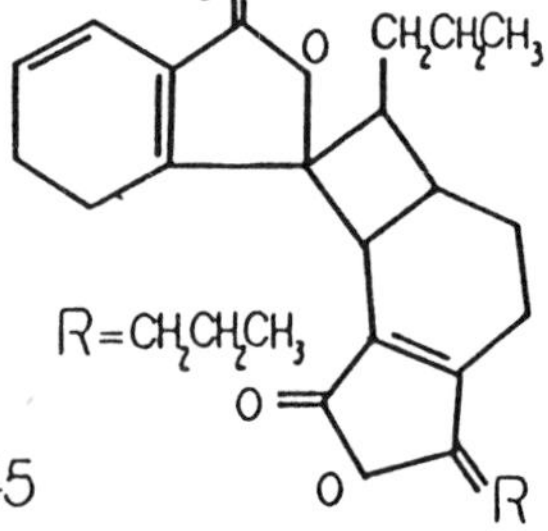
45 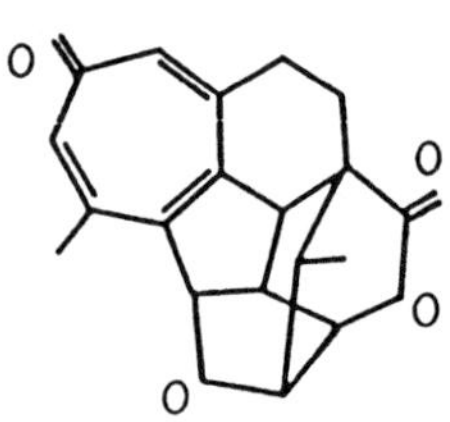
R=CH₂CH₂CH₃
46

47 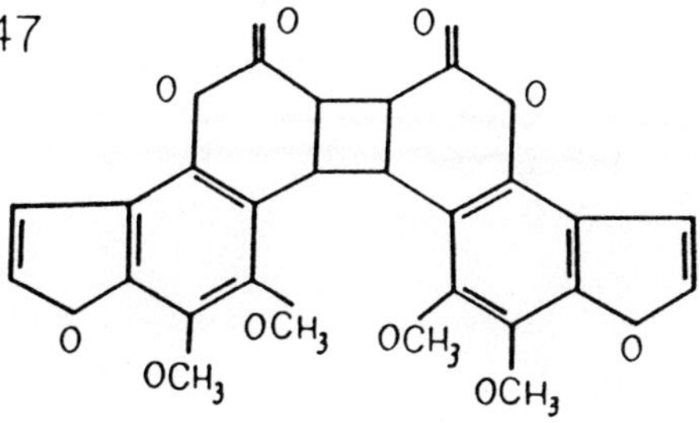
48 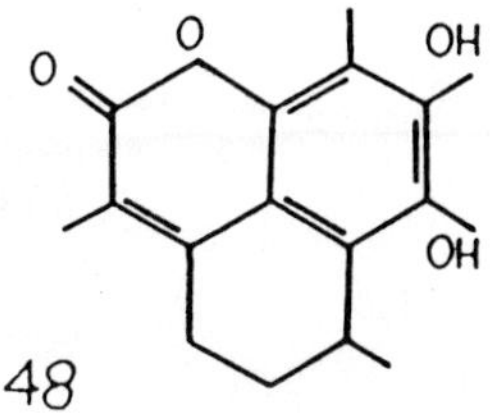
49
50 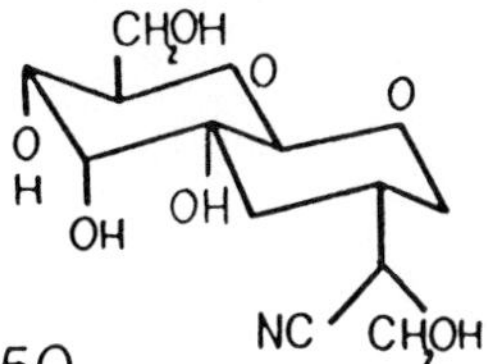
51
52 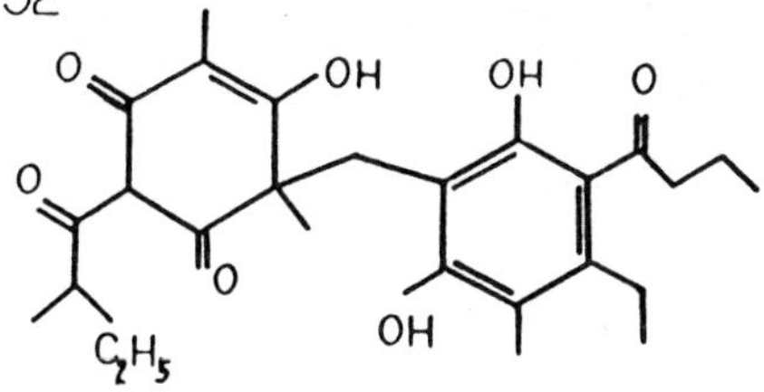
53 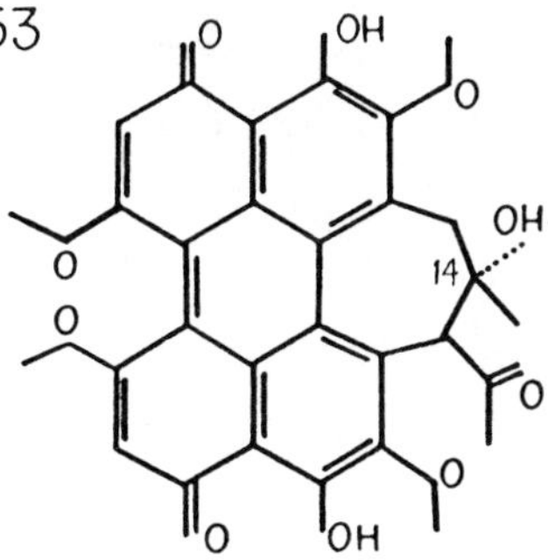
54 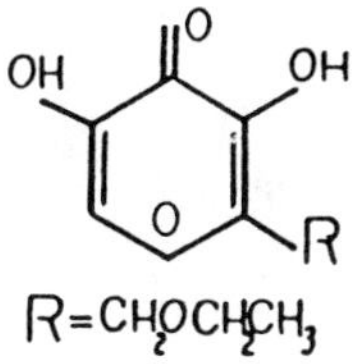
R=CH₂OCH₂CH₃
55
14

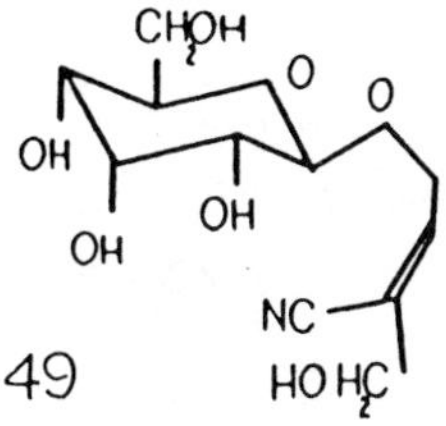

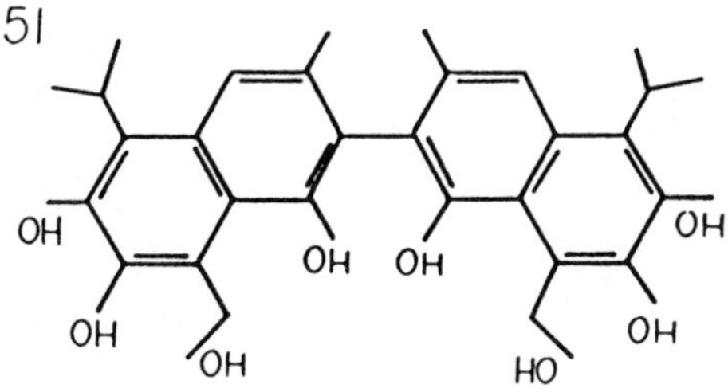

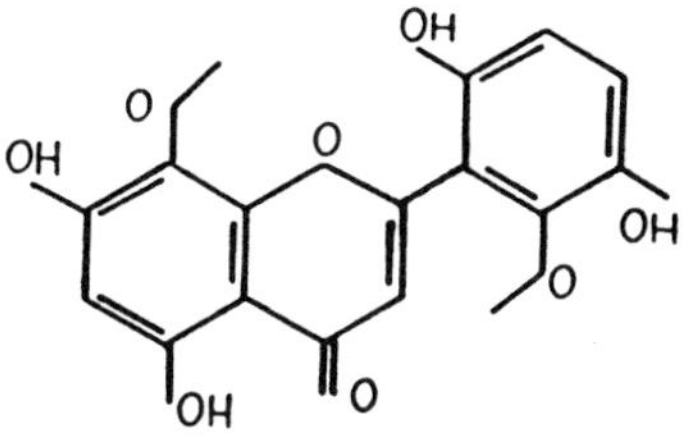

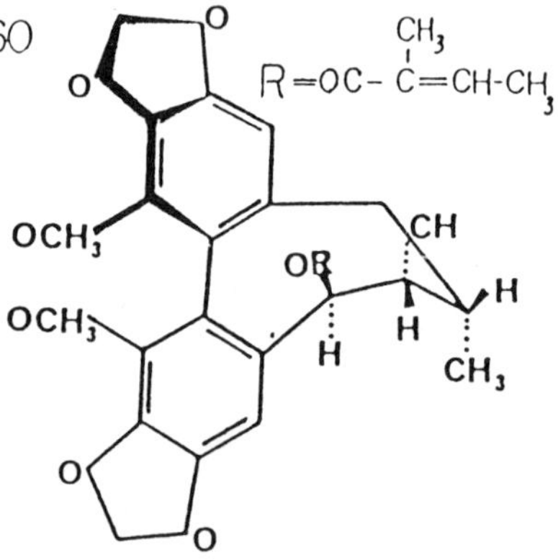

SCHEME 1

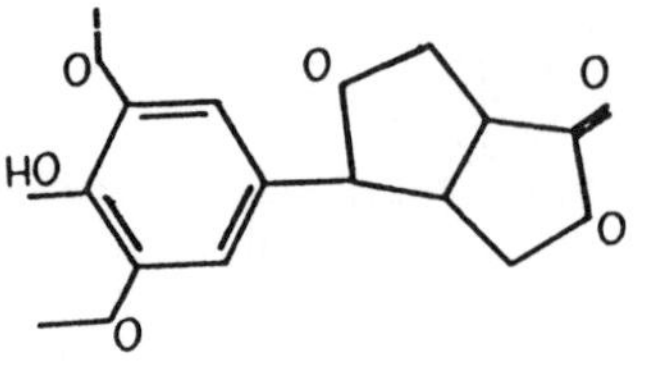

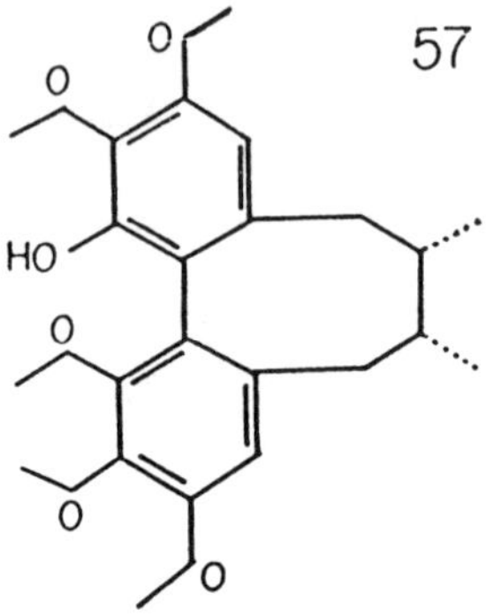

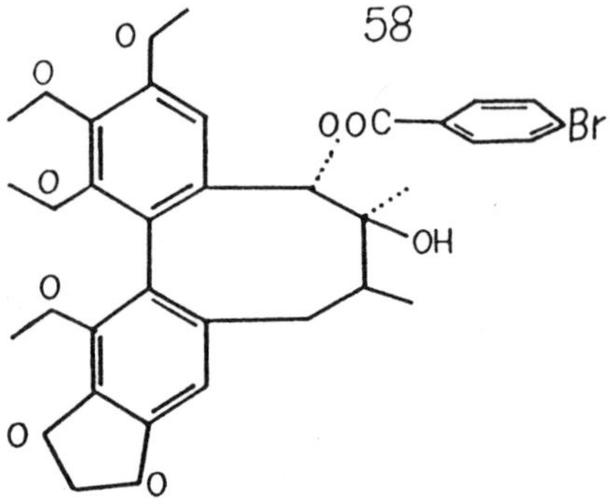

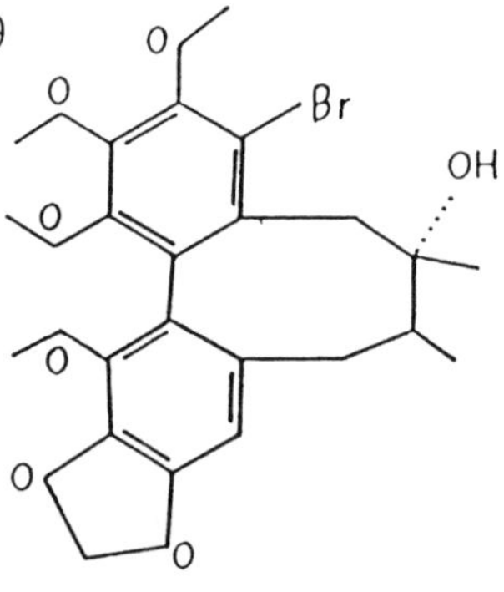

Bromoqinghaosu, **18**, is a derivative of **17**; its absolute configuration was determined using Mo K_α radiation.

Suberogorgia SP. acid, **20**, is isolated from <u>Suberogorgia SP.</u>; the absolute configuration was determined by anomalous scattering techniques.

Compound **21** is a caplellene derivative extracted from Soft Coral of the South China Sea.

Baimuxiangic acid, **22**, is isolated from Baimuxiang and acts to inhibit paranoic psychosis.

Diterpenes

Robdocalin B, **23**, consists of three six-membered rings (A, B, and C), two of which (A and B) are in chair conformations while the third has a twist-chair conformation; side chains α-CH_3 and β-CH-CH_2 are on C_{13}.

Rabdocalin A, **24**, a Hela cell inhibitor, is an effective anticancer drug extracted from a Chinese traditional herb medicine.

Rabdophylin G, **25**, belongs to spirosekaurene and is an effective anticancer drug isolated from <u>Rabdosin macrophylla</u> (<u>Migo</u>); robdophylli H, **26**, is also an effective anticancer drug isolated from the same source.

Zujiunsu, **27**, also has antitumor activity. Its chemical structure has three six-membered rings (A-C) and one five-membered ring (D).

Praelolide D, **28**, is isolated from <u>Chanxiaoyue-Suberogorgia SP.</u>

Pseudoclaric acid, **29**, isolated from the bark of <u>Pseudolorix Kaempferi Gorden</u>, has been used for treatment of skin diseases caused by fungi.

Torreyagrandate, **30**, is endemic in China and belongs to <u>Taxaceae</u> of <u>Gymuospermae</u> whose edible seeds are a famous nut.

Triterpenes

Isochuanliansu, **31**, is isolated from the bark of <u>Melia toosendan</u> and <u>Melia azedarach</u> which is used to treat ascarid. Its toxicity is higher than Chuanliansu. The positions of the side chains are: C_1-αOH, C_3-αOAc, C_4-αCH$_3$, C_5-αH, C_7-αOH, C_8βCH$_3$, C_{12}-αOAc, C_{14}-αH and C_{17}-α-furan. Rings A-E are chair conformers.The junctions between A/B, B/C and B/E, C/D are cis-fusions. There is an oxygen bridge between atoms C_{19} and C_{28}.

Chuanliansu, **32**, (isolated from the bark of <u>Melia toosendan S. et Z.</u>) is an active antilimintic Chinese folk medicine.

Anemosupogenin G, **33**, is isolated from the Chinese traditional herb medicine Anemosupogenin.

Picfeltarragenin A, **34**, is isolated from <u>Picriafeltarrae Lour</u> which belongs to <u>Scrophulariaceae</u>. The glycosides of picfeltarragenin were shown to possess significant antitumor activity. The sample for the X-ray study was picfeltarragenin A epoxide.

Tylophorin, **35**, was isolated from <u>Tylophora</u> which is another Chinese folk medicine.

Tetraterpene

Methylsartortuoate A, **36**, is a unique tetraterpenoid that was isolated from Chinese soft coral (<u>Sarcophyton Tixier-Durivault</u>).

NUCLEOTIDE

Uridine 3',5'-cyclic phosphate, **37**, was obtained as colorless and transparent crystals.

AMINO ACIDS

Cucurbitine, **38**, is a natural amino acid recently found to be the effective compound in the crude drug. The crystal structure was done for the perchlorate salt.

Quisqualic acid, **39**, is the anthelmintic principle of a chinese traditional herb drug (Shijunzi). It was isolated from the seeds and leaves of Quisqualis indica L.

L-Citrulline, **40**, and α-hydoxymethylserine were isolated from a solution of Tianhuafun. Their crystal structures were determined.

STEROID

Tenacissigennin, **41**, was isolated from Marsdenia Tenacissima (Roxb) Wight et Arn. The molecular structure contains only cis-junctions for rings.

ESTERS

Armillarisin A, **42**, is used in the treatment of cholecystitis in Chinese fork medicine.

Arteanomalactone, **43**, was isolated from Artemisia anomala S. Moore, which is a traditional Chinese folk medicine.

Glucopyranosyl-deoxyandrographolide, **44**, isolated from Andrographics paniculata Nees, is used to alleviate pain and is antiphlogistic.

Riligustilide, **45**, was isolated from the root of Ligusas acutiloburn, grown in Gansu Minxian province.

Chufeineizhi, **46**, (isolated from Hainanchufei) has anti-tumor activity.

COUMARINS

Moellendorffiline, **47**, was isolated from Heracleum moellendorffii Hance Var. paucivitatum Shan et Wang which is used in the treatment of malignant boils. The molecule appears to be symmetrical; planes A-B-C and A'-B'-C' are nearly vertical and the D-ring is a twist square.

Heliclactone, **48,** was isolated from the roots of <u>He-licteres angustifolia L.</u> which is grown in Guansi, Guandun and Fujian provinces. It is used to treat hemorrhoids and as an antiviral for influenza.

GLYCOSIDES

Sarmentosin, **49,** was isolated from the whole herb of Sarmentosum Bunge which belongs to the family <u>crassulaceae</u>. The compound showed considerable effect in lowering the serum glutamic transaminase level of patients with chronic viral hepatitis. The sample used for the crystal structure was the pentaacetyl derivative.

Isosarmentosin, **50,** was isolated from the whole herb of <u>Sedum Sarmentosum Bunge</u>. This compound also showed effects in lowering serum glutamic transaminase level.

Gossypol, **51,** is used as a male infertility agent. It is isolated from the oil of cotton seed.

Agrimophol, **52,** was isolated from <u>Agrimonia pilosa Ledeb.</u> It is used to treat anthelmintic.

QUINONE

Hypocrellin B, **53,** was isolated from the <u>Shiraia Bambosicola</u>, which is a <u>Saprohpytic Ascomycetes</u> on bamboo in southern China. Its main pharmacological action is as a new phototherapy drug. The structure was determined by X-ray methods in order to confirm the chemical structure and to determine possible differences in isomers. The results indicate that the isomerism difference is an α,β-configuration for the CH_3 and OH in position C_{14}.

FLAVONOIDS

Argutone, **54**, was isolated from <u>Incarvillea arguta</u> (Royle). It is used to inhibit mollified action.

LIGNANS

Glaberide A, **56**, was isolated from <u>ILex pubescens Hock et Arn Var. glaber Chang</u>. It is used to cure heart disease.

Schisanhenol, **57**, was isolated from the kernels of <u>Schisandra rubriflora Rhed et wils</u> and has been used in folk medicine to reduce serum glutamic pyruvic transaminase. Rub-schisantherin A, **58**, was isolated from the same source and has similar application.

Wuweizichun B, **59**, is a constituent of the kernels of <u>Schizandra Chinese Baill</u>. 4-Bromowuweizichun B is a bromination product of **59**. Its structure determination showed that the positions of methylenedioxy groups at C_{10} and C_{11} conform to the p-bromobenzoyloxy ester of **59**.

Angeloylgomisin, **60**, was isolated from the roots and stems of <u>Kadsura Longipedunculata Finet et Gagnep (Schisandraceae)</u> which finds use in Chinese folk medicine for treatment of rheumatoid arthritis and gartric and duodenal ulcers. The conformation of **60** differs from that of **59** in the eight-membered ring. In **60** the twist boat conformation is found.

References

BAO, G.H., HE, C.H. and XU, C.F. (1983), <u>Acta Academiae Medicinae Sinicae</u> **5**, 89.

CHEN, S.Z. and MA, X.Q. (1984), <u>Acta Chimica Sinica</u> **42**, 109.

CHEN, Y.Z., WU, Z.W. and CHEN, P.Y. (1984), <u>Acta Chimica Sinica</u> **42**, 645.

CHEN, Y.Z. and CHEN, P.Y. (1985a), Acta Chimica Sinica 43, 1190.

CHEN, Y.Z. and CHEN, P.Y. (1985b), Kexue Tongbao, 1709.

DAI, J.B., WAN, Z.L., RAO, Z.H., LIANG, D.Z., LU, Y.K. and LONG, K.H. (1985), Scientia Sinica B 220.

FAN, H.F. and LIN, Z.J. (1965), Acta Physica Sinica 21, 253.

FAN, H.F. (1975), Acta Physica Sinica 24, 57.

FAN, H.F. and ZHENG, Q.T. (1978), Acta Physica Sinica 27, 169.

FAN, H.F. and ZHENG, Q.T. (1980), Acta Physica Sinica 29, 1342.

GU, Y.X., ZHENG, Z.D., QING, J.Z. and ZHENG, Q.T. (1981), Acta Physica Sinica 30, 383,

GUA, F., HOU, Y.G., ZHU, N.J. and FU, H. (1984), J. Struct. Chem. 3, 91.

HE, C.H., XU, C.F. and BAO, G.H.(1977) Kexeu Tongbao 22, 504.

HE, C.H., XU, C.F., BAO, G.H. and MU, S.T. (1982), Acta Physica Sinica 31, 832.

HE, C.H., BAO, G.H., XU,C.F. and MU, S.T. (1982) Acta Pharmaceutica Sinica 17, 597.

HUAN, F.S., ZHENG, Q.T. and FAN, H.F. (1981),, Acta Physica Sinica 30, 1141.

JIN, X.L., PAN, Z.H., GUA, R.H. and HUAN, Q.C. (1985), Acta Chimica Sinica 43, 5.

KONG, Y.C., LAN, C.H., CHEN, K.F. and MAK, T.C.W. (1985), J. Struct. Chem. 4, 30.

LI, G.P. and TANG, Y.Q. (1982), Kexeu Tongbao 27, 1113.

LIN, S.X. and HUAN, J.L. (1984), J. Struct. Chem. 3, 226.

LIN, X.Y. and ZHANG, S.D. (1985), Acta Chimica Sinica 43, 724.

MA, X.Q. and CHEN, S.Z. (1983), Acta Chimica Sinica 41, 663.

MAK, T.C.W., SHE, K.L. and LONG, K.H. (1985), Acta Scientiarum Naturalliuw, Universitatis Sunyatseni, 22.

MUN, Y.M., WANG, Q.C., ZHANG, H.D., CHEN, Y.Z., FAN, Y.G. and WAN, F.S. (1983, Acta Lanzhou Univer. 19, 75.

NEU, L.W., DAI, J.B., WAN, Z.L., LIANG, D.Z., WU, Z.D. and YAO, Z.N. (1985), Scientia Sinica B, 709.

QINGHAOSU GROUP (1977), Kexeu Tungbao, 142.

QINGHAOSU GROUP (1979), Kexeu Tungbao, 1114.

RAO, Z.H., WAN, Z.L. and LIANG, D.Z. (1982), Acta Physica Sinica 31, 547.

RAO, Z.H., DAI, J.B., WAN, Z.L., LIANG, D.Z. and WAN, D. (1984), Scientia Sinica, 217.

SHU, J.Y., LONG, H., PENG T.S., ZHENG, Q.T. and LIN, X.Y. (1985), Acta Chimica Sinica 43, 796.

TANG, Y.Q., LI, G.P., JIN, X.L. and HAO, R.R. (1982), Acta Chimica Sinica 40, 193.

WANG, S.D. (1978), Kexeu Tungbao 23, 670.

WANG, Z.T., LENG, H.J. and LIN, J.Y. (1984), Journal Molecular Science 4, 385.

XU, Z.B., ZHENG, Q.T. and FAN, H.F. (1982), Acta Physica Sinica 31, 1064.

XU, C.F., HE, C.H., BAO, G.H. and ZHENG, Q.T. (1981), Kexeu Tungbao 26, 467.

XU, C.F., HE, C.H., and BAO, G.H. (1982), Scientia Sinica B, 636.

XU, C.F., BAO, G.H. and HE,C.H. (1982), Kexeu Tungbao, 1081.

XU, C.F., HE, C.H., BAO, G.H. and MU, S.T. (1985), Acta Academiae Medicinae Sinicae 7, 187.

XU, X.J., HUA, Z.Q., ZHOU, G.D. and TANG, Y.Q. (1985), Scientia Sinica B, 494.

YANG, R.Z., ZHOU, J., ZHENG, Q.T. and CLARDY, J., (1983), Acta, Botanica Yunnanica 5, 213.

YAO, J.X. and LIN, X.Y. (1982), Acta Chimica Sinica 40, 385.

ZHANG, S.D., ZHANG, J.P., WU, B.M., YAO, J.X. and LIN, X.Y. (1981), Acta Physica Sinica 30, 976.

ZHENG, C.D., QING, J.Z., GU, Y.X. and ZHENG, Q.T. (1981), Acta Physica Sinica 30, 242.

ZHENG, Q.T., GU, Y.X., WANG, L.L. and LIU, J.Y. (1980) Acta Physica Sinica 29, 1014; (1981), Acta Physica Sinica 30, 1593.

ZHENG, Q.T., DOU, S.Q. and GU, Y.Z. (1981), <u>Acta Physica Sinica</u> **30**, 1369.

15. Towards a molecular model for the large ribosome particle

A. Yonath and H. G. Wittmann

Abstract

Large crystals of 50S ribosomal subunits, diffracting to 6Å have been grown. Crystal of the same particles from a mutant which lacks protein L11 are isomorphous to those of the wild type. An undecagold cluster was prepared and bound to a -SH group on isolated protein L11. This cluster was also used for soaking experiments. Two-dimensional sheets from 50S ribosomal subunits have been subjected to three-dimensional image reconstruction studies at 30Å resolution. The resulting model contains several projecting arms, arranged radially around a cleft which turns into a tunnel which may be the path taken by nascent protein.

Crystallographic Studies

To shed light on the role of ribosomes in protein biosynthesis, we undertook crystallographic studies. As an object for such studies, ribosomal particles are of enormous size, with no internal symmetry. Nevertheless, procedures for _in vitro_ growth of crystals of intact ribosomal particles have been developed. These led to the production of crystals and sheets of whole ribosomes (Wittmann et al, 1982, Piefke et al, 1986) as well as of 50S ribosomal subunits (Yonath et al, 1984, 1986a,b, Yonath and Wittmann, 1987, Arad et al, 1984,

Makowski et al, 1987). Only active particles could be crystallized and the crystalline material retains its biological activity for long periods, in contrast to the short life time of isolated ribosomes. The best crystals are of 50S ribosomal subunits from <u>Halobacterium marismortui</u> and <u>Bacillus stearothermophilus</u>. The 50S particles are of molecular weight of 1,600,000 daltons with approximate size of 150x170x180 Å^3. They are composed of 30-35 different proteins and 2 RNA chains (Wittmann, 1983). Synchrotron radiation is essential for crystallographic studies of these crystals due to their fragility, their short life time and the large unit cell dimensions.

Halophilic ribosomal particles are stable and active at high salt concentrations. Taking advantage of the fact that these ribosomes maintain their activity over a wide range of salt concentrations, as well as of the major role played by the Mg^{++} in crystallization of ribosomal particles (Arad et al, 1984), we could obtain large, well ordered crystals from the 50S ribosomal particles from <u>H</u>. <u>Marismortui</u>. These (0.6x0.6x0.2mm^3) are monoclinic (P2$_1$), diffract to 6Å (Fig. 1), and have relatively small, compactly packed unit cells of 182x186x584 Å, γ = 109° (Makowski et al, 1987). Although up to 15 rotation photographs can be taken from an individual crystal, the high resolution terms appear only on the first diffraction patterns. Hence over 260 crystals were irradiated in order to obtain a complete data set.

Because ribosomes from <u>eubacteria</u> fall apart at high salt concentrations, they were crystallized from organic solvents. Currently crystals of the 50S subunits of <u>B</u>. <u>stearothermophilus</u> (2.0x0.3x0.3 mm) are obtained directly in X-ray capillaries, at 4°C, from mixtures of methanol and ethylene glycol (Yonath et al, 1984, 1986a,b). The crystals of space group P2$_1$2$_1$2 are loosely packed in a unit cell of 360x680x921 Å. Since most of the crystals adhere to the walls of the capillaries in different directions, it was possible to obtain

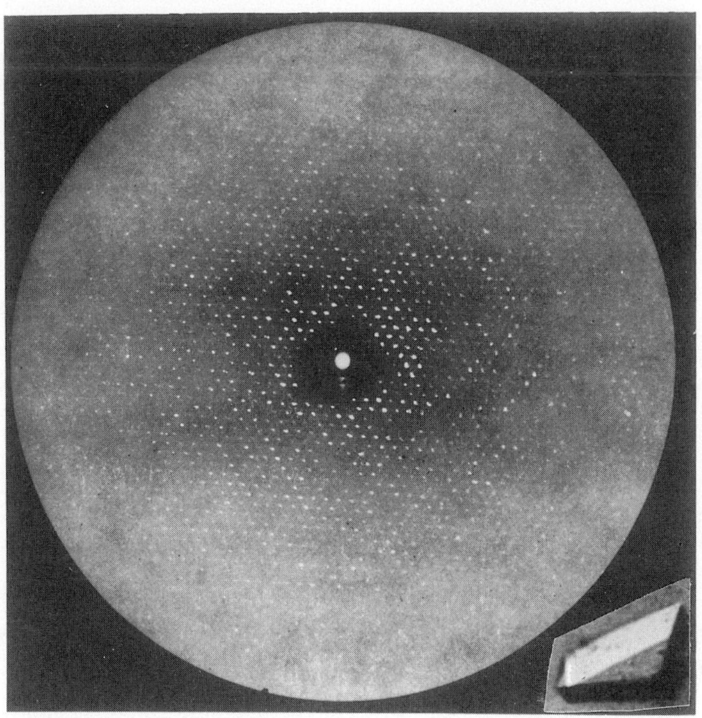

Figure 1. A 1° rotation pattern from the crystal in the insert obtained in 2 min. at 4°C with synchrotron radiation {(XII/EMBL/DESY) operating at 5 GeV, 34 mA. $\lambda=1.51$Å; slit collimation (0.3×0.3mm^2), distance=235mm}. Insert: a crystal of the 50S subunits of H. marismortui (Makowski et al, 1987).

diffraction patterns from all zones without manipulation. For an object as large as the 50S ribosomal subunit, it is necessary to use extremely heavy and dense compounds for derivation. An example is a gold cluster:

$$Au_{11}(CN)_3[P(C_6H_4CH_2NH_2)_3]_7$$

with a diameter of 8.5Å (Yang and Frey, 1984). Preliminary experiments show that crystals of H. marismortui soaked in this cluster are isomorphous to the native crystals and their X-ray patterns show intensity differences.

Because the surface of the particles has a variety of

potential binding sites, in parallel to soaking experiments attempts to bind the cluster covalently to the particles are in progress. A model compound radioactive N-ethylmaleimide was interacted with exposed -SH groups on the entire 50S ribosomal subunit. For both bacteria only two ribosomal proteins participted in the reactions (Weinstein, unpublished). A mutant of B. stearothermophilus which lacks protein L11 was obtained. Crystals of the 50S mutated ribosomal subunits are isomorphous to those of the wild-type (Yonath et al, 1986b). Protein L11 has only one sulfhydryl group to which the cluster, perpared as a monofunctional reagent, was bound.

Three Dimensional Image Reconstruction

The large size of ribosomal particles, which is an obstacle for crystallographic studies, permits direct investigation by electron microscopy. Two-dimensional sheets from 50S ribosomal subunits from B. stearothermophilus (148x370 Å γ = 109°) have been subjected to reconstruction studies at 30 Å resolution (Yonath et al, 1987). The resulting model (Fig. 2) shows several projecting arms arranged radially around the presumed interface with the 30S subunit. A cleft is formed between the projecting arms. This turns into a tunnel of a diameter of up to 25 Å and a length of 120 Å. The tunnel branches, forming a Y shape. An elongated region of low sensitivity was also detected in ribosomes from lizard (Milligam and Unwin, 1986).

The functional role of the tunnel is still to be determined. However, originating at the presumed site for actual protein biosynthesis, terminating on the other end of the particle, (Bernabeau and Lake, 1982), and being of a diameter large enough to accomodate every amino acid and of a length which could provide protection to a 40mer peptide in an extended conformation (Malkin and Rich, 1967, Blobel and Sabatine, 1970, Smith et al, 1978), this tunnel appears to be the path

taken by nascent protein. The reconstructed model shows higher detail than the image seen by electron microscopy due to
the inherently more objective character of the analysis by diffraction methods.   However, it can be positioned so that its
projected view resembles the latter (Fig. 2).

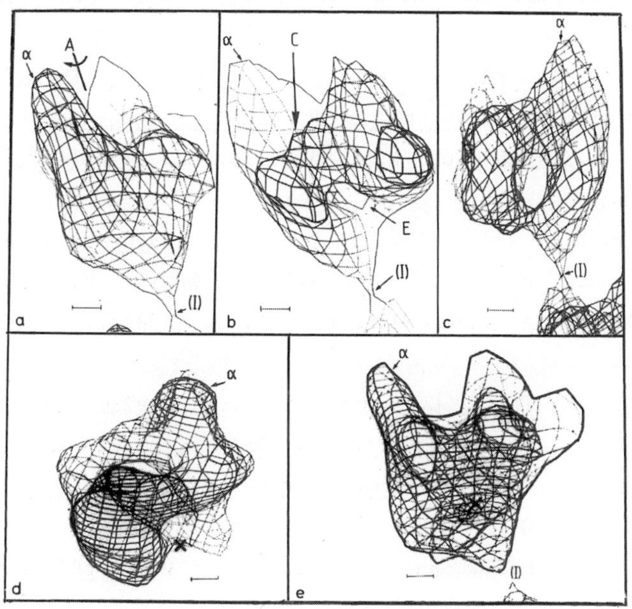

Figure 2.   Computer graphics display of the outline of the
reconstructed model of the 50S ribosomal subunit at 30 Å resolution.   Heavy lines are in front, lighter ones in back. Image (a): A side view of the model, (I) is the contact between
the two particles; (A) the approximate axis around which the
model was turned to obtain the view in (b). Image (b): (C) is
the cleft between the projecting arms, at the site it turns
into the tunnel. The exit point of the tunnel is marked (E).
Image (c) is a view into the branch of the tunnel from the
exit point, (d) a view into the tunnel from the cleft. In
image (e) the model is viewed in a projection which resembles
images vizualized by electron microscopy.

Acknowledgements

    This work was supported by Bundes Ministerium für For-
schung und Technologie, National Institutes of Health and
Minerva Research grants

References

ARAD, T., LEONARD, K.R. WITTMANN, H.G. and YONATH, A. (1984). EMBO Journal 3, 127.

BERNABEAU, C. and LAKE, J.A. (1982). Proceedings of the National Academy of Sciences, USA 79, 3111.

BLOBEL, G. AND SABITINE, D.D. (1970). Journal of Cell Biology 45, 130.

MAKOWSKI, I., FROLOW, F., SAPER, M.A., WITTMANN, H.G. and YONATH, A. (1987). Journal of Molecular Biology in press.

MALKIN, L.I. and RICH, A. (1967). Journal of Molecular Biology 26, 329.

MILLIGAN, R.A. and UNWIN, P.N.T. (1986). Nature 319, 693.

PIEFKE, J., ARAD, T., MAKOWSKI, I., GEWITZ, H.S., HENNEMANN, B., YONATH, A. and WITTMANN, H.G., (1987). FEBS Letters in press.

SMITH, W.P., TAI, P.C. and DAVIS, B.D., (1978). Proceedings of the National Academy of Sciences, USA 75, 5922.

WITTMANN, H.G. (1983). Annual Reviews of Biochemistry 52, 35.

WITTMANN, H.G., MÜSSIG, J., GEWITZ, H.S., PIEFKE, J., RHEINBERGER, H.J., and YONATH, A. (1982) FEBS Letters 146, 217.

YANG, H. and FREY, P.A. (1984) Biochemistry 23, 217.

YONATH, A. and WITTMANN, H.G. (1986). Methods of Enzymology, in press.

YONATH, A., BARTUNIK, H.D., BARTELS, K.S. and WITTMANN, H.G. (1984). Journal of Molecular Biology 177, 210.

YONATH, A., SAPER, M.A., MAKOWSKI,I., MÜSSIG, J., PIEFKE, J., BARTUNIK, H.D., BARTELS, K.S., and WITTMANN, H.G. (1986a), Journal of Molecular Biology 187, 633.

YONATH, A., SAPER, M.A., FROWLOW, F., MAKOWSKI,I. and WITTMANN, H.G. (1986b), Journal of Molecular Biology 192, 161.

YONATH, A., LEONARD, K.R. and WITTMANN, H.G. (1987). submitted.

16. Structural aspects of the tetrahedral intermediate in the carboxypeptidase A mechanism

David W. Christianson and William N. Lipscomb

The metalloenzyme carboxypeptidase A (CPA; peptidyl-L-amino acid hydrolase, EC 3.4.17.1) is an exopeptidase of molecular weight 34,472 containing one zinc ion bound to a single polypeptide chain of 307 amino acids. Its biological function is the hydrolysis of the C-terminal amino acid from polypeptide substrates, although it exhibits equally potent activity toward ester and depsipeptide substrates. The enzyme displays a marked preference toward substrates which possess large hydrophobic C-terminal side chains such as phenylalanine. Although CPA has been the subject of nearly fifty years of research efforts in many laboratories around the world, considerable uncertainty and debate persist as to its particular mechanism(s) in the hydrolysis of peptide and ester substrates (for recent reviews, see: Lipscomb, 1982; Lipscomb, 1983; Vallee, _et al_., 1983; Vallee and Galdes, 1984). Chemical and X-ray crystallographic studies have implicated Arg-145, Tyr-248, Glu-270, Arg-127, Zn, and the zinc-bound water molecule as important in substrate binding and catalysis. Furthermore, the results of recent crystallographic studies suggest that Glu-270, Arg-127, Zn, and the zinc-bound water molecule form a "catalytic tetrad" which may be conserved in other zinc proteases (Christianson

and Lipscomb, 1986a). Various conformations and configurations
of this catalytic tetrad have been observed through the study
of enzyme-transition state analogue complexes by X-ray
crystallography (Christianson and Lipscomb, 1985;
Christianson and Lipscomb, 1986a,b; Christianson, et al.,
1986). In particular, the differing structures of these
complexes, even though they model the "same" tetrahedral
intermediate resulting from the Glu-270 and/or zinc-promoted
attack of water at a substrate carbonyl, suggest that the
actual tetrahedral intermediate of a hydrolytic reaction may
proceed through a well-defined and necessary sequence of
orientations, which cause its transient stabilization and
destabilization, in order to complete the catalytic event.

The role of zinc as a possible electrophile at the
active site of CPA is supported by the observation that the
carbonyl (and the N-terminal amino group) of glycyl-L-
tyrosine coordinated to zinc in its complex with the
holoenzyme (Lipscomb, et al., 1968; Rees, et al., 1981;
Christianson and Lipscomb, 1986c). However, more recent
studies of glycyl-L-tyrosine with the apoenzyme (Rees and
Lipscomb, 1983), and the ketonic substrate analogue
(—)-3-(p-methoxybenzoyl)-2-benzylpropanoic acid with the
holoenzyme (Christianson, et al., 1985), have shown that
the carbonyl of the substrate (or ketone) can receive a
hydrogen bond from the positively-charged guanidinium
moiety of Arg-127. The flexible arginyl moiety moves closer
to the reacting centers at the enzyme active site. Hence,
there are two electrophiles present at the active site of
CPA: the divalent zinc ion and enzyme residue Arg-127.
Since chemical and X-ray methods identify Glu-270 and the
zinc-bound water molecule as catalytically important species
(as base and nucleophile in a promoted-water mechanism, or as
nucleophiles in separate steps of the anhydride mechanism),

the whole group can be classified under a more general
structural scheme which appears to be conserved among other
zinc proteases such as carboxypeptidase B (Schmid and
Herriott, 1976), thermolysin (Holmes and Matthews, 1982), and
peptidase G (Dideberg, et al., 1982). The elements of this
catalytic tetrad are depicted in Figure 1. In these enzymes,
residues Glu or Asp (or incipient alkoxide of Ser or Thr)
might serve as a base in the promoted-water mechanism, and
the electrophile might be present as Arg or His. The
presence of two electrophiles in an active site (for CPA,
zinc and Arg-127) of a protease could stabilize a tetrahedral
intermediate or the nearby transition states which flank it
if the Hammond Postulate applies (Hammond, 1955).

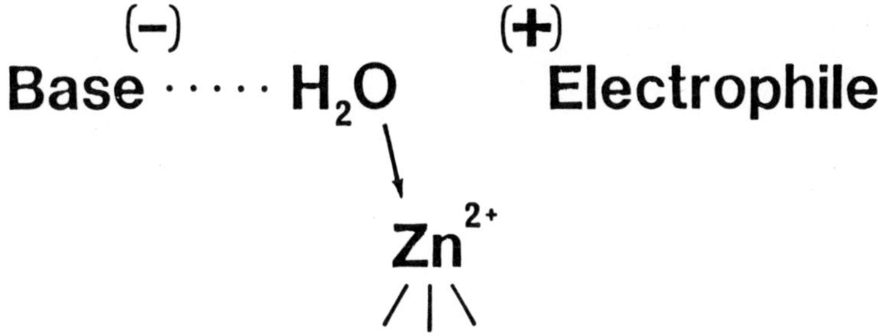

FIGURE 1. The catalytic tetrad which may be conserved among
many zinc proteases. An important feature is the presence of
an auxiliary electrophile in addition to the divalent zinc ion.

The structure of a tetrahedral intermediate can be
reasonably approximated by a variety of chemical groupings on
an enzyme inhibitor. We have investigated the complexes of CPA
with four tetrahedral transition state analogues: (1) the
hydrated aldehyde 2-benzyl-3-formylpropanoic acid (BFP,
Figure 2); (2) the hydrated ketone 2-benzyl-4-oxo-5,5,5-tri-
fluoropentanoic acid (TFP, Figure 3); (3) the cleaved

phosphonamidate (i.e., the corresponding phosphonic acid of)
N-[[[(benzyloxycarbonyl)amino]methyl]hydroxyphosphinyl]-
L-phenylalanine (hydrolyzed ZGP', Figure 4); and (4) the
hydrated ketone 5-benzamido-4-oxo-2-benzylpentanoic acid
(BOP, Figure 5). Each enzyme-inhibitor complex resembles the
tetrahedral intermediate of the promoted-water mechanism.

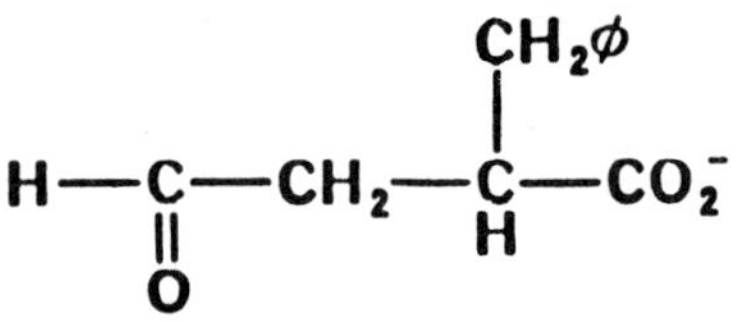

$$H-\underset{\underset{O}{\|}}{C}-CH_2-\underset{\underset{H}{|}}{\overset{\overset{CH_2\phi}{|}}{C}}-CO_2^-$$

FIGURE 2. The CPA inhibitor 2-benzyl-3-formylpropanoic acid
(BFP), which binds to CPA as the covalent hydrate adduct at
the aldehyde carbonyl.

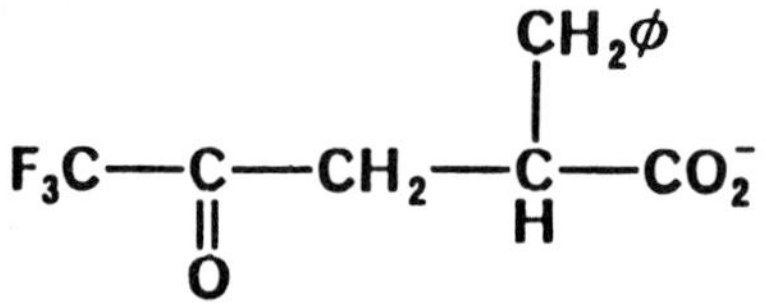

$$F_3C-\underset{\underset{O}{\|}}{C}-CH_2-\underset{\underset{H}{|}}{\overset{\overset{CH_2\phi}{|}}{C}}-CO_2^-$$

FIGURE 3. The CPA inhibitor 2-benzyl-4-oxo-5,5,5-trifluoro-
pentanoic acid (TFP), which binds to CPA as the covalent
hydrate adduct at the ketone carbonyl.

For bound BOP, the enzyme has probably facilitated the
addition of a water molecule to an otherwise unactivated
(i.e., not greatly electrophilic) ketone carbonyl. The
results of these X-ray crystallographic studies can be
classified according to the contacts of the atoms configuring
the tetrahedron with the zinc ion, Glu-270, and Arg-127.
The distances of these particular interactions for BFP, TFP,

hydrolyzed ZGP', and BOP are summarized in Scheme I. A considerable variation of contacts can occur between the tetrahedral intermediate and the enzyme, as represented by these four transition state analogues. Furthermore, both enzyme residues and the zinc ion move in order to accommodate specific binding modes. The observed rotational flexibility

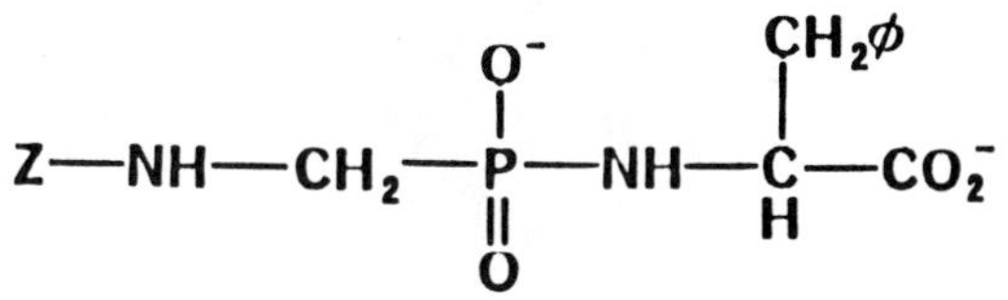

$$Z—NH—CH_2—\overset{\overset{\displaystyle O^-}{|}}{\underset{\underset{\displaystyle O}{||}}{P}}—NH—\overset{\overset{\displaystyle CH_2\phi}{|}}{\underset{\underset{\displaystyle H}{|}}{C}}—CO_2^-$$

FIGURE 4. The CPA inhibitor N-[[[(benzyloxycarbonyl)amino] methyl]hydroxyphosphinyl]-L-phenylalanine (ZGP'), which was observed to bind to CPA as the corresponding phosphonic acid plus phenylalanine. Although as such it would qualify as a product complex, the phosphonic acid moiety which binds to zinc can still be considered a tetrahedral transition state analogue. Note that Z = benzyloxycarbonyl.

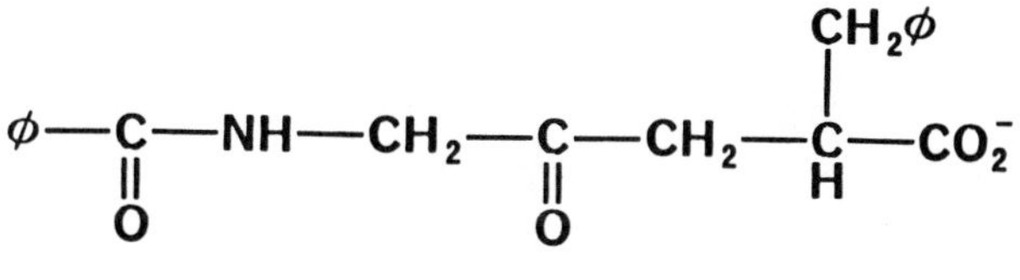

$$\phi—\overset{}{\underset{\underset{\displaystyle O}{||}}{C}}—NH—CH_2—\overset{}{\underset{\underset{\displaystyle O}{||}}{C}}—CH_2—\overset{\overset{\displaystyle CH_2\phi}{|}}{\underset{\underset{\displaystyle H}{|}}{C}}—CO_2^-$$

FIGURE 5. The CPA inhibitor 5-benzamido-4-oxo-2-benzyl-pentanoic acid (BOP), which binds to CPA as the covalent hydrate adduct at the ketone carbonyl.

of the tetrahedral moiety implies that there is flexibility in the attack of the scissile carbonyl group by zinc-bound

water, stabilization of the tetrahedral intermediate, and
dispersal of products through progressive reorientation away
from Arg-127 toward Glu-270. This method toward the study of
enzyme mechanism through the study of enzyme-inhibitor
complexes is an example of the reaction coordinate approach
used in small molecules by Slater (1959) and Slater (1959),
and by Bürgi, Dunitz, and Shefter (1973), and has been
reviewed by Bent (1968).

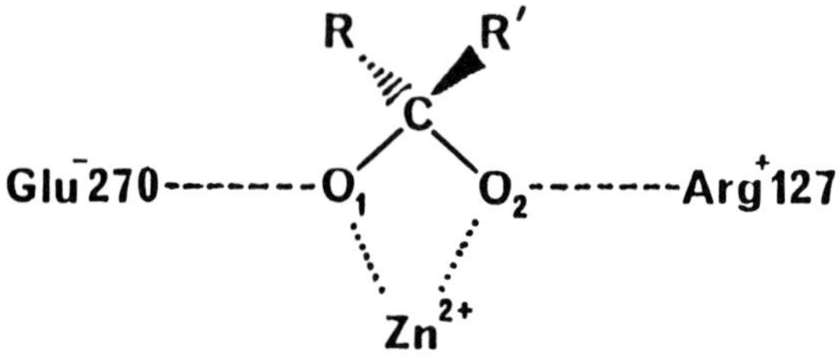

$$\text{Glu}^-\text{270} \text{------} O_1 \qquad O_2 \text{-------} \text{Arg}^+\text{127}$$

with R and R' on carbon C above, and Zn^{2+} below bridging O_1 and O_2.

Distance (Å)

Interaction	CPA-BFP	CPA-TFP	CPA-BOP	CPA-hydrolyzed ZGP'
Glu-270--O1	2.4	2.6	2.6	3.4
O1--Zn	2.7	3.3	2.9	2.2
O2--Zn	2.5	2.6	2.5	3.3
O2--Arg-127	**3.6**	3.2	3.2	2.7

SCHEME I. The results of current enzyme-transition state
analogue structures can be summarized in terms of the contacts
made by the oxygens of the tetrahedral moiety in each case with
Glu-270, zinc, and Arg-127. Both zinc and Arg-127 can
stabilize the oxyanion of an actual tetrahedral intermediate.
A well-defined sequence of orientations of the tetrahedron
may be required for catalysis, especially with regard to the
two active site electrophiles.

A tetrahedral intermediate, as represented by the four enzyme-inhibitor complexes described, can require certain, discrete orientations and contacts depending upon its position along the reaction coordinate, and whether it is forming or collapsing. Additionally, an oxyanion would probably be drawn closer to zinc than would the hydroxyl moiety of a hydrate. Interactions within the S_1' hydrophobic pocket, and in other subsites on the enzyme, are known to be of kinetic significance (Abramowitz, et al., 1967; Kuo, et al., 1979) and may influence the structure, stability, and control of the tetrahedral intermediate as well as its flanking transition states. Bearing these considerations in mind, a reasonable promoted-water mechanism might involve the initial hydrogen bonding of the substrate carbonyl to Arg-127. The positively-charged guanidinium moiety of this residue could polarize the scissile carbonyl, making it more susceptible to nucleophilic attack by zinc and/or Glu-270-promoted water. Formerly, the role of carbonyl polarization was traditionally ascribed to the zinc ion. However, such a step would increase the pK_a of zinc-bound water, making it less nucleophilic. By letting this water molecule attack the scissile carbonyl polarized by Arg-127, the potent nucleophilicity of zinc-bound water is preserved (and enhanced by its hydrogen bond with Glu-270). As the tetrahedral intermediate is formed, or immediately subsequent to its formation, the developing oxyanion could shift its association from Arg-127 to zinc. This would involve a clockwise "rotation" of the tetrahedral center depicted in Scheme I. The oxyanion contact with Arg-127 might be retained as an additional stabilizing factor; however, the positively-charged zinc ion would probably favor a close coordination interaction with the negative oxyanion. Also facilitated by this rotation would be the

transfer of a proton from the former zinc-bound water molecule
to Glu-270. Hence, the nucleophilic promotion of the water
molecule is a task which could be shared between the zinc ion
and Glu-270. The hydroxyl of the tetrahedral intermediate
could retain a hydrogen bond with the carboxylic acid
carbonyl of the now-protonated Glu-270. In addition to a
hydrogen bond with the backbone carbonyl of Ser-197, this
hydroxyl could retain some coordination to zinc for
additional stabilization. Hence, such a tetrahedral
intermediate could be stabilized through a pentacoordinate
zinc ion (counting the three enzyme residues Glu-72, His-69,
and His-196 as single ligands). The collapse of the
tetrahedral intermediate would then involve a proton transfer
from the newly-protonated Glu-270 to the leaving amino group.
This step is illustrated in Figure 6 for an extended
substrate, in which all postulated contacts with the enzyme
are depicted in the S_1' and S_1 subsites.

It is possible that some ester substrates proceed
through the anhydride mechanism, which would involve a mixed
anhydride intermediate between the substrate and enzyme
residue Glu-270 (Makinen, et al., 1979; Kuo and Makinen,
1982; Kuo and Makinen, 1985; Suh, et al., 1985). Many of the
structural aspects of the tetrahedral intermediate of a
promoted-water mechanism, outlined above, are applicable
to the tetrahedral intermediates of this alternative
pathway. Glu-270 would initially attack the scissile
carbonyl of the substrate, instead of a Glu-270 and/or zinc-
promoted water molecule. The resulting tetrahedral
intermediate could be stabilized by both zinc and Arg-127,
while the zinc-bound water molecule remains coordinated to
zinc. The collapse of this intermediate would yield the
metastable anhydride, labile to zinc-bound water. The
hydrolysis of the anhydride by zinc-bound water might be

facilitated by shifting the scissile carbonyl from zinc to
Arg-127. These mechanistic alternatives have been discussed
previously (Lipscomb, 1982).

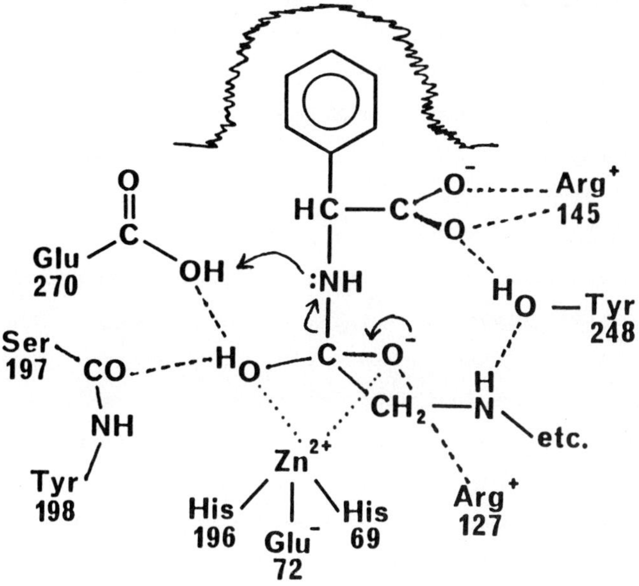

FIGURE 6. Tetrahedral intermediate for the promoted-water
mechanism, illustrating all possible enzyme-substrate
contacts in the S_1' and S_1 subsites on the enzyme for a
peptide substrate. The great number of such contacts serves
to stabilize this postulated tetrahedral intermediate and,
by extension, the flanking transition states along the
reaction coordinate. The collapse of the tetrahedral intermediate is depicted, which involves a proton transfer from
Glu-270 to the leaving amino group.

It is interesting to note that the role proposed for Arg-127 is analogous to that of hydrogen bond donors which comprise the "oxyanion hole" of the serine proteases (for an excellent review see Ferscht (1977)), and is analogous to that proposed for His-231 of the related zinc protease thermolysin (Hangauer, et al., 1984). In fact, such hydrogen bond donors/electrophiles may be a common feature of many other zinc proteases in their interactions with the carbonyl moieties of substrates or oxyanion intermediates. For CPA, the electrophile of the catalytic tetrad outlined in Figure 1 (Arg-127) may be necessary for the catalytic hydrolysis of peptide substrates. Further three-dimensional X-ray studies of these and other zinc proteases may uncover the presence of the catalytic tetrad in other metalloproteases. If such an electrophilic residue is present in these enzymes, a new class of metalloprotease inhibitors might be targeted for specific interaction with this residue (as well as with other groups at or near the active site). These inhibitors would bind strongly to metalloenzymes such as angiotensin converting enzyme, collagenase, and enkephalinase, and their use toward such pharmaceutical targets might be exploited for human benefit.

ACKNOWLEDGEMENT: We thank the National Institutes of Health for Grant GM 06920 in support of this research. Additionally, D.W.C. thanks AT&T Bell Laboratories for a doctoral fellowship.

REFERENCES

BENT, H. A. (1968). Chemical Reviews 68, 587-648.

BURGI, H. B., DUNITZ, J. D., and SHEFTER, E. (1973).
Journal of the American Chemical Society 95, 5065-5067.

CHRISTIANSON, D. W., KUO, L. C., and LIPSCOMB, W. N. (1985).
Journal of the American Chemical Society 107, 8281-8283.

CHRISTIANSON, D. W. and LIPSCOMB, W. N. (1985). Proceedings
of the National Academy of Science USA 82, 6840-6844.

CHRISTIANSON, D. W., DAVID, P. R., and LIPSCOMB, W. N. (1986).
Journal of the American Chemical Society, submitted.

CHRISTIANSON, D. W. and LIPSCOMB, W. N. (1986a). Journal of
the American Chemical Society 108, 4998-5003.

CHRISTIANSON, D. W. and LIPSCOMB, W. N. (1986b). Journal of
the American Chemical Society 108, 545-546.

CHRISTIANSON, D. W. and LIPSCOMB, W. N. (1986c). Proceedings
of the National Academy of Science USA 83, 7568-7572.

DIDEBERG, O., CHARLIER, P., DIVE, G., JORIS, B., FRERE, J. M.,
GHUYSEN, J. M. (1982). Nature (London) 299, 469-470.

FERSCHT, A. Enzyme Structure and Mechanism. W. H. Freeman:
San Francisco, 1977.

HANGAUER, D. G., MONZINGO, A. F., and MATTHEWS, B. W. (1984).
Biochemistry 23, 5730-5741.

HOLMES, M. A. and MATTHEWS, B. W. (1982). Journal of
Molecular Biology 160, 623-639.

KUO, L. C. and MAKINEN, M. W. (1982). Journal of Biological
Chemistry 257, 24-27.

KUO, L. C. and MAKINEN, M. W. (1985). Journal of the
American Chemical Society 107, 5255-5261.

LIPSCOMB, W. N. (1982). Accounts of Chemical Research 15,
232-238.

LIPSCOMB, W. N. (1983). Annual Review of Biochemistry 52,
17-34.

LIPSCOMB, W. N., HARTSUCK, J. A., REEKE, G. N., JR., QUIOCHO,

F. A., BETHGE, P. H., LUDWIG, M. L., STEITZ, T. A., MUIRHEAD, H., COPPOLA, J. (1968). Brookhaven Symposium of Quantitative Biology 21, 24-90.

MAKINEN, M. W., KUO, L. C., DYMOWSKI, J. J., and JAFFER, S. (1979). Journal of Biological Chemistry 254, 356-366.

REES, D. C., LEWIS, M., HONZATKO, R. B., LIPSCOMB, W. N., and HARDMAN, K. (1981). Proceedings of the National Academy of Science USA 78, 3408-3412.

SLATER, J. C. (1959). Acta Crystallographica 12, 197-200.

SLATER, R. C. L. M. (1959). Acta Crystallographica 12, 187-196.

SUH, J., CHO, W., and CHUNG, S. (1985). Journal of the American Chemical Society 107, 4530-4535.

VALLEE, B. L., GALDES, A., AULD, D. S., RIORDAN, J. F. In Metal Ions in Biology. Spiro, T. G., Ed.; Wiley: New York, 1983; Vol. 5, pp. 25-75.

VALLEE, B. L. and GALDES, A. (1984). Advances in Enzymology 56, 283-430.

17. Structural studies on cytochrome P-450 enzyme inhibitors

Miriam Rossi

Structural Studies on Cytochrome P-450 Enzyme Inhibitors

Miriam Rossi

1. Introduction

Cytochrome P-450 enzymes are membrane-bound, inducible heme proteins that function as the effecters of the metabolic oxidation of widely different exogenous and endogenous compounds via a variety of redox reactions. This overall apparent lack of specificity is explained by the existence of families of different isozymes with explicit substrate specificities. The two-electron transfer in the enzymatic cycle requires two other components for activity: NADPH cytochrome P-450 reductase and a lipid fraction (*e.g.*, phosphotidylcholine).

In particular, two of these isozyme families have been extensively studied, the *PB-inducible* and *3-MC-inducible*, indicating two of the compounds which are responsible for their induction (PB, phenobarbital and 3-MC, 3-methylcholanthrene). The PB-inducible forms occur in relatively few tissues and metabolize non-planar compounds with a pronounced three-dimensional shape (*e.g.*, phenobarbital) and some hydrophobic character. The 3-MC forms, instead act upon primarily planar, aromatic, hydrophobic compounds (*e.g.*, 3-methylcholanthrene). The complete nucleotide sequences of the two forms from the rat liver differ, although there is enough sequence homology to indicate that they evolved from a common source (Sogawa, *et al.*, 1984). Environmental factors such as diet and exposure to chemicals (Ryan *et al.*, 1982) as well as genetic factors such as age and sex (Larrey *et al.*, 1984) control the distribution of the different isozymes. Most compounds predominantly induce one form of the P-450 enzymes (Ryan *et al.*, 1982). The two isozyme forms have noticeably different substrate specificities which reflects the experimentally observed dissimilar substrate binding sites (Phillipson *et al.*, 1982; Dus, 1982). Despite this, evidence for a common chemical environment in the active site of all P-450s exists (Tarr *et al.*, 1983).

Recently, a report by Poulos *et al.*, (1985) on the 2.6Åcrystal structure of the substrate (camphor) bound form of *pseudomonas putida* cytochrome P-450 has made clear the structure of the active site. The heme group, iron (III) protoporphyrin IX, is pentacoordinate: the four pyrrole nitrogens and an axial cysteinyl sulfur which links the heme group to the protein in a manner similar to, but not exactly that found in other heme proteins. The heme is buried in a flexible hydrophobic pocket which can adapt to the presence of different substrates. In the structure, the camphor is held in place at the active site by a series of hydrophobic interactions and a hydrogen bond with neighboring residues; it sits adjacent to the oxygen binding site, ready for its stereoselective hydroxylation.

2. Cytochrome P-450 (3-MC-inducible) and Chemical Carcinogenesis

This family of P-450 isozymes is responsible for the solubilization and activation of polycyclic aromatic hydrocarbons (PAH) into chemical carcinogens. Examination of the structures of known PAHs (Glusker, 1985) reveals a phenanthrene structure that can be defined with a bay region and an opposite K region, used in categorizing the carcinogenic potential of compounds. After PAH's enter the body, they are oxidized in the endoplasmic reticulum of many mammalian tissues (liver, skin, *etc.*) to hydroxylated products (Gillette *et al.*, 1972). Activation results from formation of an intermediate epoxide derivative at stereospecific sites in the previously unreactive molecules. These epoxides are hydrolyzed by the enzyme epoxide hydrase; the resulting diol is again a substrate for the P-450 system. The new product is a reactive diol epoxide - an alkylating agent capable of covalent interaction with cell macromolecules, such as DNA. (Sims *et al.*, 1974; Brookes and Lawley, 1964). This is the initiation of carcinogenesis.

2.1 Inhibitors

Flavonoids are a class of compounds commonly found in plants and vegetables that display a wide range of biological activity (Cody *et al.*, 1986). In particular, several flavones are known to be inhibitors of the 3-MC-inducible form of cytochrome P-450. The nature of this inhibition is not well known (Glusker and Rossi, 1986). 7,8-Benzoflavone (I) is a potent inhibitor (Gelboin *et al.*, 1970; Kinoshita and Gelboin, 1972) of P-450 and prevents carcinogenesis by 7,12-dimethylbenz[a]anthracene. An isomer, 5,6-benzoflavone (II) is a weaker inhibitor but a strong inducer of the 3-MC-inducible P-450. Quercetin (III), a naturally occurring flavonoid found in a wide variety of foods, is also an inhibitor of P-450 (and a variety of seemingly disparate enzyme systems). Some of these inhibitory activities are due to covalent interaction between quercetin and the diol-epoxides, rather than with P-450 itself (Huang *et al.*, 1983). A weaker inhibitor is the flavonoid, naringenin (IV), found in grapefruit.

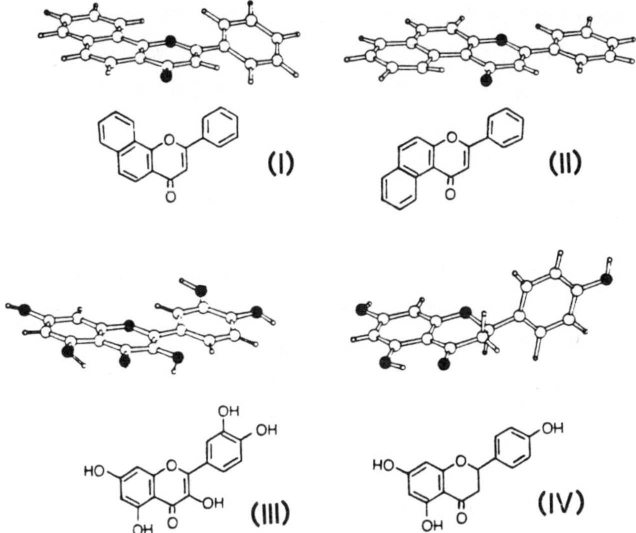

Figure 1.7,8-Benzoflavone,(I);5,6-Benzoflavone,(II); Quercetin,(III);Naringenin,(IV).

It is found, from X-ray crystal structure determinations (Rossi *et al.*, 1980) that the 7,8-BF is nonplanar with an angle of 23.8° between the phenyl ring and the rest of the molecule, while 5,6-BF and quercetin are approximately planar (2° and 7°). The molecular conformations can be described as the result of two opposing effects: steric hindrance and added conjugation. Planarity increases the existence of repulsive interactions from the steric hindrance between α-hydrogen atoms. On the other hand, enhanced stability would result from the added conjugation that planarity would impart. With planarity, the exocyclic oxygen has added partial negative charge and is more susceptible to intra and intermolecular interactions (which stabilize this added charge). These favorable interactions in 5,6-BF are the strong hydrogen bond between the exocyclic oxygen and a solvent water molecule' and another favorable close contact (2.05 Å) involving the exocyclic oxygen and a neighboring hydrogen atom. In quercetin, the exocyclic oxygen appears to be the focal point of cohesive forces in the crystal structure, participating in extensive hydrogen bonding. (Rossi *et al.*, 1986). Naringenin, a flavonone, has a saturated C2-C3 bond and has its heterocyclic ring buckled and the phenyl group lies 113° out of the plane with the rest of the molecule. Theoretical potential energy calculations (Rossi *et al.*, 1980) indicate that the maximum energy barrier to rotation is small. 0.7 Kcal/mol must be overcome for the flavone skeleton to be in the range 0-40° out of the plane with the phenyl ring. Minimum energy was obtained at a conformation corresponding to a 22.8° torsion angle about the bond connecting the phenyl and pyrone rings (reflecting an equilibrium between the two opposing forces).

The pivotal role of the exocyclic oxygen in the molecular structures is in agreement with the description of P-450 enzyme inhibitors by Testa and Jenner (1981) in that potentially active functional groups are needed. Additionally, the structural data reveal the structures of the flavones to resemble those of the PAHs whose metabolism they inhibit, as well as having ring areas similar to that occupied by PAHs. These structural similarities allow the flavones to mimic the PAHs at the active site (Lewis *et al.*, 1986).

3. Cytochrome P-450 (PB-inducible)

This family of P-450 isozymes is involved in the metabolism and hydroxylation of non-planar compounds (*e.g.*, DDT) and for the endogenous cholesterol hydroxylation reactions in the adrenal cortex that are responsible for steroid synthesis. That the overall shape of the substrate/inducer is important in determining which P-450 is responsible for activation was shown by Kaminsky, *et al.*, (1981). Experimental studies on the metabolism of dichlorinated biphenyls with different chlorine substitution patterns showed the involvement of both 3-MC inducible and PB inducible P-450s. Halogen substitution at sites that resulted in more planar biphenyl structures were metabolized via the 3-MC inducible P-450; more twisted biphenyl derivatives were metabolized by PB-inducible P-450.

3.1 Inhibitors

Two inhibitors of this family of isozymes include metyrapone (V) and proadifen hydrochloride (VI), shown in Figure 2. From spectroscopic studies, it appears that metyrapone covalently interacts with the heme iron at the enzyme active site (Peterson *et al.*, 1971). It is used as a diagnostic and therapeutic tool because of its selective inhibition of adrenal steroid 11-β-hydroxylase in the two reactions: the 11-β-hydroxylation of deoxycorticosterone to corticosterone and deoxycortisol to cortisol. Synthesis of aldosterone and cortisol, then, is prevented. Proadifen hydrochloride has been known to decrease the metabolism of other pharmaceuticals and thereby, increase their effectiveness. The P-450 mediated metabolic reactions inhibited by SKF-525A and its metabolites include N-dealkylation, O-demethylation and deamination (Testa and Jenner, 1981).

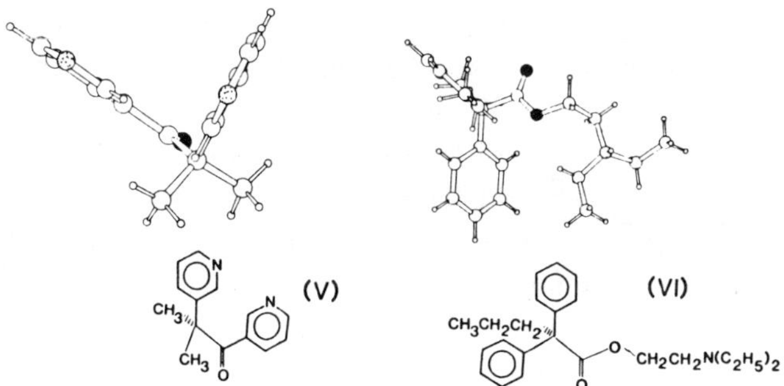

Figure 2. Metyrapone (V) and Proadifen Hydrochloride (VI).

The molecular structure of metyrapone is characterized by a twist of the molecule about the central two-carbon link (C7-C10) connecting the two pyridine rings (Rossi, 1983). The torsion angle is 59.4°. The observed conformation is such that the three heteroatoms form a triangle with the two nitrogen atoms on the same side of the molecule but opposite to the carbonyl oxygen. The intramolecular distances are: N5-N14, 4.347Å, N5-O11, 5.850Å, N14-O11, 4.750Å. The absence of hydrogen bond donors in the molecule and crystal precludes hydrogen bond formation and the only intermolecular interactions are van der Waals forces.

Proadifen hydrochloride has a highly twisted shape reflecting its chemical makeup (Rossi, *et al.*, 1987). The angle between the two phenyl rings is 70.7° and the two heteroatoms in the structure N1 and O2 are 4.762Åapart and point away from each other. In the crystal structure the electronegative chlorine takes part in a hydrogen bond with the protonated N1, 3.090Å, and is surrounded by hydrogen atoms. There is clear separation of the interactions that the polar and non-polar regions of the molecule exhibit.

Substrates whose metabolism is inhibited by both compounds include DDT, phenobarbital and camphor. Once again, it is clear that the molecular structures of inhibitors resemble those of the substrates whose metabolism they inhibit (Rossi, *et al.*, 1987). It is evident that hydrophobic interactions play an important role in the inhibition of PB-inducible P-450, as does the overall molecular shape of the molecule.

4. Summary

Inhibition of P-450 enzymes involves molecular recognition, both structural and electronic: the overall shape of the molecule must be apparent to the enzyme, the atomic charges and hydrophobicity must be complementary to the macromolecule. This forms the basis for the identification of the pharmacophore. In the case for the 3-MC inducible P-450, it is clear that more planar, hydrophobic substrates, inducers and inhibitors are preferred. Inhibitors appear to have functional groups (carbonyl or hydroxyl) capable of interaction with the enzyme. For the PB-induced P-450, it is the rotationally flexible molecule, the presence of hydrophobic groups (*e.g.*, phenyl or pyridyl) and a negatively charged heteroatom that are requirements for successful interaction. These latter molecular markers are consistent with the view of the active site noncovalent interactions between substrate-enzyme as seen in the X-ray study of P-450$_{cam}$ (Poulos *et al.*, 1985). Clearly, better definition of P-450 active site geometry awaits further data.

References

BROOKES, P. and LAWLEY, P.D. (1964). *Nature* (London) **202**, 781.

CODY,V., MIDDLETON JR., E. and HARBORNE, J.B. (eds.) (1986). *Plants Flavonoids in Biology and Medicine*, Alan R. Liss, Inc. New York.

DUS, K.M. (1982). *Xenobiotica* **12**, 745.

GELBOIN, H.V., WEIBEL F. and DIAMOND, L. (1970). *Science* **170**, 169.

GILLETTE, J.R., DAVIS, D.C. and SASAME H.A. (1972). *Annual Reviews in Pharmacology* **12**, 57.

GLUSKER, J.P. (1985). In *Polycyclic Aromatic Hydrocarbons and Cancer* (ed. Harvey, R.) American Chemical Society, Washington, D.C., Chapter 7, pp. 125-185.

GLUSKER, J.P. and ROSSI, M. (1986). In *Plant Flavonoids in Biology and Medicine* (eds. Cody, V., Middleton, Jr., E. and Harborne, J.) Alan R. Liss, Inc. NY pp. 395-410.

HUANG, M.T., WOOD, A.W., NEWMARK, H.L., SAYER, J.M., YAGI, H., JERINA, D.M. and CONNEY, A.H. (1983). *Carcinogenesis* **4**, 1631.

KAMINSKY, L.S., KENNEDY, M.W., ADAMS, S.M., and GUENGERICH, F.P. (1981). *Biochemistry* **20**, 7379.

KINOSHITA N. and GELBOIN, H.V. (1972). *Cancer Research* **32**, 1329.

LARREY, D., DISTLERATH, L.M., DANNAN, G.A., WILKINSON, G.R. AND GUENGERICH, F.P. (1984). *Biochemistry* **23**, 2787.

PETERSON, J.A., ULRICH, V., and HILDEBRANDT, A.G. (1971). *Archives in Biochemistry and Biophysics* **145**, 531.

LEWIS, P.F.V., IOANNIDES, C., PARKE, D.V. (1986). *Biochemical Pharmacology* **35**, 2179.

PHILLIPSON, C.E., IOANNIDES, C., DELAFORGE, M. and PARKE, D.V. (1982). *Biochemical Journal* **207**, 51.

POULOS, T.L., FINZEL, B.C., GUNSALUS, I.C., WAGNER, G.C. and KRAUT, J. (1985). *The Journal of Biological Chemistry* **260**, 16122.

ROSSI, M., CANTRELL, J.C., FARBER, A.J., DYOTT, T. CARRELL, H.L. and GLUSKER, J.P. (1980). *Cancer Research* **40**, 2774.

ROSSI, M. (1983). *Journal of Medicinal Chemistry* **26**, 1246.

ROSSI, M., RICKLES, L.F. and HALPIN, W.A. (1986). *Bioorganic Chemistry* **14**, 55.

ROSSI, M., CALLAHAN, T.C. and MARKOVITZ, S.B. (1987). *Carcinogenesis* accepted for publication.

RYAN, D.E, THOMAS, P.E., REIK, L.M and LEVIN, W. (1982). *Xenobiotica* **12**, 727.

SIMS, P., GROVER, P.L., SWAISLAND, A., PAL, K. and HEWER, A. (1974). *Nature* (London) **252**, 326.

SOGOWA, K., GOTOH, KAWAJIRI, K. and FUJII-KURIYAMA, Y. (1984). *Proceedings of the National Academy of Sciences U.S.A.* **81**, 5066.

TARR, G.E., BLACK, S.D, FUJITA, V.S and COON, M.J. (1983). *Proceedings of the National Academy of Sciences, U.S.A.* **80**, 6552.

TESTA, B. and JENNER, P. (1981). *Drug Metabolism Reviews* **12**,1.

18. Dehydrogenase binding by the cofactor analogue TAD

Barry M. Goldstein, J. Ellis Bell, Victor E. Marquez and Roland K. Robins

1.   Introduction

Tiazofurin (2-β-D-ribofuranosylthiazole-4-carboxamide; Fig. 1) is a novel C-glycosyl thiazole which has demonstrated significant antitumor activity in a number of model tumor systems. Tiazofurin is essentially curative *in vivo* for the Lewis lung carcinoma in mice (Robins et al, 1982). It has also demonstrated *in vitro* activity against both human lymphoid and lung tumor cell lines (Earle and Glazer, 1983; Carney et al, 1985). Tiazofurin is now undergoing clinical trials in the United States.

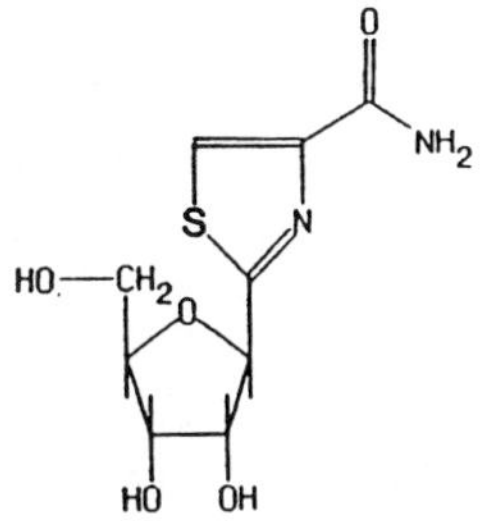

Fig. 1. Tiazofurin

Work by several groups suggests that the oncolytic activity of tiazofurin is related to its ability to create a state of cellular guanine nucleotide depression with a subsequent interruption of DNA and RNA synthesis. This is a result of a decline in the guanosine precursor xanthosine monophosphate due to the inhibition of inosine monophosphate dehydrogenase (**IMPd**) (Jayaram et al, 1982). The major IMPd inhibitor is an anabolite of tiazofurin. *In vivo* tiazofurin is incorporated into an analogue of the cofactor nicotinamide adenine dinucleotide (**NAD**) in which the nicotinamide ring is replaced by the thiazole-4-

carboxamide group. This NAD analogue, called thiazole-4-carboxamide adenine dinucleotide (**TAD**, Fig. 2) is a potent IMPd inhibitor (K_i ~0.5 μM) (Cooney et al, 1982).

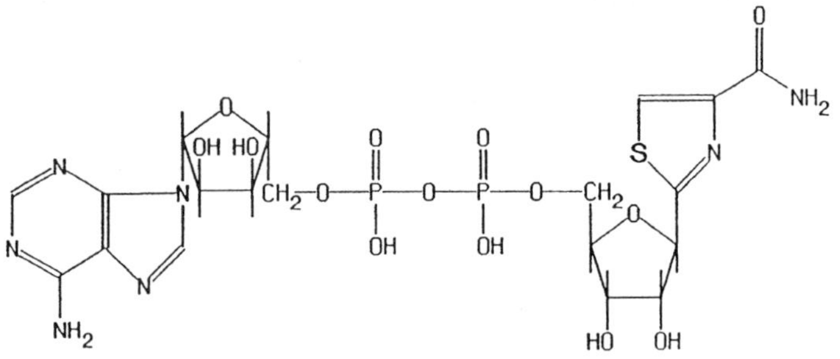

Fig. 2. Thiazole-4-carboxamide adenine dinucleotide (TAD)

One model for the mechanism of TAD inhibition of IMPd is that this molecule binds at the NAD(H) binding site, the heterocyclic base occupying the pocket normally filled by the nicotinamide ring (Goldstein et al, 1983; 1985). However, no X-ray structure of IMPd has been obtained, thus the specific stereochemical requirements of the nicotinamide end of the cofactor binding site(s) on IMPd are unknown. Nevertheless, many features of NAD binding are common to a number of dehydrogenases of known structure (Grau, 1982). Thus, TAD might be expected to inhibit other dehydrogenases in addition to IMPd. In order to test this hypothesis, enzyme inhibition and modeling studies have been initiated.

2.   Enzyme Inhibition and Modeling Studies

Enzymes examined to date have been bovine liver glutamate dehydrogenase (**GDH**) and equine liver alcohol dehydrogenase (**ADH**). Inhibition assays have been carried out using TAD as the inhibitor. Rates of NADH formation in the presence and absence of inhibitor were examined as a function of concentration of both coenzyme (NAD) and substrate (glutamic acid or ethanol). Initial rates of NADH formation were followed by monitoring characteristic fluorescence emission of the reduced coenzyme (LiMuti and Bell, 1983). Lineweaver-Burk plots for GDH and ADH with [NAD] varied in the

presence and absence of TAD are shown in Figure 3. It is apparent that TAD acts
as a competitive inhibitor of both enzymes with respect to NAD. Comparison of

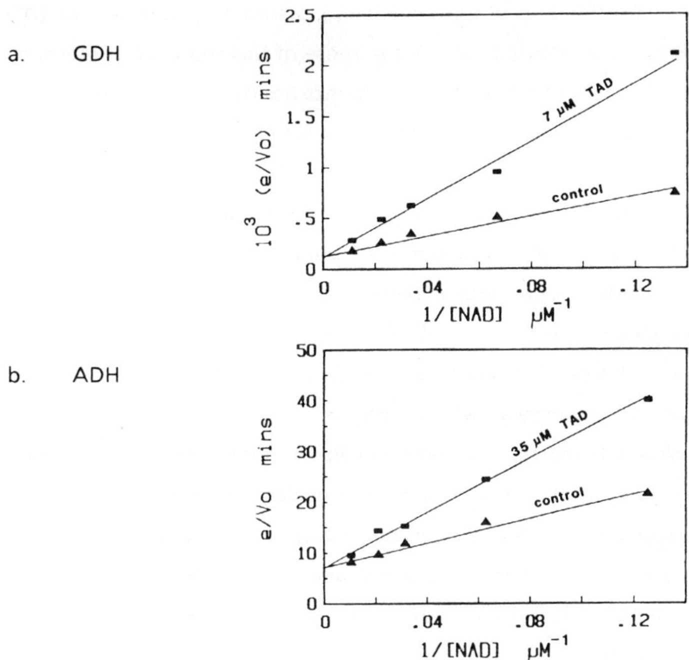

Fig. 3. Double reciprocal plots showing inhibition of (a) GDH and (b) ADH by TAD
with NAD as the variable substrate. (a) K_i(TAD) = 5 µM; K_m(NAD) = 20 µm; pH
7.0. (b) K_i(TAD) = 34 µM; K_m(NAD) = 16 µM; pH 8.0.

K_i for TAD and K_m for NAD in GDH indicates that TAD binds more tightly to this
enzyme than does NAD. Comparison of K_i for TAD with the independently
determined K_d for NAD binding to ADH (~51 µM, Dalziel, 1975) indicates that
binding of TAD is comparable to, but somewhat tighter than, binding of NAD to
ADH. TAD also showed noncompetitive inhibition of GDH with respect to
glutamic acid (K_i ~17 µM) but no inhibition of ADH with ethanol as the variable
substrate.

The structure of GDH is unknown. It is also not known whether TAD is
binding at one or several of the six high affinity binding sites on this enzyme

(Dalziel and Egan, 1972). However, the structure of equine liver ADH has been extensively studied and the cofactor binding site well characterized (Eklund et al, 1984). Only one NAD binding site exists per subunit, hence TAD presumably binds to this site. On the basis of this knowledge, preliminary modeling of TAD binding to ADH has been carried out. The purpose of this study is to determine the potential steric and hydrogen-bonding interactions between the thiazole-4-carboxamide moiety on TAD and the side chains forming the cofactor binding site.

Figure 4a shows the nicotinamide end of the cofactor binding site on equine liver ADH. Coordinates were obtained, via the Protein Data Bank, from the crystallographic study of a ternary complex of equine liver ADH with NADH and dimethylsulfoxide (Eklund et al, 1984). This structure is isomorphous to three additional ADH complexes with NAD, indicating that the features of the cofactor binding site remain relatively unchanged.

The nicotinamide moiety is anchored in an interior pocket of the enzyme by three hydrogen bonds (dotted lines) to the carboxamide group. The carboxamide oxygen acts as a hydrogen bond acceptor from the main chain nitrogen of residue Phe-319. The carboxamide amino group donates H-bonds to carbonyl oxygens on Ala-317 and Val-292 (Eklund et al, 1984). H-bond distances are given in angstroms.

Figure 4b shows a model of TAD binding to ADH. This was obtained by performing a least-squares fit between the nicotinamide moiety and the thiazole-4-carboxamide ring from the tiazofurin crystal structure. Only distances between atom pairs were minimized. No energy terms were used. Potential hydrogen bonds to the thiazole carboxamide group are shown. These bonds are quite similar to those formed by the nicotinamide moiety. Further, the carboxamide NH_2 is able to remain in its putative low energy conformation, on the same side as the thiazole ring nitrogen (Goldstein et al, 1983; 1985). The thiazole ring itself fits well into the nicotinamide binding pocket, forming no unfavorable contacts with neighboring side chains.

In order to achieve the fit shown in figure 4, the thiazole ring must be rotated about the C-glycosyl bond by about 50° relative to the conformation observed in the crystal structure. However, the sulfur remains *cis* to the furanose ring oxygen, as observed in the crystal structures of a number of thiazole nucleosides (Goldstein et al, 1983; 1985). Allowing small changes in side chain conformations using more extensive modeling techniques may modify this requirement.

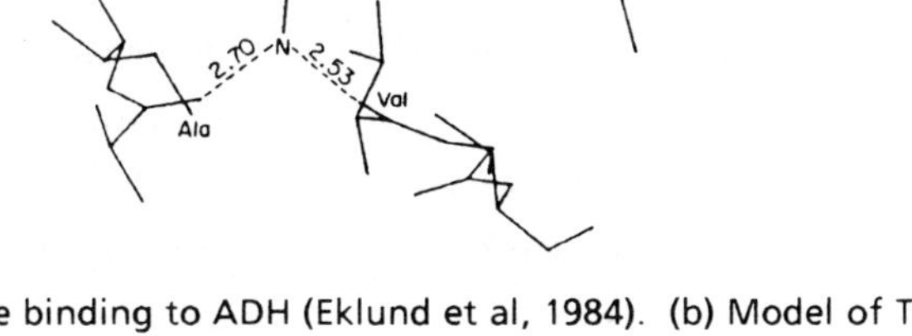

Fig. 4. (a) Nicotinamide binding to ADH (Eklund et al, 1984). (b) Model of TAD binding to ADH.

3.   Summary

The above studies indicate that it is likely that TAD inhibits ADH by mimicking NAD binding at the catalytic binding site. The thiazole-4-carboxamide group is capable of simulating the steric and hydrogen bonding properties of the nicotinamide moiety in this system. In particular, the location of the H-bonding carboxamide groups is similar in both rings. Hydrogen bonding to the nicotinamide carboxamide groups is seen in all dehydrogenase structures. Tiazofurin requires a carboxamide moiety, specifically in the 4 position of the thiazole ring, for activity (Jayaram et al, 1982). This suggests that the thiazole-4-carboxamide moiety on TAD may mimic nicotinamide binding to IMPd in a manner similar to that proposed for ADH.

ACKNOWLEDGEMENTS: The authors thank Pamela Wade and Evelyn Bell for their technical assistance. This work was supported in part by a grant from the James P. Wilmot Foundation.

REFERENCES

Carney, D.N., Ahluwalia, G.S., Jayaram, H.N., Cooney, D.A., Johns, D.G. (1985). *Journal of Clinical Investigation* **75**, 175-182

Cooney, D.A., Jayaram, H.N., Gebeyehu, G., Bets, C.R., Kelley, J.A., Marquez, V.E., Johns, D.G. (1982). *Biochemical Pharmacology* **31**, 2133-2136

Dalziel, K. (1975). *The Enzymes* **11**, 1-60

Dalziel, K., Egan, R.R. (1972). *Biochemical Journal* **126**, 975-998

Earle, M.F., Glazer, R.I. (1983). *Cancer Research* **43**, 133-137

Eklund, H., Samama, J.-P., Jones, T.A. (1984). *Biochemistry* **23**, 5982-5996

Goldstein, B.M., Takusagawa, F., Berman, H.M., Srivastava, P.C., Robins, R.K. (1983). *Journal of the American Chemical Society* **105**, 7416-7422

Goldstein, B.M., Takusagawa, F, Berman, H.M., Srivastava, P.C., Robins, R.K. (1985). *Journal of the American Chemical Society* **107**, 1394-1400

Grau, U.M. (1982). Structural Interactions with Enzymes. In *The Pyridine Nucelotide Coenzymes* (ed. B.M. Anderson, J. Everse and K.-S. You) pp 135-187. Academic Press, New York

Jayaram, H.N., Dion, R.L., Glazer, R.I., Johns, D.G., Robins, R.K., Srivastava, P.C., Cooney, D.A. (1982). *Biochemical Pharmacology* **31**, 2371-2380

LiMuti, C., Bell, J.E. (1983). *Biochemical Journal* **211**, 99-107

Robins, R.K., Srivastava, P.C., Narayanan, V.L., Plowman, J., Paull, K.O. (1982). *Journal of Medicinal Chemistry* **35**, 107-108

19. Conformational analysis of antifolate drugs: structure–activity relationships, electronic properties and enzyme active-site modelling studies

Vivian Cody, Paul A. Sutton and William J. Welsh

1. INTRODUCTION

The enzyme dihydrofolate reductase (DHFR), present in all cells and a necessary component for all cell growth, is responsible for maintaining intracellular folate pools in their biochemically active reduced state. Therefore, by tight binding to DHFR, folic acid (FA) antimetabolites such as methotrexate (MTX) deplete intracellular reduced folates required for single carbon transfer reactions and interfere with de novo thymidine and purine synthesis as well as intermediary amino acid metabolism. Because of the important metabolic role of this enzyme, its inhibition by these potent antifolates has been the focus of chemotherapy of infectious and neoplastic diseases (Blakely et al,1984; Hitchings,1983). Since there are a number of folate related cofactors which are capable of carrying different kinds of one-carbon groups, inhibition of the DHFR pathway will have significant effect on these subsequent products.

Structure-activity studies have shown that a 2,4-diaminopyrimidine, s-triazine, quinazoline or pteridine ring structure is sufficient for tight binding to DHFR from any species (Blakely, et al,1984; Hitchings,1983; Hitchings et al,1965). Thus, the antimetabolite analogues of folic acid can be grouped into four classes defined by variations in the parent ring structure. It has also been shown that the folate substrate analogues (classical antifolates) require transport for entry into

cells whereas lipophilic antifolates without the acidic side
chain groups ("nonclassical" antifolates) enter cells by diffusion. Furthermore, many of these antifolates have an inhibitory action which is specific for different species of DHFR or
is greatly modulated by only small structural changes. For example, methotrexate (MTX) (Fig. 1), is a potent inhibitor of
all species of DHFR and is also the most widely used anticancer

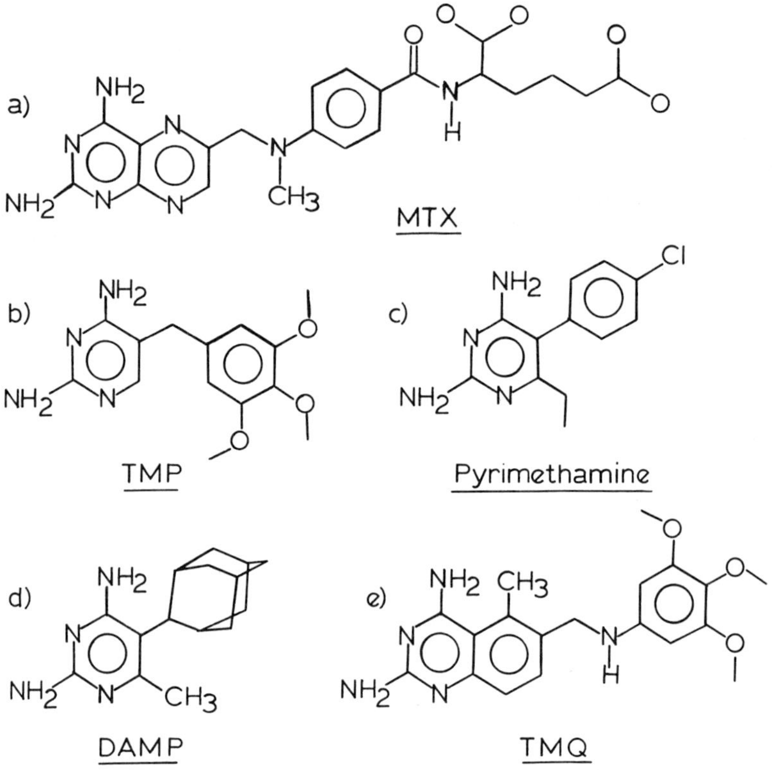

FIG. 1. Structures of DHFR inhibitors.

agent. On the other hand, trimethoprim (TMP) is so highly specific for bacterial sources of DHFR that it is the most widely
used antibacterial antifolate. Similarly, primethamine is specific for plasmodial sources of DHFR and is a widely used anti-

malarial drug (Hitchings,1983). Furthermore, inclusion of a
lipophilic group on these antifolates enhances their inhibitory
activity by their more rapid diffusion into cells as demonstrated by the antifolates 2,4-diamino-5-(1-adamantyl)-6-methylpy-
rimidine (DAMP) (Jonak et al,1971; Greco et al,1980) and trime-
trexate (TMQ). In general the triazine and quinalzoline antifolates are effective inhibitors of mammalian DHFR.

Knowledge of the structural basis of effective analogue
interaction at both the transport and target enzyme sites permits the rational design of new analogues with greater therapeutic effectiveness. Therefore, to delineate the structural,
conformational and electronic properties of these dihydrofolate
reductase inhibitors which are important for the species specificity and selectivity, X-ray crystallographic and molecular
orbital calcultions were carried out (Cody, et al,1982;1984;
Cody,1983;1984a; Sutton et al,1986) and computer graphic modeling of their binding interactions within the active site of
chicken liver DHFR were investigated (Cody,1984b;1986).

2.   CRYSTALLOGRAPHIC ANALYSIS
Crystallographic analyses of 5-(1-adamantyl) diaminopyrimidine antifolates (Fig. 2a-d) revealed that within a series of
6-substituted analogues (DAHP, DAMP, DAEP, DAPP), the pyrimidine ring in each structure is distorted from planarity with
its substitutents, with the exception of DAHP. As a result of
the steric strain placed upon the system by the close intramolecular interactions of the admantyl hydrogen atoms and those
of the 4,6-substituents, the molecules have a bowed shape.
Since the high binding affintiy of antifolates for the enzyme
active-site is assumed to involve strong multiple hydrogen
bonding interactions (Cody,1984b;1986), these pyrimidine ring
distortions can influence the placement of the diamino groups
and affect the strength and directionality of their hydrogen
bonds in the enzyme active site.

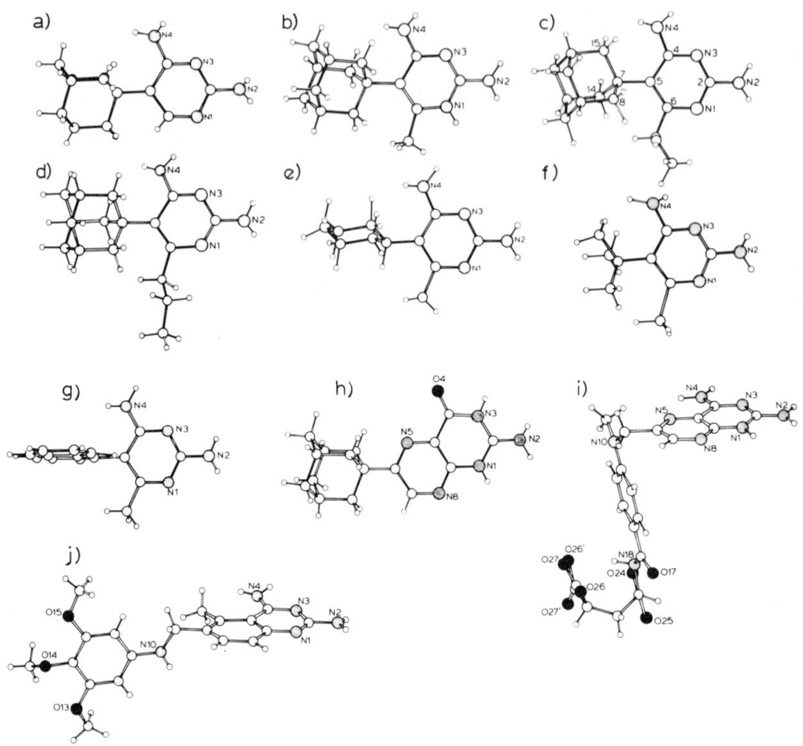

FIG. 2. Crystal structures of antifolates.

To determine the influence of hydrophobicity and pyrimidine ring pertubations on antifolate potency, the crystal structures of the 5-(1-naphthyl), 5-cyclohexyl and 5-t-butyl analogues (Fig. 2e-g) were also investigated. The naphthyl, cyclohexyl and adamantyl rings have similar hydrophobicities and the t-butyl group provides a crowded group at position five. These structures have planar pyrimidine rings.

The structures of pteridine ring antifolates were also investigated (Fig. 2h-j) and their properties compared with the pyrimidines (Cody,1986;Cody et al,1984). The inhibitory potency of the 6-adamantyl analogue (Fig. 2h) is comparable to MTX.

The molecular conformation of MTX.4H$_2$O is folded with a 97° angle between the plane of the pteridine ring and the p-aminobenzoyl ring. The structure is a zwitterion with protonation at N(1) (Sutton et al,1986). This conformation differs from that observed for enzyme-bound MTX and from the crystal structures reported for folic acid and quinespar (QU) (Mastropaola, et al,1980;1986) (Table 1). The largest conformational

TABLE 1.

Conformational Comparison of Pteridine Antifolates

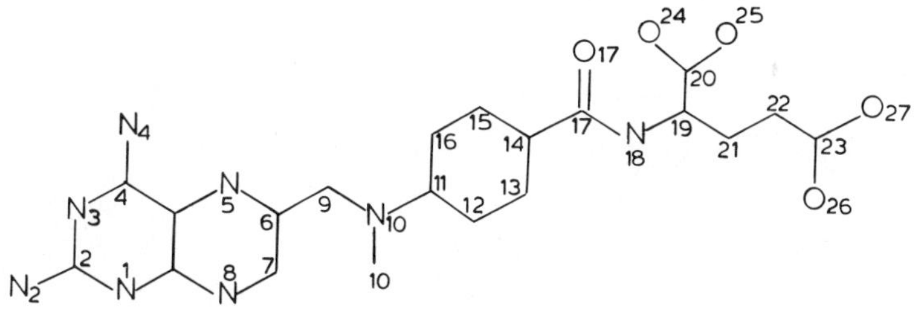

Angle	MTX[*]	TMQ	FA	QU	L. casei
N5–C6–C9–N10	39°	79°	31°	80°	–146°
C6–C9–N10–C11	–63	178	180	179	57
C9–N10–C11–C12	165	170	174	169	175
C13–C14–C17–N18	7	–	–7	–2	31
C17–N18–C19–C21	79	–	77	158	123
N18–C19–C21–C22	56	–	59	–68	–68
C19–C21–C22–C23	–83	–	–73	–	136

[*]MTX (Sutton, et al 1986); FA (Mastropaola et al,1980); QU (Mastropaolo et al, 1986); L. casei (Matthews, et al, 1985).

differences involve the torsion angles about the pteridine-p-aminobenzoyl bond and the glutamate side chain. The structure of TMQ, a potent lipophilic antifolate with a potency greater than MTX, has also been determined. The crystal data for TMQ show that the molecule is a free base and that it has an extended conformation similar to that of quinespar (Table 1).

3.   HYDROGEN BONDING AND GEOMETRY

The high affinity of antifolates for the enzyme active site is assumed to involve multiple hydrogen bonding interac-

tions with the enzyme side chain residues. Examination of
intermolecular interactions of MTX shows that its hydrogen
bonding pattern is remarkably similar to that observed in the
enzyme-bound structure (Fig. 3). Analysis of lipophilic ant-

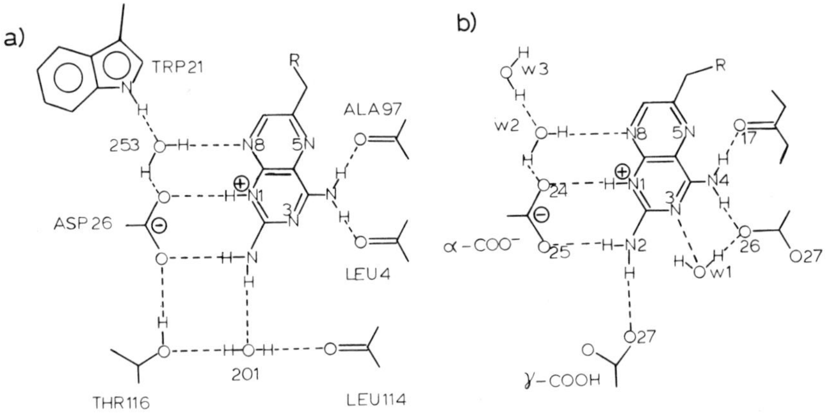

FIG. 3. Hydrogen bonding in (a) crystal structure of MTX·4H$_2$O
and (b) DHFR (Matthews et al,1985).

folates in the solid state reveals the characteristic formation
of N...N base-pair hydrogen bonds involving N(4) hydrogen dona-
tion to N(3) of an adjacent molecule (Cody et al,1982; Schwalbe
et al,1983). However, in the structures of MTX and TMQ, there
are no N...N hydrogen bonds.

     Comparison of average geometric parameters of 11 pyrim-
idine rings protonated at N(1) with 17 free base antifolates
reveals characteristic differences that undoubtly affect the
hydrogen bonding strength of the other nitrogen substitutents
(Cody,1986). These data also indicate considerable double bond
character in the exocyclic nitrogen bonds as the exocyclic C-N
bond distances are shorter than those found in other diamino
compounds (Schwalbe et al,1983). These differences are also
observed in MTX (Sutton et al,1986).

4. MOLECULAR MODELING STUDIES

In order to assess the influence of ring substitution and
N(1) protonation on pyrimidine structure, CNDO/2 geometry-op-
timized molecular orbital (Welsh et al,1983) and molecular
mechanics force field (MM2p) energy minimization calculations
were carried out. These crystallographic data were used to
derive several MM2p force field parameters, in particular those
associated with the exo- and endocyclic nitrogen atoms in the
diaminopyrimidine ring system.

CNDO/2 calculated values of atomic fractional charges for
these molecules were nearly identical for common atoms, averag-
ing -0.352, -0.308, -0.360, and -0.312 (relative to -1.000 for
an electron charge) for N(1), N(2), N(3) and N(4), respectively. The charges on the endocyclic N(1) and N(3) are uniformly
more negative than those on N(2) and N(4), reflecting the former's greater bascity and proton-binding affinities.

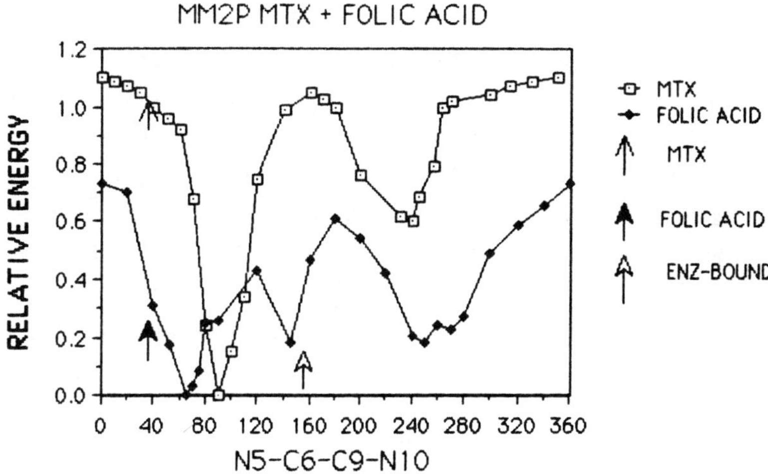

FIG. 4. Relative energy barrier to rotation about C(9)-N(10).

The results of these derived force field parameters show
that the MM2p calculated geometry accurately reflects the pyrimidine nonplanarity and the short exocylic C-N bond lengths

observed in these crystal structures (Welsh et al, in press).
MM2p energy minimized structures were calculated for MTX and
folic acid and the energy barrier to rotation about the C(9)–
N(10) bond in these structures calculated (Fig. 4). These data
show that there is only a small barrier to rotation and that
the energy of binding of these molecules to the enzyme could be
easily overcome to change the structure to that required by the
enzyme active site.

5. DHFR ACTIVE-SITE MODELING

Crystallographic data delineating the DHFR enzyme structure and inhibitor/substrate complexes are available from two
bacterial and one avian source (Bolin et al,1982; Matthews et
al,1985) and show that the active site is located within a 15Å
cavity cutting across one face of the enzyme. Structural data
show that DAMP also binds to chicken liver DHFR in a similar
manner (Bolin et al,1982). The interactions of DAMP in the
DHFR active site show that the adamantyl ring fits tightly into
the hydrophobic pocket near that occupied by the p-aminobenzoyl
ring of MTX, and that the 6-methyl is in a hydrophobic pocket
surrounded by the residues Trp-24, Glu-30, and Tyr-31.

Investigation of the 6-ethyl environment of DAEP shows
that, while there is a reasonable fit, a 110° rotation of the
ethyl side chain moves it into a more suitable pocket (Cody et
al,1984). A similar comparison of the propyl side chain indicates that it also fits into this space. In its observed
orientation, the propyl side chain atom C(63) is 2.38Å from the
hydroxyl oxygen of Tyr-31. A 60° rotation of the side chain
increases this distance to 3.64Å (Cody,1986).

These observations suggest that other 6-substituted lipo-
hilic antifolates can be proposed which could take advantage of
specific interactions with these residues. Accordingly, com-

puter generated models were made for other 6-substituted 5-adamantyl antifolates and their fit tested in this pocket.

6. SUMMARY

The results of these structural and molecular orbital energy studies show that the pyrimidine ring of lipophilic 5-adamantyl antifolates become more distorted from planarity as the size of the 6-substituent increases.   These modeling studies show that other parameters besides bulk and hydrophobicity are of importance to the binding affinity of these lipophilic antifolates which could better explain the optimal activity of DAEP within this series.

Structural data for MTX show that although its conformation differs from that bound in the enzyme, the hydrogen bonding pattern observed for the diaminpteridine ring is identical to that found in the enzyme structure, including the placement of key water molecules to stabilize the enzyme-bound structure. The conformational flexibility of the pteridine ring and p-aminobenzoyl ring, as shown by MM2p energy minimization calculations, indicates a low energy barrier to rotation and suggest that enzyme interactions dictate the conformational preferences.

7. ACKNOWLEDGEMENTS

This research was supported in part by grants from NCI CA-34714, FRA-287 from the American Cancer Society Faculty Research Award, and the Buffalo Foundation.

8. REFERENCES

BLAKELY, R.L. (1984). in: Folates and Pterins, Blakely, R. L. and Benkovic, S. J. (eds), John Wiley & Sons, New York, pp 191-253.
BOLIN, J. T., FILMAN, D. J., MATTHEWS, D. A., HAMLIN, R. C., & KRAUT, J.

(1982). J. Biol. Chem., **257**, 13650-13662.

CODY, V. (1983). Cancer Biochem. Biophys. **7**, 173-177.

CODY, V. (1984a). Acta Cryst., **C40**, 1000-1004.

CODY, V. (1984b). in: Molecular Basis of Cancer,Part B: Macromolecular Recognition, Chemotherapy and Immunology, Rein, R. (ed), Alan R. Liss, New York, pp. 275-284.

CODY, V. (1986). Mol. Graphics, **4**, 69-73.

CODY, V., OPITZ, S. & ZAKRZEWSKI, S. F. (1984). Fed. Proc. **43**, 1730.

CODY, V., WELSH, W. J., OPITZ, S. & ZAKRZEWSKI, S. F. (1984). in: QSAR in Design of Bioactive Compounds, Kuchar, M. (ed), JR Prous Science, Barcelona, Spain, pp. 241-252.

CODY, V. & ZAKRZEWSKI, S. F. (1982). J. Med. Chem. **25**, 427-431.

GRECO, W.R. & HAKALA, M. T. (1980). Mol. Pharmacol. **18**, 521-528.

HITCHINGS, G. H. (1983). Inhibition of Folate Metabolism in Chemotherapy (1983). Springer-Verlag, New York.

HITCHINGS, G. H. & BURCHALL, J. J. (1965). Adv. Enzymol. **27**, 417-468.

JONAK, J. P., ZAKRZEWSKI, S. F. & MEAD, L. H. (1971). J. Med. Chem., **14** 408-411.

MASTROPAOLO, D., CAMERMAN, A. & CAMERMAN, N. (1980). Science, **210**, 334-336.

MASTROPAOLO, D., SMITH, H. W., CAMERMAN, A. & CAMERMAN, N. (1986). J. Med. Chem., **29**, 155-158.

MATTHEWS, D. A., BOLIN, J. T., BURRIDGE, J. M., FILMAN, D. J., VOLZ, K W., KAUFMAN, B. T., BEDDELL, C. R., CHAMPNESS, J. N., STAMMERS, D. K. & KRAUT, J. (1985). J. Biol. Chem., **260**, 381-391.

SCHWALBE, C. H. & CODY, V. (1983). in: Chemistry and Biology of Pteridines, Blair, J. A. (ed), Walter de Gruyter, Berlin, pp. 511-515.

SUTTON, P. A., CODY, V. & SMITH, G. D. (1986). J. Amer. Chem. Soc., **108** 4155-4158.

WELSH, W. J., CODY, V., MARK, J. E., ZAKRZEWSKI, S. F. (1983). Cancer Biochem. Biophys., **7**, 27-36.

WELSH, W. J. & CODY, V. (in press). in: Pteridines and Folic Acid Derivatives, Cooper, B. A. & Whithead, V. M. (eds)., Walter de Gruyter, Berlin.

20. A structure–function study of elastase receptor + substrate interactions derived from high-resolution X-ray crystallography

Edgar F. Meyer, Jr., Lori Takahashi and R. Radhakrishnan

ABSTRACT

The high-resolution structure analysis of native and complexed porcine pancreatic elastase reveals the dominant role of weak interactions, leading to anomalous binding modes. The absence of strong (e.g., Coulombic) interactions makes this system appropriate for the study of subtle molecular interactions. The structural identification of water channels between the active site and the exterior of the enzyme suggests several functional roles of water in binding and catalysis.

The crystallographic method remains the definitive method for advancing the frontiers of knowledge of the structures of biological macromolecules, as evidenced by over 200 entries in the Protein Data Bank (Bernstein, et al., 1977), initiated by one of us (E.M.) in 1968. Less well recognized is the ability of the crystallographic method to make lateral explorations involving structure-function relationships such as enzyme mechanisms, receptor + substrate binding, and other types of molecular interactions. Our studies explore the chosen enzyme, porcine pancreatic elastase (PPE), in native and complexed forms.

PPE is a member of the serine protease family (Bode & Huber, 1986); as such it catalyzes the hydrolysis of ester and amide bonds. Uncontrolled elastases have been linked to degenerative diseases such as pancreatitis, emphysema, and forms of arthritis. Due to the ubiquitous nature of the serine proteases, it is a major challenge to molecular modelling and drug design efforts to create molecules specifically targeted to a given enzyme that would have minimal effect on the other members of the family.

The structure of tosyl elastase, refined to 2.5Å resolution, was published by Sawyer, et al. (1978). This structure showed that His 57 can rotate "out" into solution, breaking the well-recognized catalytic triad: Ser 195-His 57-Asp 102 (chymotrypsinogen numbering system). The catalytic role of these amino acids is a subject for debate and experimentation. A trifluoroacetylated dipeptide (TFAI) complex with PPE has been refined to 2.5Å resolution by Hughes, et al. (1982). We first studied native PPE using data to 1.65Å resolution and refined to R=0.16 (Meyer, et al., 1986b); this structure serves as our reference point for subsequent binding studies and was also the basis for predicting and modelling the structure of a homologous (58%) rat elastase II structure (Carlson, et al., 1986).

Next, a crystal of PPE was reacted in a capillary tube with an excess of the chromogenic substrate, APA-pNA, Acetyl-Ala-Pro-Ala-para-nitroaniline (Fig.1). <u>These</u> <u>studies</u> <u>demonstrate</u> <u>that</u> <u>no</u> <u>structural</u> <u>artifacts</u> <u>were</u> <u>introduced</u> <u>by</u> <u>analyses</u> <u>at</u> <u>pH</u> <u>5.0.</u> <u>They</u> <u>also</u> <u>demonstrated</u> <u>that</u> <u>minimal</u> <u>conformational</u> shifts <u>occur</u> <u>upon</u> <u>ligand</u> <u>binding</u>, compared to native PPE (a classical, Fisher (1894) "lock and key" enzyme). Most significantly, <u>they</u> <u>demonstrate</u> <u>the</u> <u>rather</u> <u>unspecific</u> <u>binding</u> <u>exhibited</u> <u>by</u> <u>the</u> <u>elastases</u> to small peptide (usually <6) substrates. This was further borne out by a study of PPE in the presence of an excess of a tetrapeptide, Acetyl-Pro-Ala-Pro-Tyr, using both 2-d NOE NMR and 1.7Å resolution X-ray data (Clore, et al., 1986). Here again, PPE demonstrated weak, backwards binding of the

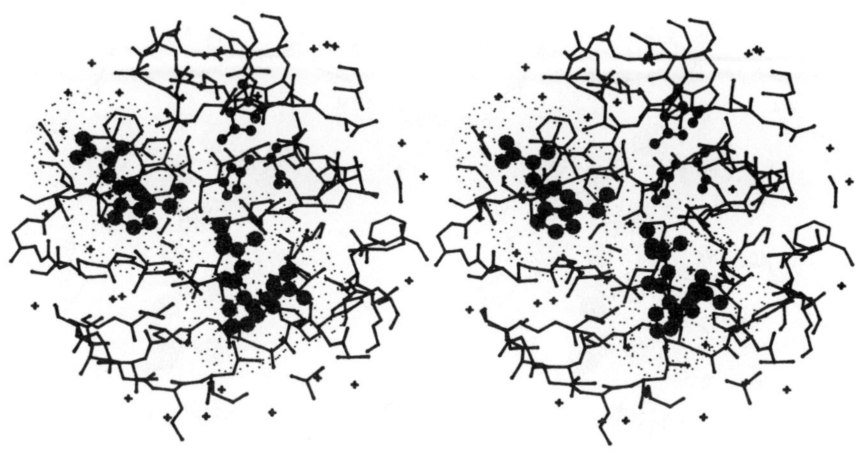

Fig.1. Stereoview of PPE + 2APA complex at 1.65Å resolution, R=0.018 (Meyer, et al., 1986a). In this and all subsequent plots, a "double van der Waals surface" (3.3Å) is used to depict bulk interactions. Crosses represent water molecules. Two APA molecules are located adjacent to the active site of PPE; both are bound backwards with respect to the established peptide binding mode (e.g., trypsin + trasylol, Huber & Bode, 1978) in the S and S1 regions of the extended binding site (notation of Schechter and Berger, 1967). The experiment was repeated at the physiological pH of PPE, 7.5, and refined to R=0.19 (Meyer, et al., 1986c), showing minimal structural shifts compared to the pH 5.0 study.

small peptide. This study provided indirect evidence in support of the hypothesis of Bizzozero and Dutler (1981) that stereochemical inversion of the amide N atom of the scissile bond is a crucial, committing step in peptide bond hydrolysis.

We next undertook the study of the binding of an isocoumarin inhibitor. A crystal of PPE was soaked in the presence of the inhibitor, 3-methoxy-4-chloro-7-amino-isocoumarin, and diffraction data collected out to 1.7Å resolution and refined to R=0.18 (Meyer, et al., 1985).

Two related studies of PPE bound to the benzoxazinones pointed out an additional challenge to the crystallographic

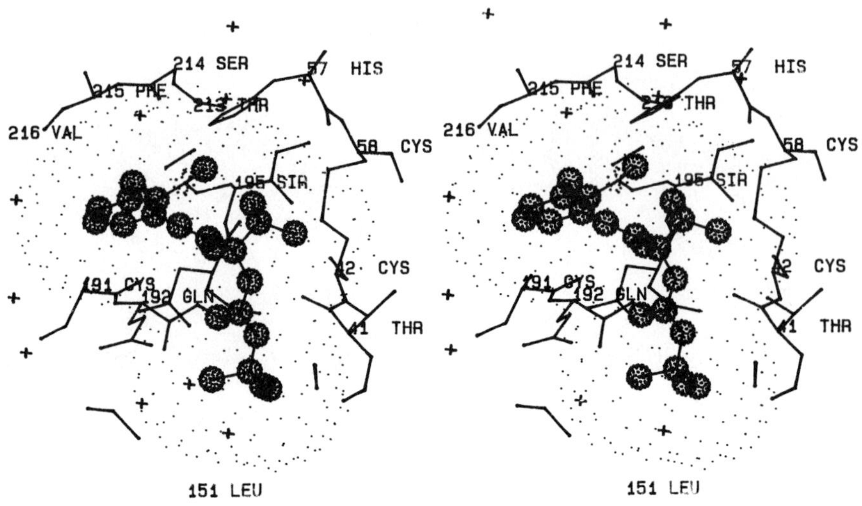

Fig.2. Stereoview of PPE + 5-chloro-benzoxazinone. Similar to the other complexes studied, there was minimal structural change in the enzyme, compared to native PPE, except that His 57 was rotated into the "out" position, making van der Waals contact with the surface of the bound inhibitor. A water molecule (406) replaces His 57 and is 3.00Å from the product carbonyl O atom and 2.88Å from Asp 102 ODI, potentially ready to assist in product hydrolysis.

method: the optimal chrystallographic pH corresponded with an "on-time" of ca. 3 days. They show how <u>both the area detector and conventional film techniques, properly adapted, can achieve significant accelerations in data collection rates, making it possible to study labile complexes.</u> In a total of 8 days, PPE was co-crystallized with the 5-methyl benzoxazinone inhibitor, at pH 5.0, and a 2.1Å data set taken at the National Area Detector Facility in the laboratory of Prof. Robert Kretsinger, R=0.17. As expected from modelling studies (Presta & Meyer, 1986d), the inhibitor was covalently bound to Ser 195 with the methyl group lodged in the S1 primary specificity pocket. A separate study with the 5-chloro-benzoxazinone was made at pH 5.0; film data were collected out to 1.75Å resolution, R=0.17, Fig. 2.

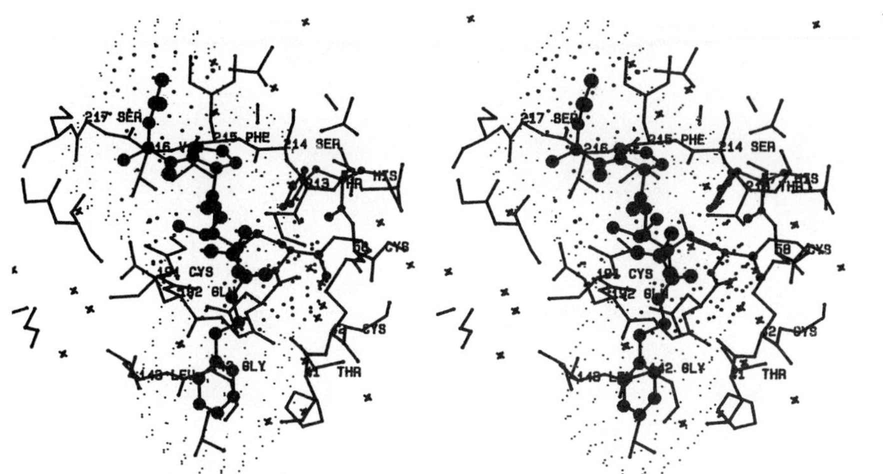

Fig. 3. Stereoview of the difluro inhibitor bound to PPE. The refined structure (R=.16) shows the expected, <u>anti-parallel β-sheet</u> <u>binding of Ala-Pro-Val in the</u> S3-S1 region with Ser 195 <u>bound as an hemiketal.</u>

Recently a novel class of the transition state analogs was studied. First, our Datex-Syntex diffractometer was used to collect data of a complex of PPE co-crystallized with acetyl-Ala-Pro-Val-trifluoromethylketone. Positive, forward binding could be established by a difference Fourier map. Subsequently, a large crystal of PPE was soaked with an excess Acetyl-Ala-Pro-Val-difluoro-β-ketophenethylamide and film data to 1.7Å resolution measured over a total of 6 days (Fig.3).

A preliminary 4Å resolution data set of PPE soaked with CBZ-Ala-Ile-boronic acid indicates forward binding in the S3-S1 region and tetrahedral geometry inherent to transition state inhibitors. Also, there is complete density for isoleucine in S1, whereas previous studies have suggested only small amino acids such as alanine can be accommodated.

THE CONSERVED STRUCTURE OF INTERNAL WATER MOLECULES

The 1.65Å resolution structure analysis of native PPE established the existence of 32 internal water molecules, most

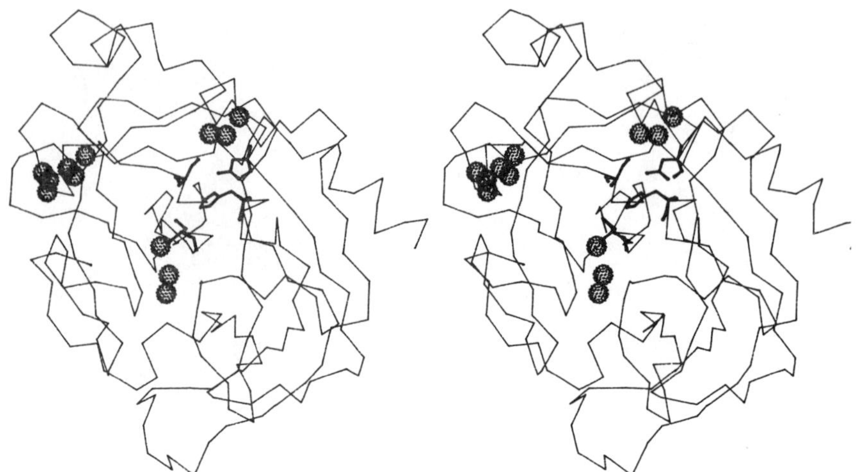

Fig. 4. Stereoview of Cα backbone of PPE; Buried water channels are drawn at the center (Fig. 5), left (Fig. 6) and top (Fig. 7). The first two of these channels link the inner portions of the primary (S1) specificity site with the surface of the enzyme. A plausible explanation suggests that these two gated canals (Fig.5,6) could play a crucial role in helping equalize "hydrostatic pressure" that must arise at the binding (and release) steps along the catalytic pathway when the hydrated P1 side chain is inserted into the hydrated S1 specificity pocket.

(23) of which had been located in the 2.5Å resolution study (Sawyer, et al., 1978). However, it became clear from a geometric analysis of the environments of these internal water molecules that <u>three clusters actually formed</u> (<u>gated</u>) <u>canals from the enzyme</u> surface directly into buried sections of the active site of PPE (Fig.4). This suggests that enzymes specific for large side chains in S1 (trypsin, kallikrein, chymotrypsin, etc.) would need these channels much more than PPE; we find buried water molecules conserved in those structures where water coordinates have been reported. Experiments are being initiated to test the catalytic role of these channels; <u>our X-ray structure here merely establishes their existence and</u>

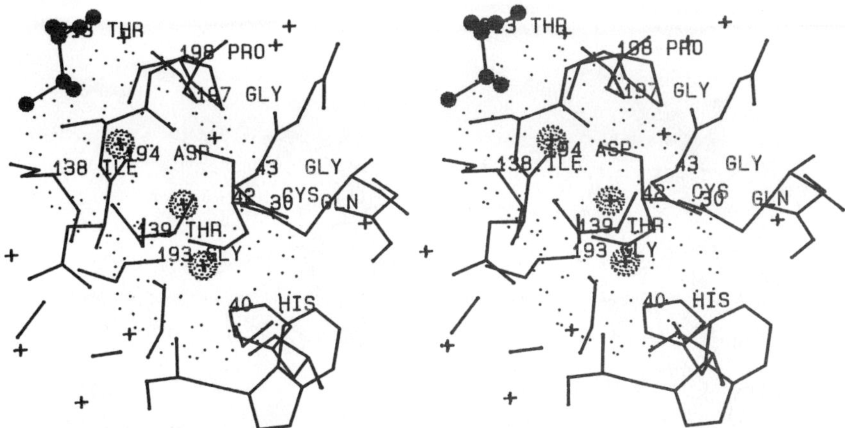

Fig.5. Water channel in PPE between Thr 213 (S1 site) and His 40 (surface). The variability of the side chain of residues 213 and 216 play a major role in defining the shape, and level of specificity of the primary (S1) specificity site in the serine proteases. Dynamically, both Thr 213 and His 40 are able to pivot, exposing the buried water channel to the S1 pocket and bulk solution, respectively.

homology; it further suggests a functional role that previously had been overlooked.

A third water channel (Fig.7) is likewise conserved in the eucaryotic and trypsin-like bacterial serine proteases.

Our observation of a conserved water channel of three tightly held water molecules linking Ser 214 with the surface of the enzyme takes on importance only when an extended substrate + receptor (Michaelis) complex is formed, ejecting all water from the immediate active site. The H-bonded triad is effectively planar. A crowded face of His 57 may still be partially hydrated. More likely, the chain of 6 H-bonds from Ser 195 to Sol 321 could provide a site for removal of a H ion at the surface of the enzyme; this "basic potential" could be transmitted, via H-bond switching (Saenger, 1979) all the way back to Ser 195 Oγ. Experiments are now being designed to test the merits or limits of this hypothesis; here we simply report

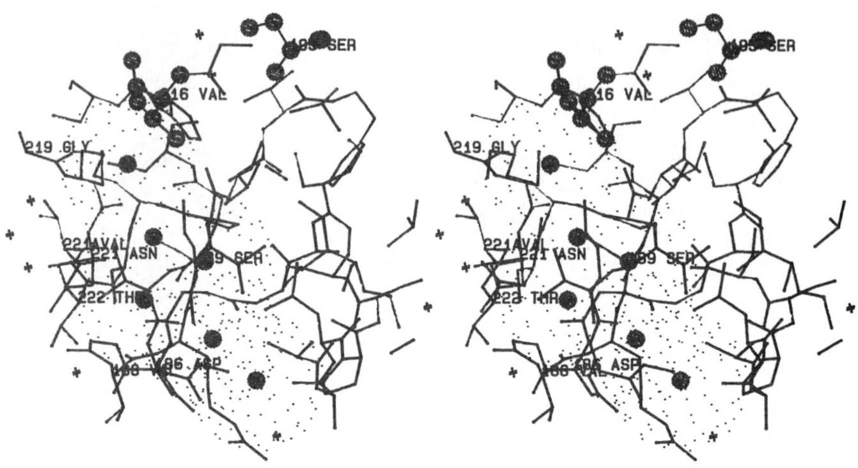

Fig.6.   15Å water channel between the surface and Val 216 (S1 site); dynamic pivoting could provide a mechanistic role for this buried water channel in substrate binding and release.

the structural results, the homology of this channel, and a plausible functional mechanism for the observed structure.

OVERVIEW

The temptation in structure:function studies is to assume that, by obtaining the structure of the receptor, that modelling and binding studies will quickly and directly lead to a full comprehension of binding modalities and thence to the design of effective and specific inhibitors. PPE may be taken as an archetypical example of a receptor with weak and mostly non-specific binding affinities. The design of ligands fully exploiting the sum of these weak interactions is an unique challenge. Unobscured by stronger formal Coulombic interactions, the weak interactions in elastase present a good starting point for systematically modelling and testing receptor + substrate affinities, with the support of computational methods and

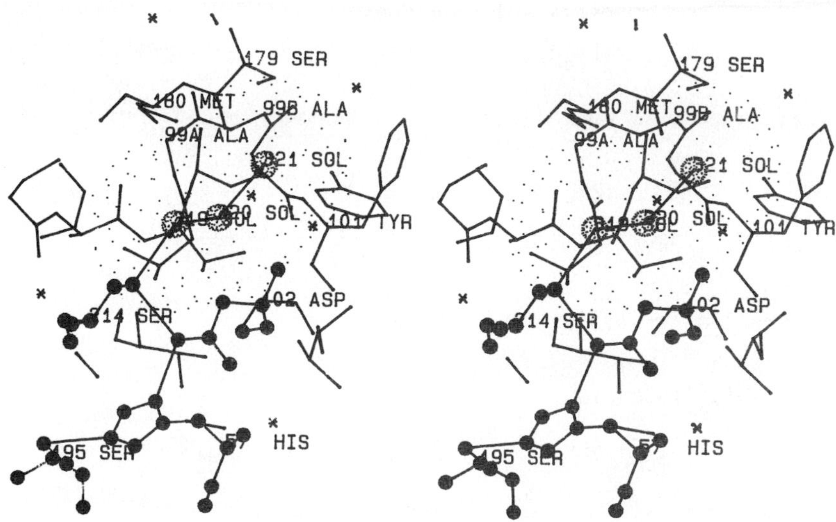

Fig.7. H-bond linkage from Asp 102-Ser 214 via 3 water molecules (319-321) to the surface of PPE. All known primary sequences of the above classes of serine proteases have the tetrad conserved, as supported by available structural data. This
tetrad is linked to the hydrated surface of the enzyme by a
conserved channel with (three) tightly bound water molecules.

especially the precise results of high-resolution crystallography.

We recognize the dedicated efforts of previous (Drs. Greg
Cole, Leonard Presta) and current (Drs. Stanley Swanson and
Dick Rosenfield) colleagues and the hospitality of Prof. Robert
Huber and congenial collaboration of Dr. Wolfram Bode and col-
leagues. We acknowledge the chemists (Dr. Morris Zimmerman,
Prof. James Powers, Drs. Amy Trainor, Richard Wildonger, Ross
Stein, Prof. John Katzenellenbogen) and the financial support
of The Robert A. Welch Foundation, ICI Americas, The Office of
Naval Research, The Council for Tobacco Research, U.S.A., The
National Science Foundation, and The Texas Agricultural Experi-
ment Station. Mrs. Sharyll Pressley assisted with the typing.

REFERENCES

BERNSTEIN, F.C., KOETZEL, T.F., WILLIAMS, G.J.B., MEYER, E.F., BRICE, M.C., RODGERS, J.R., KENNARD, O., SHIMANOUCHI T. and TASUMI, M. J. Mol. Biol. (1977) 112, 535-543, and Eur. J. Biochem. (1977) 80, 319-324.

BIZZOZERO, S.A. and DUTLER, H., Bioorg. Chem. (1981) 10, 46-62.

BODE, W. and HUBER, R., in Desnuelle, Sjöström, and Noren, (eds.) Molecular and Cellular Basis of Digestion, Elsevier (1986) 213-234.

CARLSON, G.M., MCDONALD, R.J. and MEYER, E.F., Jr., J. Theoretical Biology (1986) 119, 107-124.

CLORE, G.M., GRONENBORN, A.M., CARLSON G. and MEYER, E., J. Mol. Biol. (1986) 190, 259-267.

FISCHER, E., Ber. Deutsch. Chem. Ges. (1894) 27, 2984-2993.

HARPER, J.W., and POWERS, J.C., Biochem. (1985) 24, 7200-7213.

HUBER, R. and BODE, W., Ac. Chem. Res. (1978) 11, 114-122.

HUGHES, D.L., SIEKER, L.C., BIETH J. and DIMICOLI, J.-L., J. Mol. Biol. (1982) 162, 645.

MEYER, E.F., PRESTA, L.G. and Radhakrishnan, R., J. Am. Chem. Soc. (1985) 107, 4091-4093.

MEYER, E.F., RADHAKRISHNAN, R., COLE, G.M. and PRESTA, L.G., J. Mol. Biol. (1986a) 189, 553-539.

MEYER, E., COLE, G. and RADHAKRISHNAN, R., Acta Cryst. B., submitted (1986b).

MEYER, E., OSBORNE-CROOK, R. and RADHAKRISHNAN, R., Acta Cryst. B., submitted (1986c)

PRESTA, L.G. and MEYER, E.F., Biopolymers, submitted (1986d).

SAENGER, W., Nature (1979) 279, 343-344.

SAWYER, L., SHOTTON, D.M. CAMPBELL, J.W. WENDELL, P.L. MUIRHEAD, H., WATSON, H.C., DIAMOND R. and LADNER, R.C., J. Mol. Biol. (1978) 118, 137-208.

SCHECHTER, I. and BERGER, A., Biochem. Biophys. Res. Comm. 27 (1967) 157-162.

21. Nucleic acid junctions. A successful experiment in macromolecular design

Nadrin C. Seeman and Neville R. Kallenbach

ABSTRACT

Stable branched nucleic acid structures, called junctions, can be formed from nucleic acids with specific sequences. The selection of these sequences is described, and their preliminary characterization is reviewed.

1. INTRODUCTION

Control of the architecture of matter is one of the major aims of the structural sciences. Biological macromolecules offer an important arena for experiments in macromolecular design: their structures reflect a subtle interaction between the forces which shape them. We have chosen to work with nucleic acids in pursuit of architectural control, because of the relatively good understanding of the influences on their structure. The specific goal of our work has been to perturb the interactions which result in double helical molecules, in order to construct stable branched structures, termed junctions, from oligonucleotides. These junctions, such as "J1", shown in Figure 1, are analogs of unstable naturally occurring intermediates in the process of recombination (Holliday, 1964), and constructing them from oligonucleotides renders them tractable to physical characterization by crystallography, NMR, calorimetry and spectroscopy.

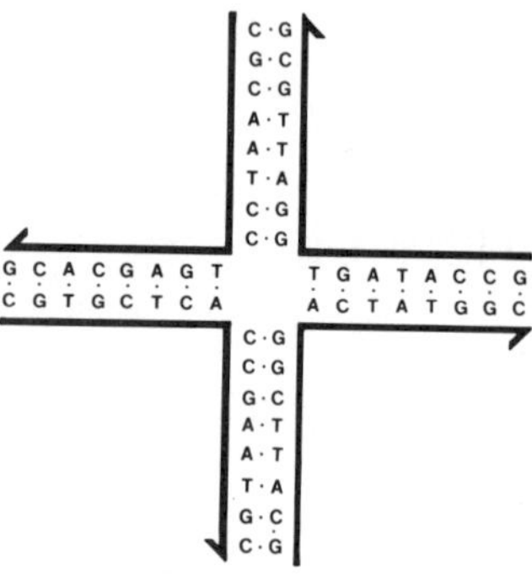

FIGURE 1.   J1, a stable branched DNA junction.

     Four principal influences have been identified as being
relevant to nucleic acid structure: three favor forming a double helix from isolated strands, and one is unfavorable.   The
stabilizing interactions are (1) Watson-Crick hydrogen bonding
of complementary base pairs, (2) base stacking interactions
between the base pairs adjacent to each other in a duplex and
(3) backbone conformational preferences.   The unfavorable
interaction is the electrostatic repulsion between the negatively charged phosphates of the strand backbones in a double
helix.   The interplay of these four interactions in linear
duplex DNA results in the delicate balance necessary for it to
function as genetic material.   We have altered the relationship
of these effects to create novel structures, the DNA junctions.

     DNA junctions are not normally seen in relaxed DNA,
because the site of the branching is about 17 kcal/mol less
favorable than an ordinary double helical site (Courey and
Wang, 1983).   This represents a combination of unfavorable
factors, including backbone conformations, stacking, and
phosphate-phosphate repulsion.   The only interaction potentially unaffected by junction formation is the base pairing in
the arms which flank it.   Thus, our strategy for forming this
"high energy" structural intermediate is based on designing the
sequence so that base pairing will be maximized only if the

junction is formed. It is also necessary to prevent a
symmetry-based isomerization process called branch point
migration (Thompson, Camien and Warner, 1976).

2. THE DESIGN OF NUCLEIC ACID JUNCTIONS

The basic idea involved in nucleic acid junction design is that
oligonucleotides can be constructed which will preferentially
associate to form all of the arms of junctions via Watson-Crick
base pairing, while the sequences of these molecules lack the
symmetry necessary to permit branch point migration. At the
same time, the overall sequence symmetry is minimized so that
the pairing configuration corresponding to the desired junction
architecture is the most probable one (Seeman, 1982; Seeman and
Kallenbach, 1983). Since even the mildest perturbation of
double helical structure can result in altered interactions
(e.g. Quigley, et al., 1986), a drastic change, such as a
junction, must be "over-engineered" until we know more about
the thermodynamics of sub-optimal pairing structures.

The essential feature of the "sequence symmetry minimization" procedure is that no pairing which interferes with the
designed architecture is permitted. Operationally, this is
accomplished by assuming that a stretch of contiguous base
pairs constitutes a single, favorable interacting unit; the
longer the stretch, the more favorable. In order to achieve
unique pairing, one must minimize the length of redundant
stretches. Each of the four hexadecameric strands of J1 (Figure 1) may be treated as a series of 13 overlapping tetramers.
The architecture of the junction dictates 52 of these tetramers. In this junction, all the tetramers are unique, and
redundancy is only seen at the level of trimers, for example
the G.G.A sequences seen at the middle of the third and fourth
strands. While it is desirable to limit redundancy to the
level of dimers only, it is not possible to choose the 56
required trimers from the 64 possible trimers.

Besides the uniqueness of each tetramer, it is important
to forbid pairing around the bends. Thus, the 4-base comple-
ments to the sequences which span a bend are also forbidden.
For example, in the upper-left strand of J1, the sequence
T.C.C.T.G.A surrounds the bend. The 4-residue sequences complementary to this are T.C.A.G, C.A.G.G and A.G.G.A, and these
must all be forbidden as well. Self-complementary sequences of
the chosen length (or one longer for odd lengths) are implicit-
ly forbidden. This means that only those restriction endonu-
cleases which recognize asymmetric sites (e.g. Mbo II) or
odd-sized sites (e.g. Eco RII) may be used with the junction
complexes.

Branch point migration results from symmetry across the
junction. This may be eliminated for the conventional Watson-
Crick base pairing partners (A-T, T-A, G-C and C-G) in the following way. On simple counting criteria, insisting that each
of the base pairs which flanks the junction be unique, allows
the maximum number of junction arms to be 4. However, applying
the geometrical criterion that a self-complementary dinucleotide cannot self pair, the same base pair can flank the
junction twice, if both occurrences are on adjacent arms
(Seeman, 1982). With this looser criterion, the maximum number
of junction arms may be as high as 8. It is possible that
immobile junctions containing more arms can also be formed:
The effective size and charge on a 9-arm junction might
disfavor migrational isomerization across three base pairs.
The limits, if any, on the number of arms which can emanate
from a junction will have to be determined empirically.

This discrete procedure for minimizing sequence symmetry
must be supplemented by equilibrium stability calculations, in
order to maximize the probability of achieving this junction
architecture, with the redundant trimers chosen (Seeman and
Kallenbach, 1983). A first-order treatment has been adequate
to generate successful junctions, when all desired pairing

partners are present. Only competing Watson-Crick base pairs are considered in this procedure. Nevertheless, it is advisable to eliminate as many potential non-Watson-Crick interactions as possible: In the region near the junction, the double helical constraints are not strong, and non-Watson-Crick interactions might be more favorable than they are within double helical domains. For example, in the design of J1, stretches of G's longer than two are forbidden, to prevent G-G pairing.

We routinely calculate a thermal transition profile for each of the sequences which passes the other criteria. These calculations indicate that junctions with arms 3 or 4 base pairs long were not likely to be stable at reasonable concentrations and temperatures. For this reason, junctions with arms of 6-10 base pairs have been prepared for all studies to date.

3. DEMONSTRATION AND CHARACTERIZATION OF DNA JUNCTIONS

Direct evidence for the formation of a 1:1:1:1 stoichiometric complex involving all four strands of J1 has been obtained from inspection of polyacrylamide gel electrophoretic patterns (Kallenbach, Ma and Seeman, 1983). UV absorbance monitoring of the order-disorder transition in this junction shows profiles in good qualitative agreement with those estimated during the design process.

Other physical experiments, including ^{1}H NMR, ^{31}P NMR and Circular Dichroism spectroscopy, have been performed on junctions. The circular dichroism spectrum indicates that the junction does not perturb the structure of the arms significantly: they remain B-DNA (Seeman et al., 1985). At low temperatures, the ^{31}P NMR spectrum resolves to give a small upfield peak which has properties consistent with the behavior anticipated for a set of phosphates with distinct chemical shifts, possibly representing those at the branch site (Kallenbach and Seeman, 1986). ^{1}H NMR of J1 suggests that the bases which flank the junction are indeed paired (Wemmer et al., 1985). ^{1}H NMR of a smaller junction at very high resolution suggests that there may be more than a single conformer stable in solution (Kallenbach et al., 1983a). Further physical studies in progress, involving calorimetry, 2-D NMR and crystallography can be expected to add to our knowledge of junction structure and dynamics within the next few years.

ACKNOWLEDGEMENTS: This work has been supported by Grants GM-29554, ES-00117 and CA-24101 from the National Institutes of Health. N.C.S. is an NIH Research Career Development Awardee.

REFERENCES

COUREY, A.J. AND WANG, J.C. (1983). Cell 13, 817-829.

HOLLIDAY, R. (1964). Genetics Research 5, 282-304.

KALLENBACH, N.R., MA, R.-I., and SEEMAN, N.C. (1983). Nature 305, 829-831.

KALLENBACH, N.R., MA, R.-I., WAND, A.J., VEENEMAN, G.H., VAN BOOM, J.H. and SEEMAN, N.C. (1983a). J. Biomolecular Structure and Dynamics 1, 159-168.

KALLENBACH, N.R. and SEEMAN, N.C. (1986). Comments on Cellular and Molecular Biophysics 4, 1-16.

QUIGLEY, G.J., UGHETTO, G., VAN DER MAREL, G.A., VAN BOOM, J.H. WANG, A.H.-J. AND RICH, A. (1986). Science 232, 1255-1258.

SEEMAN, N.C. (1982). J. Theoretical Biology 99, 237-247.

SEEMAN, N.C. and KALLENBACH, N.R. (1983). Biophys. J. 44, 201-209.

SEEMAN, N.C., MAESTRE, M.R., MA, R.-I., and KALLENBACH, N.R. (1985). in Molecular Basis of Cancer, ed. by R. Rein, Alan R. Liss, Inc., New York, pp. 99-108.

THOMPSON, B.J., CAMIEN, M.N., and WARNER, R.C. (1976). Proc. Nat. Acad. Sci. (USA) 73, 2299-2303.

WEMMER, D.E., WAND, A.J., SEEMAN, N.C. and KALLENBACH, N.R. (1985). Biochemistry 24, 5745-5749.

22. A kinked model of a DNA tridecamer with an unpaired adenosine: energy mimimization and X-ray structural studies

Joel L. Sussman, Leemor Joshua-Tor, Miriam Hirschberg, Mark A. Saper, Felix Frolow, Håkon Hope and Etorre Appella

ABSTRACT

X-ray crystallographic studies have been carried out on the DNA tridecamer d(CGCAGAATTCGCG) (Saper et al., 1986). Previous NMR studies (Patel et al., 1982) predict that this tridecamer forms a duplex similar to the B-DNA dodecamer, d(CGCGAATTCGCG) (Wing et al., 1980), except for an extra adenosine residue that is stacked within the helix but remains unpaired:

```
5' C-G-C-A-G-A-A-T-T-C---G-C-G 3'
   : : :   : : : : : :   : : :
3' G-C-G---C-T-T-A-A-G-A-C-G-C 5'
```

Initially we collected only low resolution X-ray diffraction data as crystals of this tridecamer decayed in the X-ray beam in a few hours, even when cooled to 4°C. By cooling the crystals to ~-150°C, it was possible to preserve their lifetime practically indefinitely, thus permitting accurate X-ray data collection to high resolution. We have employed methods of energy minimization and molecular dynamics (Levitt 1978, 1983) resulting in a symmetrically kinked model of this tridecamer.

1. Introduction

Recent advances in DNA synthetic methods (Itakura et al., 1975) have made it possible to analyze by X-ray crystallography single crystals of DNA with predefined sequences, 4-12 nucleotides long (Dickerson et al., 1982). Until now, apart from one example (McCall et al., 1986), only self-complementary DNA sequences or those containing mismatched base-pairs (Brown et al., 1985) have been examined. From these studies, DNA appears to exhibit enormous conformational flexibility.

The conformation of extrahelical bases in nucleic acid duplexes has been a problem of long standing interest in light of their relevance to frame shift mutagenesis (Streisinger, _et al_., 1966; Drake _et al_., 1983).   The extrahelical base could stack into the helix or loop-out depending on the base type (purine or pyrimidine) and the flanking base-pairs in the sequence.   This could result in an insertion or deletion mutation due to partial misalignment of strands during replication.

NMR and calorimetric studies showed that the extrahelical adenosine, in an otherwise self-complementary duplex of 13 base-pairs, stacks into the duplex (Patel _et al_., 1982; Hare _et al_., 1986).   In contrast 1D and 2D NMR studies on extrahelical cytidine (Morden, _et al_., 1983) and thymidine (Patel _et al_., 1987) have shown that these bases tend to loop-out into solution.   They generalized that purine bases tend to stack in the duplex while pyrimidine bases loop-out.   NMR studies of nucleic acids in solution and X-ray studies that can be done at atomic resolution provide complementary structural information.

We are studying by X-ray crystallography, together with theoretical energy calculations, the three-dimensional structure of the tridecamer, d(CGCAGAATTCGCG).   Except for the inserted adenosine at position 4, it has a sequence identical to the previously determined B-DNA dodecamer, d(CGCGAATTCGCG) (Wing _et al_., 1980)).   As discussed above, high resolution NMR analysis demonstrated that the extra adenosine residue is not _looped-out_, but instead appears to be stacked into the helix and remains unpaired (Patel _et al_., 1982).

The X-ray structural analysis will help determine if, indeed, the extra residue is unpaired, whether the extra base is stacked inside the helix as predicted by NMR or looped-out, and if insertion causes a kink in the helical geometry.

2.   METHODS AND STARTING MODEL
2.1. X-ray Crystallographic Studies
Details of the synthesis, purification and preliminary crystallization conditions have been reported (Saper _et al_., 1986). Large crystals of the tridecamer are now grown in small drops (4-8ul) on washed, plastic coverslips.   Each drop typically contains 2.4 mg/ml tridecamer, 1.5 mM spermine-HCl, 15 mM $MgCl_2$, 10% (v/v) 2-methyl-2,4-pentanediol (MPD), and 10 mM- -sodium cacodylate pH 7.0.   Drops were equilibrated at 4°C by the vapor diffusion technique against a solution of 30% MPD +

0.2 M NaCl.   Elongated triangular prisms, up to 0.3 x 0.2 x 0.2 mm appear within 7 - 10 days.

We attempted to collect X-ray diffraction data on crystals at room temperature and at 4°C, however they decomposed in the X-ray beam within a few hours, and showed significant shrinkage in one of the unit cell dimensions. Using techniques of Hope (Hope and Nichols, 1981), for extreme low temperature X-ray data collection, we were able to preserve the lifetime of the crystals practically indefinitely in the X-ray beam. This me-thod consists of first coating the crystal with a viscous oil (STP) in the crystallization droplet, and then removing all mo-ther liquor solution covering it. The crystal is then picked up with a thin glass fiber putting it directly under a lamellar stream of boiled liquid N_2 ~-150°C; n.b. the crystal was <u>not</u> mounted in a capillary when exposed to X-rays.

The space group and unit cell as determined at this cryo-genic temperature on a Rigaku AFC5-R Rotating Anode Diffrac-tometer operated at 15kW are: Space Group C2: <u>a</u> = 78.48 Å, <u>b</u> = 42.84 Å, <u>c</u> = 25.16 Å and β = 99.36°. These are all within 1% of the unit cell parameters as measured at 4°C (Saper <u>et al</u>., 1986). At ~-150°C the crystals diffract to approximately 1.8 Å resolution.

The unit cell dimensions of the B-DNA dodecamer studied by Dickerson are closely related to those of the tridecamer (Saper <u>et al</u>., 1986).   There is an obvious similarity between the <u>a</u> and <u>b</u> dimensions of B-DNA and the <u>c</u> and <u>b</u> of the tridecamer. Furthermore, the ratio between lengths of the long axes is 14:12. This suggests a similar, but stretched helix that con-tains extended linkages opposite the extra adenosine residues as predicted by Patel <u>et al</u>. (1982). We have collected a com-plete set of X-ray data, from <u>a single crystal</u>, out to 2.5 Å resolution, on the Rigaku AFC5-R at ~-150°C.

We plan to determine the three-dimensional structure of the tridecamer via molecular replacement techniques (Crowther, 1972; Harada <u>et al</u>., 1981), followed by Constrained-Restrained Least-Squares (CORELS) refinement (Sussman <u>et al</u>., 1977).

2.2. Energy Minimization and Molecular Dynamics

In order to use molecular replacement methods a good starting model is required. A model of the tridecamer sequence with two unpaired adenosines was constructed using energy minimization and molecular dynamics techniques. Using these methods, one attempts to derive the equilibrium structure by minimizing the potential function.   The potential energy was calculated as a function of positions of all atoms in the structures.   The potential function contains terms for bond stretching, bond angle bending, hindered bond twisting, van der Waals interactions and hydrogen bond interactions.   The electrostatic interactions were neglected and no solvent was included.   The potential function and the various empirical parameters used, are described in detail in previous works (Levitt 1978, 1983).

Based on the assumption that the unpaired bases are stacked into the helix, a standard B-form double helix (Arnott and Hukins, 1972) was generated for the 14-mer sequence, a sequence with two extra thymidines opposite to the two unpaired adenosines (see Fig. 1).   The two thymidines were then removed and the structure was allowed to relax by energy minimization using the program ENCAD (Levitt, 1986 private communication). The temperature of the relaxed system was then raised up to 300K (room temperature) and the system was subjected to 3000 cycles of dynamic minimization with time step of 0.002 psec., followed by energy minimization till convergence.   Several starting models were generated using the described procedure. These models differ from one another in their initial rotational angles per base-pairs around the unpaired base (Fig. 1).

$$
\begin{array}{l}
\qquad\qquad\quad \alpha \quad \beta \qquad\qquad\qquad\qquad\qquad \beta \quad \alpha \\
5' \;\; C_1-G_2-C_3-A_4-G_5-A_6-A_7-T_8-T_9-C_{10}-\;t-\;G_{11}-C_{12}-G_{13}\;\;3' \\
3' \;\; G_{13}-C_{12}-G_{11}-\;t-C_{10}-T_9-T_8-A_7-A_6-\;G_5-\;A_4-\;C_3-G_2-C_1\;\;5' \\
\qquad\qquad\quad \alpha \quad \beta \qquad\qquad\qquad\qquad\qquad \beta \quad \alpha
\end{array}
$$

Fig. 1.   Generation of the starting models of DNA tridecamer for energy minimization and molecular dynamics calculations. The residues label "t" opposite the extrahelical A's were removed, before minimization. α and β are the turns per base-pair for the 5 models that were generated: (α,β) = [$(12°,24°)$, $(15°,21°)$, $(18°,18°)$, $(21°,15°)$, $(24°,12°)$].

4.   RESULTS AND DISCUSSION

The most stable structure was obtained from a starting model with initial helical rotations of $(\alpha,\beta)=(18°,18°)$ (see Fig. 1). Introducing an unpaired base to a regular B-form double helix results in local changes in the angle between the base-pair planes of the neighboring base-pairs: C_3-G_{11} and G_5-C_{10}. Consequently, the global structure has two kinks at the unpaired sites.   It thus consists of 3 fragments of B-DNA: a central segment six base-pairs long, and two outer segments 3 base-pairs long each (see Fig. 2).   The angle between the helical axes of the central part and each of the outer parts is 34.5°.

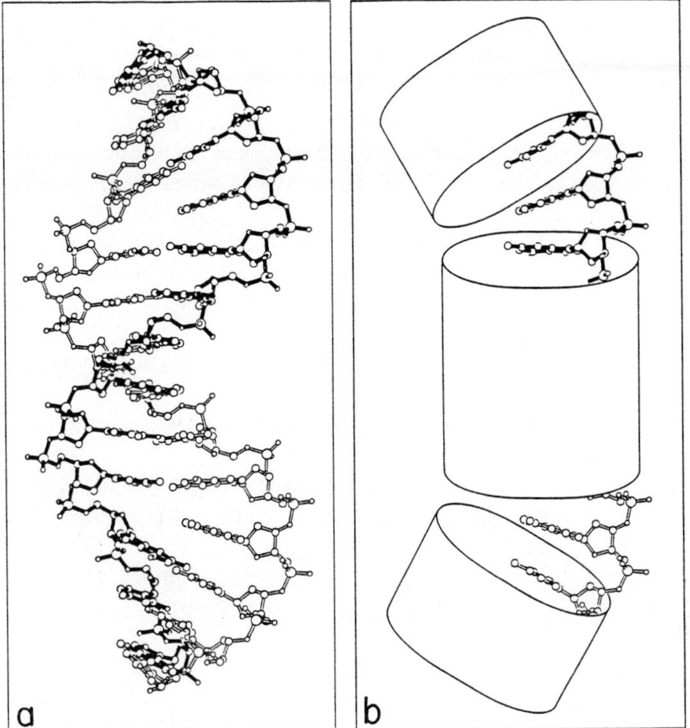

Fig. 2.   DNA tridecamer structure d(CGCAGAATTCGCG) following
energy minimization. a) Pluto (Motherwell, 1979) drawing of the
structure; b) Schematic representation of the structure, symbo-
lized by 3 cylinders; n.b. although the extra adenosines are
from different strands they fall on the same side of the model.

The overall structure consists of two outer segments on
the same side of a central region. This is due to the symmetry
in the sequence with exactly six base-pairs in the central segment, 1/2 a helical turn, and that the two extrahelical bases
are on opposite strands.   We await completion of the X-ray
structure determination to confirm this model.

ACKNOWLEDGEMENTS: We would like to thank our colleagues H.
Eisenberg, W. Traub, M. Levitt, A. Yonath, M. Shoham, B. Shaa-
nan, Y.-Y. Chiu, S. Weinstein, A. Wlodawer, and M. Miller for
encouragement and suggestions throughout this work.   This research was supported in part by grants from the Gutwirth and
Minerva Foundations and the U.S. Army Research Office to J.L.S.

REFERENCES

ARNOTT, S. and HUKINS, D.W.L. (1972). _Biochem. Biophys. Res. Commun._ 47 1504-1509.

BROWN, T., KENNARD, O., KNEALE, G. and RABINOVICH, D. (1985). _Nature_ (London) 315, 604-606.

CROWTHER, R.A. (1972). The Molecular Replacement Method, M.G. Rossmann, ed., Gordon and Breach, Science Publishers, N.Y. 173-178.

DICKERSON, R.E., DREW, H.R., CONNER, B.N., WING, R.M., FRATINI, A.V. and KOPKA, M.L. (1982). _Science_ 216, 475-485.

DRAKE, J.W., GLICKMAN, B.W. and RIPLEY, L.S. (1983). _American Scientist_ 71, 621-630.

HARADA, Y., LIFCHITZ, A. and BERTHOU, J. (1981). _Acta Cryst._ A37, 398-406.

HARE, D., SHAPIRO, L. and PATEL, D.J. (1986). _Biochemistry_ 25, 7456-7464.

HOPE, H. and NICHOLS, B.G. (1981). _Acta Cryst._ A37, 158-161.

ITAKURA, K., KATAGIRI, N., BAHL, C.P., WIGHTMAN, R.H. and NARANG, S.A. (1975). _J. Amer. Chem. Soc._ 97, 7327-7332.

LEVITT, M. (1978). _Proc. Natl. Acad. Sci._ (USA) 75, 640-644.

LEVITT, M. (1983). _Cold Spring Harbor Symp. Quant. Biol._, 47, 251-262.

McCALL, M., BROWN, T., HUNTER, W.N. and KENNARD, O. (1986). _Nature_ (London) 322, 661-664.

MORDEN, K.M., CHU, Y.G.,MARTIN, F.H. and TINOCO, Jr., I. (1983). _Biochemistry_, 22, 5557-5563.

MOTHERWELL, S. (1979). A Program for Plotting Molecular and Crystal Structures, University Chemical Lab., Cambridge, UK.

PATEL, D.J., KOZLOWSKI, S.A., MARKY, L.A., RICE, J.A., BROKA, C., ITAKURA, K. and BRESLAUER, K.J. (1982). _Biochemistry_, 21, 445-451.

PATEL, D.J., SHAPIRO, L. and KALNIK, M. (1987). _J. Biol. Chem._ (submitted for publication).

SAPER, M.A., ELDAR, H., MIZUUCHI, K., NICKOL, J., APPELLA, E. and SUSSMAN J.L. (1986). _J. Mol. Biol._, 188, 111-113.

STREISINGER, G., OKADA, Y., EMRICH, J., NEWTON, J., TSUGITA, A., TERZAGHI, E. and INOUYE, M. (1966). _Cold Spring Harbor Symp. Quant. Biol._ 31, 77-84.

SUSSMAN, J.L., HOLBROOK, S.R., CHURCH, G.M. and KIM, S.-H. (1977). _Acta Cryst._ 33, 800-804.

WING, R., DREW, H., TAKANO, T., BROKA, C., TANAKA, S., ITAKURA, K. and DICKERSON, R.E. (1980). _Nature_ (London) 287, 755-758.

23. Carcinogens and carcinogen–DNA interactions

Jenny P. Glusker

The carcinogenic effects of certain compounds were
recognized in 1775 when Sir Percivall Pott, in England, noted
that young chimney sweeps acquired a particular form of scrotal
cancer, referred to as "soot wart" (Pott, 1775). He deduced
that this disease resulted from their constant exposure to
soot. Later studies by his grandson, Earle (1823) and by
Curling (1856) led to the notion that an inherited disposition
to cancer may be significant in determining who acquired this
fatal disease; not all chimney sweeps were afflicted with it.
It was noted that soot generally only caused cancer in a
specific site in the body and that there was often a time lag
of many years between exposure and subsequent acquisition of
the disease.

However, it was many years before the active, carcinogenic
ingredients of soot were identified. Yamagiwa and Ichikawa,
in 1916, reported that rubbing coal tar on the ears of rabbits
caused malignant tumors to result (Yamagiwa and Ichikawa,
1916). Passey (1922) extracted a single chemical from soot and
showed that it was carcinogenic. The various carcinogens in
soot and coal tar were eventually identified as polycyclic
aromatic hydrocarbons (PAHs) by Kennaway, Cook and co-workers
(Kennaway, 1955). These active carcinogens have formulae such
as those of (benzo[a]pyrene (B[a]P, I) and 7,12-dimethyl-

benz[a]anthracene (DMBA, II). They are formed as a result of
pyrolysis such as in cigarette smoking and barbecuing meat.

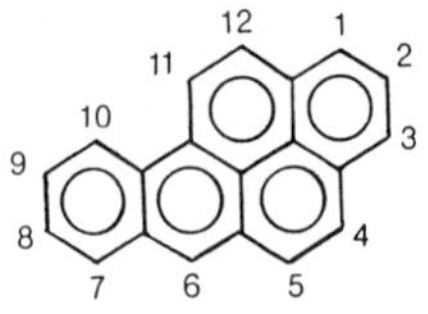

I 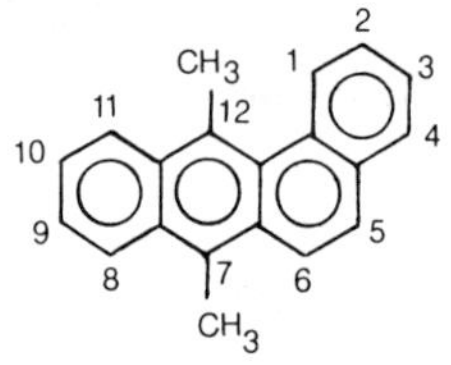   II

The manner by which the human body reacts to carcinogenic
PAHs is important to this story. PAHs are oleophilic,
hydrophobic compounds that, if they enter the body, settle in
fatty tissues. Excretion can occur readily only if the PAHs
are rendered water-soluble. This is done by an enzyme system,
cytochrome P-450, which adds hydroxyl and epoxide groups to
the PAHs. Many different PAH metabolites are formed and often,
if they become covalently attached ("conjugated") to peptides
such as glutathione, they are excreted. This results in no
more problem for the human. However occasionally a potentially
lethal, precarcinogenic event, termed "activation," occurs
(Miller, 1970). For carcinogenic PAHs, the "activated"
metabolite is a diol epoxide that will react with biological
macromolecules and, in so doing, may initiate the carcinogenic
process. Four diol epoxides of B[a]P are III-VI. The
biological macromolecule that is attacked by the diol epoxide
must be one that is involved in the genetic apparatus in some
way since the resulting lesion is inherited; current thinking
identifies it with DNA. This process of activation can be
modified by inhibition of the cytochrome P-450 and other enzyme
systems involved. This may be effected by naturally-occurring
compounds such as flavonoids, by certain synthetic chemicals
and even by other weakly carcinogenic PAHs; such modification
is termed "chemoprevention."

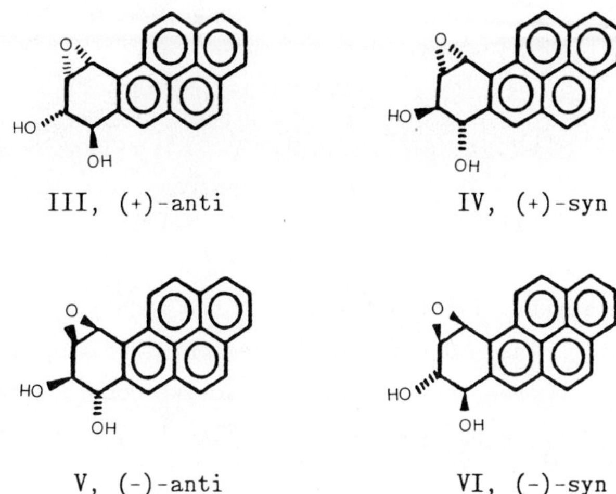

III, (+)-anti   IV, (+)-syn

V, (-)-anti   VI, (-)-syn

There appear to be a group of PAHs of similar size and
somewhat similar shape that are carcinogenic; generally they
contain four or five aromatic rings arranged with a similar
disposition to the ring systems of steroids (Huggins and Yang,
1962) and base pairs in DNA (Haddow, 1957). Such structural
similarities have intrigued scientists since Cook and Hazelwood
converted the steroid, deoxycholic acid, to the carcinogen,
3-methylcholanthrene in 1933 (Cook and Hazelwood, 1933). For
example, Coombs has demonstrated the carcinogenicity of 11-
methyl-15,16-dihydrocyclopenta[a]phenanthrene (VII) which has
an aromatized steroid-like nucleus (Coombs et al, 1979; Kashino
et al, 1986). These ideas suggest some mode of activity of
carcinogenic PAHs that interferes with the normal functioning
of steroids and/or base pairs in DNA. Carcinogenic PAHs also
have similar shapes to aflatoxin B_1 (VIII) which is one of the
most potent human carcinogens; it is found in moldy peanuts and
grain. Aflatoxin B_1 is activated by epoxidation of a double
bond (Swenson et al, 1973) in an analogous position to that at
which B[a]P or DMBA are activated. This, again, emphasizes the
importance of shape in carcinogenesis by PAHs.

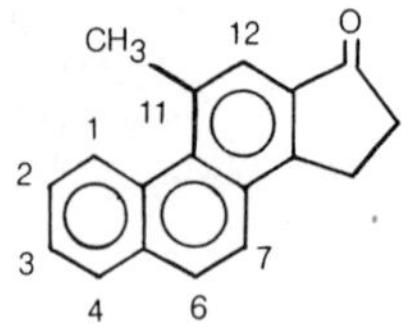

VII 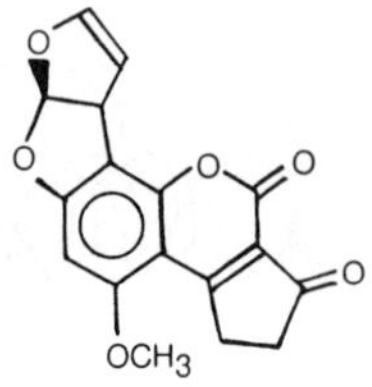   VIII

There are two important areas in the structures of
carcinogenic PAHs. One is the "K-region" (Pullman and Pullman,
1955) which corresponds to the very active double bond between
atoms 9 and 10 in phenanthrene (see Fig. 1). Bonds in the

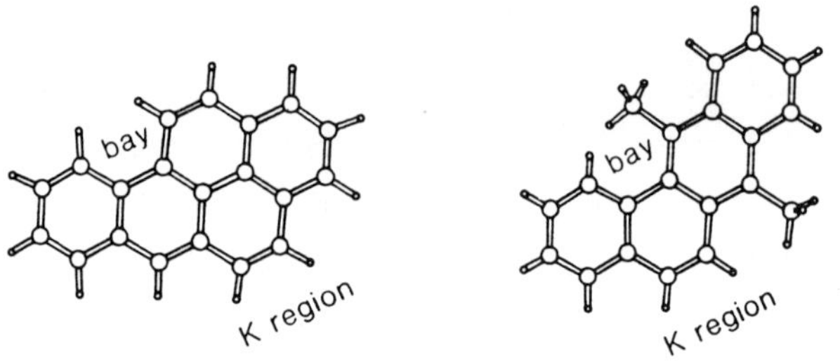

FIG. 1. K- and bay-regions of B[a]P (left) and DMBA (right).

"K-region" are usually near values for a pure double bond
(1.34-1.35 Å); most other C-C bonds in PAHs are in the range
1.36-1.44 Å (compared with 1.39 Å in benzene). The second area
of interest is the "bay region" (Jerina and Daly, 1977). This
corresponds to the hindered region between the 4- and 5-
positions of phenanthrene. While it was originally thought
that the condition for carcinogenicity of a PAH was the
presence of a K-region, it is now clear that the presence of a
bay-region, particularly if there is a methyl group attached to
the side opposite the benzo ring, is a better criterion (Jerina
and Daly, 1977).

Early work on the crystal structures of carcinogenic PAHs was done by John Iball at the University of Dundee, Scotland (Iball, 1936) and included the crystal structure of B[a]P (Iball and Young, 1956; Iball et al, 1976). DMBA was studied by Sayre and Friedlander (1960), refined by Iball (1964) and remeasured at low temperature by Klein et al (1986). The crystal structures of many other carcinogenic PAHs have since been determined.

Three-dimensional structural studies show that distortions from planarity occur in many PAH systems; these are generally caused by repulsions between hydrogen atoms that are not bonded to adjacent carbon atoms. These distortions are further increased on methyl substitution adjacent to the bay-region. In the bay-region the C-C bond length may be increased to from 1.39 Å to 1.48 Å, and it is usual to see large increases of interbond angles from the normally expected values of 118-122°. The buckling of the molecule from this type of steric hindrance arises from torsion about the bonds rather than from distortions from planarity of sp^{2}-hybridized carbon atoms. For example, in B[a]P (I), there is a slight overcrowding between hydrogen atoms in the bay region that is relieved by a twist of approximately 2° about a bond in the bay region; as a result these two hydrogen atoms each lie 0.04 Å away from the molecular plane, above and below respectively. However, DMBA (II) is much more buckled because of the overcrowding between a hydrogen atom of the methyl group on C(12) and a hydrogen atom on C(1) of the benzo ring. In this case the strain is relieved by a twist of 23° about one bond in the bay region, as well as twists about other bonds in that area. The angle between the planes of the two outer rings of DMBA becomes 24.0° as a result of this twisting. In 1,12-dimethylbenz[a]anthracene (Jones et al, 1978) this value is increased to 36°. Similar effects are seen in 5-methylchrysene, the only carcinogenic monomethylchrysene, and its derivatives (Hoffmann et al, 1974; Kashino et al,

1984; Zacharias <u>et al</u>, 1984). Here the ring system is more
amenable to in-plane distortions and therefore torsion angles
in the bay region are not as great as for DMBA, but angular
distortions are greater. Thus we have a picture of a bay
region that, on substitution with a methyl group, may become
distorted by in-plane or out-of-plane distortions, as shown in
Fig. 2.

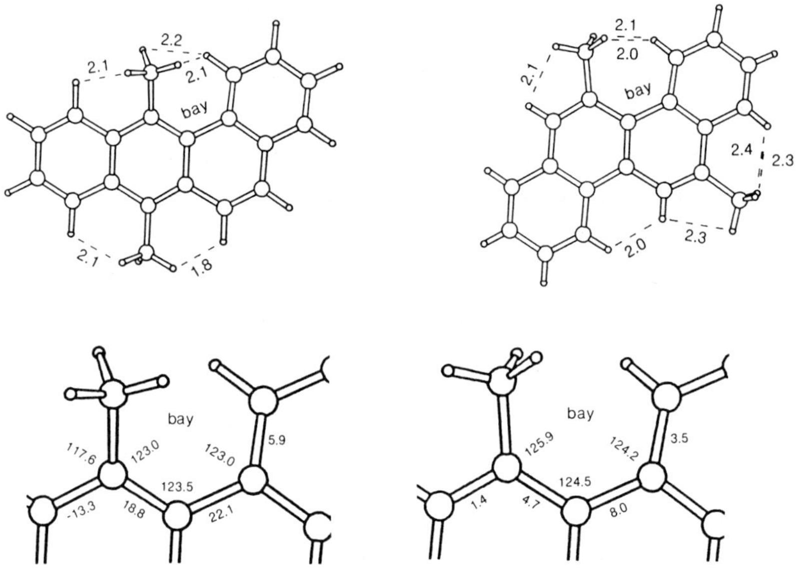

FIG. 2. Bay region distortions in DMBA (left) and 5,12-
dimethylchrysene (right). Short H...H distances, in Å, are
listed in the upper diagram. Angles and torsion angles (listed
along the bond) are shown below.

We next might ask what happens during the activation
process when diol epoxides (Sims and Grover, 1974) are formed?
The bay-region hypothesis (Jerina and Daly, 1977) states that
a diol epoxide, with an epoxide group located adjacent to the
bay-region, is the activated metabolite of a carcinogenic PAH
that is involved in alkylating the "critical target." The

angle between the outer rings of B[a]P is increased on epoxidation at the K-region from 1° to 5°, while the angle between the
outer rings of DMBA is, with K-region epoxidation, increased
from 24° to 35°. Epoxide bond C-O bond lengths are in the
region of 1.45-1.48 Å. In the K-region oxide of DMBA epoxide
the C-O bonds are unequal, probably due to steric interactions
with the methyl group on C(7) in this area. It would be
expected that the longer C-O bond, that is, C(6)-O, would be
cleaved more readily than C(5)-O, and this is, in fact, the
case. Relevant to this are structural studies on a 5,6-cis-
diol of DMBA (Zacharias et al, 1977) in which it was found that
the conformation of the 6-hydroxyl group (nearest the 7-methyl
group) is axial, while that of the 5-hydroxyl group (further
from the 7-methyl group) is equatorial. If the 6-hydroxyl
group were equatorial it would "bump" into the 7-methyl group.
Since the ring bearing the cis-diol group is more saturated,
and hence more flexible than the ring bearing the 7-methyl
group, the strain is accommodated by distortions at C(5) and
C(6) in the more saturated ring. Thus, distortions that occur
in response to steric overcrowding respond in a manner that
costs least in energy. This is shown in Fig. 3.

FIG. 3. Steric hindrance in a) the K-region oxide (epoxide)
of DMBA and b) the 5,6-cis-diol of DMBA. The preferred site
of C-O cleavage in the oxide is C6-O (the longer bond).

Our understanding of the importance of steric factors was furthered by structure determinations by X-ray diffraction techniques of two <u>trans</u>-diols of benz[<u>a</u>]anthracene (Zacharias <u>et</u> <u>al</u>, 1979) (IX and X). The crystal structures showed diequatorial conformations for the hydroxyl groups in the unhindered compound (IX), and diaxial conformations in the hindered compound (X) which had a hydroxyl group adjacent to the bay-region. In X, the hydroxyl group on C(1) would bump into the

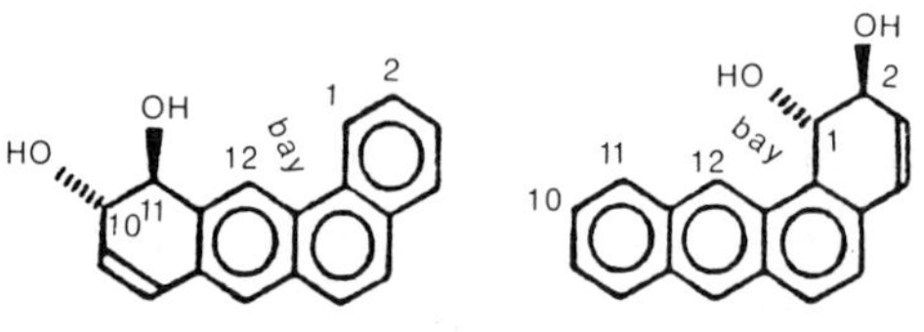

IX,diequatorial   X, diaxial

hydrogen atom on C(12) if the hydroxyl group conformation were diequatorial. Further, NMR studies in solution showed that in X the hydroxyl groups were still 100% diaxial in solution, but that for the less hindered compound, IX, the hydroxyl groups were 30% diaxial and 70% diequatorial in solution. Thus, similar conclusions, derived from a consideration of steric interactions between non-bonded hydrogen atoms, apply to solid state and solution data.

Two of the diol epoxides of B[a]P (III and IV) have the epoxide oxygen atom below the plane of B[<u>a</u>]P and the other two (V and VI) have the oxygen above this plane. In addition the 7-hydroxyl group may be on the same (<u>syn</u>) side (IV and VI) as the epoxide oxygen atom or on the opposite side (<u>anti</u>) (III and V). Also the pairs of <u>trans</u> hydroxyl groups may be either diaxial or diequatorial. The structure of the <u>anti</u>-diol epoxide of B[<u>a</u>]P (III) was determined by Neidle and co-workers (Neidle <u>et</u> <u>al</u>, 1980). The two hydroxyl groups in the diol

epoxide are diequatorial (as shown to be the case in solution
by NMR studies) and the epoxide ring lies in a plane nearly
perpendicular to the PAH system.

The first syn-compound to be studied was a naphthalene
derivative with a methyl group in place of one of the hydroxyl
groups (Glusker et al, 1982). Here the formation of an
internal hydrogen bond, as predicted by Hulbert (1975) and
illustrated in Fig. 4, is observed. However, this hydrogen
bond is not the cause of the presence of axial groups, but

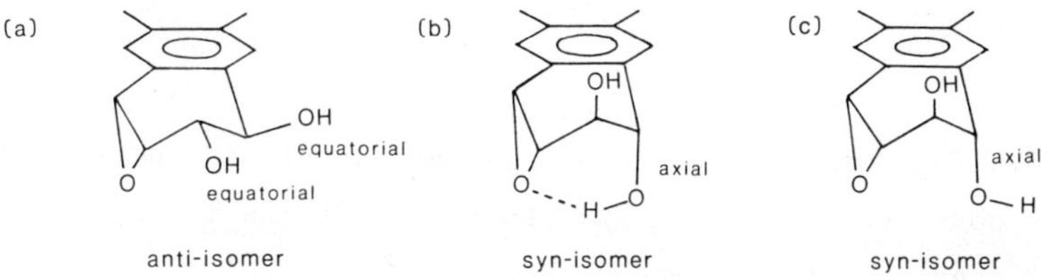

FIG. 4. Anti- and syn-conformations of diol epoxide. In (b)
the internal hydrogen bond predicted by Hulbert and observed by
us is shown.

rather the result, since if the 7-hydroxy group is replaced by
a methoxy group, the conformation of the hydroxyl groups is
also diaxial (Klein and Stevens, 1984). A syn-diol epoxide of
B[a]P (VIII), studied by Neidle (Neidle and Cutbush, 1983), was
shown to have equatorial hydroxyl groups, even though potential
energy calculations had indicated that this might be a higher
energy form. As discussed by Whalen and co-workers (Sayer et
al, 1982; Sayer et al, 1984), in the anti- and syn-diol epoxides of B[a]P, the major differences in the conformations of
diol epoxides involve orientation of the epoxide group with
respect to the planar PAH part of the molecule; they used the
relative orientation of the benzylic C-0 bond to the p-orbitals
of the aromatic rings (Sayer et al, 1982) as a measure
(referred to as "aligned" or "nonaligned"). In the syn-isomer

("aligned") the O-C(10) bond is nearly perpendicular to the
plane of the PAH (torsion angle 98°); in the anti-isomer
"nonaligned" it is inclined to this plane with a torsion angle
of 49°, as shown in Fig. 5. We conclude from these studies
that the conformation of hydroxyl groups in a diol epoxide may

FIG. 5. Two possible relative orientations of PAH and epoxide
groups observed in crystal structures. Left, anti-isomer
(nonaligned), right, syn-isomer (roughly aligned).

be important in determining how the diol epoxide might interact
with a biological macromolecule since diaxial and diequatorial
hydroxyl groups must form hydrogen bonds to groups that are
vastly different in location with respect to the PAH group that
bears the epoxide group. The bay-region methyl group may play
a significant role in determining this hydroxyl group conformation since the hydrogen atoms of the methyl group may interact
sterically with a hydrogen atom of the epoxide group carbon
atoms and so influence the conformation of the ring containing
the hydroxyl groups.

The effect of alkylation of a biological macromolecule
by an activated PAH may be drastic, since, in solution, the
hydrophobic aromatic group of the alkylating agent tries to
avoid the aqueous environment. This we demonstrated (Glusker
et al, 1977) in a structure determination of a tripeptide
alkylated by an activated PAH, together with the structures
of the alkylating PAH derivative and of the tripeptide

(sarcosylglycylglycine). This is a model of alkylation of a
protein. The crystal structure of the tripeptide was dominated
by a three-dimensional network of hydrogen bonds. However the
conformation of the peptide was changed on alkylation from that
of an α-helix to that of a β-pleated sheet. Of particular
interest was the distortion of a peptide group distant from the
site of alkylation. In the crystal packing, the hydrophobic
(PAH) and hydrophilic (peptide and water) areas are segregated,
as shown in Fig. 6. The structure now included water of crystallization and the peptide conformation had, on alkylation,
changed. Presumably the non-planar peptide group ($\omega = 159^{\circ}$)
resulted from the packing; a planar peptide group, as shown in
Fig. 6, would have caused a carboxyl group to lie in a
hydrophobic area.

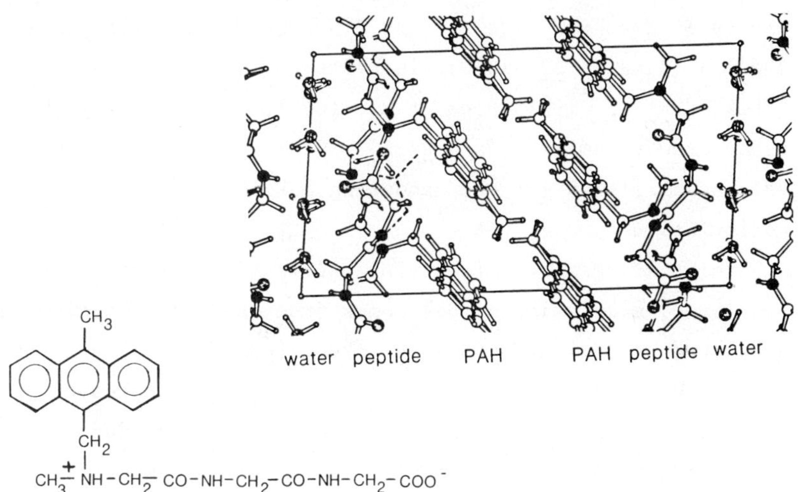

FIG. 6. Packing in the crystal structure of a tripeptide that
has been alkylated by a halomethyl PAH.

These structural results can now be used to consider what
happens when a diol epoxide attacks DNA. The epoxide group

will open and alkylate an amino base on DNA forming a new C-N
bond. The PAH will, as a result, have DNA substituted in it
adjacent to the bay-region. The DNA will be trans to the
remaining C-0 bond of the original epoxide group and will lie
axial to the PAH ring system. This means that the plane of the
PAH and the alkylated base of DNA must have a perpendicular
relationship to each other. The exact nature of the lesion in
DNA, caused by an activated carcinogen, is unknown, and so is
the conformational type of DNA that is attacked. The amino
groups N2 on guanine and N6 on adenine are the commonest sites
of interaction with halomethyl PAH alkylating agents (Dipple
et al, 1971; Dipple and Slade, 1970); this implies attack of
guanine in the minor groove and adenine in the major groove of
B-DNA. However the groove size varies with other conformers
of nucleic acids (Dickerson et al, 1983). In B-DNA the major
groove is deep and wide and the minor groove is narrower and
shallow (Wing et al, 1980); in A-DNA the major groove is deeper
(Shakked et al, 1981; Drew et al, 1980); in Z-DNA only the
minor groove remains (Wang et al, 1979). So there are different possibilities for accommodating a bulky PAH group for each
conformation type of DNA.

Very interesting information relevant to the stereochemical
results of alkylation of DNA comes from studies of nucleosides
alkylated by activated PAHs (Glusker, 1985). Nine such structures have been reported (Carrell et al, 1981; Stezowski et al,
1984; Zacharias et al, manuscript in preparation; Parthasarathy
and Fridey, 1986; Kuroda et al, 1984). Three are products of
the interaction of chloromethyl PAHs with N6 of adenosine (XI),
two with deoxyadenosine, three products of alkylation of
guanosine at 0-6 and one is an acetylaminofluorene derivative
of guanosine, alkylated at C(8).

In the structure of the alkylated adenosine and
deoxyadenosine derivatives, the adenine and PAH residues lie
nearly perpendicular to one another. Alkylation changes the

XI

sugar-base conformation (distant from the site of alkylation)
from <u>anti</u> (as in deoxyadenosine and in B-DNA) to <u>syn</u> (as in
alkylated deoxyadenosine and in some bases in Z-DNA). In these
crystal structures the more planar portion of the PAH is
stacked between adenine residues of other molecules throughout
the crystal. The highly buckled region of the PAH does not
take part in this stacking, but is positioned out of the way,
as shown in Fig. 7.

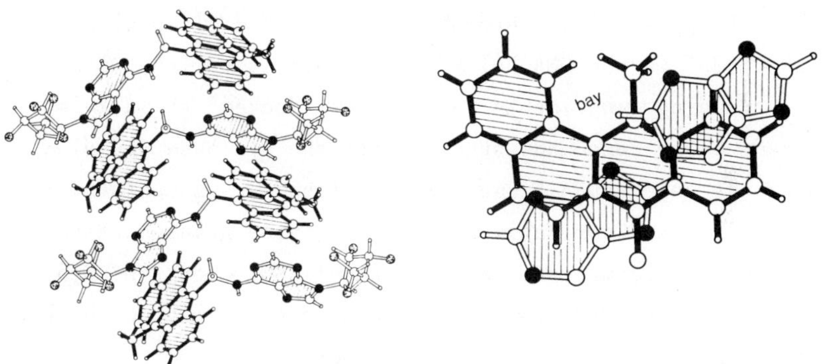

FIG. 7. Stacking of bases and PAH groups in an alkylated
adenosine. Left, packing of four molecules in the crystal
showing interleaving of adenine and PAH groups. Right, overlap
of adenines on a PAH group; note that the buckled bay region of
the PAH does not intercalate between two adenine groups.

This general scheme of packing is found in all five adenine
derivatives studied by diffraction methods; the PAH and adenine
are interleaved, but the hydrogen bonding arrangements in the
crystals of different derivatives are different. This led us
to suppose that this stacking is a significant interaction.
These structural data also indicate that the formation of an
internal hydrogen bond is not necessary for a syn-conformation
of the molecule. Finally there seems to be a weak (but possibly significant) interaction between the ribose ring oxygen
atom of one molecule and the carbon atom of a 12-methyl group
in the 12-methylbenz[a]anthracenyl derivative; such interactions have been noted in other crystal structures (Drew et al,
1980; Wang et al, 1979; Brennan et al, 1983). Such an interaction may be important in aiding in the positioning of the PAH
portion of the adduct within DNA.

There are two general classes of models for DNA alkylated
by a diol epoxide. In the first the PAH lies in one of the
grooves of DNA but on the perimeter of the helix (Beland, 1978;
Geacintov et al, 1978); this is referred to as "external binding." The pyrene-like chromophore of B[a]P diol epoxide lies
at about 35° to the DNA helix axis. This is in line with our
model building experiments in which we "docked" the alkylated
deoxyadenosine compounds studied onto the syn-base (guanine) in
Z-DNA. We used Z-DNA because we had found a syn-conformation
of the alkylated derivatives; Z-DNA contains guanine residues
in the syn-conformation. A rotation of 180° about the C(6)-
N(6) bond to the PAH (which is free to rotate) placed the PAH
on the surface of the helix and did not disrupt normal inter-
base hydrogen bonding within the helix in any way. The interesting feature of this rotated model is that the curvature of
the DMBA anthracene moiety conforms to some extent to the shape
of the helix. In an alternative model the PAH is internally
bound ("quasi-intercalated") (Pulkrabek et al, 1977). Our
X-ray results on the interleaving of adenine and PAH are also

in line with this model, and suggest that there are forces that
favor at least partial intercalation.

Presumably, however, some distortion of the DNA helix must
occur if a PAH is _both_ covalently bound to a base _and_ lies
between that base and its neighboring base up the DNA helix
axis. This is illustrated in Fig. 8 which first shows a PAH
diol epoxide being docked (by computer) into B-DNA and then
the effect of trying to bring the PAH group between the bases;
because the PAH group is now covalently attached to the DNA,
some distortion of bases from planes approximately perpendicular to the helix axis must occur. One attempt to resolve this
problem led to the suggestion that the DNA becomes kinked at
this point (Taylor _et al_, 1983).

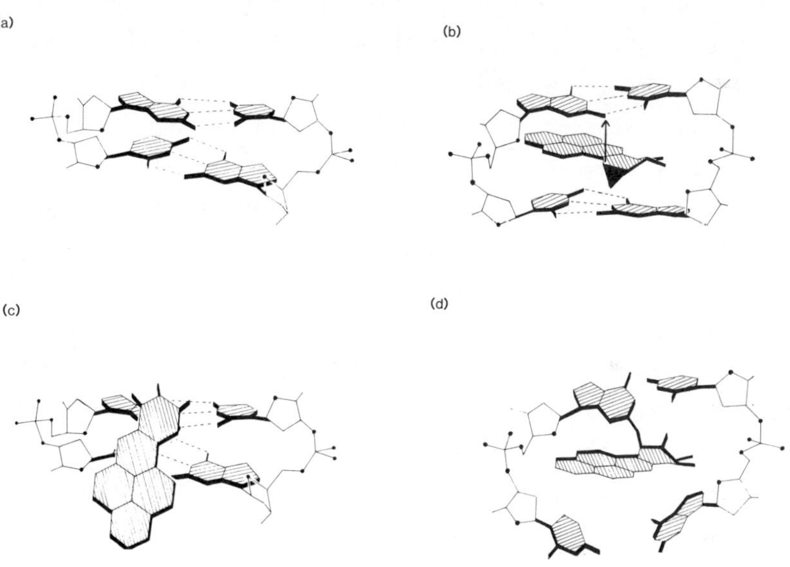

FIG. 8. Model of interaction of an activated PAH with a) DNA.
b) Diol epoxide approaches and partially intercalates in DNA.
c) Diol epoxide alkylates DNA and the PAH group swings into a
groove of DNA. d) The PAH group moves into partial
intercalation between the bases of DNA, but this causes local
denaturation.

To date we do not have enough information to be able to
tell what the lesion in DNA would be on alkylation in the
appropriate position of DNA (that is, the one that is carcino-
genically active). We are now studying a synthetic heptameric
duplex d(GTCA*GAC):d(GTCTGAC), prepared by Stezowski and
colleagues (Stezowski et al, manuscript in preparation), which
has an adenosine group (A*) in the center of one strand alkylated by a PAH group. Attempts are also in progress to make a
minor groove adduct. The report of the crystal structure of a
synthetic oligonucleotide duplex with a T:T mismatch (Hunter et
al, 1986), a model for a mutagenic event, shows no loss of
polynucleotide helicity, but modified (wobble) base-pairing.

So we are still left with two models of the stereochemistry
of DNA alkylated by a PAH diol epoxide; the PAH either lies in
a groove of DNA or else tries to intercalate between the bases
of DNA. Since it is covalently bonded to a base it must cause
considerable distortion if it tries to lie between the bases of
helical DNA. However, the stacking observed in the crystalline
state seems to argue for partial intercalation. We will need
crystal structures of some appropriately alkylated polynucleo-
tides to determine the structural consequences of a DNA of
alkylation by an activated PAH. And when these are completed
it will be just the beginning of the answer to the problem of
the structural significance of alkylation of DNA by activated
carcinogens to the carcinogenic process. The subsequent
question is, what is the lesion in DNA that is important in
carcinogenesis, and then what does it cause to happen so that
tumor formation is initiated?

ACKNOWLEDGEMENTS: I thank Drs. H. L. Carrell, A. Dipple,
R. G. Harvey, S. Hecht, R. Moschel, J. J. Stezowski, and D. E.
Zacharias for many helpful discussions and collaborations.
Most of the diagrams were drawn using the computer programs
VIEW (Carrell, 1976) and DOCK (Badler et al, 1982). The work

of the author was supported by grants CA-10925, CA-22780, CA-06927, RR-05539 from the National Institutes of Health, BC-242 from the American Cancer Society, and by an appropriation from the Commonwealth of Pennsylvania.

REFERENCES

BADLER, N., STODOLA, R.K. and WOOD, W. (1982). Program DOCK. Philadelphia, The Institute for Cancer Research, The Fox Chase Cancer Center.

BELAND, F.A. (1978). Chemical-Biological Interactions 22, 329.

BRENNAN, R.G., PRIVE, G.G., BLONSKI, W.J.P., HRUSKA, F.E. and SUNDARALINGAM, M. (1983). Journal of Biomolecular Structure and Dynamics 1, 939.

CARRELL, H.L. (1976). Program VIEW. Philadelphia: The Institute for Cancer Research, The Fox Chase Cancer Center.

CARRELL, H.L., GLUSKER, J.P., MOSCHEL, R., HUDGINS, W.R. and DIPPLE, A. (1981). Cancer Research 41, 2230.

COOK, J.W. and HAZELWOOD, G.A.D. (1933). Chemistry and Industry (London) 11, 758.

COOMBS, M.M., BHATT, T.S. and YOUNG, S. (1979). British Journal of Cancer 40, 1914.

CURLING, T.B. A Practical Treatise on the Diseases of the Testes and of the Spermatic Cord and Scrotum. Second edition: Blanchard and Lea: Philadelphia, 1856.

DICKERSON, R.E., KOPKA, M.L. and DREW, H.R. (1983). in "Structure and Dynamics: Nucleic Acids and Proteins"; Clementi, E. and Sarma, R.H., Eds.; Adenine Press: New York, pp. 149-179.

DIPPLE, A. and SLADE, T.A. (1970). European Journal of Cancer 6, 417.

DIPPLE, A., BROOKES, P., MACKINTOSH, D.S. and RAYMAN, M.P. (1971). Biochemistry 10, 4323.

DREW, H.R., WING, R.M., TAKANO, T., BROKA, C., TANAKA, S., ITAKURA, K. and DICKERSON, R.E. (1980). Proceedings of the National Academy of Sciences, U.S.A. 78, 2179.

EARLE, H. (1823). Medico-Chirurgical Trans. 12, part II, pp. 296-307.

GEACINTOV, N.E., GAGLIANO, A., IVANOVIC, V. and WEINSTEIN, I.B. (1978). Biochemistry 17, 5256.

GLUSKER, J.P. (1985). in "Polycyclic Hydrocarbons and Carcinogenesis"; Harvey, R.G., Ed.; A.C.S. Symposium Series No. 283, Ch. 7, pp. 125-185.

GLUSKER, J.P., CARRELL, H.L., BERMAN, H.M., GALLEN, B. and PECK, R.M. (1977). Journal of the American Chemical Society 99, 595.

GLUSKER, J.P., ZACHARIAS, D.E., WHALEN, D.L., FRIEDMAN, S. and POHL, T.M. (1982). Science 215, 695.

HADDOW, A. (1957). Proceedings of the Canadian Cancer Research Conferences 2, 361.

HOFFMANN, D., BONDINELL, W.E. and WYNDER, E.L. (1974). Science 183, 215.

HUGGINS, C. and YANG, N.C. (1962). Science 137, 257.

HULBERT, P.B. (1975). Nature (London) 256, 146.

HUNTER, W.N., BROWN T., ANAND, N.N. and KENNARD, O. (1986). Nature (London) 320, 552.

IBALL, J. (1936). Zeitschrift für Kristallographie 94, 7.

IBALL, J. and YOUNG, D.W. (1956). Nature (London) 177, 985.

IBALL, J. (1964). Nature (London) 201, 916.

IBALL, J., SCRIMGEOUR, S.N. and YOUNG, D.W. (1976). Acta Crystallographica B32, 328.

JERINA, D.M. and DALY, J.W. (1977). In "Drug Metabolism - From Microbe to Man"; Parke, D.V. and Smith, R.L., Eds.; Taylor and Francis: London, p. 13.

JONES, D.W., SOWDEN, J.M., HAZELL, A.C. and HAZELL, R.G. (1978). Acta Crystallographica B34, 3021.

KASHINO, S., ZACHARIAS, D.E., PECK, R.M., GLUSKER, J.P., BHATT,
    T.S. and COOMBS, M.M. (1986). Cancer Research **46**, 1817.

KASHINO, S., ZACHARIAS, D.E., PROUT, C.K., CARRELL, H.L. and
    GLUSKER, J.P. (1984). Acta Crystallographica **C40**, 536.

KENNAWAY, E. (1955). British Medical Journal **2**, 749.

KLEIN, C.L. and STEVENS, E.D. (1984). Acta Crystallographica
    C40, 315.

KLEIN, C.L., STEVENS, E.D., ZACHARIAS, D.E. and GLUSKER, J.P.
    (1986). Carcinogenesis (in press).

KURODA, R., NEIDLE, S., EVANS, F.E., BROYDE, S. and HINGERTY,
    B.E. (1984). International Union of Crystallography Meeting,
    Hamburg, Germany, 8-18 August. Abstract 03.2-3.

MILLER, J.A. (1970). Cancer Research **30**, 559.

NEIDLE, S., SUBBIAH, A., COOPER, C.S. and RIBEIRO, O. (1980).
    Carcinogenesis **1**, 249.

NEIDLE, S. and CUTBUSH, S.D. (1983). Carcinogenesis **4**, 415.

PARTHASARATHY, R. and FRIDEY, S.M. (1986). Carcinogenesis **7**,
    221.

PASSEY, R.D. (1922). British Medical Journal **2**, 1112.

POTT, P. (1775). "Chirurgical Observations Relative to the
    Cataract, the Polypus of the Nose, the Cancer of the Scrotum,
    the Different Kinds of Ruptures and the Mortification of the
    Toes and Feet"; Hawes, Clarke, and Collins: London.

PULKRABEK, P., LEFFLER, S., WEINSTEIN, I.B. and GRUNBERGER, D.
    (1977). Biochemistry **16**, 3127.

PULLMAN, A. and PULLMAN, B. (1955). Advances in Cancer
    Research **3**, 117.

SAYER, J.M., YAGI, H., SILVERTON, J.V., FRIEDMAN, S.L., WHALEN,
    D.L. and JERINA, D.M. (1982). Journal of the American
    Chemical Society **104**, 1972.

SAYER, J.M., WHALEN, D.L., FRIEDMAN, S.L., PAIK, A., YAGI, H.,
    VYAS, K.P. and JERINA, D.M. (1984). Journal of the American
    Chemical Society **106**, 226.

SAYRE, D. and FRIEDLANDER, P.H. (1960). Nature (London) 187, 139.

SHAKKED, Z., RABINOVICH, D., CRUSE, W.B.T., EGERT, E., KENNARD, O., SALA, G., SALISBURY, S.A. and VISWAMITRA, M.A. (1981). Proceedings of the Royal Society of London B213, 479.

SIMS, P. and GROVER, P.L. (1974). Nature (London) 252, 326.

STEZOWSKI, J.J., STIGLER, R-D., JOOS-GUBA, G., KAHRE, J., LOSCH, G.R., CARRELL, H.L., PECK, R.M. and GLUSKER, J.P. (1984). Cancer Research 44, 5555.

STEZOWSKI, J.J., JOOS-GUBA, J., GLUSKER, J.P., REILY, M.D. and MARZILLI, L.G. (manuscript in preparation).

SWENSON, D.H., MILLER, J.A. and MILLER, E.C. (1973). Biochemical and Biophysical Research Communications 53, 1260.

TAYLOR, E.R., MILLER, K.J. and BLEYER, A.J. (1983). Journal of Biomolecular Structure and Dynamics 1, 883.

WANG, A. H-J., QUIGLEY, G.J., KOLPAK, F.J., CRAWFORD, J.L., VAN BOOM, J.H., VAN DER MAREL, G. and RICH, A. (1979). Nature (London) 282, 680.

WING, R., DREW, H., TAKANO, T., BROKA, C., TANAKA, S., ITAKURA, K. and DICKERSON, R.E. (1980). Nature (London) 287, 755.

YAMAGIWA, K. and ICHIKAWA, K. (1918). Journal of Cancer Research 3, 1 (1916 Mitt. Fak. Univ. Tokyo, 15, 296).

ZACHARIAS, D.E., GLUSKER, J.P., HARVEY, R.G. and FU, P.P. (1977). Cancer Research 37, 775.

ZACHARIAS, D.E., GLUSKER, J.P., FU, P.P. and HARVEY, R.G. (1979). Journal of the American Chemical Society 101, 4043.

ZACHARIAS, D.E., KASHINO, S., GLUSKER, J.P., HARVEY, R.G., AMIN, S. and HECHT, S.S. (1984). Carcinogenesis 5, 1421.

ZACHARIAS, D.E., GLUSKER, J.P., MOSCHEL, R. and DIPPLE, A. (manuscript in preparation).

24. Molecular strain and carcinogenic activity: the shape of methylated benz[A]anthracenes

Derry W. Jones, Clive E. Briant and Julian D. Shaw

1   INTRODUCTION

Of all the classes of known chemical carcinogens, the polycyclic aromatic hydrocarbons (PAH), which are widely distri-
buted in the environment, were among those identified
earliest and studied most, both _in vivo_ and _in vitro_ (Dipple
_et al_. 1984).   The monomethylbenz[a]anthracenes (MBA) and
dimethylbenz[a]anthracenes (DMBA), formerly regarded as
benz-1:2-anthracenes with differently numbered substitution
sites, represent series of isomers and closely related compounds embracing a wide range of carcinogenic activity (Jones
& Matthews 1974; Wislocki _et al_. 1982) and somewhat different
molecular shapes and dimensions (Jones _et al_. 1978; Briant
& Jones 1985).   Their structure-activity relationships have
thus been subjected to considerable study.

The carcinogenic activity of benz[a]anthracenes (BA) can
be modified by substituting with electron-donating or
electron-withdrawing groups. Methyl substitution of BA at
positions 7, 8 or 12 is particularly effective at enhancing
activity (Jones & Matthews 1974), while substitution at
1,2,3 or 4 in the A or benz ring yields compounds of low
activity, whether MBA or DMBA (except for 4,5-DMBA) (Dipple
_et al_. 1984).

Changes in substituent position will evidently influence

electronic molecular properties (Lowe & Silverman 1984),
especially in the so-called phenanthrenic K region.   In
solution, ^{1}H NMR coupling constants can probe electron distributions, while chemical shifts of what we have described
(Bartle et al. 1966;   Bartle & Jones 1972) as bay and penin-
sular aromatic and methyl hydrogens can sense confrontation
effects across the angular bay region of BAs.   For PAH,
several aspects of molecular shape and dimensions, including
degree of planarity, overall encumbrance area, thickness, and
analogy with steroids, have been proposed as contributory
factors in determination of carcinogenic potency.   Accordingly, diffraction studies of crystal structures can aid
investigations of the mechanism of chemical carcinogenicity
among PAH (Glusker 1985).

We have determined by direct methods the crystal structures of a number of MBA, including those with substituents
at the K-region carbons C(5) & (6) (the moderately carcinogenic 5-MBA and 6-MBA), and also the strongly carcinogenic
(Cavalieri and Rogan 1985) related compound 3,6-dimethylcho-
lanthrene (3,6-DMC);   thus previous comparisons of the steric
effect of substituents on the shapes and dimensions of MBA
and DMBA molecules (Briant et al. 1985) can be extended.

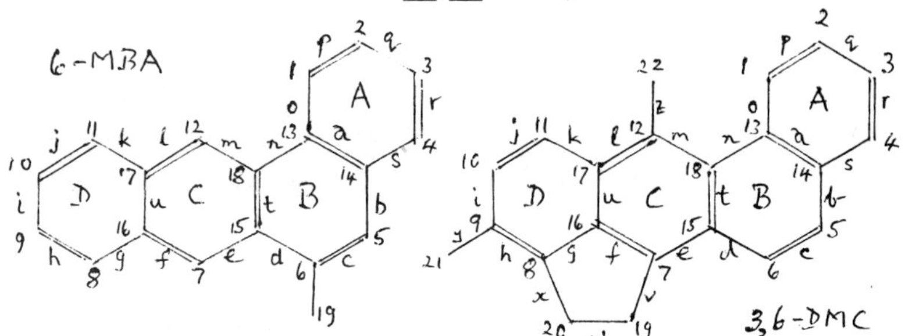

2.   RESULTS AND DISCUSSION

2.1   Molecular Strain

The overall molecular shapes in methyl-substituted BA structures can be divided into two main types (Briant et al. 1985)

<u>I</u>:   BA that have no substituent in the bay region and so are almost planar (i.e. 2-, 5-, 6-, 8-, 11-MBA, 5-MeO,7-MBA, 3,9-DMBA; and

<u>II</u>:   BA with substituents at positions 1 and/or 12, and thus sterically hindered and non-planar (i.e. 1-, 12-MBA, 7,12-, 1,12-DMBA and 3,6-DMC), with the benzo(A) ring bent out of the BCD anthracene plane.   For type I, angle A/D, a measure of the extent of buckling, ranges from about 2°(2-MBA, 11-MBA) to 7°(3,9-DMBA, 5-MBA), whereas for type II angle A/D ranges from about 21°(1-MBA, 12-MBA) through 24°(7,12-DMBA) to 29°(1,12-DMBA).   Some of these are shown in Table 1 in

TABLE 1

Angles($^{\circ}$) between least-square planes of individual benzo-rings in some type I and II BAs (e.s.d.s typically 0.3°)

Compound	Type	A/B	A/C	A/D	B/C	B/D	C/D
5-MeO,·7-MBA	I	1.3	1.5	0.8	0.5	1.0	0.9
11-MBA	I	0.8	0.8	1.8	0.1	1.5	1.4
6-MBA	I	1.4	1.6	1.7	0.5	0.7	0.3
8-MBA	I	1.9	4.6	5.1	3.2	4.1	1.5
		1.8	1.9	2.4	0.6	3.6	2.3
5-MBA	I	2.4	4.6	6.2	2.9	4.6	1.7
		2.0	3.1	3.4	2.6	3.3	0.8
3,6-DMC	(II)	8.9	16.3	15.7	7.9	7.4	0.8
1-MBA	II	8.9	18.7	21.8	10.2	13.0	4.5
12-MBA	II	10.7	19.2	20.1	8.7	9.5	3.7
7,12-DMBA*	II	10.9	21.2	24.0	10.5	13.2	5.0
1,12-DMBA	II	14.8	28.9	29.3	13.5	14.6	2.6

*D.E. Zacharias <u>et</u> <u>al</u>. personal communication

approximate sequence of increasing angle A/D.   When the methyl substituent is one atom distant from the bay region, deviations from planarity are already small.   Thus the distances of the methyl carbon from the mean BA plane in 5-, 6- and

TABLE 2

Bond Lengths (Å) of some benz[a]anthracene (BA) structures

(e.s.d.s typically 0.005 Å)

	Bond	mean 5-MBA	6-MBA	8-MBA		5-MeO, 7-MBA	Type-I mean	Type-II 3,6-DMC
K	a	1.420	1.402	1.399	1.397	1.409	1.409	1.415(6)
	b	(1.473)	1.450	1.439	1.445	(1.448)	1.440	1.446(7)
	c	(1.328)	(1.337)	1.345	1.345	(1.354)	1.338	1.331(7)
	d	1.477	(1.450)	1.434	1.440	1.445	1.448	1.445(6)
	e	1.368	1.380	1.402	1.393	(1.394)	1.384	1.380(6)
	f	1.396	1.402	1.390	1.390	(1.420)	1.395	(1.385(6))
	g	1.431	1.411	(1.443	1.439)	1.428	1.424	(1.412(6))
	h	1.360	1.342	(1.353	1.348)	1.356	1.354	(1.362(7))
	i	1.400	1.414	1.410	1.402	1.414	1.411	(1.426(7))
	j	1.359	1.359	1.361	1.368	1.365	1.359	1.366(7))
	k	1.418	1.432	1.415	1.407	1.436	1.421	1.438(6)
	l	1.421	1.402	1.411	1.419	1.401	1.409	(1.421(6))
	m	1.408	1.391	1.372	1.379	1.394	1.391	(1.418(6))
beach	n	1.475	1.461	1.476	1.471	1.457	1.467	1.480(6)
	o	1.383	1.473	1.410	1.394	1.422	1.406	1.413(7)
N	p	1.417	1.372	1.389	1.393	1.371	1.387	1.380(7)
	q	1.363	1.378	1.397	1.369	1.393	1.380	1.383(8)
M	r	1.425	1.405	1.361	1.362	1.364	1.382	1.369(8)
	s	1.392	1.396	1.405	1.414	1.410	1.402	1.420(7)
	t	1.413	1.444	1.421	1.417	1.440	1.427	1.457(6)
	u	1.427	1.438	1.423	1.422	1.418	1.422	1.409(6)
C-Me		1.511	1.533	1.493	1.500	1.520	1.507	1.522(8)
							1.507	1.517(6)
C-O						1.374	1.374	
						1.428	1.428	

BA bonds in parentheses are at substituent points and excluded from type I means

8-MBA and 5-MeO, 7-MBA are, respectively, 0.12, 0.05, 0.22 and 0.03Å, while the carbon atom skeleton in these compounds is more nearly planar than reported in the BA:PMDA complex (Foster et al. 1976).

Both molecules in the asymmetric unit of 8-MBA are slightly bowed, as in 3,9-DMBA (Briant and Jones 1985), with

A/D 5.1° and 2.4°; 5-MBA is marginally more bowed, with inter-
plane inclination A/D more pronounced (A/D 6.2° and 3.4°).
Evidently, the combination of crystal packing variations and
small strains arising from differences in substitution position
can cause 4-5° bowing in nominally planar BAs. 3,6-DMC (A/D
15.7°), expected to be an intermediate type-I/type-II because
of the inclusion of the 5-membered ring, has an overall shape
closer to that of 12-MBA (A/D 20.1°) than 7,12-DMBA (A/D 24.0°).

2.2   Molecular Dimensions in Regions Relevant to
       Carcinogenesis

Bond lengths (typical e.s.d.s about 0.05Å) and angles (e.s.d.s
about 0.3°) for 5-, 6- and 8-MBA, 5-MeO,7-MBA (all type I) and
3,6-DMC are given in Tables 2 and 3. Substitution in BA of
methyl at one or both bay-region sites (1 & 12) not only
distorts the BA skeleton but concomitantly causes enlargement
of the beach bay bond, n (1.484 Å for 1,12-DMBA compared with
the mean n for type I of 1.467 Å) and its torsion angle and
also of the distance(s) of 1- and/or 12-methyl carbons from
the mean molecular BA plane.

     K-region bonds C(5)-C(6) (c) in 2-, 5-,6-, 12-MBA, and
3,9- and 7,12-DMBA are all within 1.332-1.336 Å; those in 6-MBA
(1.332(6)Å) and 5-MBA (1.320(6)Å) are the shortest C-C bonds in
each molecule, shorter than in phenanthrene but much as in
other MBA. Bonds C(8)-C(9) = 1.339(7)Å and C(10)-C(11) =
1.360(7)Å in ring D of 6-MBA are also quite short.

     In the bay region, the long beach bonds C(13)-C(18) (n) of
5-, 6-, and 8-MBA are all very close to the mean (1.467 Å);
1,12-DMBA, with two bay-region methyl substituents, has the
longest (0.013 Å more than the mean) of any MBA or DMBA, just
as long as in 3,6-DMC. Comparison of bond lengths in the two
K-region structures 5- and 6-MBA reveals a surprisingly large
difference for one of the bay-region bonds, C(13)-C(1) (o); in

METHYLATED BENZ[A]ANTHRACENES

TABLE 3

Bond angles ($^\circ$) of some BA structures (e.s.d.s typically 0.3°)

Bond angle	5-MBA		6-MBA	8-MBA		5-MeO 7-MBA	Mean type-1	Type II 3,6-DMC
op	120.8	120.8	117.1	120.0	120.8	120.9	120.9	121.9
pq	120.0	119.6	124.1	119.1	119.8	121.3	120.8	119.7
qr	120.7	121.1	119.0	121.9	120.8	119.8	120.5	120.9
rs	119.0	118.6	119.9	119.5	120.6	119.9	120.7	120.0
bc	(120.4	120.5)	123.2	121.0	121.0	(122.4)	121.9	120.2
cd	122.8	121.8	(120.3)	121.1	121.1	121.7	121.4	122.3
ef	121.5	121.5	123.3	122.0	122.4	(120.0)	122.1	120.0
gh	120.0	119.7	121.7	(118.9	118.8)	122.1	122.3	120.1
hi	121.6	121.1	120.3	122.3	122.6	120.7	121.3	(117.3)
ij	120.5	120.4	120.6	120.4	120.2	119.5	120.8	123.5
jk	121.0	120.7	121.0	119.9	119.7	121.0	120.7	120.3
lm	120.4	120.2	121.8	122.2	122.2	121.3	121.7	(119.5)
on	121.3	121.4	121.5	121.7	122.1	123.1	122.4	123.5
ao	118.8	119.3	118.3	119.4	118.9	116.6	118.2	117.0
na	119.9	119.4	120.2	119.0	118.9	120.2	119.4	119.4
sb	(120.4)	(119.5)	120.2	119.6	120.6	120.6	120.6	119.3
sa	120.7	120.6	121.4	120.0	119.1	121.5	120.0	120.0
ab	119.0	120.0	118.4	120.4	120.3	117.9	119.4	120.7
de	120.6	120.3	(122.3)	120.3	120.6	122.1	120.9	120.8
dt	118.7	119.4	118.9	120.1	119.7	117.4	119.1	119.8
et	120.7	120.3	118.7	119.5	119.7	120.5	119.9	119.3
fg	122.3	121.1	123.3	123.2	123.1	123.3	122.7	113.3
fu	119.4	119.7	117.6	118.1	118.2	119.5	118.7	122.9
gu	118.3	119.2	119.1	118.7	118.7	117.2	118.7	123.8
kl	122.3	122.7	123.0	120.8	121.0	120.5	121.9	126.8
ku	118.6	119.0	117.4	119.9	120.0	119.5	118.9	114.9
lu	119.1	118.3	119.6	119.3	118.9	120.0	119.2	118.2
mn	122.0	121.2	122.2	123.1	122.6	121.1	122.4	124.1
mt	118.8	119.9	118.9	118.6	118.5	118.7	118.6	119.5
nt	119.2	118.9	118.9	118.3	118.9	120.2	119.0	116.3

Angles in 3,6-DMC involving bonds e,f,g,h,k,l,u,v,w,x are particularly affected by five-membered ring. BA angles in parentheses are at substituent points and are excluded from BA type I mean.

6-MBA it is 1.473 Å, appreciably larger than the mean for type
I (1.406 Å) and much larger than the $\sim$1.383 Å in 5-MBA.   The
adjacent bond C(1)-C(2) (p) at 1.372 Å is rather smaller than
the mean (1.387 Å) but appreciably less than 1.417 Å in 5-MBA.
^{1}H NMR chemical shifts of H(1) measured in solution (Jones and
Mokoena 1982) are not very different for the two compounds.
Corresponding bond lengths in 5- and 6-MBA typically differ by
0.01-0.02 Å but by $\sim$0.03 Å for C(5)-C(15) (d) and C(15)-C(18)
(t).   In 5-, 6-, and 8-MBA, all mn bay bonds fall marginally
below the mean (122.4°) while the no bonds in 5- and 6-MBA are
close to the mean; they are slightly larger in 8-MBA and over
3° larger in 1,12-DMBA.

At the M-region, C(3)-C(4) is longer in both 5- and 6-MBA
(1.424/1.426 and 1.407 Å) than the mean for type I; bonds in
8-MBA and 1,12-DMBA (1.361/1.362 and 1.360 Å) are shorter than
the mean of 1.38 Å.   For the N-region, C(1)-C(2) in 5-MBA is
long (1.421/1.412 Å) and bonds in 6- and 8-MBA and in 1,12-DMBA
fall closer to the mean (1.387 Å).   3,6-DMC has an acenaphthene moeity similar to that in acenaphthene (Hazell et al.
1986); away from the regions of methyl-substitution and pent-
acyclic-ring formation the bond lengths in 3,6-DMC conform
surprisingly closely to those of an MBA.   Thus, the K-region
bond C(5)-C(6) = 1.331(7) Å is typical of those in both type I
(nearly planar) and type II (non-planar) MBA.

At the sites of methyl substitution in MBA, interior ring
C-C-C angles are generally 2° (1° for gh in 8-MBA) less than
120°, while adjacent bond angles are increased by $\sim$1.0-1.6°
(Briant et al.1985) (122.5-123° for hi and fg in 8-MBA (Walker
et al. 1986)). For 5- and 6-MBA, however, the endocyclic
angles bc and cd, respectively, exceed 120° while ef at the
7-methyl site in 5-MeO,7-MBA is 120°.   These relatively large
C-C-C angles may be associated with greater rigidity near the
higher-bond-order K-region.

In most MBA, the C-C bonds to the methyl carbon are very close to 1.50 Å long, marginally longer (1.51 Å in 1,12-DMBA) if the methyl group is hindered.   Unhindered C-C methyl bonds tend to be relatively long in 5-MBA, 5-MeO,7-MBA and, especially, 6-MBA (1.53 Å).

We thank Professor M.S. Newman for samples, Dr. M.B. Hursthouse and the S.E.R.C. for data, and the Yorkshire Cancer Research Campaign for financial support.

REFERENCES

BARTLE, K.D. and JONES, D.W. (1972). Advances in Organic Chemistry 8, 317-423.

BARTLE, K.D., HEANEY, H., JONES, D.W., and LEES, P. 1966. Spectrochimica Acta 22, 941-951.

BRIANT, C.E. and JONES, D.W. (1985).   Cancer Biochemistry and Biophysics 8, 129-136.

BRIANT, C.E., JONES, D.W. and SHAW, J.D. (1985). Journal of Molecular Structure 130, 167-176.

CAVALIERI, E.L. and ROGAN, E.G. (1985).   In Polycyclic hydro-carbons and carcinogenesis, A.C.S. Symposium Series No. 283 (ed. R.G. Harvey) pp. 289-305 Amer. Chem. Soc., Washington, D.C.

DIPPLE, A., MOSCHEL, R.C., and BIGGER, C.A.H. (1984).   In Chemical carcinogenisis, Vol. 1. (2nd edn.) A.C.S. Monograph 182, (ed. C.E. Searle), pp. 47-174. Amer. Chem. Soc., Washington, D.C.,

FOSTER, R., IBALL, J., SCRIMGEOUR, S.N., and WILLIAMS, B.C. (1976). Journal of the Chemical Society, Perkin 2, 682-685.

GLUSKER, J.P. (1985).   In Polycyclic hydrocarbons and carcino-genesis. A.C.S. Symposium Series No. 283 (ed. R.G. Harvey) Amer. Chem. Soc., Washington, D.C., pp. 125-185.

HAZELL, A.C., HAZELL, R.G., NØRSKOV-LAURITSEN, L., BRIANT, C.E., and JONES, D.W. (1986). Acta Crystallographica C42, 690-693.

IBALL, J. (1938) Zeitschrift für Kristallographie 99, 230-231

JONES, D.W. and MATTHEWS, R.S. (1974). Progress in Medicinal Chemistry 10, 159-203.

JONES, D.W. and MOKOENA, T.T. (1982). Spectrochimica Acta 38A, 491-4.

JONES, D.W., SOWDEN, J.M., HAZELL, A.C., and HAZELL, R.G. 1978). Acta Crystallographica B34, 3021-3026.

LOWE, J.P. and SILVERMAN, B.D. (1984) Accounts of Chemical Research 17, 332-338.

MASON, R. (1956). Naturwissenschaften 43, 252-253.

WALKER, N.P.C., BRIANT, C.E., JONES, D.W. and SHAW, J.D. Acta Crystallographica C42, 1392-1395.

WISLOCKI, P.G., FIORENTINI, K.M., FU, P.P., YANG, S.K. and LU, A.Y.H. (1982). Carcinogenesis 3, 215-217.

25. Steric and electronic features of antitumor anthracyclines

Graziella Ranghino, Camillo Tosi, Vincenzo Malatesta and Nereo Sacchi

1.   INTRODUCTION

The anthracyclines comprise a small family of antibiotics that were found to be active as antineoplastic drugs some fifteen years ago and are used in the treatment of a wide spectrum of human tumours (Arcamone, 1981).

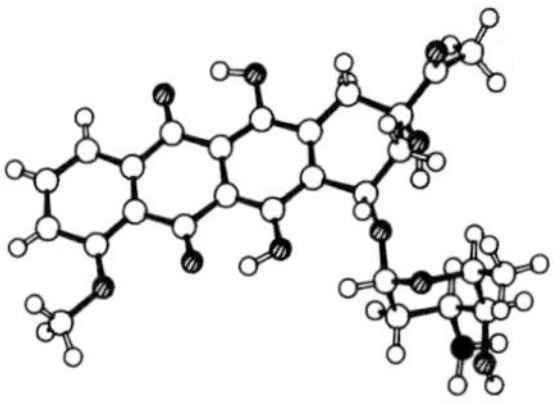

FIG. 1. Computer-drawn projection (with the program SCHAKAL, courtesy of E. Keller, University of Freiburg, FRG) of dauno-mycin, in the α-conformation of the A ring.

The structure of a representative molecule (daunomycin) is shown in Fig .1: it has a bulky group of four fused rings (daunomy-cinone) and an amino sugar (daunosamine). The daunomycinone is

almost flat and is able to intercalate into the DNA double helix, as it has been shown by many binding studies and X-ray
crystallography (Neidle, 1978; Quigley _et al_, 1980). The mode
of action in vivo is far from being established, although it is
clear that the ability of binding to DNA and the biological activity are somehow related. An important point in understanding
structure - activity - reactivity relationships is that the
molecule loses its activity if the glycosidic bond C7-O7 splits.
The mechanism of detachment of the daunosamine is closely related to the reductive activation of this kind of molecules, and
in fact electrochemical experiments have shown that anthracyclines undergo reductive splitting of the sugar according to
the scheme of Fig. 2.

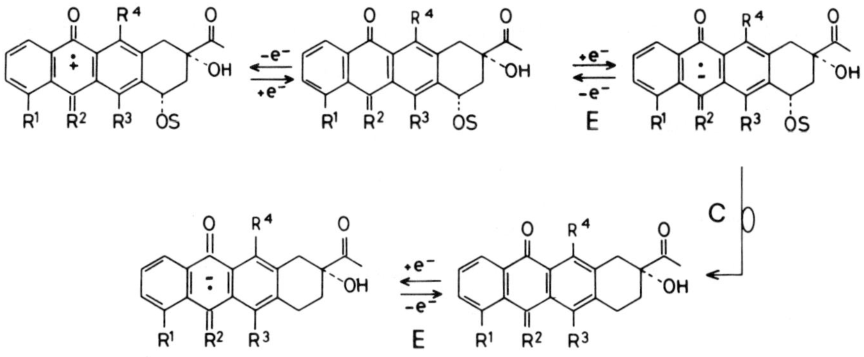

FIG. 2.   Schematic representation of the reductive activation
process of anthracyclines.

This redox pathway suggests a correlation between toxicity (in
particular cardiotoxicity) and oxidation - reduction properties.
It has been pointed out (Malatesta _et al_, 1984) that the an-

thracyclines can be easily reduced and in this way interfere with the electron transfer processes (e.g. respiratory chain), and also that the easy formation of radicals by these molecules can lead to the formation of toxic species such as the OH radical (Davies and Doroshow, 1986; Doroshow and Davies, 1986). The goals of our study at this stage are the following:

a) To find a correlation between substitution on the four fused rings and conformations of the A ring and the daunosamine on one hand and redox potentials on the other.

b) To predict the occurrence of reductive deglycosidation.

c) To find a rationale that could possibly explain activity and toxicity in terms of electronic properties.

2.   RESULTS AND DISCUSSION

We have performed an extensive conformational analysis study on daunomycin and a small series of substituted analogs (Tosi et al, 1986; Penco et al, 1986). The reductive splitting appears to be under stereoelectronic control, i.e. a repulsive interaction between orbitals on C6a and O7, which facilitates the splitting, is observed only for a particular pucker (α) of the partially saturated A ring and not for the other (β) (Fig. 3).

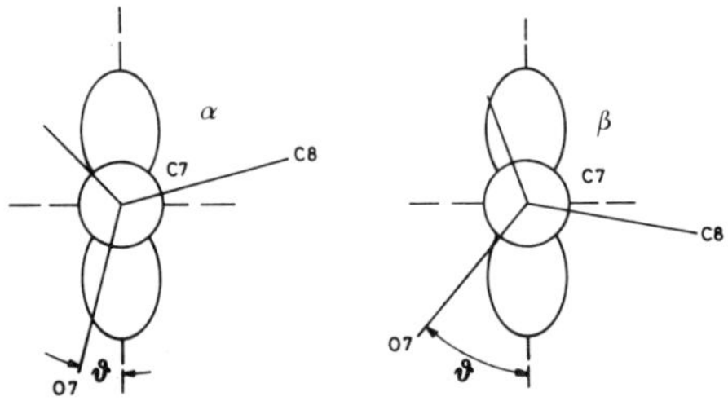

FIG. 3. Newman projections along the C7-C6a bond, showing the dihedral angle ϑ between the C7-O7 bond and the π-orbitals.

Moreover, the conformation of the A ring depends on the substituents on the other three rings, particularly on the presence of both the hydroxyl group O-H at position 6 and the carbonyl group C=O at position 5 (cf. Tosi et al, 1986, Table 1). Another important observation is that, because of the extensive delocalization of the anthraquinone π-system, daunomycinone is very sensitive to any substitution, in the sense that the whole electronic structure of the fused rings is perturbed.

If we thus consider the frontier molecular orbitals, we have that ca. 80% of the electronic cloud of the HOMO and the LUMO comes from the atomic functions shown in Fig. 4 .

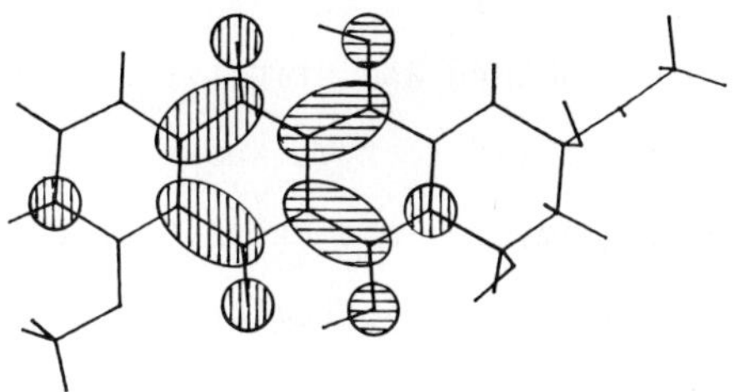

FIG. 4. Composition of the HOMO (horizontal hatching) and the LUMO (vertical hatching) of daunomycin.

We can correlate the eigenvalues of HOMO and LUMO with oxidation and reduction potentials, resp. As shown in Tables 1 to 3 of Ranghino et al (1987), the reduction potentials and the HOMO -LUMO gap correlate quite well: this parameter might give useful information on the level of insertion of a molecule into a given electron transfer process . The oxidation potential of 5-iminodaunomycin does not fit into the experimental observations, but this has an explanation in terms of different protonated or H-bonded species which have different stabilities and different redox potentials according to the pH of the solution.

The removal of oxygen at position 4 allows an easier detachment

of the sugar, due to the enhancement of the partial charge on

C6a that, in turn, enhances the repulsive interaction between

orbitals on C6a and O7. The removal of O6, on the contrary,

facilitates the retention of the sugar, since it stabilizes the

β conformation, as seen from semiempirical and molecular me-

chanics calculations (Tosi et al, 1986). Also the energy of

the glycosidic bond C7-O7 is consistent with the experimental

observations, as seen in Table 3 of Ranghino et al (1987).

In order to shed some light onto the problem of toxicity, in

particular cardiotoxicity, we focussed our attention on the dif-

ferences, both in structure and stability, between daunomycin

and its 5-imino derivative that is known to be active (though

less potent than daunomycin) and exhibits little or no tendency

to undergo reduction during biochemical experiments (Davies and

Doroshow, 1986; Doroshow and Davies, 1986). The main observa-

tion on the electronic structure is that N5 contributes both to

the LUMO and the HOMO, thus indicating a different reactivity

of this portion of the aglycone; moreover, probably different

forms are stable at different pH values and these appear to have

different redox potentials. In order to evaluate the stability

of the fully reduced species, we also carried out ab initio cal-

culations on the doubly charged anions of daunomycinone and its

5-imino derivative (Malatesta et al, 1987). The two species dif-

fer by only 35 and 46 kcal/mol from the parent compounds. The

HOMO of the fully reduced species coincides almost completely

with the LUMO of the fully oxidized species .Furthermore, the

fact that the hydroquinone from the 5-iminodaunomycinone is much less stable than the parent compound, i.e .daunomycinone, adds further theoretical rationale to the fact that the daunomycinone is more prone to undergo reductive activation.

REFERENCES

ARCAMONE, F. (1981), in "Doxorubicin .Anticancer Antibiotics" (Medicinal Chemistry, Vol. 17), Academic Press, New York.

DAVIES, K.J.A. and DOROSHOW, J.H. (1986) J. Biol. Chem. 261, 3060.

DOROSHOW, J.H. and DAVIES, K.J.A. (1986) J. Biol. Chem. 261, 3068.

MALATESTA, V., PENCO, S., SACCHI, N., VALENTINI, L., VIGEVANI, A. and ARCAMONE, F. (1984) Can. J. Chem. 62, 2845.

MALATESTA, V., RANGHINO, G. and TOSI, C. (1987) to be published.

NEIDLE, S. (1978), in "Topics in Antibiotic Chemistry", Vol. 2, Part D, P. Sammes Ed., Ellis Horwood, Chichester.

PENCO, S., VIGEVANI, A., TOSI, C., FUSCO, R., BORGHI, D. and ARCAMONE, F. (1986) Anti-cancer Drug Design 1, 161.

QUIGLEY, G.J., WANG, A.H.J., UGHETTO, G., VAN DER MAREL, G., VAN BOOM, J.H. and RICH, A. (1980) Proc. Natl. Acad. Sci. USA 77, 7204.

RANGHINO, G., TOSI, C., MALATESTA, V. and SACCHI, N. (1987) Theochem, in press.

TOSI, C., FUSCO, R., RANGHINO, G. and MALATESTA, V. (1986) Theochem 134, 341.

26. Small molecules as potential anticancer agents

V. L. Narayanan

1. INTRODUCTION

The discovery of new small molecular weight anticancer drugs
is a very challenging task involving multiple approaches--
rational design, screening and analog synthesis (Narayanan,
1984). The majority of the new anticancer leads are discovered through screening of compounds of a wide variety of
structural and biological types against animal tumor systems.

2. ANTITUMOR SCREENING EFFICIENCY

Since screening to discover new anticancer leads is a
low-yielding process, several strategies have been developed
to improve screening efficiency.

2.1. Preselection

We use a variety of criteria to select compounds for
screening. We extensively scan the literature to identify a
rational basis for selecting compounds for screening, e.g.,
antitumor activity reported in other screening programs,
compounds reported to exhibit relevant biochemical and other
biological activities and structural features suggestive of
effects on enzymatic processes relevant to cancer chemo-

therapy. However, by and large, the majority of compounds come only with structural information. We have developed a computerized model that aids us in selecting compounds for screening based on structural information alone (Hodes, 1981). The model prospectively assigns "activity" and "novelty" scores to compounds utilizing "training sets" tested against both leukemias and solid tumors.

2.2. Structure-Activity Analysis

Detailed structure-activity analyses based on our large chemical-biological database are an integral part of our acquisition and screening activities. Such analyses allow us to select compounds for screening maximizing structural uniqueness and anticancer potential. It also allows us to identify gap areas for further research (Nasr, 1985).

2.3. Congener and "Pro-Drug" Synthesis

Often the development of a screening lead is hampered by problems of lack of water-solubility, stability, bioavailability and potency. These problems can be solved through the judicious synthesis of congeners and "pro-drugs" of the parent lead. The parameters being changed are molecular size, steric shape, bond angles, hybridization, electron distribution, lipid solubility, water solubility, pKa, the chemical reactivity to cell components and metabolizing enzymes, and the capacity to undergo receptor interactions through hydrogen bonding.

3. SMALL MOLECULAR WEIGHT NEW ANTICANCER LEADS

Figure 1 illustrates examples of anticancer leads that

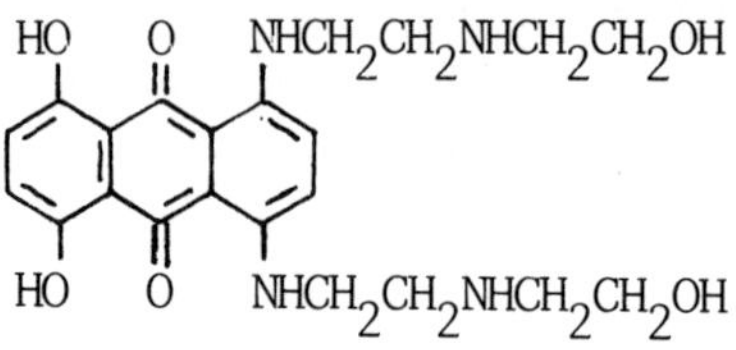

Mitoxantrone

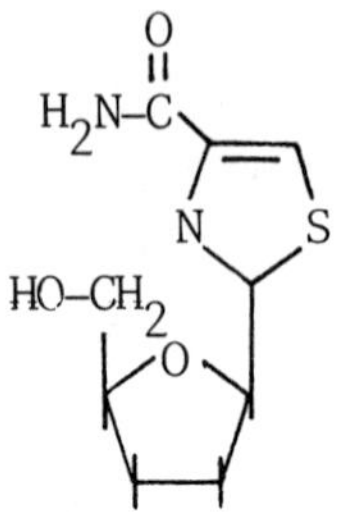

Tiazofurin

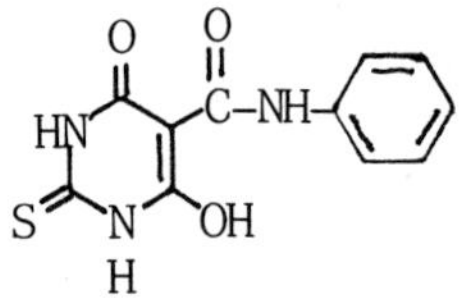

Merbarone

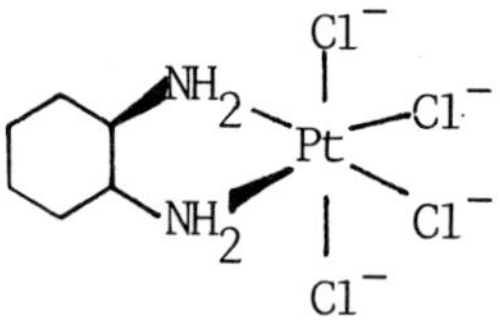

Tetraplatin

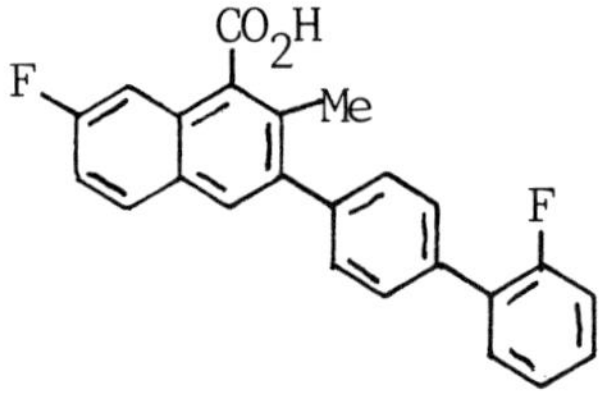

Biphenquinate

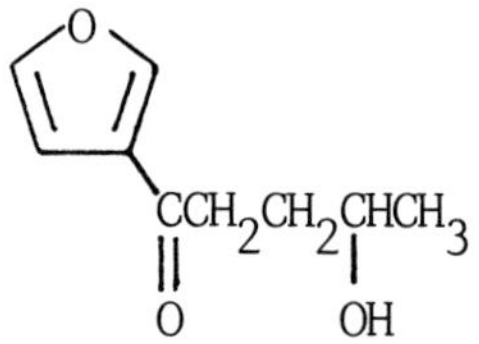

Ipomeanol

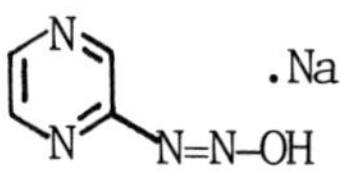

Pyrazine diazohydroxide

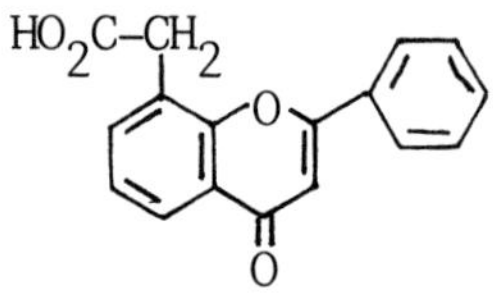

Flavoneacetic acid

Fig. 1 Chemical Structures of New Anticancer Leads

are under active development. A brief description of the
compounds are given below:

<u>MITOXANTRONE (DHAQ, DIHYDROXYANTHRACENEDIONE)</u>. The high
antineoplastic activity of the lead compound, 1,4-bis(2-[(2-
hydroxyethyl)amino]ethylamino)-9,10-anthracenedione, made it
an obvious candidate for analog studies, from which emerged
dihydroxyanthracenedione (DHAQ). DHAQ demonstrated outstand-
ing anticancer activity against P388 and L1210 leukemias, B16
melanocarcinoma and colon 26 in experimental animals. Excellent tumor responses have been achieved in clinical trials
against metastatic breast cancer and adult and pediatric
leukemias.

<u>TIAZOFURIN</u>. Tiazofurin has shown remarkable activity against
Lewis Lung Carcinoma in mice producing long-term survivors at
several dose levels. The compound is of considerable
biochemical interest since it has been shown to be a specific
inhibitor of IMP-dehydrogenase. Tiazofurin is in Phase I/II
clinical trials.

<u>MERBARONE</u>. In contrast to other barbiturates, Merbarone has
shown exceptional antitumor activity against the L1210
leukemia as well as good activity against the B16 melanoma
and M5076 sarcoma. However, the compound is only marginally
active against intracerebrally implanted tumor; the absence
of 5,5-disubstitution contributes to a pKa of 4.0 and renders
it highly ionized at physiological pH. The compound is in
Phase I clinical trials.

<u>TETRAPLATIN</u>. Tetraplatin is a new platinum IV coordination
complex with octahedral geometry in contrast to the square
planar cisplatin. The compound, unlike several other

platinum complexes, is stable, water-soluble, can be obtained
in good purity, and is equal in potency to cisplatin.
Interestingly, tetraplatin shows activity against a subline
of L1210 leukemia which has acquired resistance to cisplatin
even to the point of producing long-term survivors. Further
studies are underway.

BIPHENQUINATE. Biphenquinate is being developed jointly by
Dupont Co. and NCI. The compound is active in the subrenal
capsule assay against the human LX-1 lung, MX-1 mammary,
BL/STX-1 stomach tumors and the human HCT-15 colon tumor
implanted subcutaneously in nude mice. Biphenquinate is in
Phase I clinical trials.

IPOMEANOL. Ipomeanol is a natural product isolated from
moldy sweet potatoes. This furan derivative undergoes
metabolic activation by the P450 system of the pulmonary
clara cells, covalently binds to these cells and acts as a
specific toxin to the lung clara cells. Ipomeanol is being
developed as a specific anticancer agent for tumors derived
from the clara cells which are the stem cells for the
bronchio-alveolar tissue.

PYRAZINE DIAZOHYDROXIDE. From several congeners of the lead
compound, pyridine diazohydroxide, that were synthesized and
evaluated, pyrazine diazohydroxide was selected for further
development on the basis of 1) extended chemical stability
(t 1/2 = ca. 100 min at pH 7.4 in contrast to 1.7 min for the
lead compound), 2) good antitumor activity, and 3) ease of
synthesis and purification. The compound showed good
antitumor activity against two human tumor xenografts and
four murine tumors. Further studies are underway.

FLAVONE ACETIC ACID. Flavone acetic acid inhibited tumor
growth completely in 60-80% of mice with early stage colon
adenocarcinoma 38. The compound also caused regression of
advanced (500 mg) colon 38 tumors. Based on the excellent
activity observed against the highly refractory colon 38
model, flavone acetic acid is undergoing clinical trial.

4.   FUTURE DIRECTIONS

     Recently, the NCI has undertaken a major restructuring
of its anticancer drug screening program in order to enhance
its effectiveness (Boyd, 1984). The goal is to discover
agents which demonstrate selective cytotoxicity towards human
tumor cell lines in order to identify new prototype agents
with high selectvity towards specific human solid tumors.

REFERENCES

BOYD, M.R. (1984). In _Proceedings of the General Motors
Cancer Research Foundation Conference_, Jackson Hole, Wyoming,
September 14-15, Frei and Freireich, eds., Lippincott, NY.

HODES, L. (1981). _Journal of the Chemical Information and
Computer Sciences_ 21, 128-132.

NASR, M., PAULL, K.D., NARAYANAN, V.L. (1985). _Journal of
Pharmaceutical Sciences_ 74, 831-836.

NARAYANAN, V.L. (1984). In _Development in Pharmacology_, 3,
5-22, Reinhoudt, D.N., Connors, T.A., Pinedo, H.M., Van De
Pall, K.W., eds., Nijhoff, Hague.

27. Electron density distributions in covalent and ionic bonds

Jack D. Dunitz

The electron density in a molecular or ionic crystal is very closely similar to the superposition of the densities of the separated atoms. This is the reason why in conventional X-ray analysis such excellent agreement can be obtained between observed structure amplitudes and calculated ones based on standard scattering curves for isolated, spherical, neutral atoms.

Nevertheless, (F_o-F_c)-difference maps calculated at the close of a normal least-squares refinement often show features that can be interpreted, at least qualitatively, in terms of chemical bonding concepts. This is obviously of great importance to everyone interested in molecular structure. Unfortunately, such maps are easily contaminated by the effects of systematic errors and noise, and it is not always easy to distinguish these from genuine bonding effects. To define the pro-crystal (the assemblage of spherical, free-atom densities), atomic positions and displacement parameters should be determined from high-order X-ray data (or alternatively by neutron diffraction) where the scattering by the valency electrons becomes negligible. This means that the displacement parameters should not be too large, and it is therefore always advisable to carry out the diffraction measurements at as low a temperature as possible.

Even when all these conditions are fulfilled, the significant features of difference maps are expected to be only a small fraction of the total density. The experimental intensity measurements from which the F_o coefficients are estimated must therefore be made with great care, and systematic sources of error, such as extinction and multiple reflection effects, must be reduced as far as possible. Problems in the measurement of accurate intensities have been discussed recently by Seiler (1985).

As an alternative to (F_o-F_c)-difference maps, the deformation density can be modelled mathematically and the appropriate parameters refined by least-squares analysis together with the usual positional and displacement parameters. The resulting deformation density is, in principle, free from the effects of vibrational smearing and is often called a "static" deformation density.

In our work in Zürich, we have restricted ourselves to small light-atom structures measured at 80-100 K (liquid N_2). For tetrafluoroterephthalonitrile, $C_8F_4N_2$, at 98 K (Dunitz, Schweizer and Seiler, 1983) the difference map (Fig. 1) shows prominent "bonding density" peaks of 0.5 e.Å^{-3} or more near the mid-points of the C-C and C-N bonds (also for the N lone pair density) but only weak bonding density (circa 0.1 e.Å^{-3}) for the C-F bonds.

Qualitatively similar features are seen in a static deformation map obtained with the same X-ray data by Hirshfeld (1984a), who also applied that the Hellmann-Feynman theorem to show that the experimental density does not satisfy the condition of zero forces on the atomic nuclei. In particular, the

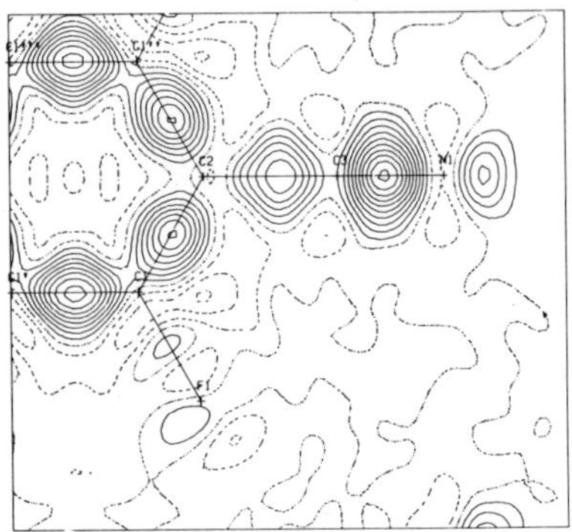

FIG. 1. Tetrafluoroterephthalonitrile: electron-density difference map in the molecular plane. Contour interval
0.075 e.$\text{\AA}^{-3}$.

electric field at the F nucleus was found to be strongly repulsive. When the deformation density was constrained to yield
vanishing electric field at the atomic nuclei, sharp dipoles
(core polarization) were produced at the atomic positions,
coupled with very small changes (less than 0.001$\text{\AA}$) in these
positions, but the rest of the deformation density was virtually unaltered (Fig. 2, Hirshfeld, 1984b). Clearly, the electric field at the nuclear positions is extremely sensitive to
features of the static deformation density close to the nuclei,
features that are not easily determined from the X-ray data
alone. The problem here is not so much the quality of the data
as the strong correlation between the core polarization functions and the atomic coordinates.

For peroxides, difference maps show a density deficit at
the mid-point of the O-O bond (Savariault and Lehmann, 1980;
Dunitz and Seiler, 1983). For example, Fig. 3 shows the difference density along the O-O bond in molecule I (in the crystal

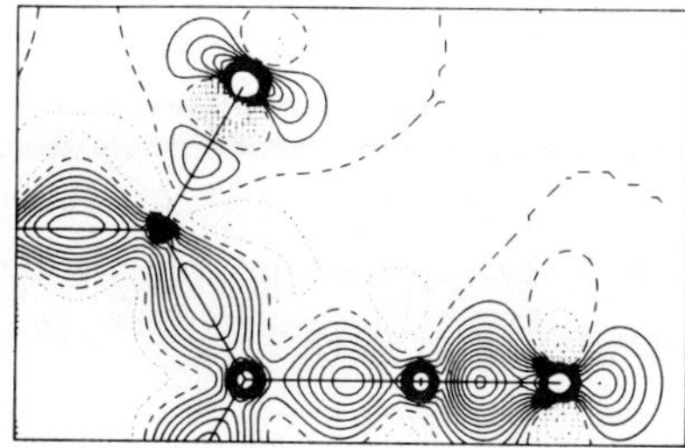 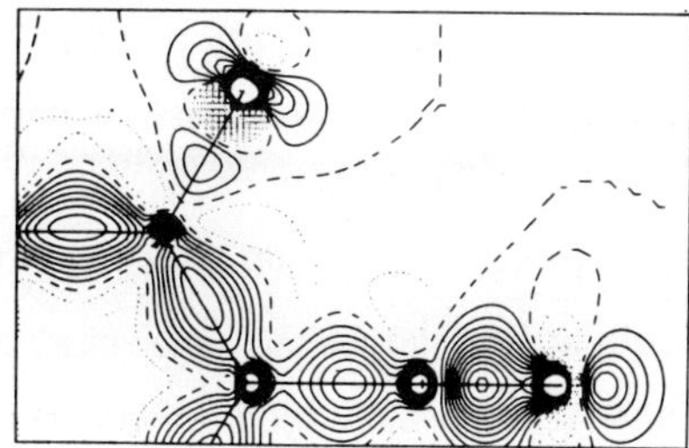

FIG. 2.   Tetrafluoroterephthalonitrile. Static deformation den-
sity in molecular plane from unconstrained (top) and
Hellmann-Feynman constrained (bottom) refinements.
Contour interval 0.1 e.Å^{-3}.

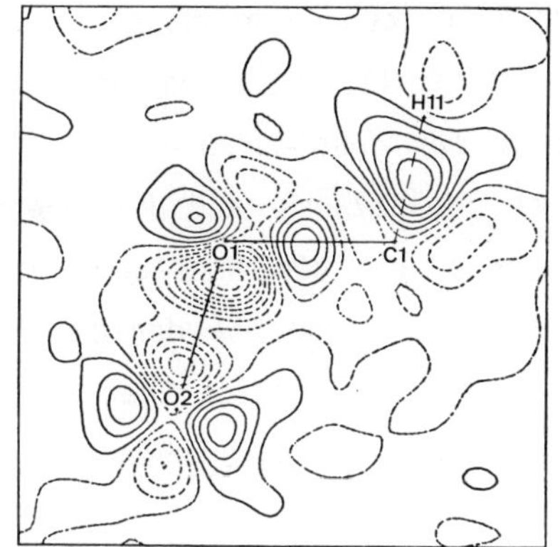 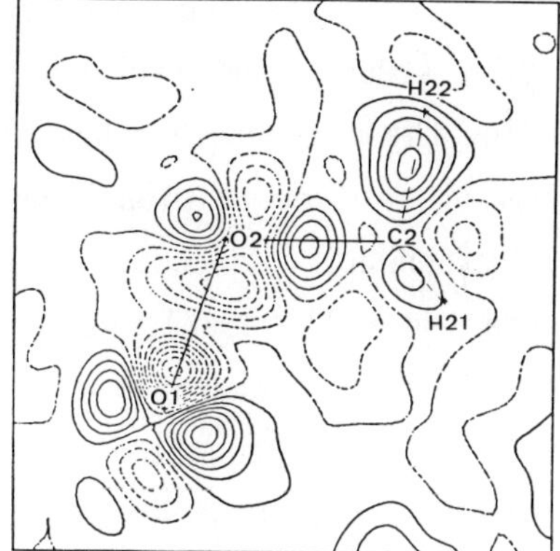

FIG. 3.   Electron-density difference maps in sections passing
through the C-O-O planes of molecule I.   Contour
interval 0.075 e.Å^{-3}, negative contours dashed.

structure of I the molecule has centrosymmetric site symmetry).
The negative density along the bond axis contrasts with the
positive density peaks corresponding to other bonding features,
such as the C-O and C-H bonds, as well as the O lone pairs.

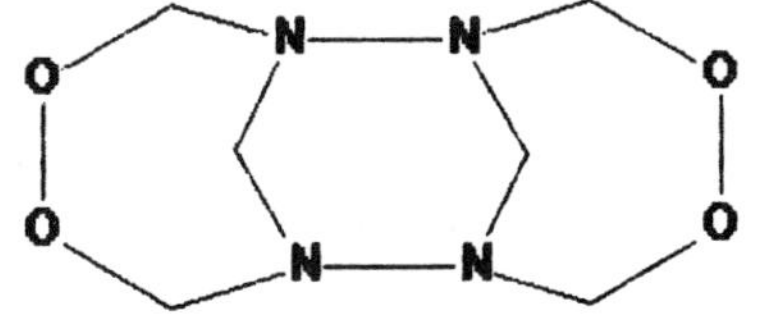

I

In fact, from these and other studies, experimental "bonding density" seems to decrease in the order $C-C > C-N > C-O \sim N-N > C-F \sim N-O > O-O$, i.e. with increase in the number of valency electrons. This trend is explained by comparing the orbital occupation numbers for the bonded atoms with those of the spherically averaged atoms that are subtracted out (Angermund, Claus, Goddard and Krüger, 1985). It is clear that the choice of reference states is to some extent arbitrary and that other kinds of trend could be produced by assuming non-spherical charge distributions for the atoms in the pro-molecule. Some aspects of these problems have been discussed by Spackman and Maslen (1985), and the use of spherically averaged atoms to define the pro-molecule has come in for considerable criticism from other theoreticians (Schwarz, Valtazanos and Rudenberg, 1985; Schwarz, Mensching, Valtazanos and von Niessen, 1986; Kunze and Hall, 1986). Nevertheless, it cannot be denied that the spherically averaged density is the only choice that allows fully for the reduction in symmetry that occurs when a free atom becomes part of a molecule.

A new, accurate analysis of lithium tetrafluoroberyllate, Li_2BeF_4, at 81 K shows that the charge density in this crystal is almost equally compatible with a superposition of spherical neutral atom charge distributions and a superposition of ionic charges (Seiler and Dunitz, 1986). Indeed, the two distributions are so similar that they can hardly be distinguished by examining the charge density in real space. Both models yield difference maps with peaks of up to 0.2 e.Å^{-3} along the Be-F bonds of the tetrahedral "anions" and still weaker ones along the Li-F directions (Fig. 4).

However, the two models are differentiated by analysis of the weak low-order reflections; the intensities of 22 low-order

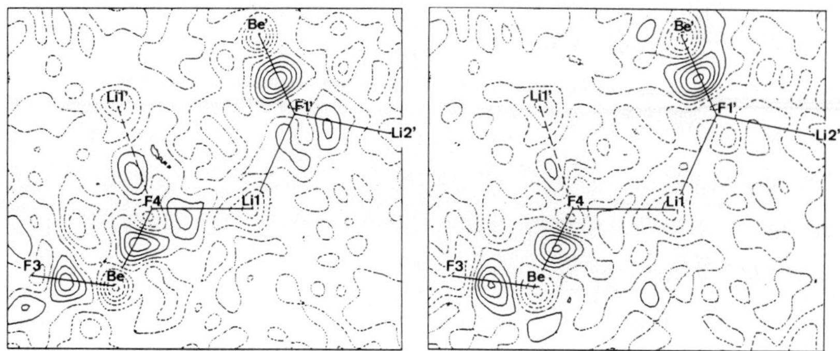

FIG. 4.   Lithium tetrafluoroberyllate; electron-density difference maps based on neutral atom (left) and conventional ionic (right) models for the pro-crystal.
Contour interval 0.045 e.Å^{-3}.

reflections (out to $\sin\theta/\lambda = 0.25$ Å^{-1}), measured with special
care, are reproduced better by calculations based on the neu-
tral-atom scattering curves.

ACKNOWLEDGEMENT: This work has been supported over the years by
the Swiss National Science Foundation.

REFERENCES

ANGEMUND, K., CLAUS, K.H., GODDARD, R. and KRUEGER, C. (1985).
Angew. Chemie. Int. Ed. Engl. 24, 237-247.

DUNITZ, J.D., SCHWEIZER, W.B. and SEILER, P. (1982). Helv.
Chim. Acta 66, 123-133.

DUNITZ, J.D. and SEILER, P. (1983). J. Amer. Chem. Soc. 105,
7056-7058.

HIRSHFELD, F.L. (1984a). Acta Crystallogr. Sect. B. 40,
484-492.

HIRSHFELD, F.L. (1984b). Acta Crystallogr. Sect. B. 40,
613-615.

KUNZE, K.L. and HALL, M.B. (1986). J. Amer. Chem. Soc. 108, 5122-5127

SAVARIAULT, J.M. and LEHMANN, M.S. (1980). J. Amer. Chem. Soc. 102, 1298-1303.

SCHWARZ, W.H.E., MENSCHING, L., VALTAZANOS, P. and VON NIESSEN, W. (1986). Intern. J. Quantum Chem. 29, 909-914.

SCHWARZ, W.H.E., VALTAZANOS, P. and RUEDENBERG, K. (1985). Theor. Chem. Acta 68, 471-506.

SEILER, P. (1985). Lecture Notes, International School of Crystallography, Static and Dynamic Implications of Precise Structural Information (Ed. Domenicano, A., Hargittai, I. and Murray-Rust, P.), Ettore Majorana Centre for Scientific Culture, Erice (Italy), pp. 79-94.

SEILER, P. and DUNITZ, J.D. (1986). Helv. Chim. Acta 69, 1107-1112. Errata, ibid, 1787.

28. The interaction between theoretical and crystallographic methods

A. Gavezzotti and M. Simonetta[§]

1. INTRODUCTION

The study of organic molecules packed in their crystal offers
a unique opportunity to both experimentalists and theoreticians.
Properties of matter can be analyzed, to a large extent, starting from the static picture provided by geometrical structure;
such is obviously not the case for gases and liquids, whose
treatment calls at once for the explicit introduction of dynamic factors. Thus, just as X-ray crystallography has allowed
giant steps in the understanding of molecular structure, there
is now the hope that structure itself can help chemists work
their way back to an understanding of crystal forces, and of
the rules that govern crystal formation and growth.

X-ray crystal structure analysis has grown into full
maturity after the advent of fast computers and direct methods,
to the point that it is now ready to face protein crystallography. It is, of course, the best method to obtain static

---

[§]Deceased on 6 january 1986.

information. But besides this well known, familiar tool, other
techniques are appearing, which provide subtler (although less
direct) information on the behaviour of condensed matter. Temp-
erature-dependent NMR spectroscopy and IQNS (Incoherent
quasielastic Neutron Scattering) are introducing chemists and
physicists to the world of large-amplitude molecular motions in
crystals. FTIR and ESR spectroscopies are used to detect product molecules after reactions in crystalline matrices. It is
also true that simple and often largely overlooked techniques,
like optical examination in polarized light, can give good
amounts of valuable (and unexpensive) evidence.

On the theoretical side, most of what is today available
for crystal structure analysis originates from simple ideas on
dispersion-repulsion non-bonded interactions between atoms in
different molecules (intermolecular interactions). There is
speculation as to whether terms in the potential, like
Coulomb-type formulas, represent truly independent physical effects or are just a way to add flexibility to the model by
incorporating more parameters. Statistical investigations, made
possible through the availability of an immense body of fully
characterized crystal structures, are yielding insight on how
minima in the crystal potential surfaces are attained, as a
function of _molecular_ structure. This correlation is searched
for mainly in terms of molecular shape and bulk. There is hope
that crystal chemistry can be explained in the mechanistic
language of steric factors, just like much of solution chemistry
can. There are good reasons to presume that inductive and electrostatic factors may be modulations, but never predominant, in

such a theory.

The following sections will give an overlook on the problematic (and therefore fascinating) subject of the interplay between experiment and theory in crystal chemistry, through a few typical (but by no means unique) examples.

2.   CRYSTAL PACKING STATISTICS

X-ray crystallographers very seldom care to analyze in detail the intermolecular part of their findings. Some statement on distances shorter than the sum of van der Waals radii is all one can expect in a standard crystallographic report. We wish to describe here some parameters that may help to characterize crystal structure types; something that can be the intermolecular equivalent of such intramolecular lables as, for instance, "strained molecule", or the like. We define the following quantities:

V_M Molecular volume

$$C_K = Z \, V_M / V_C \tag{1}$$

S_M Molecular envelope surface

$$PPE = \sum_{i=1,N} \sum_{j} \{ A \exp(-BR_{ij}) - C \, R_{ij}^{-6} \} = \sum_{i} \{ E_{A,i} \} \tag{2}$$

   Packing Potential Energy

$$PE = 1/2 \, PPE \qquad \text{Packing energy} \tag{3}$$

$E_{A,i}$ Atomic relevance of atom i

$$F_{A,i}(R) = d\{E_{A,i}(R)\} / dR \qquad \text{Atomic sensitivity} \tag{4}$$

Details for the fast and accurate calculation of these quantities can be found in Gavezzotti (1982, 1983, 1985); Gavezzotti and Simonetta (1982). Of course, V_M and C_K were originally

TABLE 1

Average packing coefficients for organic compounds, and A,B,C
for use in eq. (2) (in kcal/mole)

Class	C_K
Hydrocarbons, overall	0.715
Condensed aromatic hydrocarbons	0.754
Compounds with N,O,P,S	0.704
Compounds with N,O, halogen	0.722
Polyhalogenated compounds	0.748

Interaction	A	B	C
C...C	71600	3.68	421
C...H	18600	3.94	118
H...H	4900	4.29	29
N...N	42000	3.78	259
O...O	77700	4.18	259.4

introduced by Kitaigorodski (1961). $E_{A,i}$ had already been used
by Bernstein and Hagler (1978). The molecular volume and outer
surface can be computed from molecular geometry plus atomic
radii. Z is the number of molecules with N atoms in a cell of
volume V_C. Table 1 shows some average values for C_K and values
for A,B,C in eq. (2). $E_{A,i}$ is the amount of packing energy due
to atom i, and R is the radial distance from it; thus, $F_{A,i}$

may be interpreted as the crystal "pull" on atom i, although it lacks directional character.

Some interesting relationships exist among the above defined quantities, when statistics over a large number of compounds is carried out. S_M is related to the number of valence electrons in the molecule, Z_V, by

$$S_M = 2.601 \ Z_V + 28.13 \quad \overset{\circ}{A}{}^2 \tag{5}$$

The packing energy, PE, is related to S_M by

$$PE = 0.0767 \ S_M + 1.448 \quad kcal/mole \tag{6}$$

when a 7 $\overset{\circ}{A}$ cutoff is used in the sums in eq. (2). Fig. 1 shows plots of PE versus S_M or Z_V for some typical populations. Molec ular packing can be classified, for any crystal, according to the deviation from the averages in Table 1 and from eq. (5) and (6). It is usually true that, when an awkward molecular shape prevents close packing, C_K is below the average and PE is less than it should be from eq. (6); a comparison of PPE and C_K for isomers (Fig.2) is convincing in this respect. PE is also less than the "theoretical" value when polar substances are consider ed. In this case, one possible interpretation is that other ingredients (e.g., electrostatic forces) contribute to the cohesive energy of the crystal. Of course, hydrogen-bonded crystals give smaller PE's than predicted by eq. (6), since the model does not contain energy terms accounting for H-bonds (this provides a quick check for spotting even feeble H-bonds in organic crystals).

On the other hand, deviations from eq. (5) reveal unusual conformational properties of the molecule; "strainless" group

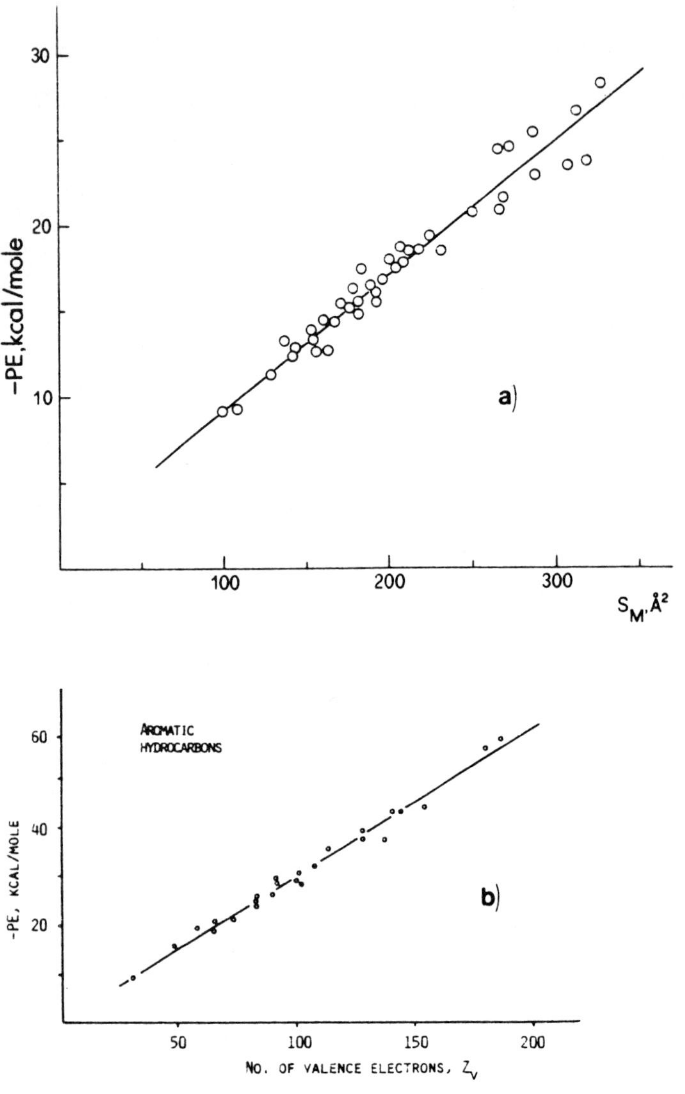

FIG. 1. a) PE (7 Å cutoff) as a function of S_M for a population of organic compounds (Gavezzotti, 1985). b) PE (10 Å cutoff) for a population of condensed aromatic compounds (Gavezzotti and Desiraju, to be published).

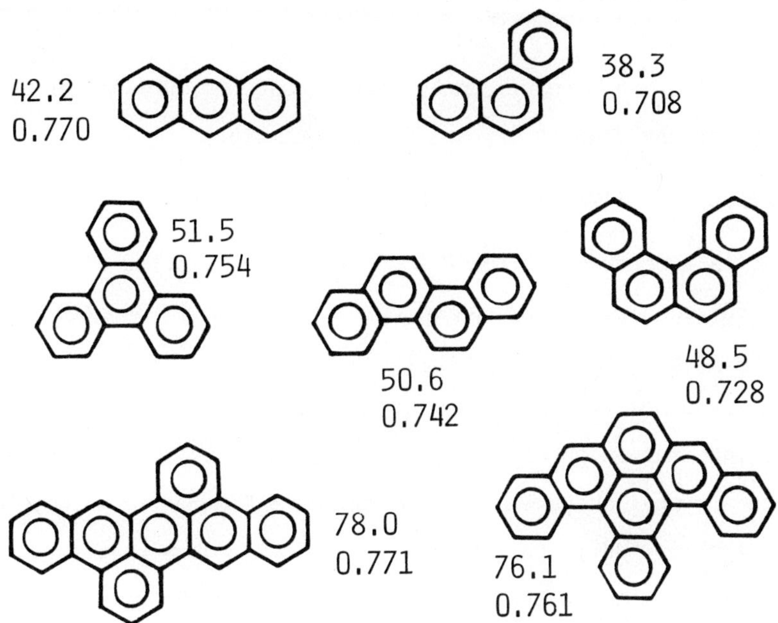

FIG. 2. Packing potential energy (kcal/mole) and packing coefficients for isomers of aromatic compounds. A.Gavezzotti and G.R.Desiraju (to be published).

contributions to S_M can be established, and overcrowding can be gauged (and located) by simply measuring the decrease in surface resulting from overlap of atomic spheres. Some caution is needed in the process (see Gavezzotti, 1985; Bianchi, Gavezzotti and Simonetta, 1985).

On the whole, it can be said that eqns. (5) and (6) describe the ideal case of a conformationally strainless, non-polar, purely van der Waals-close-packed molecule, and that

deviation from them qualifies and quantifies, if properly inter
preted, the amount of deviation from such an ideal case.

The absolute significance of the equations is doubtful,
although it can be confidently stated that the sublimation
energy of any organic crystal is approximately

$$\Delta H(\text{subl}) \sim (1/0.8)\ 0.0767\ S_M + 1.448\ \text{kcal/mole}$$

if it is accepted that the 7 $\overset{\circ}{A}$ cutoff in the lattice sums makes
them 80% convergent. By use of eq. (5), $\Delta H(\text{subl})$ can also be
estimated from Z_V.

The quantity here called atomic relevance, $E_{A,i}$, provides
more information on crystal packing. It appears that a close-
packed crystal obeys what may be called a "homomeric principle"
(Gavezzotti, 1982): the contribution of each atom to the packing energy is constant for each atom type, and does not change
with the intermolecular environment. Therefore, molecules arran
ge themselves in crystals in such a way that each atom can
reach the homomeric share of the cohesive energy which is proper
of its species. In the ideal packing, there may be no corner-
stone and no outlier atoms. Fig. 3 shows histograms of the $E_{A,i}$
for boron atoms in borane crystals (Beringhelli, Filippini,
Gavezzotti and Simonetta, 1983) and for carbon atoms in aromatic
molecules - for the latter, it is necessary to distinguish
between tertiary and secondary carbon atoms. There are also
preliminary results confirming that the homomeric principle is
obeyed also in heteroatom-containing crystals.

By use of this principle, close-packing can be tracked
down to the level of the single atom, and it is possible, if
not entirely proved, that it can be used for the construction

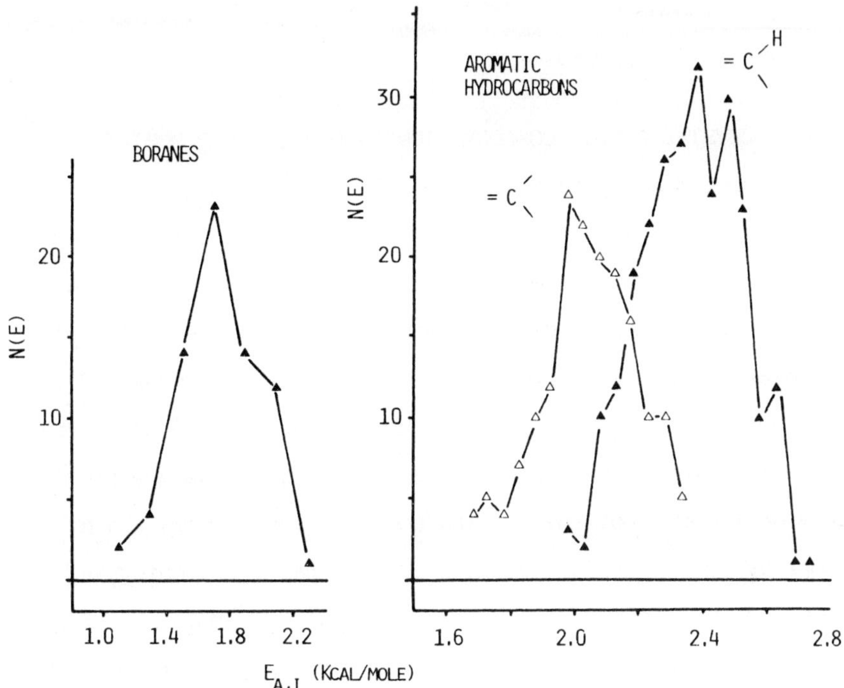

$E_{A,I}$ (KCAL/MOLE)

FIG. 3. Left: a histogram of atomic relevances (kcal/mole) for boron atoms in borane crystals (7 Å cutoff). Right: the same for carbon atoms in aromatics (10 Å cutoff).

of a theory of organic crystal packing and growth, and that deviations from its requirements can rationalize crystal dynamics and reactivity. Is there, for instance, a substituent effect in crystal packing? A reasonable working hypothesis is that atoms with $E_{A,i}$ below the average may be advantageously substituted by bulkier groups, within the same overall crystal framework; conversely, at some threshold bulk (or "shape") value the substituent steers the crystal to a different structure.

Computational tests can be carried out (see, for work along these lines, Thomas _et al._, 1985). Research on this subject is in progress in our laboratory.

3.   CRYSTAL FIELD, CONFORMATION, AND MOLECULAR MOBILITY

A molecule in a crystal is a flexible object, pulled upon (anisotropically) by intra- and intermolecular forces. It settles in an overall minimum where these forces have been satisfied at best. To quote just a simple example, in 1,1-di-_p_-tolylethylene the dihedral angles between the plane of the two rings and the ethylene plane are different in the crystal (35 and 50°). Fig. 4 explains why; the intramolecular energy rises as one ring is twisted to 50°, but the packing energy drops faster, and a compromise is reached. But this sort of naive reasoning in terms of pulls along well defined molecular coordinates often fails; for example, in compound 1 four different values of the C(1)–C(6) distance were found in the solid state, but it is not easy to tell which forces are acting at which molecular sites. The $F_{A,i}$ of atoms 1 and 6 do not correlate with bond length, so presumably it is not just a matter of intermolecular action at the bridgehead carbon atoms.

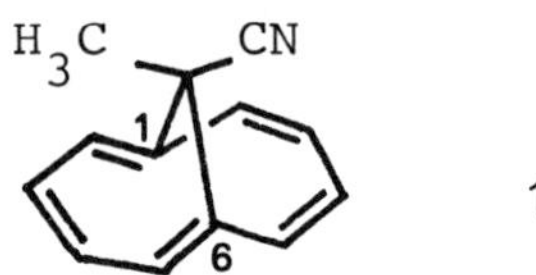

1

A closely related matter is that of molecular mobility (meaning here large-amplitude displacements other than molecular libration). The experimental side is generous of reliable results; wide-line, or even better, CP-MAS (Cross Polarization

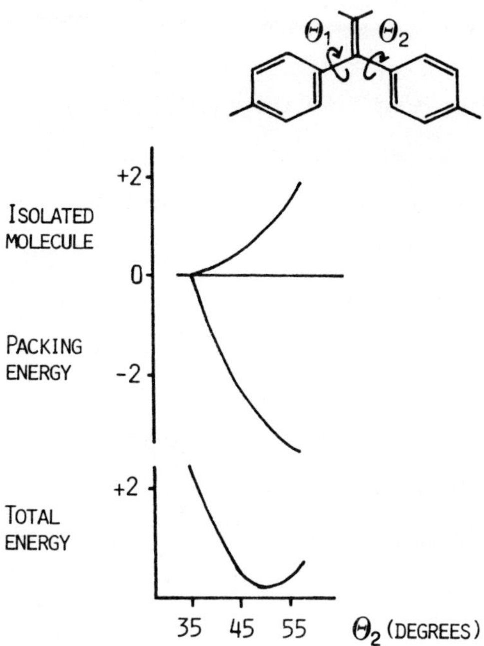

FIG. 4. 1,1'-di-p-tolylethylene; energy curves for torsion of
one ring, while keeping the second at 35° torsion. Kcal/mole
units. Gavezzotti and Casalone, 1971.

Magic Angle Spinning) NMR measurements allow a detailed descrip
tion of the type and of the energetics of molecular motion in
condensed matter. X-ray single-crystal analysis parallel to the
NMR study, would give the most complete structural picture of
the process, were it not for a certain reluctance on the part
of X-ray crystallographers (there are exceptions!) to take the
trouble of carrying out the analysis at many temperatures for
the same compound. Yet, temperature control is what makes most
of the difference between purely analytical X-ray crystallogr-

aphy and true physical chemistry of condensed matter.

The computational counterpart of these experiments on molecular mobility is the use of eq. (2) to calculate the PPE of a molecule displaced from its equilibrium position along a certain coordinate, immersed in the crystal field. This technique has been extensively used, and the reader will find key references in reviews by Gavezzotti and Simonetta (1982, and in press). Although such calculations may require some ingenuity in the geometrical modeling, they run in seconds on small (or even desktop) computers. The ratio of the information retrieved to the computer expense is perhaps one of the best in theoretical chemistry.

One thing is computing, another is understanding, however. What are the (molecular or crystal field) requirements to make large-amplitude motion possible in crystals? The simplest answer (although never complete or foolproof) is in terms of molec ular shape. Let us consider first the condensed aromatics (see Fig. 5). It is evident that an intuitive law, stated in simple words by Fyfe and Harold-Smith (1975), is obeyed: "those axes which cause the fewest atoms to move and to transcribe curves of the smallest radii will have the lowest barriers to rotation". In Fig. 5, disc-like molecules are seen to have very small rotation barriers in the molecular plane.

Another example concerns the 11,11-disubstituted annulenes. The parent hydrocarbon 2 and the difluoroderivative 3 have quite different crystal structures, but quite similar rotation barriers. The dimethylderivative 4 and the methylcyano derivative 1 are isostructural, but the barriers to rotation in the

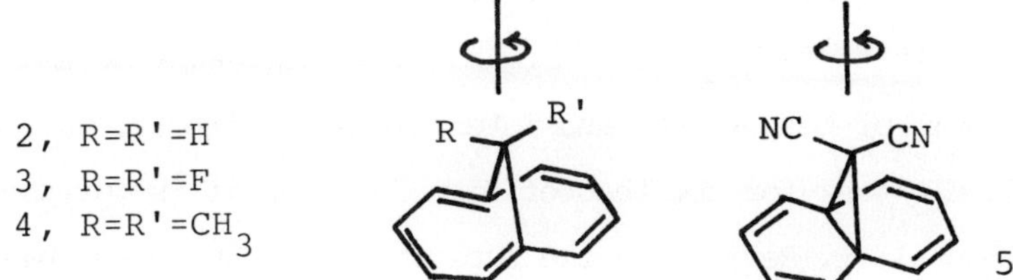

2, R=R'=H
3, R=R'=F
4, R=R'=CH$_3$

crystal are different. Thus, the ease of rotation depends on molecular shape rather than on crystal field. The shape effect at work here is apparently the bulk of the CN group sticking outwards; rotation in 5 is completely forbidden.

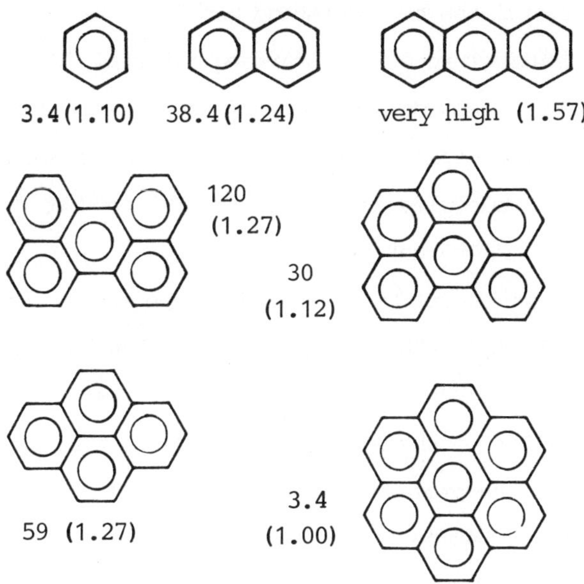

FIG. 5. In-plane rotation of aromatics. Barriers (kcal/mole) and, in parentheses, ratio of maximum to minimum molecular dimension in the molecular plane.

4.   ORGANIC SOLID-STATE REACTIVITY

The solid-state reactivity of organic molecules must be analyzed in terms of both intra- and intermolecular effects. This is
a difficult task for the theoretician, just as it is for the
experimentalist. Reviews on the experimental techniques, based
on sophisticated spectroscopic methods and, once again, X-ray
crystallography, have been given (Scheffer, 1980; McBride, 1983).
Organic reactivity in channels formed by a host substance has
been probed by experiment and calculations (Tang et al.,1985).
We will describe in the following some of our theoretical
approaches to solid-state reactivity.

The starting point is the simplifying assumption that reactivity in a crystal can be understood mechanistically in terms
of the crystal structure of the starting material. The perturbations caused by reaction are then analyzed in terms of the
availability of free space at the reaction site, since in the
very first steps molecules and fragments will tend to diffuse
towards holes in the structure.

The procedure starts by drawing a volume analysis map (see
details in Gavezzotti, 1983), which gives a three-dimensional
picture of the distribution of matter in the crystal. Quantitative information on the width and overall size of cavities
can be obtained; the procedure, if sometimes open to a rather
subjective interpretation, is at least systematic. Figure 6
shows such a map for the crystal of bis-triphenylpropanoyl
peroxide, which undergoes double decarboxylation (Walter and
McBride, 1981). The next step is the calculation, by eq. (2),
of the PPE variation for the proposed displacements. The main

$(C_6H_5)_3CCH_2\overset{\text{-C-O-O-C-}}{\underset{O\qquad O}{}}CH_2C(C_6H_5)_3$

Triphenylpropanoylperoxide

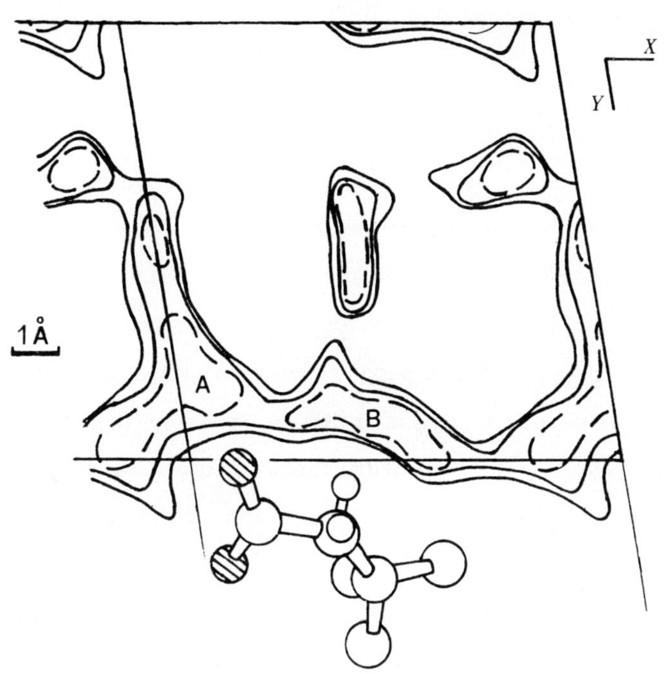

FIG. 6. Packing density map for triphenylpropanoylperoxide.
Dashed lines enclose zones of empty space. The molecular model
shows a $(C_3)CCH_2COO$ group (oxygens are dashed). The CO_2 moves
into cavity A; the CH_2 group could move to cavity B.

difficulty here is to find the detailed pathway of the molecule
and the fragments, consisting in general of rotations and
translations. Intramolecular rearrangements may also occur. For
the case of triphenylpropanoylperoxide, it was found after
optimization (Gavezzotti, 1986) that the CO_2 molecule moves
into cavity A (Fig.6) with an intermolecular activation energy

of no more than 18 kcal/mole. The remarkable result is that the
crystal lattice can accommodate two CO_2 molecules without substantial deformation.

5.   CONCLUSIONS

This paper demonstrates, it is hoped, that theoretical methods
in organic crystallography are growing to the stage where
reliable information can be obtained on the structure and dynamics of organic solids. Molecular structure can be studied by
both quantum chemistry and crystallography; but quantum chemist
ry is as yet unable to deal with crystal structure. The challenge of founding and developing a theory of crystals must
therefore rely on semiempirical methods, the key to which rests
in the large availability of databases. A screening of the Camb
ridge files for intermolecular information is a timely and
rewarding undertaking - and not only in the interest of pure
science. An understanding of the principles of crystal structure
is vital to polymer science, and to all the branches of
materials science that aim at the production of molecular
devices, molecular conductors, and other similar building blocks
of high technology in the years to come. The day must come when
the principles of the synthesis of crystals will be as plain
and clear as are now those of the synthesis of molecules.

REFERENCES

BERINGHELLI,T.,FILIPPINI,G.,GAVEZZOTTI,A. and SIMONETTA,M.
(1983). J.Mol.Struct.(TheoChem) 94,51.
BERNSTEIN,J. and HAGLER,A.T. (1978). J.Amer.Chem.Soc. 100,673.
BIANCHI,R.,GAVEZZOTTI,A. and SIMONETTA,M. (1985). J.Mol.Struct.

(TheoChem) 135,391.

CASALONE,G. and GAVEZZOTTI,A. (1971). Istituto Lombardo (Rend. Sc.) A105,824.

FYFE,C.A. and HAROLD-SMITH,D. (1975). J.Chem.Soc.Faraday II, 967.

GAVEZZOTTI,A. (1982). Nouv.J.Chimie 6,443.

GAVEZZOTTI,A. (1983). J.Amer.Chem.Soc. 105,5220.

GAVEZZOTTI,A. (1985). J.Amer.Chem.Soc. 107,962.

GAVEZZOTTI, A. (1986). Tetrahedron (submitted).

GAVEZZOTTI,A., and SIMONETTA,M. (1982). Chem.Rev. 82,1.

KITAIGORODSKI,A.I. (1961). Organic Chemical Crystallography, Consultants Bureau, New York.

MCBRIDE,J.M. (1983). Acc.Chem.Res. 16,304.

SCHEFFER,J.R. (1980). Acc.Chem.Res. 13,283.

TANG,C.P.,CHANG,H.C.,POPOVITZ-BIRO,R.,FROLOW,F.,LAHAV,M., LEISEROWITZ,L., and MCMULLAN,R.K. (1985). J.Amer.Chem.Soc. 107,4058.

THOMAS,N.W.,RAMDAS,S., and THOMAS,J.M. (1985). Proc.Roy.Soc. (London) A400,219.

WALTER,D.W. and MCBRIDE,J.M. (1981). J.Amer.Chem.Soc. 103,7074.

29. Size and softness of hydrocarbon atoms from *ab initio* SCF dimer calculations

Donald E. Williams and Karen L. Lobb

ABSTRACT

<u>Ab</u> <u>initio</u> self-consistent-field (SCF) energy calculations at
the 6-31G** level were carried out for acetylene, ethylene,
and methane dimer supermolecules. Dimer configurations were
chosen to separately define H...H and C...C nonbonded
interactions. The SCF results were corrected for basis set
superposition error. Net atomic charges for the monomers were
obtained by the potential-derived method. After subtraction
of the scaled intermolecular electrostatic energy, the
remaining energy was fitted by atom-atom exponential repulsion
functions. The atomic size of hydrogen and carbon obtained
by this method was not found to be uniform in this series of
molecules.

1.   INTRODUCTION

Because of the Pauli exclusion principle overlap of filled
orbitals on different molecules is prohibited. If the lowest
unoccupied orbitals are high in energy, the filled orbitals
must distort to avoid overlap. The repulsion energy caused
by this distortion is approximately proportional to the amount
of overlap (Kita, Noda, and Inouye, 1976). It can further be
shown (Starr, 1976) that the amount of overlap may be
represented by an exponential function.

Because of computer advances it is now possible to
calculate by _ab initio_ methods the energy of interaction of
small molecule dimers. However, these calculations are quite
time-consuming; we show that the essential results for
acetylene, ethylene, and methane dimers can be represented by
an atom-atom empirical model. The great value of the empirical model is that energies of interaction can be evaluated
very rapidly for aggregates of much larger molecules where
ab initio calculations are difficult or impossible.

Electrostatic interaction in hydrocarbon dimers, although
small, is not negligible. It has been shown (Williams, 1974)
that a model for hydrocarbon crystals using net atomic charges
gives better agreement with observed data. Net atomic charges
q may be derived from the molecular electric potential
obtaineddirectly from the monomer molecular wavefunction (Cox
and Williams, 1981). The atom-atom repulsion function (exp-1)
can then be corrected for this electrostatic effect:

$$B_{jk} \exp(-Cr_{jk}) = V_{jk} - q_j q_k r_{jk}^{-1}$$

In this equation V is the atom-atom interaction energy between
nonbonded atoms j and k at distance r; B and C are empirical
size and softness parameters for the interaction. The total
intermolecular energy is the sum of all of the V's between the
molecules of the dimer. Given the total intermolecular energy
as a function of intermolecular distance, values of B and C
for the atom types can be obtained by least-squares fitting.

2.   MONOMER AND DIMER SCF CALCULATIONS

Ab initio SCF wavefunctions were obtained at the 6-31G** level

using the GAUSSIAN-82 program. The geometry of the monomers
was fixed at the following values: acetylene, C-C=1.203 A,
C-H=1.061 A; ethylene, C-C=1.339 A, C-H=1.085 A, H-C-H=117.8
deg; methane, C-H=1.094 A. For definition of H...H repulsion
the monomers were oriented with linear C-H...H-C. To define
the C...C repulsion acetylene and ethylene monomers were
twisted by 90 deg and separated so that these contacts were
closest. The potential-derived net atomic charges of Cox and
Williams (1984) were used.

Table 1 shows energies for acetylene dimer. The energies
of the monomer vary slightly because of the basis set superposition correction (BSSC). The intermolecular energy varies
from 2.59 to 24.61 kJ/mol for the linear configuration as the
H...H distance decreases from 3.0 to 1.6 A. In the twisted
configuration the intermolecular energy varies from 3.38 to
55.96 kJ/mol as the distance decreases from 4.0 to 2.8 A.

Table 1. SCF and (exp-1) energies for acetylene dimers.

distance(a)	SCF-BSSC monomer (hartrees)	SCF dimer (hartrees)	SCF inter (kJ/mol)	(exp-1) inter(b) (kJ/mol)
linear configuration				
1.6	-76.8215818	-153.6337892	24.61	24.61
1.8	-76.8214746	-153.6371101	15.33	15.30
2.0	-76.8213830	-153.6389004	10.15	10.15
2.2	-76.8213106	-153.6399098	7.12	7.14
2.4	-76.8212565	-153.6405130	5.25	5.26
2.6	-76.8212184	-153.6409005	4.03	4.03
2.8	-76.8211933	-153.6411700	3.19	3.18
3.0	-76.8211782	-153.6413711	2.59	2.56
twisted configuration				
2.8	-76.8223346	-153.6231373	55.96	56.09

3.0	-76.8222249	-153.6312251	34.15	34.49
3.2	-76.8221153	-153.6360289	21.53	21.32
3.4	-76.8219917	-153.6388528	13.47	13.29
3.6	-76.8218484	-153.6404722	8.47	8.38
3.8	-76.8216931	-153.6413552	5.33	5.37
4.0	-76.8215422	-153.6417989	3.38	3.51

a. Nearest H...H distance for linear configuration; distance between molecular centers for twisted configuration.
b. Calculated with charge scale factor K=0.9704.

Table 2 shows energies for ethylene dimer. The intermolecular energy varies from 0.47 to 19.98 kJ/mol for the linear configuration as the H...H distance decreases from 3.0 to 1.6 A. In the twisted configuration the intermolecular energy varies from 5.45 to 77.84 kJ/mol as the distance decreases from 4.0 to 2.8 A.

Table 2. SCF and (exp-1) energies for ethylene dimers(a).
linear configuration

1.6	-78.0380187	-156.0684285	19.98	20.00
1.8	-78.0379922	-156.0719355	10.63	10.59
2.0	-78.0379709	-156.0737581	5.73	5.73
2.2	-78.0379561	-156.0747082	3.16	3.18
2.4	-78.0379474	-156.0752070	1.81	1.83
2.6	-78.0379433	-156.0754732	1.09	1.08
2.8	-78.0379419	-156.0756192	0.69	0.67
3.0	-78.0379411	-156.0757014	0.48	0.43

twisted configuration

2.8	-78.0395740	-156.0495022	77.84	77.78
3.0	-78.0394113	-156.0600956	49.17	49.35
3.2	-78.0392465	-156.0665409	31.38	31.41
3.4	-78.0390682	-156.0704483	20.19	20.09

3.6	-78.0388687	-156.0727754	13.03	12.94
3.8	-78.0386547	-156.0741046	8.41	8.42
4.0	-78.0384466	-156.0748156	5.46	5.55

a. Calculated with charge scale factor K=0.8039

Table 3 shows results for methane dimer. In this case there was no configuration which emphasized the C...C repulsion. For the linear configuration the energy varied from 0.22 to 19.99 kJ/mol as the distance decreased from 3.0 to 1.6 A.

Table 3. SCF and (exp-1) energies for methane dimers.
linear configuration

1.6	-40.2015493	-80.3954858	19.99	20.23
1.8	-40.2015161	-80.3990696	10.40	10.40
2.0	-40.2014908	-80.4009270	5.39	5.37
2.2	-40.2014735	-80.4018861	2.79	2.80
2.4	-40.2014643	-80.4023803	1.44	1.47
2.6	-40.2014609	-80.4026359	0.75	0.78
2.8	-40.2014606	-80.4027691	0.40	0.43
3.0	-40.2014606	-80.4028379	0.22	0.24

3. ATOM-ATOM MODEL FOR THE INTERMOLECULAR ENERGY

Preliminary calculations showed that the use of a single H...H and a single C...C exponential repulsion function did not well fit the SCF results for all three molecular dimers. Therefore, separate functions were derived for each molecule, except that no C...C function was obtained from methane dimer. The figures below show curves for H...H and C...C atom-atom exponential repulsion. Since previous studies indicated a possible overestimation of potential-derived net atomic charges using the 6-31G** basis set (Cox and Williams, 1981), we also optimized a charge scale factor parameter, K, for the acetylene and ethylene data. The value of K was 0.9704 for acetylene and

0.8039 for ethylene; these are in a reasonable range. A good fit of the model potential to the calculated intermolecular energies was obtained. The last columns of Tables 1, 2, and 3 show the empirical model energies compared to the SCF energies.

From the left-hand figure below it is seen that the size of the hydrogen atom increases going from acetylene to ethylene to methane. The fourth curve is one derived from hydrocarbon crystal data by Williams and Cox (1984); it depicts an even larger hydrogen atom. In the right-hand figure the carbon atom in ethylene appears smaller in size than the acetylene carbon, reversing the hydrogen sequence. The value derived from crystal data is still smaller. Even if the absolute values of the curves have some error, the variation in atomic size with bonding environment appears to be significant. In the future, accurate work on molecular aggregates will need to take these size differences into account.

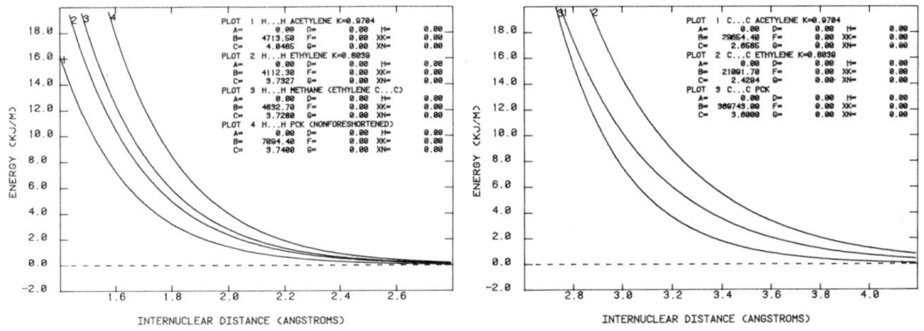

AKNOWLEDGEMENT. This work supported by National Institutes of Health Research Grant GM37453.

REFERENCES

COX, S. and WILLIAMS, D. (1981). J. Computational Chem. 2, 304.
KITA, S., NODA, K., INOUYE, H. (1976). J. Chem. Phys. 64, 3446.
STARR, T. (1976). Acta Cryst. A30, 71.
WILLIAMS, D. and COX, S. (1984). Acta Cryst. B40, 404.

30. Paradigms and rules inferred from structural organic chemistry

Alajos Kálmán

ABSTRACT

This work presents a few paradigms which have been recognized between the bonding and electronic structure of some groups of related organic compounds possessing such important hetero atoms as nitrogen and sulphur.

These paradigms (or rules) are the following:

1.) $S(IV)N_3$ pyramids as a component of various compounds have a certain capacity to buffer the electron-withdrawing or -releasing effect of the ligands bound to the N atoms in order to maintain the valence number around the S atom as near to four as possible. In the formation of the individual $S(IV)-N$ bonds the N-lone pair plays an important role, however.

2.) The orientation of the N-lone pair(s) to the $S(VI)[O,O,N,C]$ and $S(VI)[O,O,N,N]$ tetrahedra has a significant effect upon their geometry especially on the bond lengths and angles which involve N atom(s).

3.) Aromatic five-membered hetero rings built up by C and N atoms (imidazole, s- and v-triazole derivatives, etc.) exhibit only two patterns of the endocyclic bond angle magnitudes which are induced dominantly by the number and relative position of N-lone pairs.

1.   INTRODUCTION

During the last twenty years the author has been involved in
the analysis of crystal structures determined by X-ray diffraction methods of more than one hundred organic compounds mainly
heterocyclic, containing nitrogen and sulphur atoms (e.g.
sulphilimines, sulphuranes, thiazolidines, thiazines, thia-
triazines, etc.). In the course of these studies several correlations between bonding and electronic structures (tautomerism,
non-bonded interactions, etc.) of the above mentioned molecules
have been recognized, some of which have helped to elucidate
the chemical behaviour of the systems studied. From them three
rules deliberately denoted as <u>paradigms</u> are presented. It is
worth noting that they have the common feature that direct
or indirect involvement of the N-lone pair(s) is a decisive
factor.

The author intends to present evidence that in science
a paradigm is not only a simple rule or pattern, but like an
accepted judicial decision in common law is an object for further articulation and specification under new and more strin-
gent conditions. The birth of a new paradigm is generaly preceded by an early fact-gathering which is restricted to the
wealth of data available. This is followed by classification
(or systematization) in terms of trends or relationships. The
articulation of a paradigm is mainly supported by its test
on novel observations or experimental data including the resolution of apparent and/or real contradictions.

In the course of the articulation of the first two rules
the Cambridge Crystallographic Data Center files, CCDC, have
been used as a rich source of relevant structural parameters.

2.   DISCUSSION

2.1.   First statement

Any $S(IV)N_3$ pyramid has a certain capacity to buffer the
electron-withdrawing or -releasing effect of the ligands bound
to the N atoms in order to maintain the valence number around
the S atom as near to **four** as possible.

Early **fact-gathering** started with the study of S(IV)-N
bond lengths determined for different N-acyl sulphilimines
(a: Kálmán, 1967; b: Cameron et al, 1971; c: Kálmán et al,
1971a; d: Kálmán et at, 1972; e: Kálmán et at, 1971b; f: Kálmán
et at, 1973; and for some related compounds (e.g. g: Cook et
al, 1971; h: Gieren et al, 1974) depicted in Chart I.

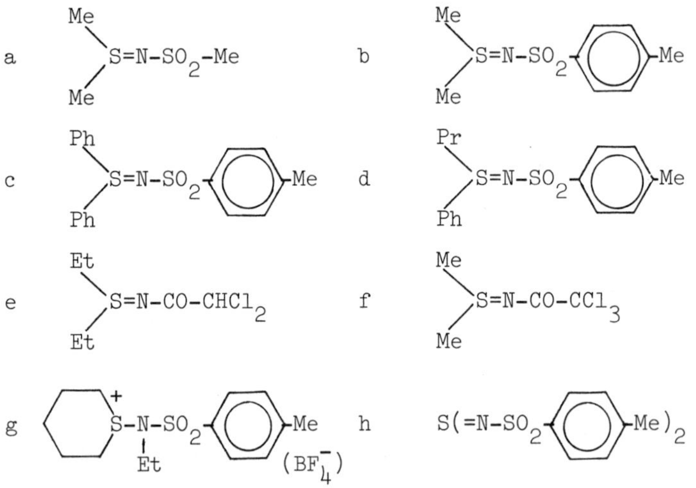

CHART I

and resulted in the systematization of various S(IV)-N multiple
bonds in the range bounded by the lengths of single and double
bonds (Table 1) with a conclusion that the $R^1R^2S(IV)N$ moieties
interact only minimally with the lone pair of the nitrogen

TABLE 1

Characteristic S(IV)-N bond lengths

N(2) and N(3) represent two- and three-coordinate nitrogen atoms, resp. δ stands for smaller, whilst Δ stands for greater polarization of the S-N bond (st = strong, w = weak)

$$S^{IV} = N(2) \qquad <1.54\ \text{Å}$$

$$S^{IV+\delta} \overset{-\delta}{-} N(2)^{-\delta}_{st} \qquad 1.63$$

$$S^{IV+\delta} \overset{-\delta}{-} N(3)^{-\delta}_{st} \qquad 1.64$$

$$S^{IV+\Delta} \overset{-\Delta}{-} N(2)^{-\Delta}_{w} \qquad 1.67$$

$$S^{IV+\Delta} \overset{-\Delta}{-} N(3)^{-\Delta}_{w} \qquad >1.68 \quad (<1.75\ \text{Å})$$

atoms (Kálmán, 1976). By the use of the pattern displayed in Table 1 the three different S(IV)-N bonds involving the same S atom in a novel compound, I, (Fig. 1) could easily be classified (Kálmán et al, 1977). It was noted simultaneously that the trigonal pyramidal $S(IV)N_3$ groups { a few were known from the literature e.g. $S_3N_4Ph_3$ (Holt and Holt, 1974), $S_3N_5PF_2$ (Weiss et al, 1974) $2K^+(C_{21}H_{21}N_3O_6S_4)^{2-}$ (Gieren and Narayanan, 1975)} could be characterized by an equilibrium distance of 1.64(1) Å. This observation served as a candidate for the paradigm to explain a discrepancy in that the exocyclic S-N bond lengths measured subsequently in two sister compounds II and III were shorter than the expected values. The similarity of the mean bond lengths of each set of three related S-N bonds suggests a certain capacity of the $S(IV)N_3$ pyramids to buffer different effects upon their delocalized electronic systems; e.g. in II two weak endocyclic S-N bonds should be balanced

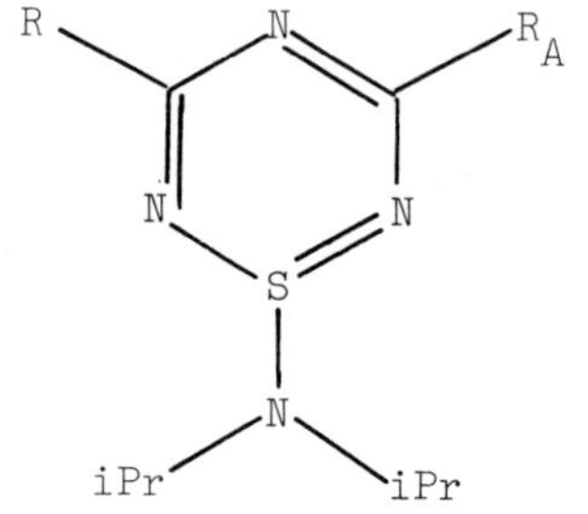

FIG. 1. General formula for $1\lambda^4,2,4,6$-thiatriazines I,II,III
and IV investigated by X-ray diffraction methods. R = Cl,
R_A = $NH-C_6H_{11}$ for I; R = R_A = Cl for II; R = R_A = $S-C_6H_5$ for
III and R = R_A = $O-CH_3$ for IV, in which the iPr (C_3H_7-) groups
are replaced by a Ts ($SO_2-C_6H_4-CH_3$) moiety.

by a relatively strong (i.e. short) exocyclic one. That is,
there is a tendency for the trigonal pyramidal $S(IV)N_3$ moieties
to maintain an equilibrium of the bonding around the sulphur
atoms that can be described best in terms of the sum of the
bond orders (σ+p) or bond numbers (n). The former (i.e. p)
was computed by the use of the modified Coulson's formula
(Liquori and Vaciago, 1956)

$$d(p) = d(1) - \frac{d(1) - d(2)}{1 + 0.6625(1-p)/p} \tag{1}$$

while the latter one was obtained by the Pauling's (1947)
equation :

$$d(n) = d(1) - c \log n \tag{2}$$

where d(1) and d(2) are single and double bond lengths. No
significant difference was observed between σ +p and n values

pertaining to the same S(IV)-N bond. Initially 13 structures
containing altogether 16 $S(IV)N_3$ groups were tested with good
results (Kálmán et al, 1979). Apart from one compound, S_4N_5Cl
(Chivers and Fielding, 1978), in which there is strong competition between the sulphurdiimide and $S(IV)N_3$ moieties no deviation greater than 5 % from $V = \Sigma n = \Sigma(\sigma+p) = 4$ was observed.
Keeping in mind the note of Johnston (1960) who assumed first
the conservation of bond order: "we feel the assumption cannot
be purely correct, but it cannot be totally wrong either" the
next test of the rule was performed on 1,4-dihydro-3,5-di-
methoxy-1-tosylimino-$1\lambda^4$,2,4,6-thiatriazine (R = 3.5 % for
2698 reflections) (see IV in Fig. 1) with good results (Kálmán
et al, 1981): S(1)-N(2) = 1.657(1), S(1)-N(6) = 1.653(1) and
S(1)-N(7) = 1.619(1) Å from which $\Sigma n = 4.001$ and $\Sigma\sigma +p = 3.999$
were obtained.

To scrutinize the "limits" of the bond order conservation
principle applied to the $S(IV)N_3$ groups a search of the CCDC
files has recently been performed. This revealed 24 structures
(including those already studied) containing a total of 34
$S(IV)N_3$ pyramids. In addition, 6 groups found in four inorganic
structures were analyzed. The CCDC REFCODES are accompanied
by the Σn values in Table 2. The numerical distribution (N)
of the S(IV)-N bond lengths is nearly continuous and Gaussian
(Fig. 2). The grand mean value of 120 observations (1.644 Å)
agrees well with the expected one. This value is a characteristic descriptor of the S(IV)-N multiple bond and confirms the
covalent radius (1.19 Å) and electronegativity (1.0) given
by Truter (1962) for sulphur in the valence state of four.

2.1.1. Further observations

This bond order conservation rule is equally valid for neutral
or charged (once or twice) $S(IV)N_3$ groups formed either

TABLE 2.

<u>V = Σ n values for 28 compounds described by their CCDC REFCODES</u>
In a few there are two or even three independent $S(IV)N_3$ groups

BADCAK10	4.22		MPSHNA	4.05	3.95	
BASVOG	3.97	4.03	MPSHNB	3.96	3.96	4.00
BINLOZ	3.98		MURTNS10	3.91	4.02	
BOJGIQ	3.93		PHTTTZ10	4.14		
BTSITZ10	4.05		PPIMTZ	3.93		
BUKHUK	4.22		SNCPDI	3.97	4.01	
CAVTIC	4.00		SOAZSN	3.91		
CHATRZ	3.98	4.02	TFSIOCO1	3.98	4.01	
EPXPTZ10	3.95		TPASSN	4.12		
HMTSAZ	4.00		TPIMSO10	3.97	4.18	
IPCLTZ10	4.12		TPPTSN10	3.90		
KTTSIM	4.00		ZEFWOW	4.06	4.07	

a) $S_3N_5PF_2$ 3.95   c) S_5N_6 3.96
b) S_4N_5Cl 3.77   3.80   d) $R(S_4N_5)$ 3.86   4.07

a) Weiss <u>et al</u>, 1974;   b) Chivers and Fielding, 1978;
c) Chivers and Proctor, 1978; d) Flues <u>et al</u>, 1976.

exclusively by bi- or tricoordinate N atoms or by a mixture
of both. The S-N distances formed by bicoordinate N atoms vary
in the full range (1.51 - 1.75 Å) while those built up by
tricoordinate nitrogens have a minimum value of <u>ca</u>. 1.59 Å
(IPCLTZ10, Kálmán <u>et al</u>, 1979).

In contrast with the $R^1R^2S(IV)N$ groups in N-acyl sulphilimines, the N-lone pairs play important role in the formation
of the $S(IV)N_3$ pyramids. Conformational analysis has shown

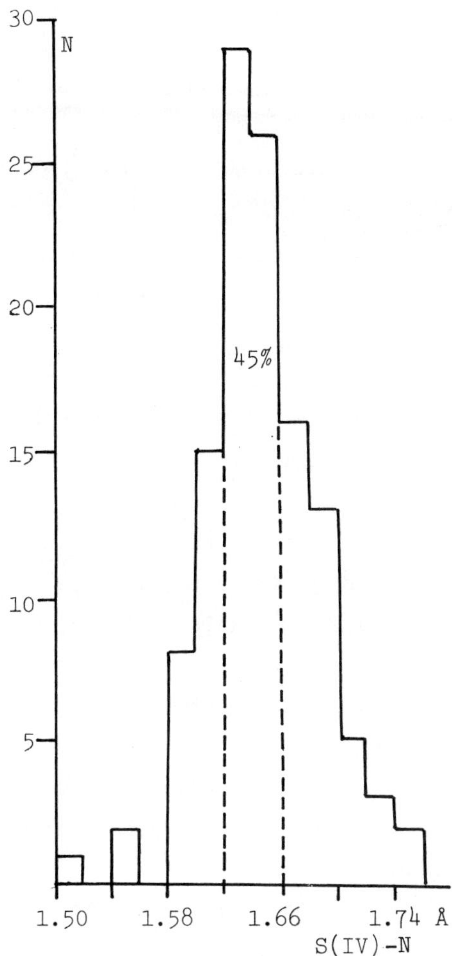

FIG. 2. Numerical distribution (N) of the S(IV)-N distances
observed for 40 S(IV)N$_3$ moieties found in 28 structures.

that the N-lone pair exerts maximum interaction with that of
the S atom when they are perpendicular which results in the
maximum shortening of the corresponding S-N bond (1.59 Å in
IPCLTZ10, Kálmán et al, 1979) while their interaction is

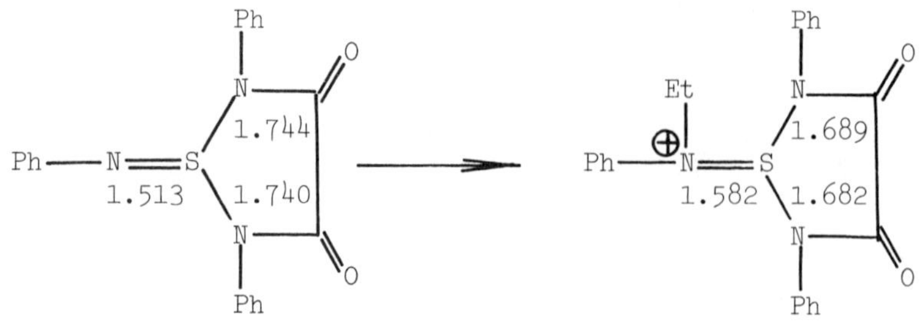

FIG. 3. The effect of the alkylation (R = Et) of the exocyclic
N atom on the bonding of the $S(IV)N_3$ moiety in 1-phenylimino-
-2,5-diphenyl-1λ^4,2,5-thiadiazolidine-3,4-dione (Gieren et
al, 1980).

minimum when they are antiperiplanar (e.g. the longest S-N bond
of 1.679 Å in BINLOZ, Romming et al, 1982).

An excellent example of the lone pair effect on the $S(IV)N_3$
moiety is given by the alkylation of the exocyclic nitrogen atom
in 1-phenylimino-2,5-diphenyl-1λ^4,2,5-thiadiazolidine-3,4-dione
(PPIMTZ, Neidlein et al, 1977) to gain 1-ethyl-1-phenyliminium-
-2,5-diphenyl-1λ^4,2,5-thiadiazolidine-3,4-dione (EPXPTZ10,
Gieren et al, 1980). With the conservation of valence number
(Σn = 3.93 vs 3.95) it considerably alters the three S(IV)-N
bond lengths as depicted in Fig. 3.

2.1.2. Summary

The 40 Σn values with a grand mean of 4.00 show a continuous
and Gaussian distribution (Fig. 4) supporting the validity of
the developed paradigm. Of course, in a few cases the shortcom-
ings of the data collections, the refinements of atomic parameters, etc. account for the observed deviations. However, there
still remain a few puzzling questions to be answered. For example, in 1,5-bis(triphenylphosphimino-cyclotetrathiazene

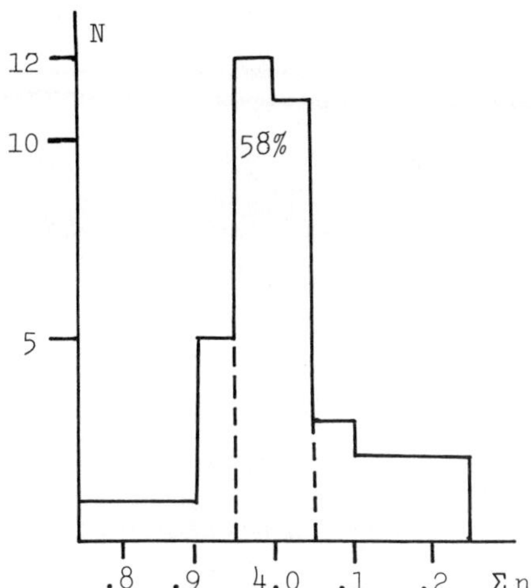

FIG. 4. Numerical distribution (N) of the V = Σn values for
40 $S(IV)N_3$ pyramids investigated in 28 crystal structures.

(TPIMSO1O, Bojes et al, 1981) there are two $S(IV)N_3$ groups
characterized by two different mean S-N bond distances:

$$< 1.647(4) \text{ Å} > \underline{vs} < 1.628(4) \text{ Å} >$$

2.2.   Second statement

The orientation of the N-lone pair(s) to the S(VI)[O,O,N,C]
and S(VI)[O,O,N,N] tetrahedra has significant effect upon
their geometry especially on the bond lengths and angles which
involve N atom(s).

Early fact-gathering also started with the study of N-acyl
sulphilimines and related compounds discussed above. From the
study of these structures it became apparent, that the S(VI)-N
bond lengths are dependent on the interactions in which the N-
-lone pairs are involved (e.g. intramolecular hydrogen bonds cf.
Kálmán et al, 1972, 1981a; N-alkylation cf. Cook et al, 1971).

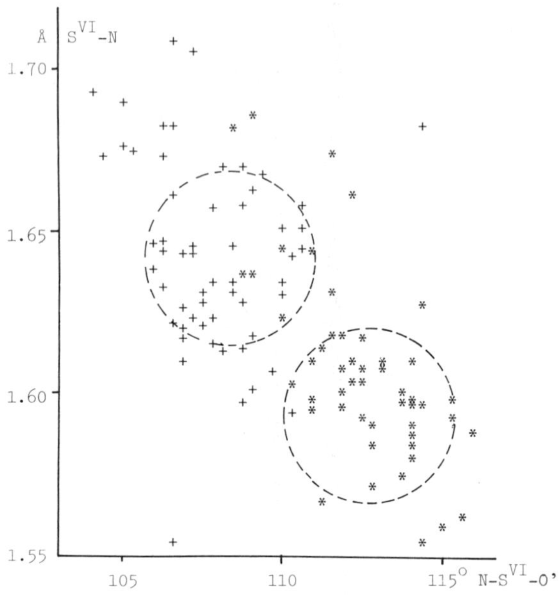

FIG. 5. A scattergram of S(VI)-N distances <u>vs</u> the greater values
of the NS(VI)O pairs. (Reproduced from the paper of Kálmán
<u>et al</u>, 1981; with the permission of Acta Crystallographica).
(+ is for arylamino, whilst * is for arylimino structures).

Novel observations have helped to articulate a paradigm during
the last few years (Kálmán <u>et al</u>, 1980, 1982). A search of
the CCDC files for compounds containing $(Aryl)[SO_2NXX']$ (X,X'
= H,C,N,S,P) fragments (121 were found) (Kálmán <u>et al</u>, 1981b)
led, among others, to a scattergram of S(VI)-N distances plotted against the greater angles of the NS(VI)O pairs (Fig. 5).
Apart from points scattered at random, two clusters may be
noted, one at the lower right corner formed almost exclusively
by structures of <u>arylimino</u> type and the other is formed predominantly by the structures of <u>arylamino</u> type. However, a few

structures violate this observation. The analysis of sulfon-amines with markedly different NS(VI)O angles showed that, in these structures, the lone-pair situated at the top of the pyramidal $N(sp^3)$ atom is antiperiplanar with O' which is in-volved in the greater NSO angle (Fig. 6). When the NSO angles are nearly equal then the lone pair bisects the O'S(VI)O angle. This suggests that the orientation of the N-lone pair has an influence on the geometry of the S(VI) O,O,N,C moiety. Indeed, according to quantum chemical studies the observed antiperipla-nar lp-oxygen arrangement permits optimum n-σ* interaction. From this it follows that, in the arylimino structures where the N-lone pair lies in the plane of the $X-N(sp^2)-S(VI)$ moiety, maximum interaction between this lone pair and the S 3d orbitals occurs when one of the $X-N(sp^2)-S(VI)-O$ torison angles is nearly zero.

A correlation between the length of the S(VI)-N bonds and the torsion angles Φ = S(IV)=N-S(VI)-O inferred from 12 cyclic sulphilimine structures (Kálmán et al, 1984a)

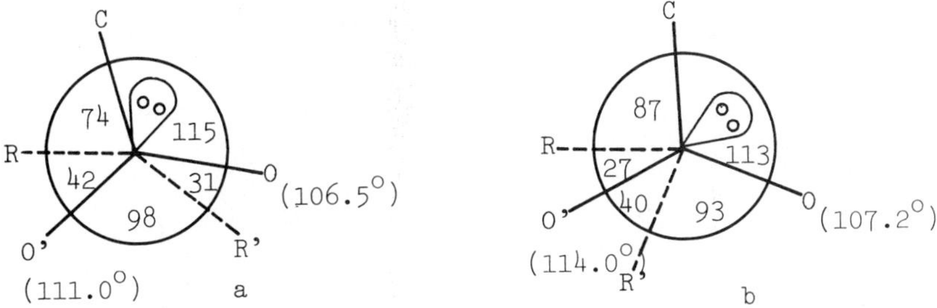

FIG. 6. Newman projections along S(VI)-N bonds for two selected structures of (Aryl)[SO_2NRR'] type: a) CXMESX (Hamodrakas and Filippakis, 1977), b) IBAZUN (Zacharis and Trefonas, 1970). The approximate orientations of the lone pairs situated at the top of the pyramidal $N(sp^3)$ atoms are also shown. The OSN angles are in parentheses.

$$\overset{\displaystyle\diagup^{R}}{C}S=N-SO_2-C_6H_4-Y \quad (Y = CH_3 \text{ or } NO_2)$$

by linear regression analysis (Kálmán et al, 1985)

$$d(S^{VI}-N) = 5.982.10^{-4}\,\Phi + 1.576 \text{ (Å)}$$

seems to substantiate this conclusion (R = 0.884). No better
correlation could be expected since in many structures there
are intra- and intermolecular close contacts (e.g. hydrogen
bonds (Kálmán et al, 1981a) which involve the N-lone pair to a
certain extent. Furthermore, only in 10 of the 121 retrieved
structures (Kálmán et al, 1981b) are there nearly synperiplanar
X-N-S(VI)-O torsion angles. In the majority (61 %) they are
about 40° which may be attributed to X...O 1,4-close contacts.

Currently, the structure determinations of two polymorphs
of famotidine a novel efficatious histamine H_2 receptor antagonist (Kálmán et al, 1986) enabled us to test our equation on
their sulfamoyl ($>C=N-SO_2NH_2$) groups. In both cases excellent
agreements were obtained:

	calc.	obs.
Polymorph A	1.610	1.608(1) Å
Polymorph B	1.609	1.609(2)

In addition to the above Newman projections perpendicular to

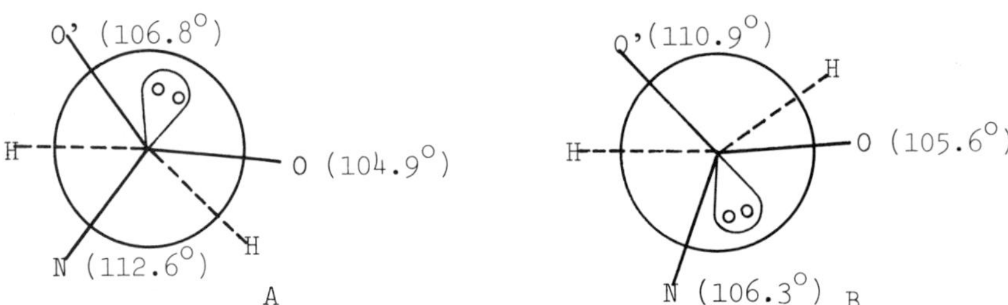

FIG. 7. Newman projections of the sulfamoyl groups along
$S(VI)-N(sp^3)$ bond observed in the polymorphs (A and B) of
famotidine. The $S(VI)-N(sp^3)$ is 1.602 for A and 1.630 Å for B.

the $S(VI)-N(sp^3)$ bond (Fig. 7) have shown different orientations of the pyramidal $N(sp^3)H_2$ group relative to the
$S(VI)[O,O,N,N]$ tetrahedra which result in different NSO and SNS
angles and $S-N(sp^3)$ bond lengths. This underscores that the
N-lone pairs, <u>via</u> n-σ* interactions, have indeed fundamental
effect on the configuration of the tetrahedra pertaining to
S(VI) atoms. Further analysis of these effects will be carried
out.

The last paradigm is fundamentally based on the importance
of the N-lone pairs in the recognized structural relationships.

2.3.   Third statement.

Hetero aromatic five-membered rings built up by C and N atoms
(imidazole, s- and v-triazole derivatives, etc.) exhibit only
two patterns for the endocyclic bond angle magnitudes which are
induced dominantly by the number and relative position of the
N-lone pairs.

Fact-gathering started with a crystallographic X-ray study
of tautomerism shown by imidazole and v-triazole derivatives of
biological importance (Lempert <u>el al</u>, 1973; Kálmán <u>et al</u>, 1974;
Kálmán and Simon, 1975). Naturally, in the absence of a
paradigm or some candidate for a paradigm, all of the facts
that could possibly pertain to the explanation of a phenomenon
are likely to seem equally relevant. Consequently, at the
beginning of this work all multiple bonds, hydrogen positions
(located in difference maps), bond angles belonging to proton-
-bearing or -free N atoms and hydrogen bonding were considered
thoroughly with equal weight. The first draft of a rule could
be set up only when enough experimental data were available,
i.e. by the late seventies (Párkányi <u>et al</u>, 1977). Further
<u>articulation</u> of the recognized <u>bond angle inequalities</u> was

done when the studies were extended to s-triazole structures
(Kálmán and Argay, 1983). It was ascertained then that (due to
the cross dependence of angular parameters ($\Sigma_{i=1}^{5} \alpha_i = 540^{\circ}$)) in
these planar hetero rings, there are only two sequences of the
bond angle magnitudes:

$$A: \sim a > b \ll c \gg d < e \sim$$
$$B: > a < b > c < d > e >$$

The angles a,b,c,d and e are denoted in Fig. 8. The first
sequence in accord with VSEPR rules (Gillespie, 1963), is
induced by two separated (1,3 = b,d) lone pairs exerting their
repulsion directly on the angles b and d. The second can either
arise from a lone pair with direct effect on the bond angle c
(subgroup B1) or from two vicinal (1,2 = a,e) lone pairs via
an indirect effect on the whole planar system (subgroup B2).
Recently, a crystal structure determination of an isomeric pair
of s-triazoles of the type depicted in Fig. 9 furnished
excellent proof of these rules (Kálmán et al, 1984b). It was
found that the sequence of the endocyclic bond angle magnitudes
for isomer A starting from the R substituted N atom (in
clockwise direction) is identical with that for isomer B except
that in the latter it follows a counterclockwise direction:

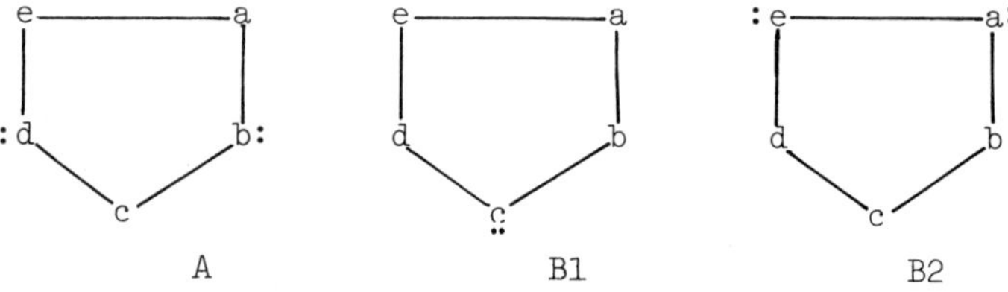

FIG. 8. The possible lone pair arrangements in imidazole, v-
and s-triazole derivatives. Letters a,b,c,d and e stand for
endocyclic bond angle magnitudes.

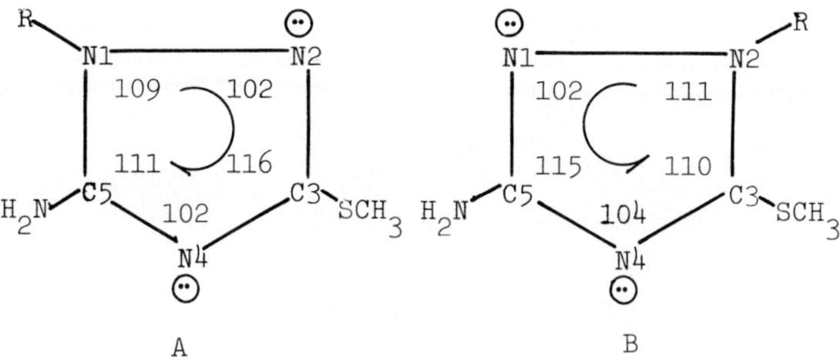

FIG. 9. The sequences of the bond angle magnitudes observed in
1-[2-(2,6-dichlorophenoxy)-ethyl]-3-methylthio-5-amino-1H-s-
-triazole (A) and 2-[2-(2,6-dichlorophenoxy)-ethyl]-3-methyl-
thio-5-amino-2H-s-triazole (B) (Kálmán et al, 1984b).

These examples not only substantiate this paradigm but
suggest that "mutatis mutandis" it may also be extended to
other heteroaromatic five-membered rings irrespective of their
nature and the type of their atoms.

REFERENCES
BOJES, J., CHIVERS, T. CORDES, A.W., MACLEAN, G. and OAKLEY,
R.T. (1981). Inorg. Chem. 20, 16.
CAMERON, A.F., HAIR, N.J. and MORRIS, D.G. (1973). J.Chem. Soc.
Perkin Trans. 2, 1951.
CHIVERS, T. and FIELDING, L. (1978). J.Chem. Soc. Chem. Commun.
212.
CHIVERS, T. and PROCTOR, L. (1978). J.Chem. Soc. Chem. Commun.
642.
COOK, R.E., CLICK, M.D., RIGAU, J.J., JOHNSON, C.R. (1971).
J. Amer. Chem. Soc. 93, 924.
FLUES, W., SCHERER, O.J., WEISE, J. and WOHLMERSHÄUSER, G.
(1976). Angew. Chem. 88, 411.

GIEREN, A. and PERTLIK, F. (1974). Abstracts of the 2nd European Cryst. Meeting, Keszthely, p. 303.

GIEREN, A. and NARAYANAN, P. (1975). Acta Cryst. A31, S 120.

GIEREN, A., DEDERER, B. and ABELEIN, I. (1980). Z. Anorg. Allg. Chem. 470. 191.

GILLESPIE, R.J. (1963). J. Chem. Educ. 40, 295.

HAMODRAKAS, S. and FILIPPAKIS, S.E. (1977). Cryst. Struct. Commun. 6, 209.

HOLT, E.M. and HOLT, S.L. (1974). J.Chem. Soc. Dalton Trans. 1990.

JOHNSTON, H.S. (1960). Adv. Chem. Phys. 3, 131.

KÁLMÁN, A. (1967). Acta Cryst. 22, 501.

KÁLMÁN, A. DUFFIN, B. and KUCSMAN, Á. (1971a). Acta Cryst. B27, 586.

KÁLMÁN, A., SASVÁRI, K. and KUCSMAN Á. (1971b). J. Chem. Soc. Chem. Commun. 144.

KÁLMÁN, A. and SASVÁRI, K. (1972) Cryst. Struct. Commun. 1, 243.

KÁLMÁN, A., SASVÁRI, K. and KUCSMAN, Á. (1973) Acta Cryst. B29, 1241.

KÁLMÁN, A., SIMON, K., SCHAWARTZ, J. and HORVÁTH, G. (1974). J. Chem. Soc. Perkin Trans. 2, 1849.

KÁLMÁN, A. and SIMON, K. (1975). Acta Cryst. A31, S 176.

KÁLMÁN, A. (1976). Abstract of the 5th Polish School or Crystal Structure Analysis, Trzebieszowice, p. 62.

KÁLMÁN, A., ARGAY, Gy., FISCHER, E., REMBARZ, G. and VOSS, G. (1977). J. Chem. Soc. Perkin Trans. 2, 1322.

KÁLMÁN, A., ARGAY, Gy., FISCHER, E. and REMBARZ, G. (1979). Acta Cryst. B35, 860.

KÁLMÁN, A., PÁRKÁNYI, L. and KUCSMAN, Á. (1980). Acta Cryst. B36, 1440.

KÁLMÁN, A., ARGAY, Gy., FISCHER, E. and TELLER, M. (1981a).
Acta Cryst. B37, 164.

KÁLMÁN, A., CZUGLER, M. and ARGAY, Gy. (1981b). Acta Cryst.
B37, 868.

KÁLMÁN, A. and ARGAY, Gy. (1983). J. Mol. Struct. 102, 391.

KÁLMÁN, A., KORITSÁNSZKY, T., JALSOVSZKY, I., RUFF, F. and
KUCSMAN, Á. (1984a). Acta Cryst. A40, S 274.

KÁLMÁN, A., PÁRKÁNYI, L. and REITER, J. (1984b) J. Mol.Struct.
118, 293.

KÁLMÁN, A., KORITSÁNSZKY, T., SVORONOS, P. and HORAK, V. (1985).
Abstracts of the 9th European Cryst. Meeting, Torino, p. 264.

KÁLMÁN, A., PÁRKÁNYI, L. HARSÁNYI, K., BOD, P. and HEGEDÜS, B.
(1986). Abstracts of the 10th European Cryst. Meeting, Wroclaw,
p. 69.

LEMPERT, K., NYITRAI, J., ZAUER, K., KÁLMÁN, A., ARGAY, Gy.,
DUISSENBERG, A.J.M. and SOHÁR, P. (1973). Tetrahedron 29, 3565.

LIQUORI, A. and VACIAGO, A. (1965). Gazz. Chim. Ital. 86, 769.

NEIDLEIN, R., LEINBERGER, P., GIEREN, A. and DEDERER, B. (1977).
Chem. Ber. 110, 3149.

PÁRKÁNYI, L., KÁLMÁN, A., ARGAY, Gy. and SCHAWARTZ, J. (1977).
Acta Cryst. B33, 3102.

PAULING, L. (1947). J. Amer. Chem. Soc. 69, 542.

ROMMING, C., NEVSTAD, G.O. and SONGSTAD, J. (1982). Acta Chem.
Scand. A36, 407.

TRUTER, M.R. (1962). J. Chem. Soc. 3400.

WEISS, J., RUPPERT, I. and APPEL, R. (1974). Z. Anorg. Allg.
Chem. 406, 32.

ZACHARIS, H. and TREFONAS, L.M. (1970). J. Heterocycl. chem.
7, 755.

31. Stereochemistry of strained, overcrowded bistricyclic ethylenes

Gil Shoham, Shmuel Cohen, Rachel Michal Suissa and Israel Agranat

Abstract

The bistricyclic ethylenes (**I** and **II**) are attractive substrates for the study of the stereochemistry of overcrowded ethylenes. Several of these systems have been shown to be thermochromic, photochromic, and piezochromic. The ethylene unit in **I** is highly overcrowded: the molecules cannot adopt a planar conformation in the ground state. The molecular structures of several representatives from the families of **I** (X=Y) and **II** (X≠Y) in the crystalline state are presented and discussed. Special attention is given to the notion of intramolecular overcrowding and to various modes of deviation from planarity: twist around the central CC double bond, *syn*- and *anti*-folding of the benzene rings, pyramidalization of the central carbon atoms, stretching of the central CC double bond, and distortions of the central double bond CCC angles. Several correlations and trends among the overcrowding parameters are revealed. The mixed structures (**II**) display an intermediate molecular conformation and local parameters corresponding to the sizes of the central rings. A certain correlation between the ethylenic twist and the degree of pyramidalization is indicated.

Introduction

The bistricyclic ethylenes (Fig. 1, where Y=X (**I**) and Y≠X (**II**)) are attractive substrates for the study of the ground-state conformation and the dynamic behavior of symmetrical (**I**) and asymmetrical (**II**) *overcrowded ethylenes* (Sandström, 1983).

STEREOCHEMISTRY OF STRAINED, OVERCROWDED BISTRICYCLIC ETHYLENES

The term "intramolecular overcrowding" was first introduced by Bell and Waring (1949) to denote aromatic systems which adopt non-planar forms to accomodate certain hydrogen atoms (e.g. hydrogen atoms at positions 4 and 5 in dibenzo[c,g]phenanthrene).

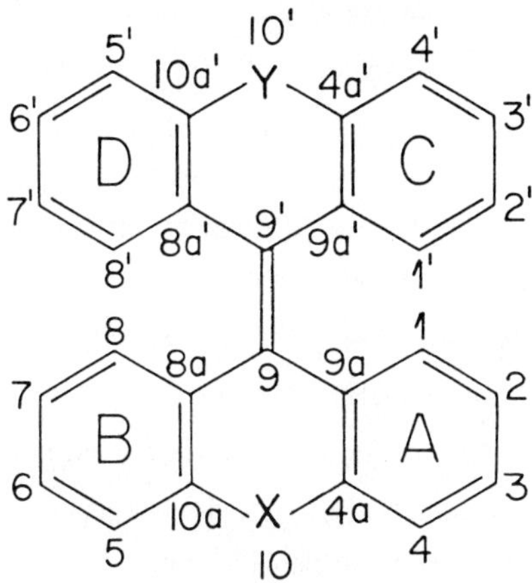

Figure 1. A scheme of Bistricyclic Ethylenes **I** and **II**.

Intramolecular overcrowding is a steric effect shown by aromatic structures in which the (intramolecular) distance of the closest approach between non-bonded atoms, calculated on the basis of conventional bond lengths and bond angles, is smaller than the sum of the van der Waals radii of the involved atoms (Harnik et al, 1951).

Why are the bistricyclic ethylenes overcrowded, or congested? An idealized coplanar bistricyclic ethylene would maintain very short nonbonded carbon-carbon and hydrogen-hydrogen distances (e.g. C1-C1' and H1-H1' in **I** or **II**), leading to a considerable overlap of the van der Waals radii in the region of the central carbon-carbon double bond (C9=C9' or the "pinch"). The associated repulsive interactions could in principle, be relieved by deviations from coplanarity and by various other distortions.

Consider the general formula of **I** with the four benzene rings labelled as *A* (C1, C2, C3, C4, C4a, C9a), *B* (C5, C6, C7, C8, C8a, C10a), *C* (C1',C2',C3',C4',C4a',C9a'), and *D* (C5',C6',C7',C8',C8a',C10a'). The following distortions of both **I** and **II** may be envisaged:

1. Stretching of C9=C9'

2. Distortions of the CCC angles around C9 and C9'.

3. Torsion around C9=C9'.

4. Folding of the benzene rings:

    a. *syn*-folding; leading to A(+), B(+), C(+), D(+) [where (+) means "up" and (-) means "down"].

    b. *anti*-folding; leading to A(+), B(+), C(-), D(-).

    c. alternate-folding; leading to A(+), B(-), C(-), D(+), or A(+), B(-), C(+), D(-).

    d. half-folding; leading to A(+), B(±), or C(+), D(±).

5. Pyramidalization of C9 and/or C9':

    a. pyramidalization of only C9 *or* C9'.

    b. pyramidalization of both C9 *and* C9':

        b1. *syn*-pyramidalization. b2. *anti*-pyramidalization.

6. Distortions of the central ring(s) of the tricyclic unit(s).

7. Distortions of the peripheral benzene rings.

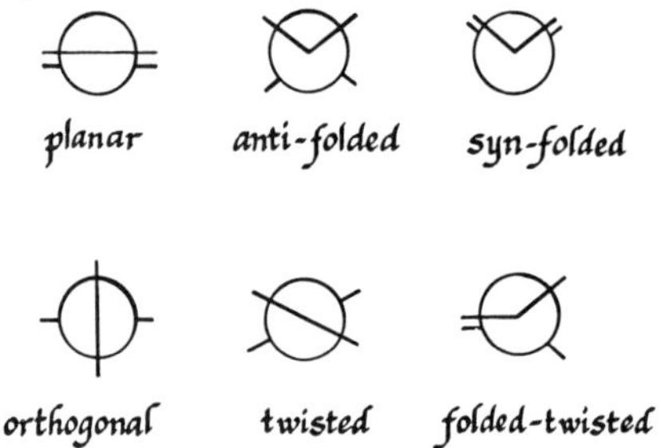

Figure 2. Newman projections along C9=C9' of classical conformations of Bistricyclic Ethylenes.

Obviously, the energetic cost of such distortions may vary considerably. The following points should be noted in the present context: Once the molecule adopts a non-planar conformation in the ground-state, it does not mean that it necessarily ceases to be overcrowded. Such a bistricyclic ethylene may still be overcrowded, but in a permissible manner. Of course, the degree of over-crowding would be smaller than in the classical planar conformation. Secondly, the molecule need not adopt just one mode of deviation from planarity to achieve the equilibrium conformation in the ground-state. Although various such modes should be considered concomitantly, in many cases it may be expected that no more than one of these modes would be

STEREOCHEMISTRY OF STRAINED, OVERCROWDED BISTRICYCLIC ETHYLENES

predominant (Greenberg and Liebman, 1978).

A schematic representation of some of the simplest overall molecular shapes of bistricyclic ethylenes is shown in Figure 2. These schemes demonstrate some of the possible molecular conformations that could result from the intramolecular overcrowding of bistricyclic ethylenes. We will refer to the definitions presented above and the schemes presented in Figure 2 later on in the discussion of the stereochemistry and deviation from planarity of overcrowded bistricyclic ethylenes.

The bistricyclic ethylenes enigma has fascinated chemists since thermochromism in bianthrone (I, X=Y=CO) was revealed by Meyer at the beginning of the century (1909). Thermochromism is the phenomenon of reversible change of color with the change of temperature. The thermochromism of bianthrones in solution was shown to result from a thermal equilibrium between two distinct and interconvertible isomeric species $(A \rightleftharpoons B)$, where A is the ground-state yellow species and B is the thermochromic green species absorbing at 650-730 nm (Bergmann, 1955; Day, 1963; Kortüm, 1974; Tapuhi et al, 1979; Fischer, 1984; Evans and Fitch, 1984). A,B type species were detected (along with other isomers) also in the photochromic and piezochromic phenomena exhibited by bianthrones. Photochromism is a phenomenon of photoinduced coloration which reverts either thermally or by irradiation Piezochromism is a phenomenon of reversible change of color with the change of pressure. It is generally accepted that the thermochromic B species and the photochromic B species are identical. Photochromism in bianthrones and related bistricyclic ethylenes has recently been reviewed (Fischer, 1984). The phenomenon of piezochromism has so far gained only little attention (Fanselow and Dickamer, 1974). The Stereochemical basis of these chromic changes is not fully understood at the present time, however we believe that as more structural information about various bistricyclic ethylenes is coming out, these interesting phenomena are going to be better understood.

In the following we review the structures of representative examples of the bistricyclic ethylene family reported in the literature, and present preliminary structural data of several other members of the same family which have been synthesized and analyzed in our laboratory in the past few years. We will look mainly at the mode and magnitude of distortion of the structure from the expected "classical" structure and especially at the deformation of the central double bond and the presumably planar central ethylene unit. We define first the parameters of distortion and then we will describe, compare and discuss the various distortions in this series of bistricyclic ethylenes.

STEREOCHEMISTRY OF STRAINED, OVERCROWDED BISTRICYCLIC ETHYLENES

Crystallographic Analyses of Bistricyclic Ethylenes

The first significant contribution of X-ray crystallography towards the elucidation of the molecular structures of the bistricyclic ethylenes was reported in 1954 (Harnik and Schmidt). They showed that yellow crystals of bianthrone (I, X=CO (1)) adopt in the ground-state a folded centrosymmetric geometry (the A form): the central rings are boat shaped and the tricyclic halves are folded in opposite directions at the olefinic termini, leading to an *anti*-folded conformation. Due to the relatively high crystallographic R factor and the relatively poor quality of the diffraction data of the original crystallographic analysis of bianthrone, this analysis has been repeated later with better data resulting in a more accurate and more reliable structure ((1'), Apgar and Wasserman, 1978). The case of another bistricyclic ethylene, 9,9'-bisfluorenylidene (I, X=— (2)), is an interesting one and illustrates the complexity of the subject. The red crystals of this hydrocarbon have been laying on the bench of crystallographers since 1877 (Arzruni, 1877; Taylor, 1936), leading to meaningful as well as erroneous conclusions pertaining to their molecular conformation. On the basis of a partial determination of the structure Fenimore (1948) first argued that the molecule is planar. Nyburg (1954), using a two-dimensional crystallographic analysis claimed that the molecule is *anti*-folded. Harnik et al (1954) reported the existence of two polymorphs, α and β, the latter being chiral crystallizing in the orthorhombic space group $P2_12_12_1$. Baily and Hull (1978) concluded, on the basis of a three-dimensional crystallographic analysis, that the molecule is twisted, with an average twist angle of 35°. They also analyzed the structure of the 1,1'-diisopropyl ester of 9,9'-bisfluorenylidene 1,1'-dicarboxylic acid (3), and demonstrated that its structure is generally similar to the parent compound. Lately, Lee and Nyburg (1985) reported the molecular structure of two forms of 2, α (achiral, (2a)) and β (chiral, (2b)), both in the twisted conformation, with an average twist angle of 41.2° and 40.1°, respectively. Very recently, Ballester et al (1985) reported the synthesis of the dark-blue, highly overcrowded and twisted, perchloro-9,9'-bisfluorenylidene and noted a twist angle of 67° (the complete structure has not yet been published). Mills and Nyburg (1963) analysed two forms of dixanthylene (I, X=O, (4)) and showed that both the yellow β-form (4b) and the thermochromic deep-blue-green α-form (4a) of dixanthylene are *anti*-folded, with an A-B dihedral angle of 38.3° and 41.1°, respectively. Dichmann et al (1974) investigated yet another interesting member of the bistricyclic ethylene family, 5,5'-bis(5H-dibenzo[a,d]cycloheptene (I, X = CH=CH, (5)), where each tricyclic system contains a central 7-membered ring. They analysed two isomers, which turned out to be the *syn*-folded and *anti*-folded conformations of this compound. The *syn* isomer crystalizes in a

STEREOCHEMISTRY OF STRAINED, OVERCROWDED BISTRICYCLIC ETHYLENES

$P2_1/c$ unit cell which contains two independent molecules (**5a** and **5b**) which are practically identical, while the *anti* isomer crystallizes in a different $P2_1/c$ cell with only one independent molecule (**5c**). The molecules of both isomers were shown to be severly overcrowded resulting in rather distorted and far from planar geometries. On the basis of the structures it has been predicted that the *anti* isomer would be the more stable of the two at all temperatures.

The numbering scheme, references and basic crystallographic data concerning the ten crystal structures mentioned above, are summarized in Table 1. It should be noted that **1** and **1'** refer to the crystal structure of the same compound in an identical space group and unit cell; however, **1'** represent the more recent and more accurate analysis of the two and therefore would be the one to be considered in the discussion below. **4a** and **4b** refer to two different crystal forms, which can possibly represent the two thermochromic forms, of dixanthylene. **5a** and **5b** refer to two independent molecules in one (*syn*-folded conformer) of the two crystal forms of 5,5'-bis(5H-dibenzo[a,d]cycloheptene). **5c** refer to the second (*anti*-folded conformer) of these crystal forms.

During the past few years we have synthesized and characterized a number of other bistricyclic ethylenes. Several of these were subjected to single crystal structural analysis utilizing X-rays. Of these we include in the present discussion the structural data of the following seven interesting molecules listed in Table 2. These are: 10,10,10',10'-tetramethyl-10,10,10',10'-tetrahydrodianthrylidene (**6**); 10,10'-bis(dicyanomethylene)-10,10,10',10'-tetrahydrodianthrylidene (**7**); 10,11,10',11'-tetrahydro-5,5'-bis(5H-dibenzo[a,d]cycloheptene) (**8**), (Schönberg et al, 1969); 10-(9-fluorenylidene)-9-anthrone (**9**), (Bergmann et al, 1952; Ismail and El-Shafei, 1957); 10-(9-fluorenylidene)-9-dicyanomethylene-9,10-dihydroanthracene (**10**); and 10-(10-xanthylidene)-9-dicyanomethylene-9,10-dihydroanthracene (**11**), of which we present two crystal forms (**11a** and **11b**). Stereoscopic drawings of **6** and **9** are shown in Figures 3 and 4, respectively.

The basic crystallographic data related to compounds **6-11** is summarized in Table 2. None of these data have so far been published. It should be noted that a number of the structures are not in the final stage of analysis, although generally they are very accurate and reliable.

Distortion Parameters

In this section we define and present various structural parameters characterizing the conformation of bistricyclic ethylenes, especially those related to modes of distortion from "classical structure", including deviations

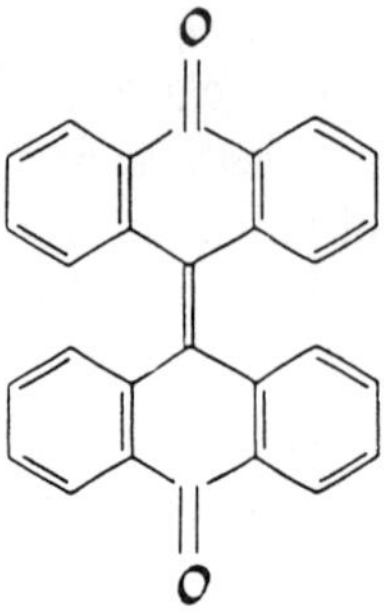

1

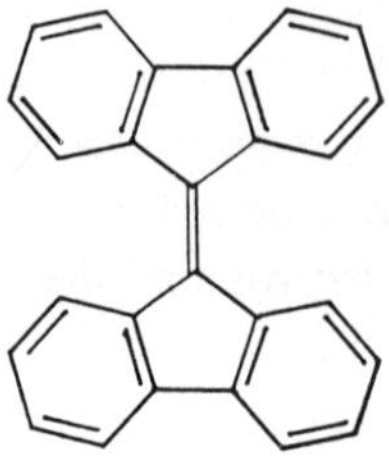

2

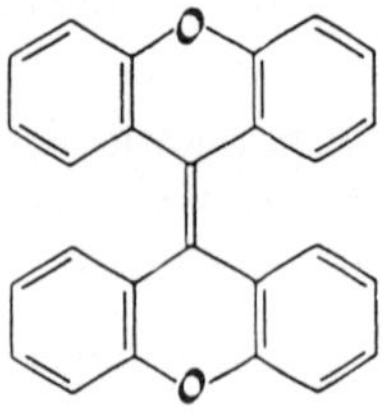

4

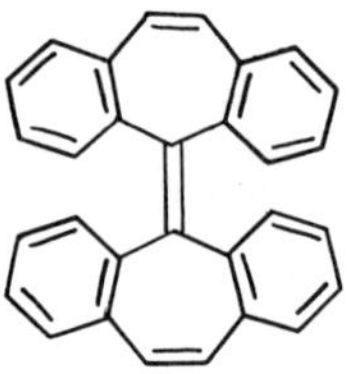

5

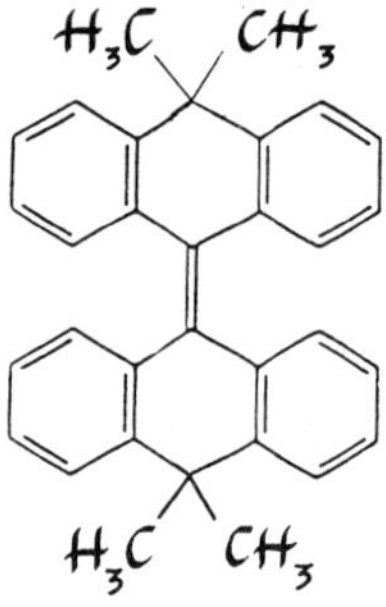

6

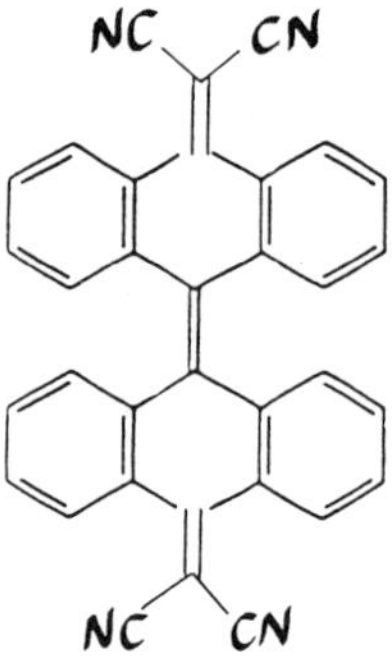

7

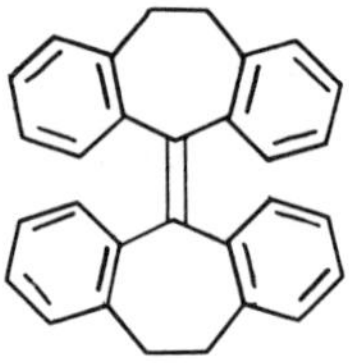

8

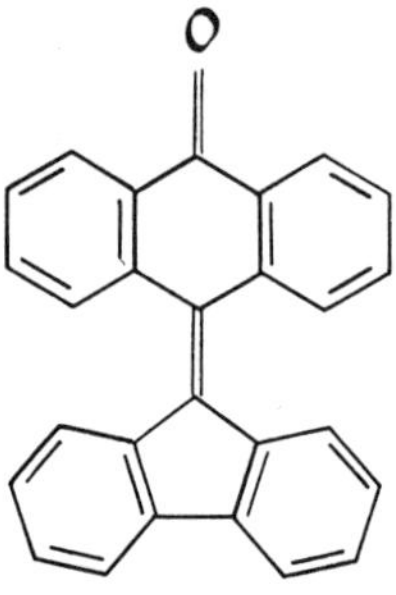

9

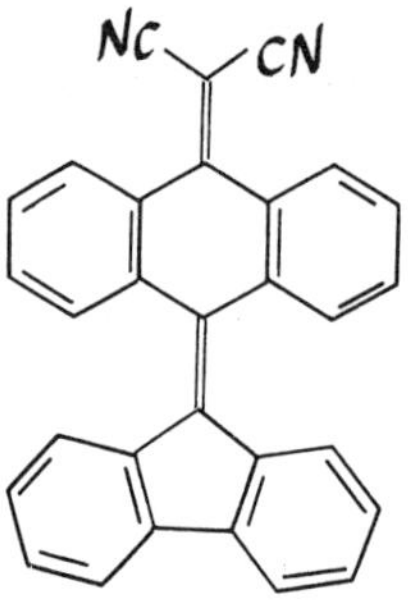

10

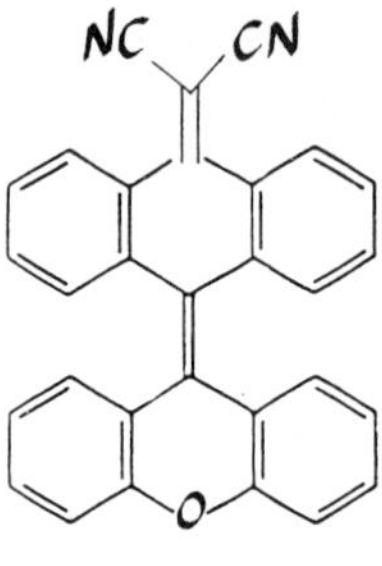

11

STEREOCHEMISTRY OF STRAINED, OVERCROWDED BISTRICYCLIC ETHYLENES

Table 1. Crystallographic Data of Bistricyclic Ethylenes (I)- Literature.

No.	X	Y	Color	Space Group[†]	R	Year
1	C=O	C=O	yellow	$P2_1/c$*	0.20	1954
1'	C=O	C=O	yellow	$P2_1/c$*	0.069	1978
2a	-	-	red	$Pbcn$	0.074	1985
2b	-	-	red	$P2_12_12_1$	0.050	1985
3‡	-	-	red	$I2/c$	0.095	1978
4a	O	O	blue-green	$A2/n$*	0.21	1963
4b	O	O	yellow	$C2/c$*	0.21	1963
5a	HC=CH	HC=CH	colorless	$P2/c$	0.059	1974
5b	HC=CH	HC=CH	colorless	$P2_1/c$	0.059	1974
5c	HC=CH	HC=CH	colorless	$P2_1/c$*	0.038	1974

Table 2. Crystallographic Data of Bistricyclic Ethylenes (I,II) - Present Work.

No.	X	Y	Color	Space Group[†]	R
6	$C(CH_3)_2$	$C(CH_3)_2$	colorless	$P2_1/c$*	0.043
7	$C=C(CN)_2$	$C=C(CN)_2$	orange-red	$P2_1/c$*	0.064
8	CH_2-CH_2	CH_2-CH_2	colorless	$P2_1/n$	0.072
9	-	C=O	yellow	$P2_1/n$	0.035
10	-	$C=C(CN)_2$	yellow	$P2_1$	0.045
11a	O	$C=C(CN)_2$	orange	$P2_1/c$	0.079
11b	O	$C=C(CN)_2$	orange	$P\bar{1}$	0.112

Notes for Tables 1 and 2:
† - a (*) next to the s.g. indicates a crystallographic $\bar{1}$ molecular symmetry.
‡ - there is an additional $-CO_2iPr$ group at the 1 and 1' positions.

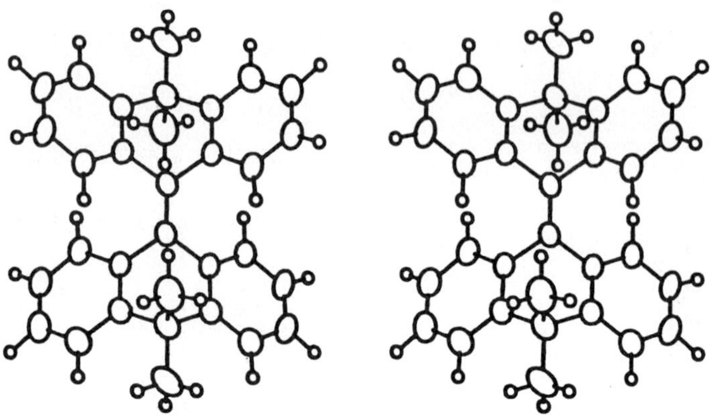

Figure 3. Stereodrawing of **6**.

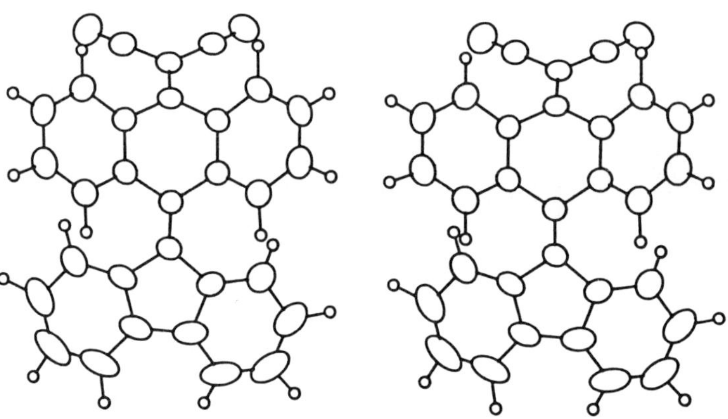

Figure 4. Stereodrawing of **9**.

STEREOCHEMISTRY OF STRAINED, OVERCROWDED BISTRICYCLIC ETHYLENES

from planarity.

Folding

Folding is the *molecular effect* describing the direction and the degree to which the *tricyclic units* of a bistricyclic ethylene are folded. We are especially interested in folding which is manifested by a symmetrical deviation from planarity of the tricyclic unit, namely, the out-of-plane bending of the two peripheral benzene rings from the center of the tricyclic system and its least squares plane. Several parameters could be used to describe this phenomenon, and in a few cases it may be necessary to use more than one parameter to fully characterize the folding. However, for simplicity and clarity we present only the dihedral angles between the least squares planes of the peripheral benzene rings of the same tricyclic unit (A-B and C-D) as a measure for folding. It should be noted that these dihedral angles describe best the folding only when there is no twist or other nonsymmetrical distortions of the tricyclic unit. In the ideal (and more common) case of symmetrical folding, however, the value of these dihedral angles describe rather well the magnitude of folding in the tricyclic systems.

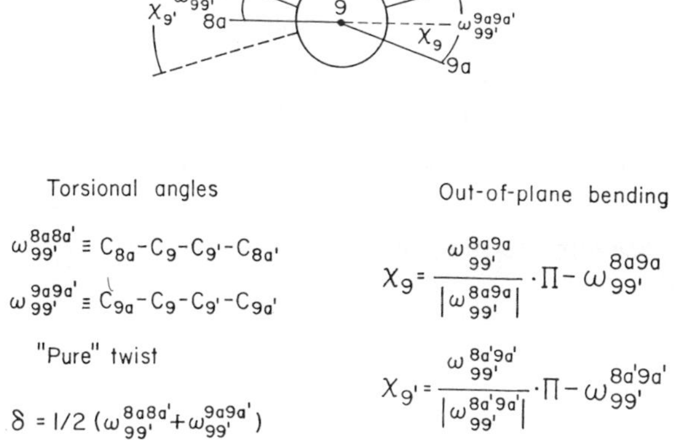

$$\omega_{99'}^{8a8a'} \equiv C_{8a}-C_9-C_{9'}-C_{8a'}$$

$$\omega_{99'}^{9a9a'} \equiv C_{9a}-C_9-C_{9'}-C_{9a'}$$

$$\chi_9 = \frac{\omega_{99'}^{8a9a}}{|\omega_{99'}^{8a9a}|} \cdot \Pi - \omega_{99'}^{8a9a}$$

$$\chi_{9'} = \frac{\omega_{99'}^{8a'9a'}}{|\omega_{99'}^{8a'9a'}|} \cdot \Pi - \omega_{99'}^{8a'9a'}$$

$$\delta = 1/2 \left(\omega_{99'}^{8a8a'} + \omega_{99'}^{9a9a'} \right)$$

Figure 5. Definitions of torsion angles and out-of-plane angles of Bistricyclic Ethylenes.

Twisting

We consider two kinds of twisting modes in bistricyclic ethylenes: the molecular twist and the twist of the ethylene unit (or the twist of the central double bond) alone. The molecular twist is the torsion effect between the two tricyclic halves of the molecules and is best described by the "torsion" angles

STEREOCHEMISTRY OF STRAINED, OVERCROWDED BISTRICYCLIC ETHYLENES

$\omega_{99'}^{11'}$ and $\omega_{99'}^{88'}$. The twist of the central double bond is the torsion effect exerted on the ethylenic unit and is best described by the torsion angles $\omega_{99'}^{8a\,8a'}$ and $\omega_{99'}^{9a\,9a'}$ (Figure 5). Pure twist of the central double bond is a measure of the net torsion effect in the ethylenic unit which may be represented by δ [where $\delta=\frac{1}{2}(\omega_{99'}^{8a\,8a'}+\omega_{99'}^{9a\,9a'})$] (Ermer, 1976). It should be emphasized that the molecular twist and the twist of the ethylenic unit should not necessarily be correlated, especially in molecules where the tricyclic systems are not planar (*vide infra*).

Pyramidalization

Pyramidalization of alkenes have attracted attention since Mock et al (1972) have indicated that out-of-plane (oop) deformation in ethylenes are not necessarily pure torsions but may be accompanied by significant bending and hence *pyramidalization* of the bonds at the double bonds carbon atoms (Mock, 1972; Radom et al, 1972; Ermer, 1974, 1976, 1977; Volland et al, 1979; Houk, 1983; Houk et al, 1983a, 1983b; Szeimies, 1983; Jeffrey et al, 1985).

Consider the central ethylenic unit C9=C9' in bistricyclic ethylenes **I** and **II**. The relevant hypothetically trigonal systems of C9,C8a,C9a,C9' and C9',C8a',C9a',C9 could be visioned as trigonal pyramids with vertexes at C9 and C9', respectively. The degree of pyramidalization of the double bond C9=C9' may be manifested by the following parameters:

1. The altitudes of the pyramids, i.e., the distances between C9 and the C8aC9'C9a plane and between C9' and the C8a'C9C9a' plane.

2. The deviations from 360° of the sum of the angles around C9 and around C9'.

3. The out-of-plane (oop) bending angles χ_9 and $\chi_{9'}$, in the Newman projection of the C9=C9' bond.

Figure 5 describes the torsional angles $\omega_{99'}^{8a\,8a'}$ and $\omega_{99'}^{9a\,9a'}$ and the out-of-plane bending angles χ_9 and $\chi_{9'}$, as defined by Ermer (1976, 1977), in the Newman projection of the C9=C9' central double bond of **I** and **II**.

While the first two pyramidalization parameters are relatively insensitive, the oop parameters χ_9 and $\chi_{9'}$ are very sensitive to the degree of pyramidalization. Moreover, the oop parameters also permit a distinction between *syn*-pyramidalization and *anti*-pyramidalization. In principle, pyramidalization of the C9=C9' bond implies a certain contamination of sp^3 hybridization character in the sp^2 hybridization character of the centers at C9 and C9'. Figure 6 present a scheme of the sp^2 and sp^3 orbital combination in the cases of *syn*- and *anti*-pyramidalization (Greenberg and Liebman, 1978). *Syn*- pyramidalization implies $\chi_9>0$ and $\chi_{9'}>0$ or $\chi_9<0$ and $\chi_{9'}<0$, while *anti*-pyramidalization

STEREOCHEMISTRY OF STRAINED, OVERCROWDED BISTRICYCLIC ETHYLENES

implies $\chi_9>0$ and $\chi_{9'}<0$ or $\chi_9<0$ and $\chi_{9'}>0$. A π-Orbital Axis Vector analysis (POAV) of conjugation and hybridization in non-planar conjugated molecules has recently been described (Haddon and Scott, 1986; Haddon, 1986).

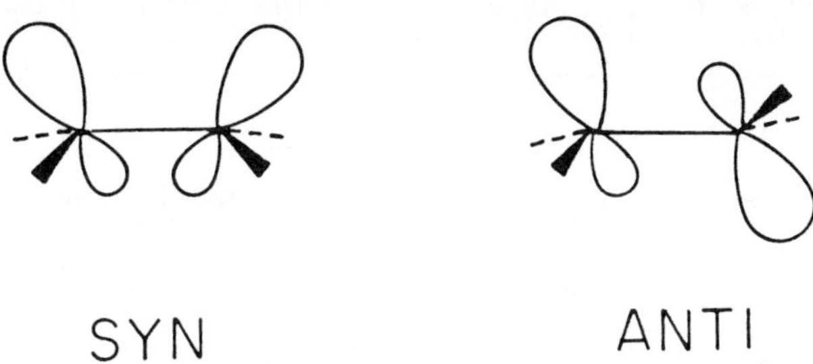

Figure 6. *Syn-* and *Anti*-Pyramidalization.

It should be noted that the distortion parameters and modes of deviation from classical structures as defined above refer only to theoretical and isolated phenomena. In the real cases, as we shall see below, the modes of folding, twisting and pyramidalization are all mixed in a wide range of ratios, so that in some cases it is difficult to isolate the net twisting or folding effects. In all cases it is desirable to examine visually the structure of the molecule concerened (e.g. stereo drawing) to fully understand the overall conformation and the way in which all the distortion parameters are combined.

Molecular Shape of Bistricyclic Ethylenes

In order to determine the general overall shape of each of the bistricyclic ethylenes presented above (**1-11**), and assign them with one of the "classical" schemes presented in Figure 2, we examine the following structural parameters: a. the dihedral angles between the peripheral benzene rings in each half of the molecule (the dihedral angles A-B and C-D), which provide information pertaining to the degree of folding (*vide supra*); b. the molecular torsion angles $\omega_{99'}^{11'}$ and $\omega_{99'}^{88'}$, which provide information pertaining to the molecular twist and to the relative direction of folding in each half of the molecule. The molecular *twisted* conformation is characterized therefore by small A-B and C-D dihedral angles (i.e. each tricyclic half is nearly coplanar) and large $\omega_{99'}^{11'}$ and $\omega_{99'}^{88'}$ torsion angles carrying the same sign. The molecular *syn*-folded conformation is characterized by large A-B and C-D dihedral angles (i.e., each half is folded) but small $\omega_{99'}^{11'}$ and $\omega_{99'}^{88'}$ torsion angles. The molecular *anti*-folded conformation is characterized by large A-B and C-D dihedral angles and large $\omega_{99'}^{11'}$ and $\omega_{99'}^{88'}$ torsion angles carrying opposite signs.

STEREOCHEMISTRY OF STRAINED, OVERCROWDED BISTRICYCLIC ETHYLENES

Table 3. Molecular Shape of Bistricyclic Ethylenes (**I, II**).

No.	Dihedral Angles(°)*		Torsional Angles(°)		Intramolecular Distances(Å)		Molecular Conformation[†]
	A-B	C-D	$\omega_{99'}^{11'}$	$\omega_{99'}^{88'}$	Cl-Cl'	C8-C8'	
1'	40.0	(40.0)	-35.0	(35.0)	2.942	(2.942)	anti-f
2a	5.2	4.2	-40.8	-41.7	3.176	3.194	twisted
2b	2.7	4.5	-39.6	-40.7	3.147	3.200	twisted
3	6.2	2.4	-48.1	-43.5	3.311	3.314	twisted
4a	41.1	(41.1)	33.8	(-33.8)	2.97	(2.97)	anti-f
4b	38.3	(38.3)	-36.3	(36.3)	3.02	(3.02)	anti-f
5a	56.6	61.8	0.5	0.9	3.252	3.315	syn-f
5b	63.2	54.5	-2.9	-5.2	3.266	3.240	syn-f
5c	55.7	(55.7)	54.1	(-54.1)	3.481	(3.481)	anti-f
6	52.9	(52.9)	42.1	(-42.1)	3.125	(3.125)	anti-f
7	47.6	(47.6)	-38.3	(38.3)	3.026	(3.026)	anti-f
8	58.5	57.0	56.1	-51.7	3.379	3.393	anti-f
9	11.8	51.2	-19.8	27.5	3.098	3.023	anti-f
10	16.1	53.1	-20.5	30.5	3.027	2.985	anti-f
11a	41.5	50.9	34.4	-38.9	3.033	3.076	anti-f
11b	36.1	48.7	34.9	-33.8	2.93	2.99	anti-f

* Values in parenthesis are those required by symmetry.
† anti-f = *anti*-folded; syn-f = *syn*-folded.

STEREOCHEMISTRY OF STRAINED, OVERCROWDED BISTRICYCLIC ETHYLENES

In Table 3 we list these parameters in each of the sixteen structures examined, together with the molecular conformation determined on the basis of the criteria described above. The overall conformation was also visually examined utilizing stero drawing of the molecules, using our coordinates and those published in the literature. Table 3 also lists the Cl-Cl', C8-C8' distances. As Cl, Cl', C8, C8' and the corresponding hydrogen atoms are involved in the intramolecular overcrowding, it is of interest to examine their proximity at the adopted conformation. For comparison, the van der Waals distance between carbon atoms in a linear C-H..H-C system is around 4.0 Å (Bondi, 1964; Baur, 1972).

Table 3 demonstrates that there is a certain correlation between the overall conformation of the molecule and the size of the central ring of the tricyclic system. In the series of symmetric bistricyclic ethylenes (**I, 1'-8**) there is a clear distinction: molecules with five-membered central rings are twisted. Even the severly overcrowded derivative of 9,9'-bisfluorenylidene with $-CO_2Pr$ groups at positions 1 and 1' (**3**), which is expected to overcome a large degree of strain, is twisted and adopts an overall conformation similar to unsubstituted 9,9'-bisfluorenylidene (**2a** and **2b**). Among the folded structures, the most common and probably the most stable form is the *anti*-folded conformation which provides a better spatial solution to the overcrowding problem. It looks as if only the molecule with seven-membered central rings is sufficiently folded to allow the *syn*-folded conformation. *Syn*-folded conformation has been observed only in one of the two crystal forms of **5** (structures **5a** and **5b**) and even then there is an additional distortion of the molecule so that the two tricyclic halves are not parrallel but rather "bent" outwards to prevent intramolecular spatial conflict between the two halves. A *syn*-conformer for the molecule with partially saturated 7-membered central rings (**8**), has not been observed. However, a disorder in one of the 7-membered rings in the crystal structure of **8** hints that this part of the molecule is sufficiently flexible to allow conformations different from the *anti*-folded form.

The degree of folding in each part of the symmetrical bistricyclic ethylenes (**I**) clearly correlated with the size of the central ring and the nature of X. The folding dihedral angles (A-B and C-D) are larger in the tricyclic systems containing a 7-membered central ring (54-63°) than those in the systems containing a 6-membered central ring (38-53°). Within the series of molecules containing 6-membered central rings the folding angle varied according to the type and hybridization of X, such that $O \approx C=O < C=C(CN)_2 < C(CH_3)_2$. This trend in the folding angle could be explained at least in part on the basis of the sp^2/sp^3 character of the atom of X involved in the central 6-membered ring. The nature of X in the molecules containing 7-membered central rings seems to be less critical for the folding angle.

In the series of mixed, nonsymmetric bistricyclic ethylenes (structures **9-11b**) the assignment of a schematic overall conformation is less obvious. The "push-pull system (**11**) adopts a distorted, yet distinct, *anti*-folded conformation in both crystal forms (**11a** and **11b**). There again, the degree of folding in the two halves correlates well with the folding angle/nature of X relationship presented above, namely $X = C = C(CN)_2$ allows a larger folding angle than $Y = O$.

The molecules containing mixed 5- and 6-membered central rings (**9** and **10**) seem to adopt an intermediate distorted conformation between *anti*-folded and twisted. The tricyclic system containing a 6-membered central ring is as folded as in the symmetrical molecules. However, the fluorenylidene half is significantly less planar than in the symmetrical molecules (A-B = 11.8° (for **9**) and 16.1° (for **10**)). The molecular torsion angles are significantly smaller than in both the twisted and the *anti*-folded symmetrical structures. It appears therefore as if these mixed bistricyclic ethylenes suffer more strain and are probably more destabilized in the ground state in comparison with the symmetrical molecules.

The Cl-Cl' and the C8-C8' distances are all in the range of 2.9-3.5 Å . It would have been rather informative to examine the distance between the corresponding conflicting hydrogen atoms; however, in most of the structures under discussion the exact positions of the hydrogen atoms are not sufficiantly defined for the examination of non-bonded interactions and it will probably be necessary to perform a neutron diffraction study for that purpose. It should be noted that when examining the overcrowding effect around C1..C1' and C8...C8', one should consider not only the C1-C1' and C8-C8' distances but also the C-H direction in each pair. When considering the distances between the carbons alone, it looks as if the overcrowding problem is conformationally best overcome only in the bistricyclic ethylenes containing a 7-membered central ring (**5a-5c**, **8**), while in the other systems the overcrowding is only partially overcome, settling for a "barely bearable" minimal C..C distance of 2.9-3.2 Å . In the exceptional case of the overcrowded derivative **3**, the molecular twist angle, and therefore the relevant C...C distances, are significantly larger than in the parent compound, because of the severe overcrowding at the C1-C1' and C8-C8' regions.

STEREOCHEMISTRY OF STRAINED, OVERCROWDED BISTRICYCLIC ETHYLENES

Table 4. Pyramidalization parameters in Bistricyclic Ethylenes (III).

No.	Central Ring Size	Bond Length (Å) C9-C9'	Internal Angles(°)* C8a C9C9a	Internal Angles(°)* C8a' C9'C9a'	Altitudes (in Å) of C9	Altitudes (in Å) of C9'	OOP† Bending Angles(°) χ_9	OOP† Bending Angles(°) $\chi_{9'}$
1'	6	1.364	112.5	(112.5)	0.024	(-0.024)	3.6	(-3.6)
2a	5	1.367	105.1	104.6	-0.037	-0.024	6.0	3.9
2b	5	1.364	104.9	105.2	0.035	-0.014	5.7	2.2
3	5	1.390	106.7	107.1	0.052	-0.064	8.2	10.1
4a	6	1.368	109.4	(109.4)	-0.008	(0.008)	-1.2	(1.2)
4b	6	1.40	110.8	(-110.8)	-0.018	(0.018)	-2.7	(2.7)
5a	7	1.344	114.2	112.4	-0.030	0.055	-4.5	-8.3
5b	7	1.341	112.1	114.1	-0.042	0.040	-6.2	-5.9
5c	7	1.348	115.0	(115.0)	0.025	(-0.025)	3.7	(-3.7)
6	6	1.345	110.1	(110.1)	0.0	(0.0)	0.1	(-0.1)
7	6	1.354	111.0	(111.0)	0.011	(-0.011)	1.7	(-1.7)
8	7	1.347	115.3	115.4	0.023	0.029	3.4	-4.2
9	5,6	1.367	104.7	110.2	0.002	0.057	-0.4	-8.6
10	5,6	1.355	104.1	109.7	0.046	0.014	7.4	-2.1
11a	6	1.355	109.6	111.1	-0.011	-0.002	-1.7	0.3
11b	6	1.34	109.8	110.2	-0.017	-0.034	-2.7	5.2

* Values in parenthesis are those required by symmetry.
† OOP = Out-Of-Plane (see text).

Pyramidalization Effects in Bistricyclic Ethylenes

As discussed above, one of the possible ways to relieve overcrowding and strain in bistricyclic ethylenes is that of partial pyramidalization of the central carbons C9 and C9'. In order to determine to what extent this mode of distortion exists in bistricyclic ethylenes, we examined several of the parameters characterizing pyramidalization in the series of structures presented above.

Of the pyramidalization criteria discussed earlier we found that the sum of angles around C9 and C9' is a very insensitive measure. All these sums in all independent C9 and C9' centers range between 359.4° to 360.0° while the standard deviation of each angle in the sum is at least 0.2°, which make this parameter insignificant for pyramidalilzation. However, the altitudes of C9 and C9' above the plane of the three atoms connected to them, and the oop bending angles χ_9 and $\chi_{9'}$, appeared to be rather sensitive and significant in detecting pyramidalization. These parameters for structures **1'-11b**, together with the central double bond length (d) and the central ring internal angle (α) are listed in Table 4. The latter two parameters represent additional interesting modes intended to overcome the overcrowding problem, and also are indirectly related to pyramidalization. Since pyramidalilztion means an increase in the sp^3 character of a formally sp^2 carbon, it causes a deviation of α from 120° (classical sp^2 carbon) and a deviation of d from 1.34 Å (classical double bond length). In a naive way, the smaller α and the longer d are, the more pronounced pyramidalization would exist at the center examined. For the purpose of the following discussion we ignore structures **4a**, **4b** and **11b** since they do not seem to be sufficiantly accurate (R > 0.10) for a significant determination of the subtle deviations we are expecting.

The central double bond length ranges from 1.34 to 1.39 Å (a typical s.d. in a bond length in the good structures we examined here is < 0.005 Å) in the molecules discussed. There is a direct correlation between d and the size of the central ring, so that it increases when the central ring size decreases. The central bond is expecially long in the severly overcrowded compound **3**, while it displays only a small deviation from normal double bond in the systems containing 7-membered central rings. In the mixed molecules (**9** and **10**) d obtains an intermediate value according to the size of the two rings involved. A similar correlation exists between α and the size of the central ring (cf. Lee and Nyburg, 1985), a phenomenon which is expected from considerations of the preferred angle in an n-membered ring. It also indicates a higher degree of pyramidalization in the smaller rings.

The altidutes of C9 and C9' above the plane of the three carbons bonded to it is well correlated with the oop bending angles. Both parameters reflect, however, a different view upon the pyramidization of the central carbons. Both parameters show rather small values corresponding to a relatively small pyramidalization effect. Yet these parameters are significant, and in contrast to the earlier two, exhibit the net pyramidalization involved. These parameters indicate a considerable pyramidalization in the symmetrical compounds containing 5-membered central rings, especially **3**, which in addition carries the bulky substituents at the 1 and 1' positions. The pyramidalization effect seems to be rather small in the systems containing 6-membered central rings and this effect diminished completely in the tetramethyl derivative **6**. It is interesting to note, though, that the systems containing 7-membered central rings unexpectedly exhibit some degree of pyramidalization, especially the *syn*-folded conformers **5a** and **5b**.

The pyramidalization in the mixed compounds with 5- and 6-membered central rings is somewhat puzzling. While compound **10** displays the excpected ratio of pyramidalization (larger at the carbon of the 5-membered ring), the ratio in compound **9** is just the opposite. We do not have a satisfactory explaination for this pattern, although one should note that the Y's in the two compounds are considerably different.

Molecular and Double Bond Twisting in Bistricyclic Ethylenes

In Table 5 we compare the torsion angles at the central double bond, from which the pure twist (δ) of the ethylenic unit, and the torsion angles between the two halves of the molecules are derived. We also list again the schematic molecular shape in order to examine a possible correlation between the molecular conformation and the ethylenic twist. The table shows clearly that there is a significant ethylenic twist ($\delta=32\text{-}38°$) only in the symmetrical compounds with 5-membered central rings (those adoptiong twisted overall conformation). In all other compounds the pure twist of the ethylenic group is quite small. Even the presence of one 5-membered central ring in the mixed structures does not seem to be sufficient to cause a significant twist of the double bond.

Another interesting observation demonstrated in Table 5 is the fact that most of the molecular torsion angles for the folded conformations (except for the *syn*-folded conformers) are rather large (20-56°) yet the corresponding ethylenic torsion angles are rather small. Hence, there is no correlation between molecular and ethylenic torsion parameters and also no correlation between molecular folding and ethylenic twisting.

Table 5. Twisting of the Ethylenic Unit in Bistricyclic Ethylenes (**I,II**).

| No. | Ethylenic Torsional Angles (°)*† | | | Molecular Torsional Angles (°) | | Molecular Conformation |
	$\omega_{99'}^{9a\,9a'}$	$\omega_{99'}^{8a\,8a'}$	Pure (δ)	$\omega_{99'}^{11'}$	$\omega_{99'}^{88'}$	
1'	-3.6	(3.6)	0.0	-35.0	(35.0)	anti-f
2a	-31.9	-34.0	32.9	-40.8	-41.7	twisted
2b	-33.7	-30.2	31.9	-39.6	-40.7	twisted
3	-38.8	-37.0	37.9	-48.1	-43.5	twisted
4a	-1.2	(1.2)	0.0	33.8	(-33.8)	anti-f
4b	2.7	(-2.7)	0.0	-36.3	(36.3)	anti-f
5a	3.0	-0.8	1.1	0.5	0.9	syn-f
5b	-4.1	-3.8	3.9	-2.9	-5.2	syn-f
5c	3.7	(-3.7)	0.0	54.1	(-54.1)	anti-f
6	0.1	(-0.1)	0.0	42.1	(-42.1)	anti-f
7	1.7	(-1.7)	0.0	-38.3	(38.3)	anti-f
8	6.8	-0.8	3.0	56.1	-51.7	anti-f
9	7.2	-1.1	3.1	-19.8	27.5	anti-f
10	9.3	-0.3	4.5	-20.5	30.5	anti-f
11a	-0.8	-1.0	2.0	34.4	-38.9	anti-f
11b	-3.6	4.3	3.9	34.9	-33.8	anti-f

* Values in parenthesis are those required by symmetry.
† For definitions of torsion angles (ω) and pure twist (δ)
  see Figure 5 and the relevant text.

STEREOCHEMISTRY OF STRAINED, OVERCROWDED BISTRICYCLIC ETHYLENES

Summary of the Structural Distortions in Bistricyclic Ethylenes

In calculating and analysing the crystallographic data we were hoping to find a simple formula correlating these different, yet related, distortion modes. As it turned out, we could not find obvious relationships between the size of the central rings and the nature of X and Y on the one hand, and the degree of pyramidalization, the extent of twisting and the molecular conformation on the other hand. However, at this stage of the investigation we wish to summarize the several correlations and trends among the parameters we found in the structures of the sixteen bistricyclic ethylenes under examination: 1) In all the bistricyclic ethylenes examined we found, as expected, a considerable overcrowding effect leading to various theoretically possible distortion modes. 2) Tricyclic halves containing 5-membered central rings tend to remain planar, while those containing 6- and 7-membered central rings tend to fold. 3) As a result of (1), symmetrical bistricyclic ethylenes containing 5-membered central rings are forced to adopt a twisted conformation in which the pure twist angle of the central double bond is close to the molecular twist angle. 4) In symmetrical bistricyclic ethylenes containing 6- and 7-membered central rings the most stable confromation is the *anti*-folded form. In the systems containing 7-membered rings (and probably larger rings as well) a *syn*-folded conformation is possible, although it is less stable than the anti-folded form. 5) The mixed structures display, in most of the cases, an intermediate molecular conformation and local parameters corresponding to the sizes of the two rings involved. 6) There is a clear correlation between the size of the central ring, the internal CCC angle and the length of the central double bond. For 5-membered rings the CCC angle is 104-108° and the C=C bond length is 1.36-1.39 Å; for 6-membered rings the CCC angle is 109-113° and the C=C bond length is 1.345-1.37 Å; and for 7-membered rings the CCC angle is 112-116° and the C=C bond length is 1.34-1.35 Å. 7) In general, the central carbon at the 5-membered rings show higher degree of pyramidalization than the central carbon at the 6- and 7-membered central rings. 8) There is a certain correlation between the ethylenic twist and the degree of pyramidalization, but it seems as there is no correlation between folding of the tricyclic halves and pyramidalization.

As indicated above, when more structures of different bistricyclic ethylenes emerge, we hope to have a better understanding of the distortion modes in overcrowded systems in general and in bistricyclic ethylenes in particular. At the present time it seems that the simple formula we were looking for, the one correlating between the various structural parameters of these systems, will have to wait untill more structural data is available.

STEREOCHEMISTRY OF STRAINED, OVERCROWDED BISTRICYCLIC ETHYLENES

The static and dynamic stereochemistry of bistricyclic ethylenes **I** has been investigated by application of DNMR spectroscopy. (Gault et al, 1970; Sutherland, 1971; Agranat et al, 1972; Rabinovitz et al, 1974; Agranat et al, 1978; Agranat and Tapuhi, 1976, 1977, 1978, 1979a, 1979b; Tapuhi, 1978; Sandström, 1983; Bindl et al, 1985). The conformational behavior of bianthrones and related bistricyclic ethylenes has recently been studied also by electrochemical methods, with special emphasis on electron transfer reactions (Olsen and Evans, 1981; Neta and Evans, 1981; Hammerich and Parker, 1981; Ahlberg et al, 1981; Evans and Xie, 1982, 1983; Olsen et al, 1982; Evans and Busch, 1982; Matsue et al, 1984; Evans and Fitch, 1984). The relationship between the intramolecular overcrowding of bistricyclic ethylenes in the ground-state and the free energies of activation for thermal E,Z isomerization and conformational inversion processes will be discussed elsewhere. The streochemistry of the bistricyclic ethylenes should also be prone to modern computational methods, including molecular mechanics (force field) calculations (Favini et al, 1982; Lenoir and Lemmen, 1980).

Acknowledgement: We thank Dr. E. Wasserman (Central Research & Development Department, E.I. duPont de Nemours & Company) for the crystal structure of bianthrone.

References

AGRANAT, I, RABINOVITZ, M., WEITZEN-DAGAN, A and GOSNAY, I, (1972). *Chem. Comm.*, 732.

AGRANAT, I and TAPUHI, Y., (1976). *J. Am. Chem. Soc.*, **98**, 615.

AGRANAT, I and TAPUHI, Y., (1977). *Nouveau. J. Chim.*, **1**, 361.

AGRANAT, I and TAPUHI, Y., (1978). *J. Am. Chem. Soc.*, **100**, 5604.

AGRANAT, I and TAPUHI, Y., (1979a). *J. Am. Chem. Soc.*, **101**, 665.

AGRANAT, I and TAPUHI, Y., (1979b). *J. Org. Chem.*, **44**, 1941.

AGRANAT, I, TAPUHI, Y. and LALLEMAND, J.Y., (1979). *Nouveau. J. Chim.*, **3**, 59.

AHLBERG, E., HAMMERICH, O. and PARKER, V.D., (1981). *J. Am. Chem. Soc.*, **103**, 844.

APGAR, P.A and WASSERMAN, E, (1978). Unpublished results (Allied Chemical).

ARZRUNI, A (1877). *Z. Kristallogr.*, **1**, 434. Cited by GROTH, P., (1906-1919). *Chem. Kristallog.*, **5**, 431.

BAILEY, N.A and HULL, S.E., (1978). *Acta Cryst.*, **B34**, 3289.

BALLESTER, M., CASTANER, J., RIERA, J., DE LA FUENTE, G. and CAMPS, M., (1985). *J. Org. Chem.*, **50**, 2287.

BAUR, W.H., (1972). *Acta Cryst., Sect. B*, **B28**, 1456.

BELL, F. and WARING, D.H. (1949). *J. Chem. Soc.*, , 2689.

BERGMANN, E.D., (1955). *Prog. Org. Chem.*, **3**, 81.

BERGMANN, E.D., HIRSHBERG, Y. and LAVIE, D., (1952). *Bull. Soc. Chim. Fr.*, **19**, 268.

BINDL, J., BURGEMEISTER, T. and DAUB, J., (1985). *Chem. Ber.*, **118**, 4934.

BONDI, A.J., (1964). *J. Phys. Chem.*, **68**, 441.

DAY, J.H, (1963). *Chem. Rev.*, **63**, 65.

DICHMANN, K.S., NYBURG, S.C., PICKARD, F.H. and POTWOROWSKI, J.A., (1974). *Acta Cryst.* **B30**, 27.

ERMER, O., (1974). *Tetrahedron*, **30**, 3103.

ERMER, O., (1976). *Structure and Bonding*, **27**, 161.

ERMER, O., (1977). *Z. Naturfosch.*, **B32**, 837.

EVANS, D.H. and BUSCH, R.W., (1982). *J. Am. Chem. Soc.*, **104**, 5057.

EVANS, D.H. and FITCH, A., (1984). *J. Am. Chem. Soc.*, **106**, 3039.

EVANS, D.H. and XIE, N., (1982). *J. Electroanal. Chem.*, **133**, 367.

EVANS, D.H. and XIE, N., (1983). *J. Am. Chem. Soc.*, **105**, 315.

FANSELOW, D.L. and DRICKAMER, H.G., (1974). *J. Chem. Phy.*, **61**. 4567.

FAVINI, G., SIMONETTA, M., SOTTOCORNOLA, M. and TODESCHINI, R., (1982). *J. Comput. Chem.*, **3**, 178

FENIMORE, C.P., (1948). *Acta. Cryst.*, **1**, 295.

FISCHER, E., (1984). *Rev. Chem. Intermed.*, **5**, 393.

GAULT, I.R., OLLIS, W.D. and SUTHERLAND, I.O., (1970). *Chem. Comm.*, 269

GREENBERG, A. and LIEBMAN, J.F., (1978). "Strained Organic Molecules", Academic Press, New York, p 91.

HADDON, R.C., (1986). *J. Am. Chem. Soc.*, **108**, 2837.

HADDON, R.C. and SCOTT, L.T., (1986). *Pure. & Appl. Chem.*, **58**, 137.

HAMMERICH, O. and PARKER, V.D., (1981). *Acta Chem. Scand., Ser. B*, **B35**, 395.

HARNIK, E., HERBSTEIN, F.H., SCHMIDT, G.M.J., (1951). *Nature*, **168**, 158.

HARNIK, E., HERBSTEIN, F.H., SCHMIDT, G.M.J. and HIRSHFELD, F.L., (1954). *J. Chem. Soc.*, 3288.

HARNIK, E. and SCHMIDT, G.M.J., (1954). *J. Chem. Soc.*, , 3295.

HOUK, K.N., (1983). in "Stereochemistry and Reactivity of Systems Containing π Electrons", WATSON, W.H. Ed., Verlag Chemie International: Deerfield Beach, FL, pp 1-40.

HOUK, K.N., RONDAN, N.G., BROWN, F.K., JORENSEN, W.L., MADURA, J.D. and SPELLMEYER, D.C., (1983a). *J. Am. Chem. Soc.*, **105**, 5980.

HOUK, K.N., RONDAN ,N.G. and BROWN, F.K., (1983b). *Israel J. Chem.*, **23**, 3.

ISMAIL, A.F.A. and EL-SHAFEI, Z.M., (1957). *J. Chem. Soc.*, 3393.

JEFFREY, G.A., HOUK, K.N., PADDON-ROW, M.N., RONDAN, N.G. and MITRA, J. (1985). *J. Am. Chem. Soc.*, **107**, 321.

JOHNSON, R.P., (1986). in "Molecular Structure and Energetics", LIEBMAN, J.F. and GREENBERG, A., Eds., VCH Publishers, Deerfield Beach, FL, Vol. 3, p. 85 and references cited therein.

KORTUM, G., (1974). *Ber. Bunsenges. Phys. Chem.*, **78**, 391.

LEE, J.S. and NYBURG, S.C., (1985). *Acta. Cryst.*, **C41**, 560.

LENOIR, D. and LEMMEN, P., (1980). *Chem. Ber.*, **113**, 3112.

MATSUE, T., EVANS, D.E. and AGRANAT, I., (1984). *J. Electroanal. Chem.*, **163**, 137.

MEYER, H., (1909a). *Monatsh. Chem.*, **30**, 165.

MEYER, H., (1909b). *Ber.*, **42**, 143.

MILLER, S.I., (1978). *J. Chem. Educ.*, **55**, 778.

MILLS, J.F.D. and NYBURG, S.C., (1963). *J. Chem. Soc.,* , 308.

MOCK, W.L., (1972). *Tetrahedron Lett.,* 475.

MOCK, W.L., (1976). *Bioorg. Chem.,* **5**, 403.

NETA, P. and EVANS, D.H., (1981). *J. Am. Chem. Soc.,* **103**, 7091.

NYBURG, S.C., (1954). *Acta Cryst.,* **7**, 779.

OLSEN, B.A. and EVANS, D.H., (1981). *J. Am. Chem. Soc.,* **103**, 839.

OLSEN, B.A., EVANS, D.H. and AGRANAT, I., (1982). *J. Electroanal. Chem.,* **136**, 139.

RABINOVITZ, M., AGRANAT, I. and WEITZEN-DAGAN, A., (1974). *Tetrahedron Lett.,* 1241.

RADOM, L., POPLE, J.A. and MOCK, W.L., (1972). *Tetrahedron Lett.,* 479.

SANDSTROM, J., (1983). *Top. Stereochem.,* **14**, 83.

SCHONBERG, A., SODTKE, U. and PRAEFCKE, K., (1969). *Chem. Ber.,* **102**, 1453

SUTHERLAND, I.O., (1971). *Annu. Rep. NMR Spectrosc.,* **4**, 71.

SZEIMIES, G., (1983). in "Reactive Intermediates", Vol. 3, ABRAMOVITCH, R.A., Ed., Plenum Press, New York, p. 299.

TAPUHI, Y., (1978). Ph.D. Thesis, The Hebrew University of Jerusalem, Jerusalem, Israel.

TAPUHI, Y., KALISKY, O., AGRANAT, I., (1979). *J. Org. Chem.,* **44**, 1949.

TAYLOR, W.H. (1936). *Z. Kristallogr.,* **A93**, 151.

VOLLAND, W.V., DAVIDSON, E.R. and BORDEN, W.T., (1979). *J. Am. Chem. Soc.,* **101**, 533.

32. Structure/function studies on substituted strained molecules

Arthur Greenberg and Joel F. Liebman

Suitably-placed substituents offer the possibilities of extending the stability range of strained molecules and selectively activating bonds as reviewed recently (Greenberg and Stevenson, 1986). We will briefly illustrate the interplay between energy and structure in such molecules using selected examples.

1. CRITERIA FOR AROMATICITY IN DIPHENYLCYCLOPROPENONE

Experimental redeterminations of the heats of combustion and sublimation of diphenylcyclopropenone (1) indicated resonance energy for the 3-membered ring of 22 kcal/mol (Steele et al, 1985; Greenberg et al, 1983), subsequently verified by two other techniques (Grabowski et al, 1984; Davis et al, 1985). It seemed self-evident on the basis of this large resonance energy and the well-known stabilities of cyclopropenones that the system is aromatic. However, a very interesting article (Childs et al, 1986) concluded, on the basis of its structure, 1 is not aromatic since it evidences very pronounced bond alternation. A very compelling contrast is provided by the salt 2 (Childs et al, 1986) in which bond alternation has disappeared. While we must remember that models must be assumed in both cases, it appears that energy and structural considerations provide opposite conclusions in this case. A plausible resolution might be offered through a study of bond ellipticity (Wiberg, 1986).

Before leaving this topic, it is worthwhile noting that the total resonance energy in hydroxytropylium ion ($\underline{3}$) is less than that in $\underline{1}$. Specifically, the enthalpy of protonation of tropone is 11.8 kcal/mol more exothermic than that of cycloheptanone (Childs et al, 1986) and comparison of $\Delta H_f^{\circ}(g)$ of tropone, cycloheptanone, cycloheptene and cycloheptane (Pedley et al, 1986) indicates a resonance energy of 8.4 kcal/mol in tropone providing a total resonance energy of 20.2 kcal/mol in $\underline{3}$. Although this is a comparison between ions and neutrals, so is the structural comparison. Comparison of gas-phase proton affinities of tropone and diphenylcyclopropenone would be of interest.

2. STRAIN ENERGIES IN FLUORINATED STRAINED MOLECULES

1,1-Difluorocyclopropane has shortened vicinal bonds and a significantly lengthened distal bond which is, in turn, relatively labile (Dolbier, 1981). The 12 kcal/mol extra strain in this molecule is largely reflected in the reduced E_a(isomerization) (Greenberg et al, 1983). The strain in hexafluorocyclopropane is 54 kcal/mol, virtually double that of the hydrocarbon, while the value for octafluorocyclobutane, 14.5-18 kcal/mol, is only half to two-thirds that of the hydrocarbon (Liebman et al, 1986). The sources of these large effects are not immediately apparent. The electronegativity of fluorine may simply destabilize the ring and, indeed, the added strain in hexafluorocyclopropane is equal to the suggested 4.5-5 kcal/mol extra strain per F on a cyclopropane (O'Neal and Benson, 1968). The reduced strain in octafluorocyclobutane could be due to reduced 1,3-nonbonded interactions (Wiberg, 1985). Alternatively, if one considers the idea of aromatic sigma delocalization in cyclopropane (Dewar, 1984; Cremer and Kraka, 1985), and presumably antiaromatic sigma deloin cyclobutane, then if the electronegative gem-difluro group were to decrease delocalization, then there should be decreased aromaticity in the cyclopropane and decreased antiaromaticity in the cyclobutane leading to the apparent strain energy changes.

One very important point is what constitutes an appropriate strain-free $-CF_2-$ increment. The $\Delta H_f^{\circ}(g)$ of the $C(F)_2(C)_2$ in a molecule such as $CH_3CF_2CH_3$ is -104.9 kcal/mol (Dolbier et al, 1982). The corresponding value in $CF_3CF_2CF_3$ can be taken to be -103.5 kcal/mol while that in $-CF_2CF_2CF_2-$ (or in cyclo- C_6F_{12}) can be estimated at -95.8 kcal/mol (Liebman et al, 1986). It is apparent that non-next-nearest neighbor interactions cannot be neglected, the "end-group" problem enters and there is no univer sal $C(F)_2(C)_2$ group (Liebman et al, 1986). While the same pro- blems should occur for hydrocarbons, the small electronegativity differences between C and H allow simple additivity to work.

A very interesting system for exploration is 2,2,4,4-tetra- fluorobicyclobutane (4). One might anticipate a long central bond by analogy to 1,1-difluorocyclopropane. Indeed, our 3-21G calculations predict a C1C3 bond length of 1.573 Å for 4 (the corresponding 3-21G value for the hydrocarbon is 1.484 Å). It has been noted that diradicaloid nature of the central bond in bicyclobutanes necessitates the use of configuration interaction (Budzelaar et al, 1986) and following these authors, we estimate a 6-31G*/GVB value of 1.600 Å for the central bond in 4. We are presently attempting synthesis of 5 for structural study.

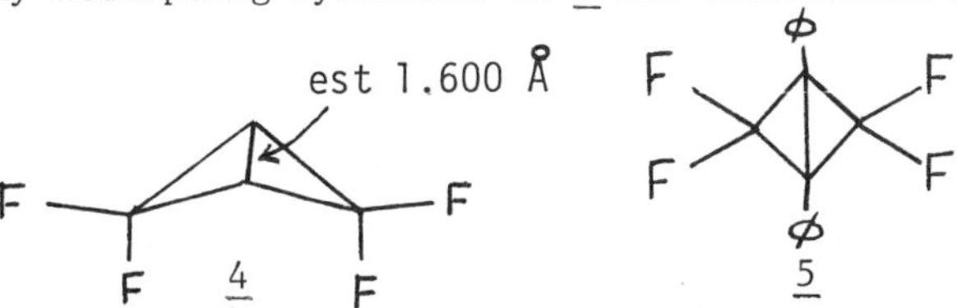

3. STRUCTURE/ENERGY RELATIONSHIPS IN BRIDGEHEAD LACTAMS

Bridgehead lactams are interesting as models for bridgehead ole- fins and distorted peptide linkages. Little structural or thermo dynamic study has been on these molecules despite their poten tial for parameterizing calculations of proteins and penicillins (Greenberg, 1987). During the past few years a number of members of this class have been reported (Uyeo et al, 1958; Pracejus, 1959; Denzer and Ott, 1969; Hall et al, 1980; Blackburn et al, 1980; Buchanan, 1981; Somajaji and Brown, 1986; Williams and Lee, 1986). We have made use of an analogy between the two major resonance contributors of amides and the corresponding exo- and methylated endo- olefins (e.g. $\underline{6}$ and $\underline{7}$). A plot of the enthalpy

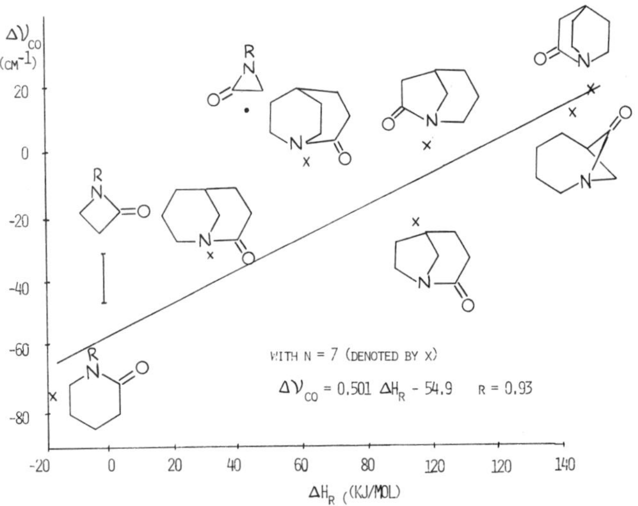

changes against $\Delta\nu_{CO}$ between the corresponding lactams and ketones for a group of bridgehead lactams and small cyclic lactams is depicted in Figure 1. The specifics of the $\Delta H_f^{\circ}(g)$ and IR data are too involved to present here and are discussed elsewhere (Greenberg, 1987). The departure of the alpha-lactam from this line may be due to its unusual geometry (Bürgi, 1986).

FIG. 1. Plot of $\Delta\nu_{CO}$ between lactams and corresponding ketones vs ΔH_{rxn} for the corresponding olefin isomerization (e.g. $\underline{6}\rightarrow\underline{7}$).

REFERENCES

BLACKBURN, G.M., SKAIFE, C.J., KAY, I.T. (1980). Journal of Chemical Research, Miniprints, 3650.

BUCHANAN, G.L. (1981), Journal of the Chemical Society,Chemical Communications, 814.

B UDZELAAR, P.H.M., KRAKA, E., CREMER, D., SCHLEYER, P.V.R. (1986), Journal of the American Chemical Society, 108, 561.

BÜRGI, H.-B. (1986), Personal communication to A. Greenberg.

CHILDS, R.F., MAHENDRAN, M., ZWEEP, S.D., SHAW, G.S., CHADDA, S.K., BURKE, N.A.D., GEORGE, B.E., FAGGIANI, R., LOCK, C.J.L. (1986). Pure and Applied Chemistry 58, 111.

CREMER, D., KRAKA, E. (1985), Journal of the American Chemical Society 107, 3811.

DAVIS, E.H., ALLINGER, N.L., ROGERS, D.W. (1985), Journal of Organic Chemistry 50, 3601.

DENZER, M., OTT, H. (1969), Journal of Organic Chemistry 34,183.

DEWAR, M.J.S. (1984), Journal of the American Chemical Society 106, 669.

DOLBIER, W.R., JR. (1981). Accounts of Chemical Research 14,195.

DOLBIER, W.R., JR., MEDINGER, K.S., GREENBERG, A., LIEBMAN, J.F. (1982). Tetrahedron 38, 2415.

GRABOWSKI, J.J., SIMON, J.D., PETERS, K.S. (1984), Journal of the American Chemical Society 106, 4615.

GREENBERG, A., STEVENSON, T.A. (1986), Molecular Structure and Energetics, Vol. 3, Liebman, J.F., Greenberg, A. (Editors), VCH Pub., Deerfield Beach, pp. 193-266.

GREENBERG, A., TOMKINS, R.P.T., DOBROVOLNY, M., LIEBMAN, J.F. (1983). Journal of the American Chemical Society 105, 6855.

GREENBERG, A., LIEBMAN, J.F., DOLBIER, W.R., JR., MEDINGER, K.S. SKANCKE, A. (1983). Tetrahedron 39, 1533.

GREENBERG, A. (1987), Molecular Structure and Energetics, Vol.7, L iebman, J.F., Greenberg, A. (Editors), VCH Pub., in press.

HALL, H.K., JR., SHAW, R.G., JR., DEUTSCHMAN, A. (1980),

Journal of Organic Chemistry 45, 3722.

LIEBMAN, J.F., DOLBIER, W.R., JR., GREENBERG, A. (1986). Journal of Physical Chemistry 90, 394.

O'NEAL, H.E., BENSON, S.W. (1968). Journal of Physical Chemistry 72, 1866.

PEDLEY, J.B., NAYLOR, R.D., KIRBY, S.P. (1986). Thermochemical Data of Organic Compounds, Chapman Hall, London.

PRACEJUS, H. (1959). Chemisches Berichte 92, 988.

SOMAYAJI, V., BROWN, R.S. (1986). Journal of Organic Chemistry 51, 2676.

STEELE, W.V., GAMMON, B.E., SMITH, N.K., CHICKOS, J.S., GREENBERG, A., LIEBMAN, J.F. (1985). Journal of Chemical Thermodynamics 17, 505.

UYEO, S., FALES, H.M., HIGHET, R.J., WILDMAN, W.C. (1958), Journal of the American Chemical Society 80, 2590.

WIBERG, K.B. (1985). Tetrahedron Letters 26, 599.

WIBERG, K.B. (1986). Personal communication to A. Greenberg based upon techniques developed by R. Bader.

WILLIAMS, R.M., LEE, B.H. (1986), Journal of the American Chemical Society 108, 6431.

33. Bent bonds in polycyclic compounds

Hermann Irngartinger, Jürgen Deuter, Reiner Jahn, Dietmar Kallfass
and Wolfgang Reimann

1. INTRODUCTION

The valence isomers of benzene, cyclobutadiene and
cyclooctatetraene are highly strained polycyclic
hydrocarbons with three and four membered ring subunits. We collected low temperature (100 K) X-ray
data from derivatives of these compounds in order to
determine the deformation densities by the X-X-method.
From these electron density distributions we obtain
experimental informations about the degree of bending
of the C-C bonds within the strained systems.

Such bonds are very sensitive against electronic
and steric influences from substituents. From the
structure determinations (bond lengths and angles)
of derivatives we obtain informations about these
effects.

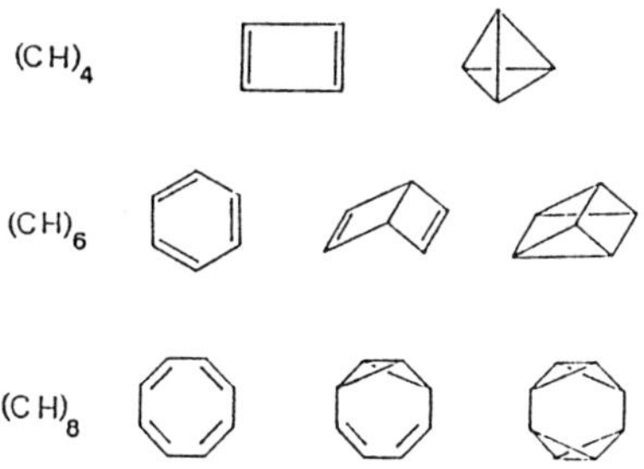

2. $(CH)_4$: TETRA-TERT-BUTYL-TETRAHEDRANE

2.1. Gas inclusion crystals

The bonds of the $(CH)_4$ compound tetra-tert-butyl-cyclobutadiene are bent (Irngartinger and Nixdorf, 1983). After the structure determination of the corresponding valence isomer tetra-tert-butyl-tetrahedrane 1 at 213 K (Irngartinger et al, 1984) we collected the data at 103 K to find the deformation densities. With these results of the low temperature form of 1 we discovered gas inclusion crystals. Nitrogen or argon are sitting on three fold inversion axes (FIG. 1). Gas clathrates hold together only by van der Waals forces were not known previously. Above 220 K the gas diffuses out of the crystals which are destroyed subsequently (Irngartinger et al, 1987 a). The C-C bonds of the tetrahedron have a distance of 1.497 Å. This is comparatively short due to the bending of these bonds and due to the unusual hybridization of the C-atoms.

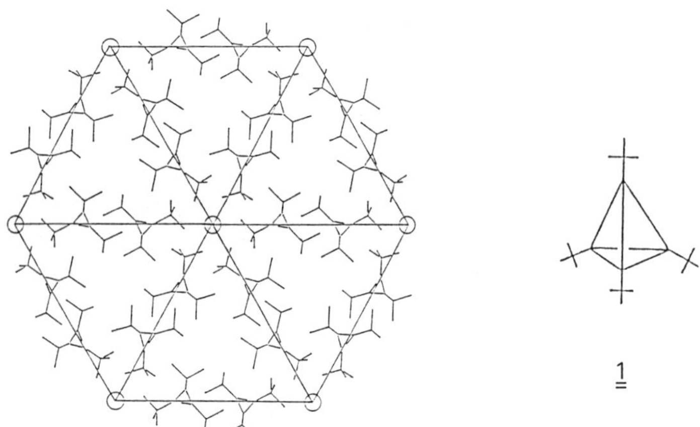

1

FIG. 1. Packing diagram of the argon inclusion crystals of 1 at 103 K.

2.2. Deformation densities of the tetrahedrane 1

The deformation densities of 1 have been obtained
from low temperature data (103 K) of the argon
clathrate by the X-X-method. Two sections shown in
FIG. 2 demonstrate the electron density distributions
on the C-C bonds of the tetrahedrane. The density
maxima of the tetrahedrane bonds are shifted by 0.37 Å
from the bond axes outwards which is equivalent to
a bending of 26°. The density maxima of the exocyclic
bonds from the tetrahedrane to the tert-butyl groups
are localized exactly on the bond axes. The hight of
the density is more pronounced in these exocyclic
bonds compared to the tetrahedrane bonds which are
weakened by bending and strain.

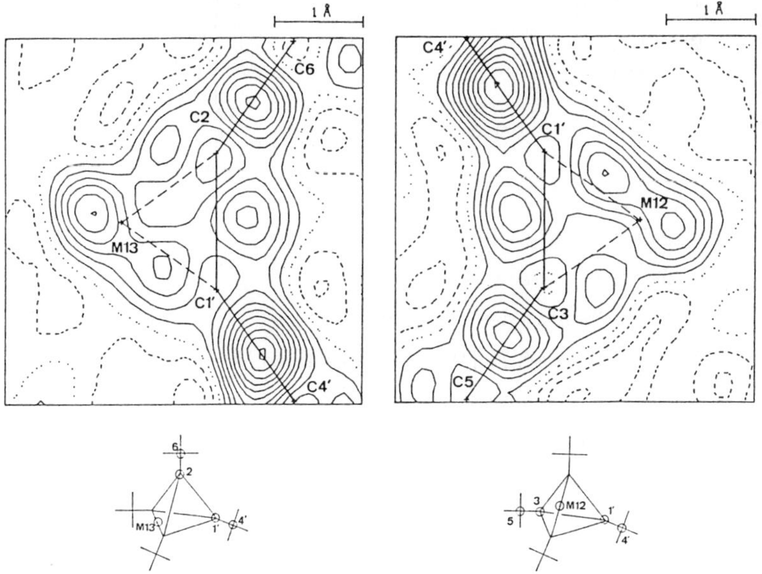

FIG. 2. Deformation density of 1. Section parallel to
one bond and perpendicular to the opposite bond of
the tetrahedrane. Contour interval 0.05 e/Å^3.

3. $(CH)_6$: PRISMANE AND DEWAR-BENZENE DERIVATIVES

3.1. Structures of prismane derivatives

The strained bonds of the prismane systems 2 - 4
(Irngartinger et al, 1987 b; Maier et al, 1986; Wingert et al, 1987) are very sensitive to electronic
effects of substituents and repulsion effects between tert-butyl substituents. Carbonyl groups in
the favourable bisecting orientation to the three
membered ring moiety of the prismane give rise to
shortenings of the distal bonds by 0.030-0.036 Å
(4 and 2). Repulsion forces between tert-butyl substituents stretch prismane bonds as far as 1.594 Å (4).

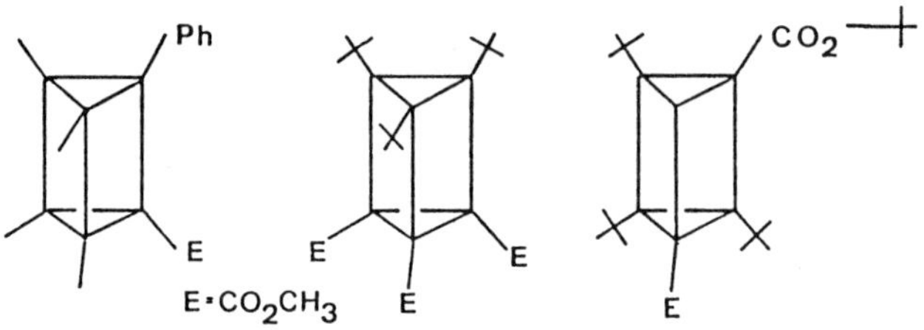

3.2 Structures and deformation densities of Dewar benzene derivatives

In the Dewar benzene derivatives 5 - 8 (Weinges et
al, 1984, 1986, and 1987) the bond lengths between
the bridge head carbon atoms vary from 1.562 (5) to
1.586 Å (8) depending on the steric influences of
the substituents.

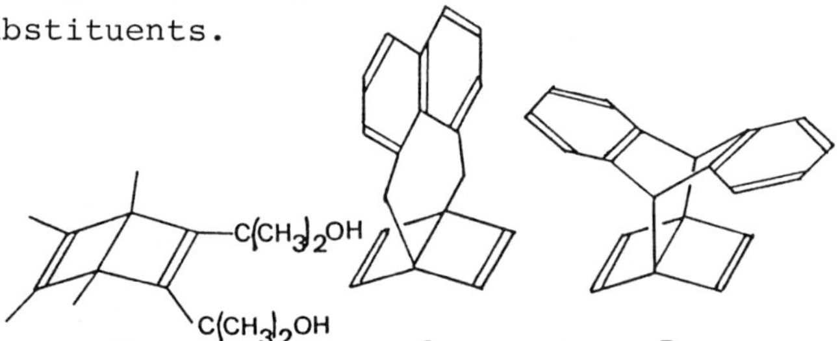

From the low temperature data of $\underline{8}$ we determined the deformation densities. The central bond between the bridge head carbon atoms is bent strongly by 28° (FIG. 3). The bendings of the remaining bonds of the four membered rings are smaller.

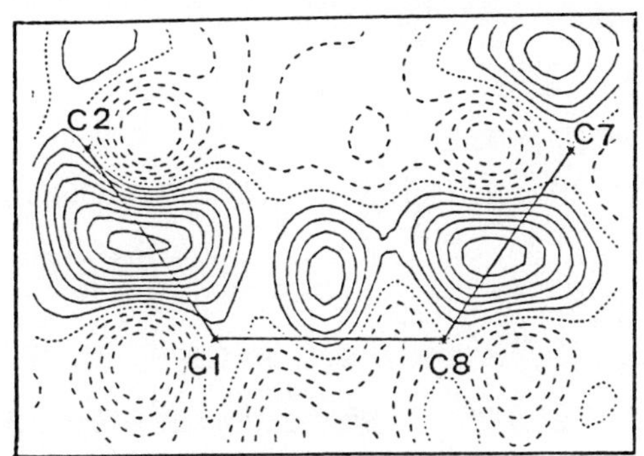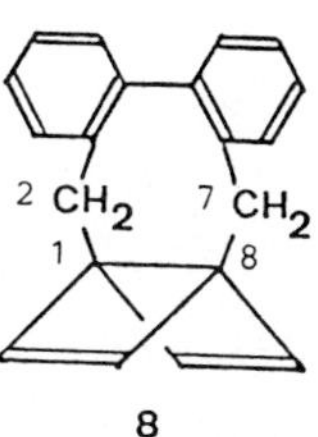

FIG. 3. Deformation density of $\underline{8}$ in a section along the central bond C1-C8 including the methylene carbon atoms C2 and C7.

4. $(CH)_8$: OCTAVALENE AND OCTABISVALENE DERIVATIVES

In the cycloaddition product $\underline{9}$ of octavalene and phenyltriazoldione (Christl $\underline{et}$ $\underline{al}$, 1984) the central bond of the bicyclo[1.1.0]butane moiety is bent by 10°, as we obtained from the deformation densities. The sulfonyl groups in the octabisvalene derivative $\underline{10}$ cause shortenings of the distal bonds by 0.017 Å (Rücker $\underline{et}$ $\underline{al}$, 1986). The bridge head carbon atoms of both compounds are inverted.

ACKNOWLEDGEMENTS : We would like to thank the Deutsche Forschungsgemeinschaft and the Fonds der Chemischen Industrie for financial support.

REFERENCES

CHRISTL, M., LANG, R., REIMANN, W. and IRNGARTINGER, H. (1984). Chem. Ber. 117, 959.

IRNGARTINGER, H., GOLDMANN, A., JAHN, R., NIXDORF, M., RODEWALD, H., MAIER, G., MALSCH, K.-D. and EMRICH, R. (1984). Angew. Chem. Int. Ed. Engl. 23, 993.

IRNGARTINGER, H., JAHN, R., MAIER, G., MALSCH, K.-D. and EMRICH, R. (1987 a). Angew. Chem. Int. Ed. Engl. 26, in the press.

IRNGARTINGER, H., KALLFASS, D., LITTERST, E. and GLEITER, R. (1987b). Acta Cryst. C43, in the press.

IRNGARTINGER, H. and NIXDORF, M. (1983). Angew. Chem. Int. Ed. Engl. 22, 403.

MAIER, G., BAUER, I., HUBER-PATZ, U., JAHN, R., KALLFASS, D., RODEWALD, H. and IRNGARTINGER, H. (1986). Chem. Ber. 119, 1111.

RÜCKER, C., PRINZBACH, H., IRNGARTINGER, H., JAHN, R. and RODEWALD, H. (1986). Tetrahedron Lett. 27, 1565.

WEINGES, K., KASEL, W., HUBER-PATZ, U., RODEWALD, H. and IRNGARTINGER, H. (1984). Chem. Ber. 117, 1868.

WEINGES, K., KLEIN, J., SIPOS, W., GÜNTHER, P., HUBER-PATZ, U., RODEWALD, H., DEUTER, J. and IRNGARTINGER, H. (1986). Chem. Ber. 119, 1540.

WEINGES, K., SIPOS, W., KLEIN, J., DEUTER, J. and IRNGARTINGER, H. (1987). Chem. Ber. 120, in the press.

WINGERT, H., IRNGARTINGER, H., KALLFASS, D. and REGITZ, M. (1987). Chem. Ber. 120, in the press.

34. Structural studies of some unsaturated eight-membered carbocycles

Thomas C. W. Mak

---

1. INTRODUCTION

Since its discovery by Willstätter and Waser in 1911 [1]
cyclooctatetraene has continued to play a dominant role in
many areas of theoretical and synthetic chemistry [2-5]. The
C_8H_8 molecule exists in a strainless non-planar D_{2d} conformation commonly described as a 'tub', and it is generally
assumed that ring inversion and bond-shift processes can
occur via planar transition states with alternating (D_{4h}
symmetry) and equal (D_{8h}) bond lengths, respectively.

In recent years our attention has been focused on
different facets of the structural chemistry of cyclooctatetraene and its derivatives, including presumably planar
fully conjugated eight-membered ring systems [6,7].

2. BENZANNELATED CYCLOOCTATETRAENES

All five possible benzannelated derivatives of cyclooctatetraene, 6A, namely benzocyclooctatetraene, 1 [8], dibenzo-
[a,c]cyclooctatetraene, 2 [9], dibenzo[a,e]cyclooctatetraene,
3 [10], tribenzo[a,c,e]cyclooctatetraene, 4 [11], and
tetrabenzo[a,c,e,g]cyclooctatetraene (commonly known as
tetraphenylene), 5 [12,13] are known. IUPAC uses the some-

what confusing 'cyclooctene' nomenclature for these compounds; for example, 2 and 3 are named dibenzo[*a,c*]cyclooctene and dibenzo[*a,e*]cyclooctene, respectively.

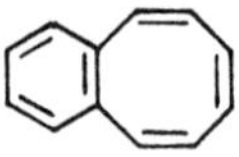

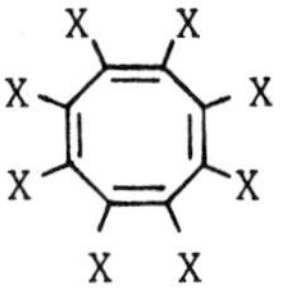

6A, X = H
6B, X = F
6C, X = Me
6D, X = Ph

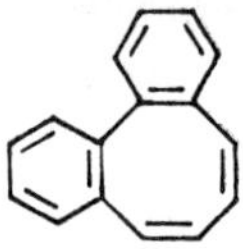

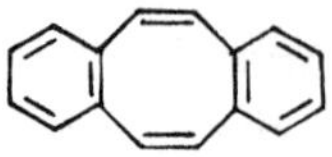

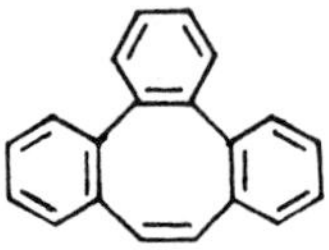

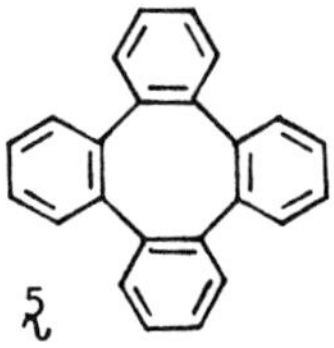

An early electron diffraction investigation of 5 indicated a tub conformation [14], which was later confirmed in an X-ray analysis by Irngartinger and Reibel, who also reported the structure of 3 [15]. Around 1981, we initiated a comparative X-ray crystallographic and semi-empirical molecular orbital investigation of the effects of benzannelation on the geometry of the cyclooctatetraene ring skeleton, as a sequel to our MINDO/3 study of cyclooctatetraenes containing annelated cyclobutene and cyclobutadiene rings [16]. As our work on 3 and 5 had been preceded, we reported only the crystal structures of 1, 2 and 4, as well as the results of a MINDO/3 and MNDO conformational study on this series of benzannelated cyclooctatetraenes and related compounds 6A and 6B [17,18].

Chemically equivalent bond lengths and angles for 1–5 are averaged in Table 1. There is remarkably good agreement between analogous bonds and angles throughout this series. The effects of benzannelation on the unsubstituted cyclooctatetraene skeleton (C=C 1.340, C–C 1.475 Å, C–C=C 126.1°) [19] may be discussed by reference to the bond labelling scheme

Table 1

Averaged bond lengths (Å) and angles (deg) in the five benzannelated cyclooctatetraenes[a]

Parameter	$1(C_s)$	$2(C_2)$	$3(C_{2v})$	$4(C_s)$	$5(D_{2d})$
k	1.449(6)	1.459	–	–	–
l	1.322(7)	1.324(1)	1.321(1)	1.316	–
m	1.479(4)	1.478(1)	1.475(2)	1.477(8)	–
n	1.399	1.400(5)	1.395(1)	1.408(3)	1.400(2)
p	–	1.492	–	1.487(2)	1.494(4)
q	1.393(5)	1.395(1)	1.398(2)	1.395(12)	1.397(2)
r	1.375(5)	1.377(1)	1.374(3)	1.381(5)	1.381(4)
s	1.376	1.378(1)	1.370(6)	1.372(6)	1.373(4)
$kl=lm$	127.2(5)	126.5(4)	127.6(4)	128.9(4)	–
$mn=np$	123.9(1)	123.5(4)	123.2(2)	123.6(8)	122.5(3)
nq	118.7(4)	118.6(3)	118.6(4)	118.5(6)	118.8(2)
qr	121.7(5)	121.9(3)	121.7(4)	121.8(5)	121.3(3)
rs	119.7(5)	119.5(3)	119.7(3)	119.7(6)	119.9(3)

[a] The scatter $s = [\Sigma(x - \Sigma x/n)^2/(n - 1)]^{\frac{1}{2}}$ given in parentheses is of the same order of magnitude as the esd's of individual measurements. The x values are the data and n is the number of data to be averaged.

adopted in Table 1. Fusion of a benzene ring to the cyclooctatetraene system results in a considerable lengthening of the shared (formally double) bond n in comparison with the free double bond l, and angle mn is significantly larger than angle lm. The observed order $p > m > k$ of $C(sp^2)$–$C(sp^2)$ single bond lengths clearly illustrates the significance of 1...3 repulsive interaction of the C atoms. Another consequence of this steric effect is that torsion angles of

the type npn', which involve two bonds shared between six-
and eight-membered rings, are about 10° greater than others
of the types lmn and lkl'. The central eight-membered ring
of 2 thus takes a more distorted conformation as compared to
that of 3. Steric repulsion due to fused benzene rings also
accounts for the fact that, within each benzene moiety,
q generally exceeds r and s, and $qr > rs > nq$ in all benzannelated cyclooctatetraenes.

Molecular dimensions of 1–5 calculated using the MNDO
procedure [20] are in very good agreement with the experimental results. The largest discrepancies are in the bond
lengths around the benzene ring(s), where the MNDO values are
consistently larger by 0.2–0.3 Å.

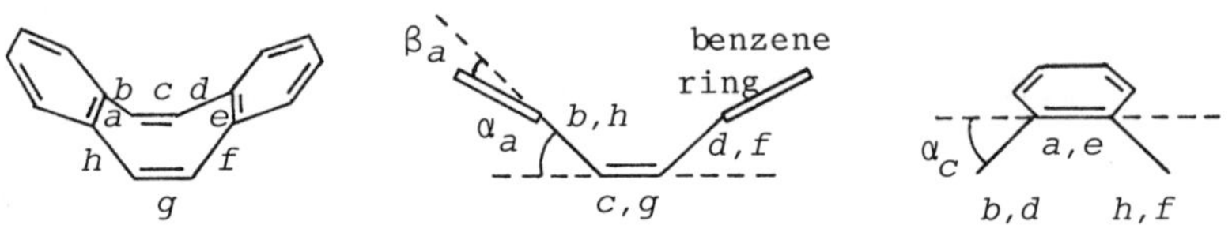

FIG. 1. Definition of fold angle α and tilt angle β in
annelated cyclooctatetraenes. The subscript refers to the
common edge of the six- and eight-membered rings. In the
illustrated example, α_a is the dihedral angle between the
plane containing bonds b, a and h, and the plane containing
bonds c and g; β_a is the dihedral angle between the left
benzene ring and the plane containing bonds b, a, and h.

The conformations of various cyclooctatetraenes may be
compared by reference to the fold angles α and tilt angles β
as defined in Fig. 1. Comparison of the calculated and
observed α values in Table 2 shows that MNDO is vastly
superior to MINDO/3 [21] in accounting for the observed
conformations of cyclooctatetraene ($6A$) and its benzannelated
derivatives, although the MNDO results still leave some room
for improvement. On steric grounds, it is expected that the
tub conformation of $6A$ is only slightly perturbed by the
successive fusion of one and two non-adjacent benzene rings,

and that repulsion between adjacent annelated benzene rings, when present, should predominate and give rise to more folded configurations. This has indeed been found experimentally and reproduced by the MNDO model. (For the same reason, MNDO predicts that 3 is thermodynamically slightly more stable than 2). On the other hand, MNDO fails dismally in gauging the more subtle effects of benzannelation such as the markedly different fold angles in 4 and the tilt angles. With the lone exception of 2, the MNDO model does not yield

TABLE 2

Comparison of conformation of benzannelated cyclooctate-traenes and related compounds[a]

Molecule		MNDO	MINDO/3	Experimental (averaged)	Ref.
1	α_a	42(0)	29	42.6(2.1)	[18]
	α_c	44	33	43.3	
	α_e	42	30	40.7	
2	α_a	47(4)	38	47.0(4.5)	[17]
	α_e	46	39	43.9	
3	α_a	44(0)	34	43.2(2.9)	[15]
	α_c	48	41	46.3	
4	α_a	50(0)		44.5(7.0)	[17]
	α_c	51(0)		50.3(0.9)	
	α_g	51		43.1	
5	α	53(0)		50.7(4.5)	[15]
6A	α	38	32	43.1	[19]
6B	α	47	26	41.4(1.9)	[22]
6C	α			50.9(1.6)	[23]
6D	α			55.5	[24]

[a] Both the fold angle α and tilt angle β (Fig. 1) are given in degrees, with the latter in parentheses. α refers to a particular double bond and its two adjacent single bonds in the eight-membered ring.

any tilting of the benzene ring(s) with respect to the eight-membered ring in the benzocyclooctatetraenes.

It seems reasonable to suggest that the series of compounds $\underset{\sim}{1}$-$\underset{\sim}{5}$ can serve as a suitable test standard for critical evaluation of other current and future semi-empirical molecular orbital models.

3.   TETRAPHENYLENE CLATHRATES

When Rapson, Shuttleworth and van Niekirk successively synthesized tetraphenylene $\underset{\sim}{5}$ in 1943, they also reported that it formed 2:1 adducts with a variety of solvent molecules [12]. In their 1944 paper on the structure determination of $\underset{\sim}{5}$, Karle and Brockway stated that 'the sample contained 10% benzene which was removed by placing the sample in the nozzle and heating it to 200° in a vacuum before electron diffraction photographs were taken' [14]. Understandably, the nature of these adducts drew no attention at that time, since the concept of molecular inclusion was unknown until the architecture of the hydroquinone (quinol) clathrates was elucidated by Palin and Powell in 1947 [25].

Several decades were to pass before we 're-discovered' the tetraphenylene clathrates [26], which constitute an isomorphous series of the general formula $2C_{24}H_{16}\cdot G$, where G is a guest species ranging in size from methylene chloride to cyclohexane (Table 3). The tetragonal clathrates crystallize in space group $P4_2/n$ (No. 86) with $Z = 2$, and are morphologically distinguishable from pure host $\underset{\sim}{5}$ belonging to $C2/c$ (No. 15) with $Z = 8$.

A stereoview of the molecular packing in a representative clathrate is shown in Fig. 2. With the unit-cell origin at $\bar{1}$, the host and guest molecules occupy Wyckoff positions $4(e)$ and $2(a)$, of point symmetries 2 and $\bar{4}$, respectively. Eight host molecules cluster around a spheroidal cavity centered at ($\frac{1}{4}\frac{1}{4}\frac{1}{4}$), which accommodates the guest molecule. The latter necessarily satisfies the $\bar{4}$ site symmetry, generally through orientational disorder. It is notable that in the carbon tetrachloride clathrate [27], the encaged CCl_4 molecule also exhibits orientational disorder, despite the fact that its size and shape ideally match the specifications for an ordered guest species. All intermolecular interactions are of the van der Waals' type.

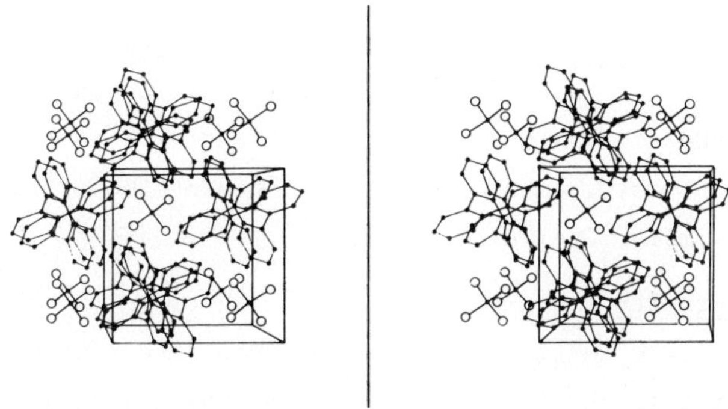

FIG. 2. Stereoview of the molecular packing in $2C_{24}H_{16} \cdot CCl_4$. The disordered CCl_4 molecule is shown in its preferred orientation, and H atoms have been omitted for clarity. The unit-cell origin lies at the upper left corner, with a pointing from left to right, b downwards, and c away from the reader.

3.1. The host molecule

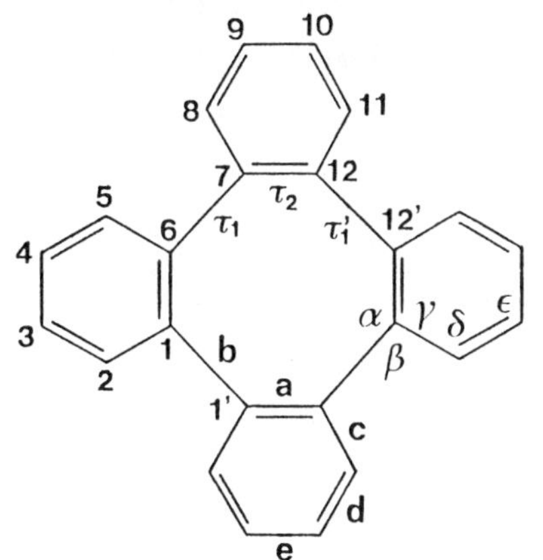

A crystallographic C_2 axis passes through the centers of the C(1)-C(1') and C(12)-C(12') bonds.

General trends:
$b \gg a \sim c \sim d > e,$
$\alpha \gg \beta \sim \gamma,$ and
$\delta > \varepsilon > \gamma$

A consistent system of numbering is used for both crystalline tetraphenylene [15] and its chloroform [26], carbon tetrachloride [27], benzene [28], and cyclohexane [28] inclusion complexes. Apart from the larger difference (up to 9.6°) between torsion angles τ_1 and τ_1' in the clathrates, there is good agreement between corresponding molecular dimensions throughout this series of related compounds. In all instances the bond lengths and bond angles follow the trends: $b \gg a \sim c \sim d > e$, $\alpha \gg \beta \sim \gamma$, and $\delta > \varepsilon > \gamma$.

Interestingly, the host molecule makes use of a dihedral C_2 axis, rather than its S_4 principal axis, in the construction of the clathration lattice. In line with this, dibenzo-[a,e]dinaphtho[c,g]cyclooctatetraene has not been found to exhibit similar inclusion behavior.

3.2. The cavity

This takes the the general shape of an oblate spheroid with its minor axis parallel to the <u>c</u> axis, the limiting shape being a sphere of free diameter 7.2–7.4 Å in the carbon tetrachloride clathrate. Table 3 presents a comparison of the volume of the cavity (calculated by subtracting the volume of the primitive cell of tetraphenylene from the clathrate cell volume and dividing by two) and the volume of the guest as reported in the crystallographic literature. Excellent correlation between the two sets of values is obtained, although the estimated volume of the guest consistently exceeds that of the cavity by 6–15%.

3.3. The guest molecule

The sizes and shapes of the guest molecules can be correlated with the two independent crystallographic axes. In Table 3, the various clathrates are divided into three groups according to the nature of the guest species [28]. In group (A) the <u>a</u> axis increases from the five-membered ring of tetrahydrofuran, through the six-membered rings of dioxan and

TABLE 3

Crystal data of tetraphenylene clathrates $2C_{24}H_{16} \cdot G$ ordered according to groups of guest molecules

Group	Guest G	$\underline{a}$ (Å)	$\underline{c}$ (Å)	Volume (Å^3)		
				Unit cell	Cavity	Guest
(A)	Tetrahydrofuran	9.906(1)	18.503(5)	1815.7	82.6	
	Dioxan	9.968(1)	18.553(5)	1843.5	96.5	
	Benzene	10.069(1)	18.431(5)	1868.6	109.0	118.5(−135 °C)
	Cyclohexane	10.073(1)	18.712(2)	1898.6	124.1	140.3(−158 °C)
(B)	CH_2Cl_2	9.892(5)	18.46(1)	1806	77.7	82.0(−120 °C)
	CH_3COCH_3	9.902(2)	18.491(6)	1813.0	81.2	
	CH_2Br_2	9.935(2)	18.546(6)	1830.6	90.0	95.1(−90 °C)
	$CHCl_3$	9.925(2)	18.593(3)	1831.5	90.5	103.8(−88 °C)
	Pr^iBr	9.973(1)	18.633(5)	1853.3	101.4	
	CCl_4	9.930(2)	18.932(6)	1866.8	108.2	116.6(10 Kbar)
(C)	Pr^nBr	10.004(1)	18.647(4)	1866.2	107.8	

benzene, to the chair structure of cyclohexane. The same variation in the $\underline{c}$ axis is found for the non-planar guests, but a significant contraction in $\underline{c}$ occurs in the case of the planar guest benzene, which has its principal molecular axis parallel to $\underline{c}$.

Several general patterns may be noted in group (B). The CH_2Cl_2 and CH_3COCH_3 clathrates have similar lattice dimensions because the size of the methyl group is close to that of chlorine, and the two hydrogen atoms in the former guest species occupy roughly the same space as the oxygen atom in the latter. In a similar way, the two bromine atoms in CH_2Br_2 have roughly the same spatial requirement as the three chlorine atoms in $CHCl_3$, so the two clathrates differ little in their lattice dimensions. The measured crystal data clearly show that the Pr^iBr guest is intermediate in size between $CHCl_3$ and CCl_4; the bulk of the bromine atom is reflected mainly in an increase in the $\underline{a}$ axis, whereas the longest $\underline{c}$ axis is observed in the CCl_4 clathrate.

The linear-chain molecule Pr^nBr which alone comprises group (C), as compared to all guest species in group (B) except CCl_4, corresponds to lengthening of both axes, the increase in $\underline{a}$ being more significant.

Taken as a whole, the measured crystallographic data are consistent with the conclusion that a given guest species tends to spread its bulk in a plane parallel to (001), and the tetraphenylene cagework readily adapts itself to accommodate the preferred orientation of the guest molecule in the cavity.

4.   METAL π-COMPLEXES OF CYCLOOCTATETRAENES

4.1. Adducts of cyclooctatetraene with silver nitrate

Cyclooctatetraene reacts with aqueous silver nitrate to form three crystalline compounds of compositions $2C_8H_8 \cdot AgNO_3$,

$C_8H_8 \cdot AgNO_3$ and $2C_8H_8 \cdot 3AgNO_3$ [29]. X-ray analysis of the 1:1 adduct provided the first elucidation of the π bonding between a cyclic polyolefin and a heavy metal ion [30,31]. All unsatisfactory aspects of this early work were rectified in a later study [32], and the coordination modes of the C_8H_8 and nitrato groups about the Ag(I) ions in the 2:1 [33] and 2:3 [34] adducts have also been determined.

The crystal structure of $C_8H_8 \cdot AgNO_3$ [32] consists of strongly-bonded $Ag-C_8H_8$ units linked by weaker metal-ligand interactions to form infinite chains parallel to the _c_ axis; neighboring chains are cross-linked by bridging nitrato groups to form corrugated sheets normal to the _a_ axis. The configuration of the ligands about Ag may be described as distorted trigonal bipyramidal, with two metal-oxygen bonds (to different nitrato groups) and the strongest metal-olefin bond lying in the equatorial plane, and the weakest π ligand in an axial position (Fig. 3). The two unequal Ag-O bonds bracket the distance of 2.384 Å observed in crystalline silver nitrate [35].

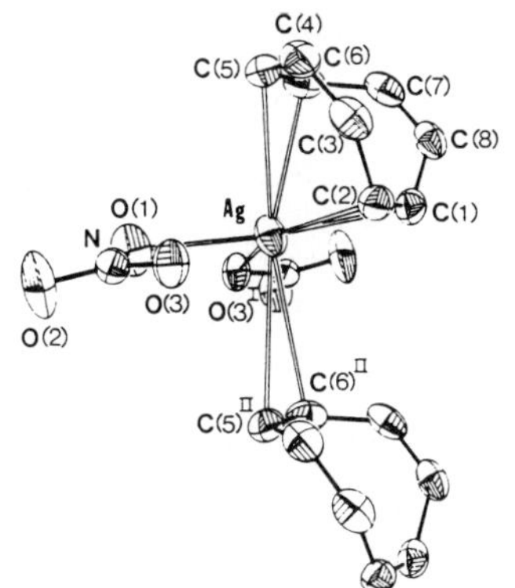

FIG. 3. Perspective view of the coordination geometry about the Ag(I) ion in $C_8H_8 \cdot AgNO_3$. Transformations: I. $\frac{1}{2}-x$, $-\frac{1}{2}+y$, $-z$; II. x, y, $-1+z$.

Ag–C(1)	2.483(5)Å
Ag–C(2)	2.475(5)
Ag–C(5)	2.787(5)
Ag–C(6)	2.805(5)
Ag–O(1)^{I}	2.431(5)
Ag–O(3)^{I}	2.368(3)
Ag...C(5)^{II}	3.284(5)
Ag...C(6)^{II}	3.222(5)

The crystal structure of the 2:1 adduct [33] is bult up by a packing of discrete molecules corresponding to the formula $(C_8H_8)_2AgNO_3$. Each molecule lies on a crystallographic diad, with the Ag(I) ion coordinated by a symmetrical bidentate nitrato group and four double bonds, two from each

C_8H_8 moiety in an unsymmetrical fashion (Fig. 4). If the bidentate nitrato group is considered to occupy one position in the coordination sphere, the configuration about Ag can be described as approximately trigonal bipyramidal: the nitrato group and the centers of the C(5)-C(6) and C(5')-C(6') double bonds lie in the equatorial plane, and the pair of C(1)-C(2) and C(1')-C(2') double bonds occupy the axial positions. This overcrowded coordination geometry leads to significantly weaker metal-ligand interactions than in the 1:1 adduct and related cyclic oligoolefin-silver(I) complexes.

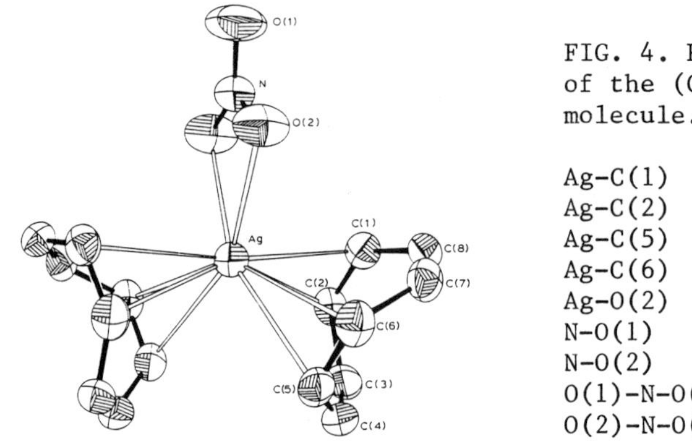

FIG. 4. Perspective view of the $(C_8H_8)_2AgNO_3$ molecule.

Ag-C(1)	2.710(4)Å
Ag-C(2)	2.694(4)
Ag-C(5)	2.737(4)
Ag-C(6)	2.731(4)
Ag-O(2)	2.518(3)
N-O(1)	1.232(7)
N-O(2)	1.239(4)
O(1)-N-O(2)	121.2(2)°
O(2)-N-O(2')	117.6(5)

X-ray analysis of $2C_8H_8 \cdot 3AgNO_3$ [34] revealed a complex three-dimensional polymeric structure in which the two C_8H_8 (COT) ligands partition the three silver(I) ions in the asymmetric unit (Fig. 5). The configuration of the ligands about each metal center can be described, to a crude approximation, as trigonal bipyramidal. The Ag(I) ion is coordinated by one monodentate and two bidentate nitrato groups, and by two non-adjacent double bonds from C_8H_8 moelcule 1; the centers of the C(5)-C(6) and O(1)-N(1)-O(2) ligands occupy the axial positions. The Ag(2) ion is coordinated by a monodentate nitrato group and four double bonds (two non-adjacent ones from each C_8H_8 moiety), with the C(7)-C(8) and C(13)-C(14) olefinic ligands at the axial positions. The

Ag(3) ion is coordinated by three monodentate nitrato groups (NG $\underline{3}$ and NG $\underline{4}$ each has a N–O bond lying on the same crystallographic C_2 axis), and by two double bonds from C_8H_8 moelcule $\underline{2}$, the axial ligands being the C(11)–C(12) double bond and O(3)^{III}. The measured silver(I)–oxygen and silver(I)–olefin distances lie in the ranges 2.384–2.894(7) and 2.500–2.969(8) Å, respectively.

FIG. 5. A perspective view of the coordination geometries about the three independent silver(I) ions in polymeric $2C_8H_8 \cdot 3AgNO_3$. Metal–ligand atom interactions are represented by broken lines.

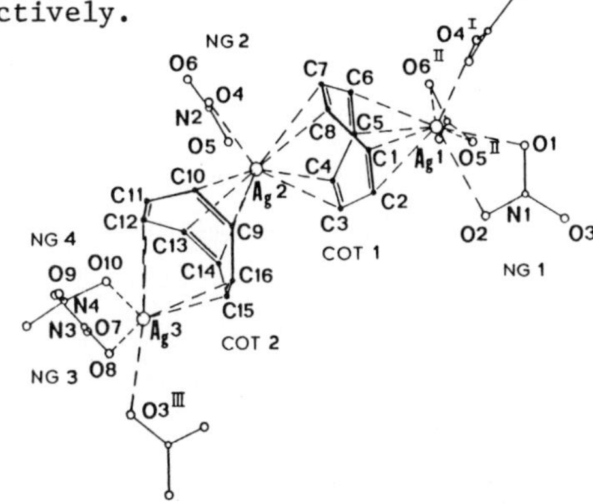

The formation and relative stability of the three adducts [29] can be rationalized in structural terms. Trigonal bipyramidal coordination of the silver(I) ion has emerged as an important common feature in all three crystal structures. The formation of discrete pentacoordinate $(C_8H_8)_2AgNO_3$ molecules is favored in the presence of a large excess of C_8H_8. A $C_8H_8/AgNO_3$ molar ratio between 1 and 2 tends to yield the 1:1 adduct with a chain–and–layer structure. Upon standing or recrystallization from ethanol, both the 2:1 and 1:1 adducts readily transform to the stable 2:3 adduct. The chemical transformation in this series is tantamount to the successive replacement of van der Waals' interactions by much stronger ionic (metal–oxygen) and covalent (metal–olefin) bonds, and is therefore energetically favorable. This series of cyclooctatetraene–silver nitrate adducts thus provides a nice example of structural transition from a simple molecular crystal, through an intermediate chain–and–layer structure, to a complex three–dimensional polymeric lattice.

4.2. Silver(I) complexes of benzannelated cyclooctatetraenes

The 1:1 adduct of 1 with $AgClO_4$ [36] provides the first example of simultaneous donor π-bonding from aromatic and olefinic ligands to a silver(I) ion (Fig. 6). With the disordered perchlorate group in its principal orientation, the configuration of the ligands about the metal center may be described as distorted trigonal bipyramidal. The silver ion is coordinated by two non-adjacent double bonds of a benzocyclooctatetraene molecule, one aromatic C–C bond from a neighbouring benzocyclooctatetraene molecule, and two oxygen atoms belonging to different perchlorate groups; atom O(1) and the center of the C(5)–C(6) olefinic ligand occupy the axial positions of the trigonal bipyramid. The silver-olefin and silver-aromatic interactions are approximately equal in magnitude. From a structural point of view, this complex is more closely related to the cyclooctatetraene adducts of $AgNO_3$, in which the metal ions have trigonal bipyramidal coordination, than to silver-aromatic complexes with a coordination number of three or four for the metal ion [37].

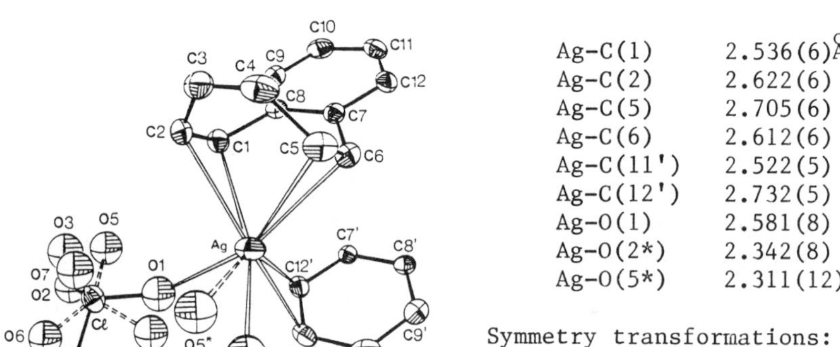

Ag–C(1)	2.536(6)Å
Ag–C(2)	2.622(6)
Ag–C(5)	2.705(6)
Ag–C(6)	2.612(6)
Ag–C(11')	2.522(5)
Ag–C(12')	2.732(5)
Ag–O(1)	2.581(8)
Ag–O(2*)	2.342(8)
Ag–O(5*)	2.311(12)

Symmetry transformations:
' $-x$, $-y$, $1-z$
* x, $\frac{1}{2}-y$, $-\frac{1}{2}+z$

FIG. 6. A perspective view of the coordination geometry about a silver(I) ion in $1.AgClO_4$. Metal-ligand atom interactions are represented by open bonds, and linkages involving oxygen atoms of the less populated orientation of the perchlorate ion by broken bonds.

Another example in which a silver(I) ion π-bonds

simultaneously to both aromatic and olefinic ligands has recently been found in a polymeric 1:4 complex of 3 with silver trifluoroacetate.

Both 1 and 3 form 1:1 adducts with $AgNO_3$. In each complex, the metal atom is tetrahedrally coordinated by two non-adjacent double bonds of an organic ligand, a bidentate nitrato group, and another bidentate nitrato group generated from the first by a symmetry operation (Fig. 7).

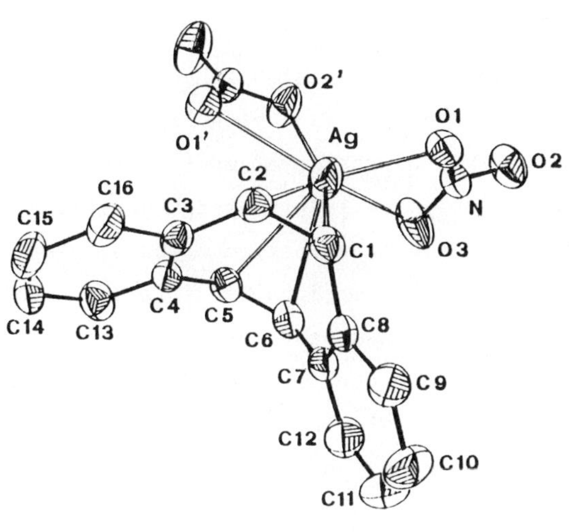

FIG. 7. Coordination geometry about the Ag(I) ion in $3 \cdot AgNO_3$. An analogous situation is found in the $1 \cdot AgNO_3$ complex.

Ag–C(1) = 2.477(6)Å
Ag–C(2) = 2.519(6)
Ag–C(5) = 2.727(6)
Ag–C(6) = 2.693(7)
Ag–O(1) = 2.311(5)
Ag–O(3) = 2.865(6)
Ag–O(1') = 2.627(5)
Ag–O(2') = 2.454(6)

Symmetry transformation:
' x, $\frac{1}{2}-y$, $-\frac{1}{2}+z$

4.3. Tetrameric complex of copper(I) chloride with 3

The cyclic $[(dibenzo[a,e]cyclooctatetraene)CuCl]_4$ molecule, which has crystallographically imposed $\bar{1}$ symmetry, consists of an eight-membered ring of alternating Cu and Cl atoms in a 'step' (or 'chair') configuration (structure I), with an organic ligand π–bonded to each metal center via a pair of olefinic double bonds [38]. The two independent copper(I) atoms are in distorted trigonal pyramidal coordination, each being bound asymmetrically to chloride and olefin ligands at distances in the ranges 2.263–2.278(2) and 2.099–2.801(6) Å, respectively.

(I)   (II)   (III)

The 'step' Cu_4Cl_4 ring is geometrically similar to, but
topologically distinct from, the well-established step M_4X_4
skeleton found in several tetrameric (pnicogen ligand)-
(coinage metal)-halogen $[LMX]_4$ cluster systems [39] (structure
II), which contains both doubly and triply bridging halides
and hence metal ions of variable coordination numbers. The
topologically equivalent but conformationally different 'tub'
ring (structure III, $\bar{4}$ symmetry) has been found in [(norbor-
nadiene)CuCl]$_4$, in which only one double bond of the diolefin
π-bonds to each metal atom in an *exo* fashion to give a
distorted trigonal planar coordination [40].

5.   PLANAR UNSATURATED EIGHT-MEMBERED CARBOCYCLES

5.1. <u>1,4,7,10-Tetramethyl-5,6-didehydrodibenzo[*a*,*e*]-</u>
     <u>cyclooctene, 9</u>

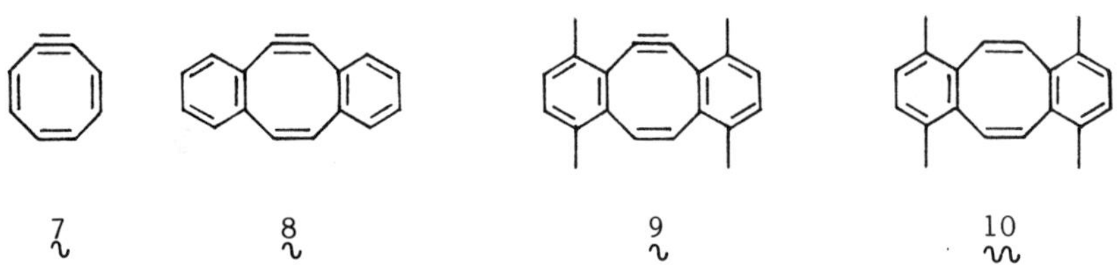

The study of planar cyclooctatetraenes was pioneered by
Krebs, who reported the fugitive existence of 1,2-didehydro-
cyclooctatetraene, 7, in 1965 [41]. The relatively more
stable dibenzo derivative 8 [13,42], which decomposes readily

at room temperature, was shown to be planar within 0.100 Å by low-temperature X-ray crystallography [43]. Subsequently we prepared the stable derivative 9, in which the methyl substituents effectively shield the otherwise rather exposed strained triple bond and kinetically stabilize it against dimerization and/or air oxidation [44]. Recent work [7] has shown that 9 adopts a 'butterfly' conformation (α_a = 11.2, β = 2.7, α_c = 21.9, α_g = 16.4°) intermediate between those of planar 8 and tub-shaped 10 (α_a = 47.0, β = 1.3, α_c = 50.7°) [cf. Fig. 1 and Table 2]. Clearly tetramethyl substitution of 3 to give 10 results in a more folded configuration, and dehydrogenation of 10 to yield 9 corresponds to a flattening of the eight-membered ring. The acetylenic and ethylenic bridges in 9 have nearly the same span between the benzene rings (Fig. 8), indicating that the difference in length of the double and triple bonds are compensated by their respective angular deviations from sp^2 and sp geometries.

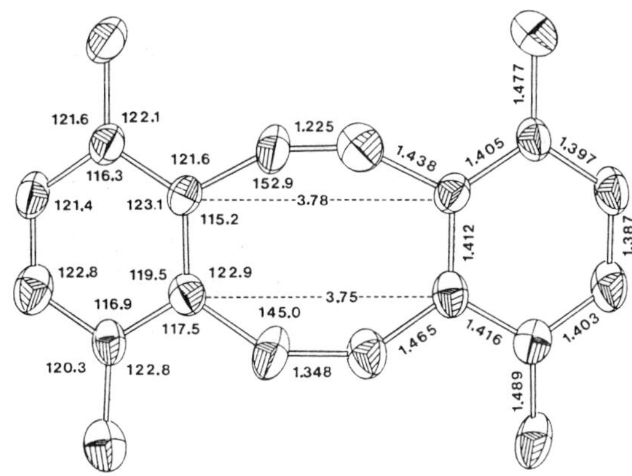

FIG. 8. Perspective view of 9 showing molecular dimensions averaged according to idealized C_s symmetry. Standard deviations: $\sim$ 0.010 Å for acetylenic and ethylenic C–C bonds, 0.008 Å for other bonds; 0.6° for bond angles.

5.2. MINDO/3 study of 7 and related planar dehydro[8]-annulenes

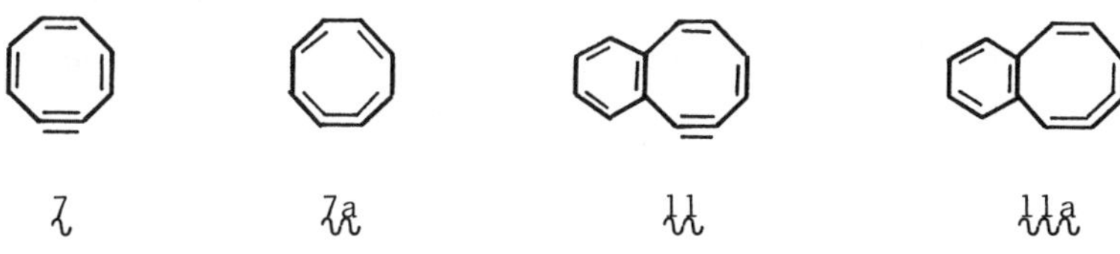

Whereas acetylenic form 7 and cumulenic form 7a have
identical molecular topologies (i.e. the same arrangement of
hydrogen atoms) and could conceivably interconvert through
readjustment of the C-C bond lengths, this is not so for the
pair of benzannelated derivatives 11 and 11a. According to
MINDO/3 calculations, there exists no minimum corresponding
to structure 7a in the potential energy surface, and acetylene 11 is thermodynamically more stable than the cumulenic
isomer 11a by about 22 kcal mole^{-1} [45].

MINDO/3 optimization indicates that a four-membered ring
fused to cyclooctatetraene helps to flatten the eight-
membered ring skeleton, exact planarity being reached with
two cyclobutene rings annelated at positions 1, 2, 5 and 6
[16]. Similarly, the annelation of cyclooctatetraene with
one or more three-membered rings converts the tub to a planar
configuration [46]. These findings are consistent with the
planar carbon skeletons found for 12 [47] and 13 [48] by
X-ray diffraction.

ACKNOWLEDGEMENT

The author gratefully acknowledges the contributions of his
collaborators, especially Dr. H.N.C. Wong (also known as N.Z.
Huang) who provided most of the compounds used in this
research.

REFERENCES

1. Willstätter, R. and Waser, E. (1911). Chem. Ber. **44**, 3432.

2. Schröder, G. (1965). _Cyclooctatetraen._ Verlag Chemie, Weinheim.

3. Röttele, H. (1972). In _Houben–Weyl Methoden der Organischen Chemie_ Band V/1d, pp. 418–525. Georg Thieme Verlag, Stuttgart.

4. Fray, I. and Saxton, R.G. (1978). _The Chemistry of Cyclo-octatetraene and its Derivatives._ Cambridge University Press, London.

5. Delganello, G. (1979). _Transition Metal Complexes of Cyclic Polyolefins._ Academic Press, London.

6. Huang, N.Z. and Sondheimer, F. (1982). Acc. Chem. Res. **15**, 96.

7. Chan, T.-L., Mak, T.C.W., Poon, C.-D., Wong, H.N.C., Jia, J.H. and Wang, L.L. (1986). Tetrahedron **42**, 655.

8. Friedman, L. and Lindow, D.F. (1968). J. Am. Chem. Soc. **90**, 2329.

9. Wong, H.N.C. and Sondheimer, F. (1980). J. Org. Chem. **45**, 2438.

10. Rabideau, P.W., Hamilton, J.B. and Friedman, L. (1968). J. Am. Chem. Soc. **90**, 4465.

11. Xing, Y.D. and Huang, N.Z. (1982). J. Org. Chem. **47**, 140.

12. Rapson, W.S., Shuttleworth, R.G. and van Niekirk, J.N. (1943). J. Chem. Soc., 326.

13. Wong, H.N.C. and Sodheimer, F. (1981). Tetrahedron **37(S1)**, 99.

14. Karle, I.L. and Brockway, L.O. (1944). J. Am. Chem. Soc. **66**, 1974.

15. Irngartinger, H. and Reibel, W.R.K. (1981). Acta Crystallogr., Sect. B **37**, 1724.

16. Mak, T.C.W. and Li, W.-K. (1982). J. Mol. Structure, Theochem **89**, 281.

17. Huang, N.Z. and Mak, T.C.W. (1983). J. Mol. Structure **94**, 135.

18. Li, W.-K., Chiu, S.-W., Mak, T.C.W. and Huang, N.Z. (1983). J. Mol. Structure **94**, 285.

19. Traetteberg, M. (1966). Acta Chem. Scand. **20**, 1724.

20. Dewar, M.J.S. and Thiel, W. (1977). J. Am. Chem. Soc. **99**, 4899, 4907.

21. Bingham, R.C., Dewar, M.J.S. and Lo, D.H. (1975). J. Am. Chem. Soc. **97**, 1285.

22. Laird, B.B. and Davis, R.E. (1982). Acta Crystallogr., Sect. B **38**, 678.

23. Bordner, J., Parker, R.G. and Sandford, R.H. (1972). Acta Crystallogr., Sect. B **28**, 1069.

24. Wheatley, P.J. (1965). J. Chem. Soc., 3136.

25. Palin, D.E. and Powell, H.M. (1947). J. Chem. Soc., 208.

26. Huang, N.Z. and Mak, T.C.W. (1982). J. Chem. Soc.,
      Chem. Commun., 543.

27. Wong, H.N.C., Luh, T.-Y. and Mak, T.C.W. (1984). Acta
      Crystallogr., Sect. C 40, 1721.

28. Herbstein, F.H., Mak, T.C.W., Reisner, G.M. and
      Wong, H.N.C. (1984). J. Incl. Phenom., 1, 301.

29. Cope, A.C. and Hochstein, F.A. (1950). J. Am. Chem.
      Soc. 72, 2515.

30. Mathews, F.S. and Lipscomb, W.N. (1958). J. Am. Chem.
      Soc. 80, 4745.

31. Mathews, F.S. and Lipscomb, W.N. (1959). J. Phys. Chem.
      63, 845.

32. Ho. W.C. and Mak, T.C.W. (1983). J. Organometal. Chem.
      241, 131.

33. Mak, T.C.W. and Ho, W.C. (1983). J. Organometal. Chem.
      243, 233.

34. Mak, T.C.W. (1983). J. Organometal. Chem. 246, 331.

35. Moyer, P., Rimsky, A. and Chevalier, R. (1978). Acta
      Crystallogr., Sect. B 34, 1457.

36. Mak, T.C.W., Ho, W.C. and Huang, N.Z. (1983). J.
      Organometal. Chem. 251, 413.

37. Silverthorn, W.E. (1975). In Advances in Organometallic
      Chemistry (eds. F.G.A. Stone and R. West) Vol. 13,
      p. 43. Academic Press, New York.

38. Mak, T.C.W., Wong, H.N.C., Sze, K.H. and Book, L.
      (1983). J. Organometal. Chem. 255, 1983.

39. Teo, B.-K. and Calabrese, J.C. (1976). Inorg. Chem. 15,
      2474.

40. Baenziger, N.C., Haight, H.L. and Doyle, J.R. (1964).
      Inorg. Chem. 11, 1535.

41. Krebs, A. (1965). Angew. Chem. Int. Ed. Engl. 4, 953;
      Krebs, A. and Byrd, D. (1967). Justus Liebigs Ann.
      Chem. 707, 66.

42. Wong, H.N.C., Garratt, P.J. and Sondheimer, F. (1974).
      J. Am. Chem. Soc. 96, 5604.

43. de Graff, R.A.G., Gorter, S., Romers, C., Wong, H.N.C.
      and Sondheimer, F. (1981). J. Chem. Soc., Perkin Trans.
      2, 478.

44. Huang, N.Z., Jia, J.H., Wang, L.L., Chan, T.-L. and
      Mak, T.C.W. (1982). Tetrahedron Lett. 23, 4797.

45. Huang, N.Z., Mak, T.C.W. and Li, W.-K. (1981).
      Tetrahedron Lett. 22, 3765.

46. Wong, H.N.C. and Li, W.-K. (1984). J. Chem. Res.(S),
      302.

47. Einstein, F.W.B., Willis, A.C., Cullen, W.R. and
      Soulen, R.L. (1981). J. Chem. Soc., Chem. Commun.,
      526.

48. Dürr, H., Klauck, G., Peters, K. and von Schnering, H.G.
      (1983). Angew. Chem. Int. Ed. Engl. 22, 332.

35. Interactions in cyclodextrin clathrates: inclusion and release of pheromones

G. Tsoucaris, N. Rysanek, G. Le Bas, F. Villain, E. Hadjoudis and I. Moustakali-Mavridis

1   INTRODUCTION

α-, β- and γ-cyclodextrins (cyd) are cyclic oligosaccharides built up from respectively 6, 7 and 8 glucopyranose units which include various guest molecules.   The toroidal cavity shows a pronounced hydrophobic behaviour, so that lipophilic and/or hydrophobic molecules can be solubilized in water.   Obviously, this opens up a major field of applications.   Crystallization of these complexes leads to clathrates, i.e. mixed crystals characterized by a complete enclosure of the guest within cavities formed by a three-dimensional framework of the host. This opens up another field of applications: retention of volatile guests and protection from oxidation, light etc.   The specific microenvironment of the guest molecule within the molecular and/or crystal cavity induces changes in guest conformation and electronic structure, and therefore in guest properties.   This leads again to other applications, namely fine tuning of molecular properties.

Cyclodextrins present an important advantage over other clathrates: the inclusion of the guest molecule occurs already in solution, and it is possible to study, for the same system, both the crystal structure and the spectroscopic properties in solution.   In our laboratory we have considered fluorescence, photochromic and thermochromic phenomena, UV and circular dichroism spectra: the latter provides a novel method of studying the chiroptical properties of conformationally-labile molecules (Hadjoudis *et al* 1986, Le Bas *et al* 1986).

Clearly, the above properties and potential applications depend strongly on the fit between host and guest.   The size, polarity and gross geometry of the guest are the principal factors presiding over the strength of the host/guest interaction.   On a finer scale, this interaction involves a feedback concerning conformational changes.

The aim of this contribution is (a) to survey critically examples of such fine interactions, and (b) to report the first results concerning cyclodextrin·pheromone systems whose practical potentiality depends sharply upon the nature and strength of the host/guest interactions.

2   CONFORMATIONAL AND ENANTIOMERIC SELECTIVITY

Enantiomeric selectivity provides a clear manifestation of host/ guest fit, resulting in the phenomena of optical rotation and circular dichroism, which are easy to detect.   In practice, despite the simplicity of the concept and the experiment, it turns out that, for every particular system, given both host and guest structures, it is very difficult to predict the preferential association or inclusion of one guest enantiomer over another. Indeed, inclusion is a complex phenomenon involving a balance between enthalpic and entropic factors, further complicated by the concomitant expulsion of water molecules located within the cyclodextrin cavity.   The enthalpy factor involves mainly van der Waals forces and, possibly, polar forces and hydrogen bonds. We shall see, with a few striking examples, that the notion of fit is very complicated when it comes down to evaluating all the interatomic forces presiding over the inclusion.

The bile pigments bilirubin and biliverdin are achiral,

but, upon interaction with cyclodextrin, they acquire a preferential chiral conformation resulting in a very strong circular dichroism spectrum ($\Delta\varepsilon > 10$). In the following scheme, we see that the size of the γ-cyclodextrin cavity is smaller than that of bilirubin.

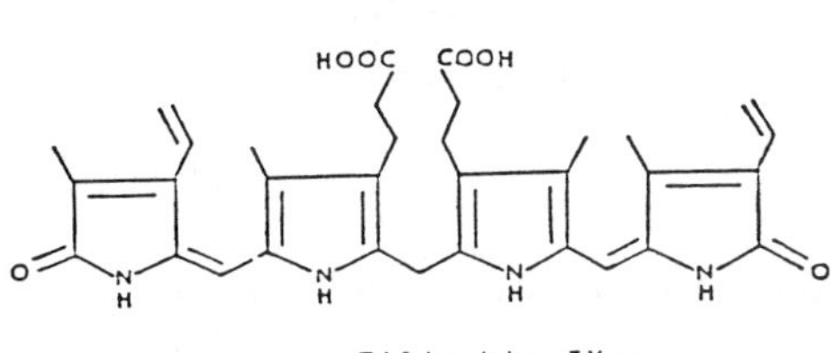

This suggests that bilirubin is 'sitting on' rather than 'included in' the cyclodextrin cavity. The molecular interaction is strong enough to discriminate between the two guest enantiomers, although the association constant is rather low. We note that no crystals could be obtained. These experiments show that, for certain systems, inclusion may not be a prerequisite for the manifestation of markedly different properties of the guest molecule upon complexation with cyclodextrin.

A different situation arises in the α-cyclodextrin·cyclopentanone complex and clathrate. This small molecule fits comfortably into the α-cyclodextrin cavity, and we have observed new phenomena upon inclusion. In the highly symmetrical P6 structural arrangement, there are two independent molecules in the cell, one on the hexagonal, and the other on the trigonal site. To our knowledge, these ideal symmetries for α-cyclodextrin molecules have not yet been found (Le Bas 1985).

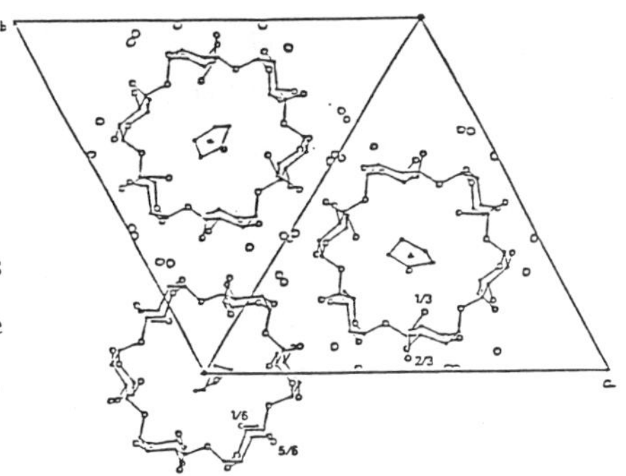

The two guest molecules in the unit cell are respectively sixfold and threefold disordered. The primary hydroxyl groups are also disordered between the gauche-gauche and the gauche-trans conformation. The occupancy factor of the gauche-trans conformation is about:

- 1/6 for the cyclodextrin molecule at the sixfold axis,
- 1/3 for the molecule at the threefold axis,

for only three primary hydroxyl groups, the other three being entirely in the gauche-gauche conformation.

Moreover, for both sites, the distances between the gauche-trans primary oxygen of the host and the carbonyl oxygen of the guest are close to that expected for an H-bond. These most remarkable facts suggest a <u>mutual dependence</u> between the disorder of the primary -OH of the host and the orientation of the guest.

In conclusion, we have to be very careful when reasoning with rigid geometrical models: a conformational feedback of statistical or dynamic origin may radically alter the expected picture of host/guest fit.

3   PHEROMONES

Pheromones, sex attractants produced by insects, can be used in agronomy to attract insects into a 'trap', i.e. a surface area containing small quantities of pesticide. Thus, pollution resulting from the massive use of these nocive substances could be avoided. The practical implementation of this process is impaired by the volatility of pheromones and their chemical instability in the air. This drawback can be greatly reduced or totally overcome by inclusion of pheromones in cyclodextrins. The major component of the sex pheromone of the olive fly *Dacus oleae*, 1,7-dioxaspiro[5-5] undecane has been included in β-cyclodextrin.

Field bioassays at the Biology Institute Demokritos in Athens showed that the clathrate is 'too stable', i.e. the release rate of the guest is too low. This is a 'molecular and crystal engineering' problem. Our programme of work comprises crystal structure determinations, jointly with the search for appropriate cyclodextrin derivatives.

The monoclinic clathrate (space group C2, Z=4, a=19.33 Å, b=24.42 Å, c=15.94 Å, β=108.72°, R=12.7% for 4400 reflections) belongs to type A of a recently-attempted classification (above references). It consists of channels where one pheromone molecule lies wihin cyclodextrin dimers, and the other in the interdimer space. The guest molecules are probably disordered around the twofold axis.

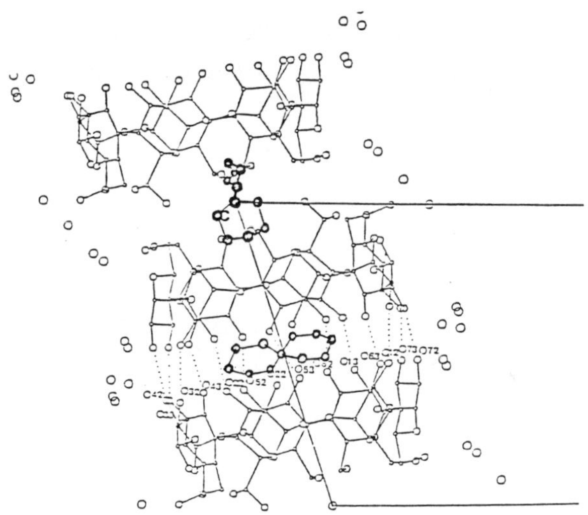

One can ask the question whether the crystal structure at this stage of the analysis already accounts for the stability of the clathrate. It may be that the continuous columns of cyclodextrin linked to pheromone molecules by several van der Waals contacts, and eventually hydrogen bonds, constitute a 'rigid pillar' of the structure.   On the other hand, guided by molecular models to obtain a less stable clathrate, we achieved inclusion in 2,6-dimethyl-β-cyclodextrin.   The first experiments have shown that this clathrate is indeed less stable than that of β-cyclodextrin (Mazomenos).

In conclusion, the notions of geometrical fit, feedback in conformational changes, and chiral discrimination upon association must be critically studied in order to understand the inclusion phenomenon and its influence on physical, chemical and biological properties.

REFERENCES

HADJOUDIS E., MOUSTAKALI-MAVRIDIS I., TSOUCARIS G. and VILLAIN F.   (1986), Mol.Cryst.Liq.Cryst., 134, 255

LE BAS G.   (1985), Thesis, Université Paris VI

LE BAS G. and TSOUCARIS G.   (1986), Mol.Cryst.Liq.Cryst., 137, 287

MAZOMENOS B.E.   Unpublished results

36. Host–guest interactions in the molecular structure of *β*-cyclodextrin inclusion complexes with some benzene derivatives having hydrophilic and/or hydrophobic side chains

Ken-ichi Tomita and Takaji Fujiwara

ABSTRACT

As the results of X-ray structure determinations, we compared
the molecular structures of β-cyclodextrin(abbr.CyD) complexes
with some benzene derivatives and also discussed how each guest
molecule is included in the hydrophobic cavity of β-CyD and how
the hydrophilic and/or hydrophobic side chains of guest molecules exert upon the host-guest interactions.

1. INTRODUCTION

The cyclic oligosaccharides, α-, β-, and γ-CyD consist respectively of six, seven and eight D-glucopyranose units connected
by an α-1,4-linkage. These cyclic oligosaccharides have a peculiar property of harbouring small guest molecules in their inner
cavities of diameter 6 to 9 Å. X-Ray structural analysis on some
α- and β-CyD complexes has revealed the existence of a hydrophobic and rather weak interaction between the guest molecules and
the host CyD (Tokuoka et al, 1981). In order to elucidate this
particularly interesting character of host-guest interaction
modes and to clarify the effects of hydrophilic and/or hydrophobic side chains on these interaction modes and on the stability
of molecular packing, an X-ray structural determination was made
on the β-CyD inclusion complexes with some benzene derivatives
having hydrophilic and/or hydrophobic side chains.

TABLE 1. Crystal data

Host	β-cyclodextrin	β-cyclodextrin	β-cyclodextrin
Guest	3,4-xylidine($\underline{1}$) (Xyl)	salicylic acid($\underline{2}$) (Sali)	acetylsalicylic acid, aspirin($\underline{3}$) (Aspi)
Formula	$(\beta\text{-CyD})_2(\text{Xyl})_3\cdot 30H_2O$	$(\beta\text{-CyD})_2(\text{Sali})_3\cdot 23H_2O$	$(\beta\text{-CyD})_2(\text{Aspi})_2(\text{Sali})\cdot 23.3H_2O$
a	19.588(7) Å	16.084(4)	19.777(5)
b	15.433(3) Å	15.538(4)	15.247(3)
c	15.431(8) Å	15.541(8)	15.475(4)
α	102.55(3) °	103.41(5)	102.63(2)
β	117.63(3) °	101.00(3)	116.96(2)
γ	104.49(2) °	101.03(4)	104.12(2)
Z	1	1	1
Space group	P1	P1	P1
V	3687 Å^3	3594	3729
Dm	1.427(1) $g\cdot cm^{-3}$	1.461(2)	1.409(2)
Dx	1.429	1.431	1.419
Mw	3173.8	3080.7	3188.2
No. of reflections $(F_0>3\sigma(F_0))$	9957	5964	9086
R factor	0.120	0.126	0.107

2. EXPERIMENTAL METHODS

3,4-Xylidine(1)(abbr.Xyl), salicylic acid(2)(abbr.Sali) and acetylsalicylic acid, aspirin(3)(abbr.Aspi) were used as guest molecules of the β-CyD inclusion complexes. These complex crystals
were prepared by heating aqueous solutions of β-CyD and each of
the guest molecules to 60-70°C and then cooling the mixture slowly to room temperature.

The crystal structure of the β-CyD inclusion complex was
solved by; a) the interpretation of the Patterson peaks for the
head-to-head dimeric β-CyD·Xyl inclusion complex as well as
those of the β-CyD·Sali and β-CyD·Aspi complex crystals. b) the
application of successive Fourier synthesis and least-squares
refinement of these β-CyD complexes with benzene derivatives.
The R factors obtained for the complexes β-CyD·Xyl, β-CyD·Sali
and β-CyD·Aspi were 0.120, 0.126 and 0.107, respectively(Tabel 1).

3. RESULTS AND DISCUSSION

The Patterson peaks observed for β-Cyd·Sali and β-CyD·Aspi were
very similar to that obtained for the β-CyD·Xyl complex. We
therefore thought of these complexes as being isomorphous to the
β-CyD·Xyl structure. The stomic coordinates obtained for the 154
atoms of the dimeric β-CyD molecules in the β-CyD·Xyl complex
were hence used for the calculation of the initial phases in the
cases of the β-CyD·Sali and β-CyD·Aspi complexes (Nishioka et al,
1984; Nishioka, 1985). The crystallographic data obtained for
these complexes are shown in Table 1.

The inside diameter of the β-CyD cavity which is about 7 Å,
is big enough to accommodate the benzene ring. Therefore, when
the benzene derivative is used as the guest molecule for the β-
CyD inclusion complex, the benzene ring may be properly fixed in
the cavity of the β-CyD molecule by rather weak hydrophobic interactions. However, the molecular association of the β-CyD complex
is influenced by the molecular dimensions of the guest molecule.
For instance, if smaller molecules such as water, methanol and

ethanol are used, the inclusion complex molecules of the β-CyD are packed into a "cage structure" in the crystalline state. The resulting crystal packing of the β-CyD complex looks like that of a herring bone type structure.

On the other hand, when the guest molecule is bigger than ethanol, e.g., n-propanol, two host β-CyD molecules interact with each other to form a dimer with a head-to-head fashion that involves three guest molecules (Stezowski et al, 1978); two in each cavity and one in the vacancy formed at the centre of the dimer. The three β-CyD inclusion complexes with benzene derivatives described in this paper belong to the above category. They form a unit with a molar ratio 2 to 3 (Table 1). Each of the three benzene derivatives is able to penetrate into the cavity of a dimer which then pile up to form a "channel structure" in the β-CyD·Xyl and β-CyD·Sali crystals or a "brick wall type structure" in the case of the β-CyD·Aspi crystal.

The orientations of the three different guest molecules (see Fig. 1) reflect the following particular features: 1) Because of the different host-guest interactions, there exist two kinds of guest molecule in each complex crystal. The first set occupies the centre of the β-CyD channel axis. In the β-CyD·Aspi complex, this guest molecule is not acetylsalicylic acid but salicylic acid, a degradated compound of acetylsalicylic acid. Although acetylsalicylic acid is only used as a guest molecule, it occupies three statistically disordered positions with van der Waals contact to the β-CyD molecule. The second set of guest molecule occupies the cavity of each β-CyD molecule in the dimer. The orientation of this second set of molecules, which is usually inclined to the β-CyD channel axis, is independent of each other. In Fig. 1(c), one of the guest molecule (lower one) is shown to be almost parallel to the β-CyD channel axis, and its molecular plane is held in place by two hydrogen bonds between the carboxyl oxygen atom and the acetyl oxygen atom of salicylic acid and a water molecule. In the case of the β-CyD complex with 3,4-xylidine

FIG. 1

(a)
(b)
(c)
3,4-xylidine complex
salicylic acid complex
aspirin complex
N
C
O

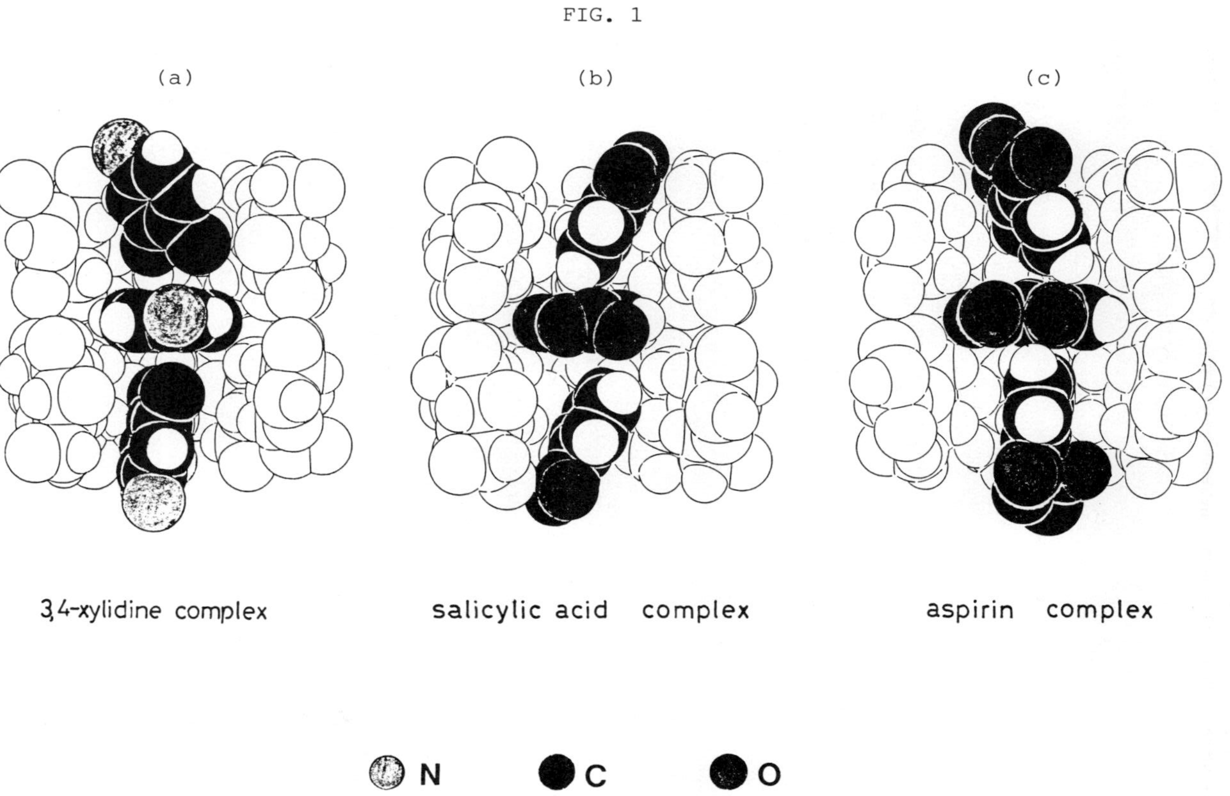

or salicylic acid, a hydrogen bond exists between the amino nitrogen atom or the carboxyl oxygen atom of the guest molecule, which is protruding out of the central cavity, and the primary hydroxyl oxygen atom of the β-CyD molecule. This bond is inclined to the molecular plane of the guest molecule. 2) The orientation of the two inclined guest molecules trapped in the cavity of the β-CyD dimer may influence the rotational disordering of the first set of guest molecule perpendicular to the β-CyD channel axis. In the case of β-CyD·Sali complex (Fig. 1(b)), the orientations of these two inclined guest molecules are almost symmetrical and the dihedral angle between the individual molecular planes of the guest molecules and that of the β-CyD molecule is also very similar (60° and 63°) where the molecular plane of the β-CyD molecule is defined by the seven oxygen atoms at the C4 position. It is therefore very probable that in the case of β-CyD·Sali complex, the orientation of the third guest molecule (first set of guest molecule) located at the centre of the β-CyD dimer is influenced by the two guest molecules (second set of guest molecule) in the cavity of the β-CyD dimer. In the other two cases (Fig. 1(a and c)), the orientations of the two inclined guest molecules varied and no correlation was found between them, hence the third molecule has a statistically disordered orientation.

REFERENCES

NISHIOKA. F., NAKANISHI. I., FUJIWARA. T. and TOMITA. K. (1984). J. Inclusion Phenomena 2, 701.

NISHIOKA. F. (1985). Master thesis, Faculty of Pharmaceutical Sciences, Osaka University.

STEZOWSKI. J. J., JOGUN. K. H., ECKLE. E. and BARTELS. K. (1978). Nature (London) 274, 617.

TOKUOKA. R., FUJIWARA. T. and TOMITA. K. (1981). Acta Cryst. B37, 1158.

37. Werner clathrates

Luigi R. Nassimbeni and Margaret L. Niven

1. INTRODUCTION

Interest in Werner Clathrates started with the work of Schaeffer and coworkers, who announced a new method of separating aromatics from petroleum fractions (Schaeffer, Dorsey, Skinner and Christian, 1957). These compounds comprise host molecules of general formula MX_2L_4 (M = Mn, Fe, Co, Ni, Cu, Zn; X = SCN, NO_2, CNO, halogen; L = substituted pyridine or 1-arylalkylamine), which entrap a wide variety of organic compounds as guest molecules. The host complex which has received the most attention is $[Ni(NCS)_2(4\text{-}MePy)_4]$. Its crystal structures with a number of guest molecules and the physico-chemical behaviour of these clathrates have been reviewed by Lipkowski (1984). The thermodynamics of clathration (Lipkowski, Starzewski and Zielenkiewicz, 1982) and its scope of application as the stationary phase in chromatography, have been studied (Kemula, Sybliska and Lipkowski, 1981). An important feature of these Werner complexes is the freedom of rotation of the pyridine rings about their Ni-N bonds. This may be the prime factor which allows the host molecules to adjust their shapes, form appropriate channels or cavities, and so capture guest molecules of different shapes and sizes. It appears that sorption of a guest molecule has the ability to transform the non-porous α-phase of the $[Ni(NCS)_2(4\text{-}MePy)_4]$ complex into a tetragonal, clathrating β-phase (Allison and Barrer, 1969). Several proposals of host-guest interactions within these clathrates have been advanced in order to account for their selectivity (Guarino, Occhiucci, Possagno and Bassanelli, 1977). On the basis of spectral data researchers have hypothesised charge-transfer interactions between host and guest molecules (de Radzitzky and Hanotier, 1962). More recent studies, however, have enabled re-interpretation of previous spectral data solely in terms of steric interactions (Borkowska, Lipkowski, Moszynska and Wolfram, 1972; Moszynska, Lipkowski and Jonowski, 1973).

2. $[Ni(NCS)_2(4\text{-}MePy)_2(4\text{-}PhPy)_2]$ STRUCTURES

We reported the first Werner clathrate having mixed pyridines in the host complex (Bond, Jackson and Nassimbeni, 1983). The clathrate with methylcellosolve as guest molecule is monoclinic, with a = 10.648(5), b = 23.005(12), c = 16.294(8)A, β = 98.04(2)°, space group C2/c. The host molecules have the central nickel atom in octahedral configuration with all like ligands <u>trans</u> to each other. There are four host molecules in the unit cell, located such that the <u>trans</u> 4-phenylpyridine moieties lie on a diad. A perspective view of the

packing is shown in Fig. 1 which clearly shows the
channel, parallel to <u>a</u>. Symmetry considerations
required the guest molecule to lie on a centre of
inversion and, as methylcellosolve, $CH_3OCH_2CH_2OH$ is
not centrosymmetric, we invoked statistical disorder
in the final model. This is a common feature of the
structure of Werner clathrates, and it reoccurred in
two other structures with $[Ni(NCS)_2(4\text{-MePy})_2(4\text{-}$
$PhPy)_2]$ as host complex, having acetylacetone and 1-
chlorobutane as guest molecules. The latter
structures are isomorphous with the methylcellosolve
one and the guest molecules are similarly disordered.

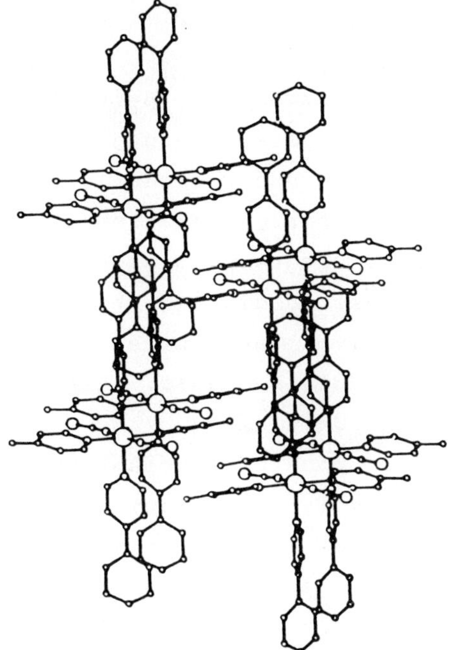

FIG. 1.   Packing of $[Ni(NCS)_2(4\text{-MePy})_2(4\text{-PhPy})_2]$
·(methylcellosolve) showing the channel.
Guest molecule omitted.

2.1. <u>Energy study</u>

The potential energy environment of the methyl cellosolve molecule in the crystal lattice was studied in order to determine possible dynamic conformation changes in the host molecules and to ascertain the nature and strength of the forces holding the guest 'captive' in the clathrate. We used the program EENY (Motherwell, 1973) to calculate the van der Waals energy using empirical atom-pair potential curves. The coefficients of the atom-atom potentials are of the form

$$U(r) = a \exp (-br)/r^d - c/r^6$$

where r is the distance between any pair of atoms and coefficients $\underline{a}$, $\underline{b}$, $\underline{c}$, $\underline{d}$, are those given by Giglio (1969). These potential curves were derived primarily to give good agreement for calculation of molecular position in crystal structure. No account was taken of partial atomic charges or dipole interactions, and the energy values derived are more useful in a comparative than an absolute sense. The summation was extended over two unit cells. We arbitrarily chose one of the four possible orientations of the disordered methyl cellosolve guest molecule, and then translated it through the channel in steps of 1Å, as shown in Fig. 2.

The energy was initially calculated on the assumption that both host and guest remained rigid yielding the energy profile '1' shown in Fig. 3. This model was then refined by allowing the guest molecule small variations in rotational parameters in order to find a local minimum in the energy profile

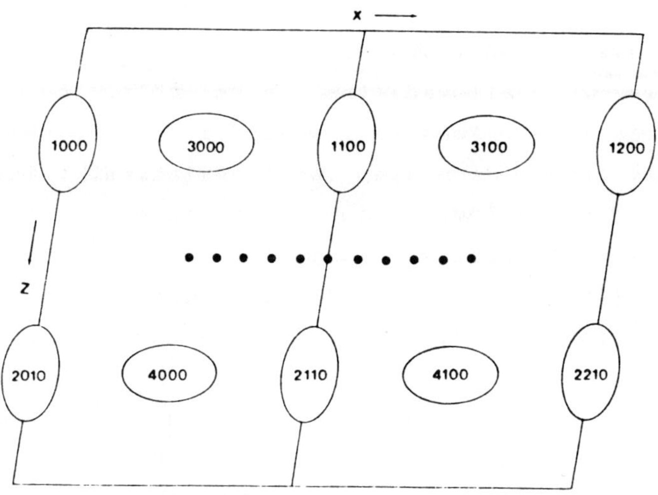

FIG. 2.   Translation of methylcellosolve molecule through the channel.

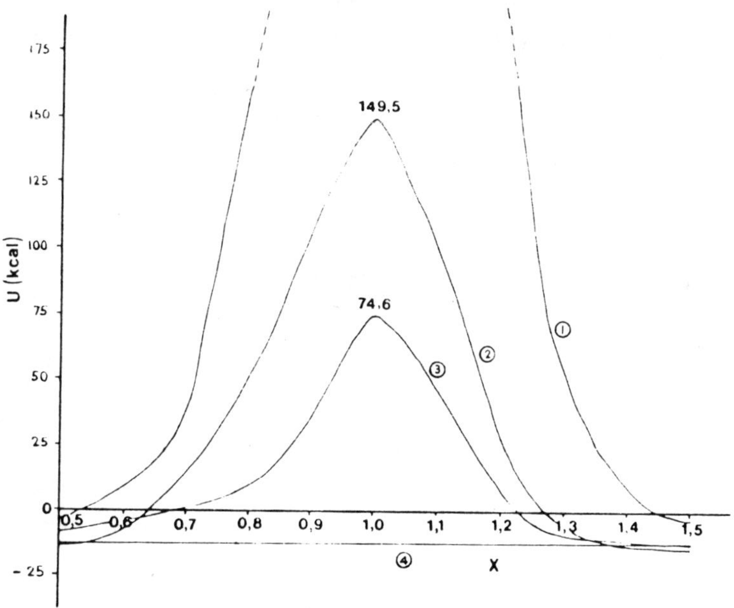

FIG. 3.   Potential energy profiles.

after every 1-A translation (Fig. 3, profile '2'). Thirdly we allowed the host molecules partial conformational freedom while holding the guest rigidly after each step of translation. This was achieved by allowing the six rings of the host molecule to twist independently and achieve minimum energy (Fig. 3, profile '3'). Finally we allowed both conformational freedom of the host molecule and rotational freedom of the guest molecule. This resulted in no energy barrier (Fig. 3, profile '4') suggesting free diffusion of the guest through the channel. This is clearly unreasonable, and we suggest that a realistic energy model lies between profile '3' and '4'. Experimentally, an activation energy for the diffusion of p-xylene through a host lattice of [Ni(NCS)$_2$(4-MePy)$_4$] has been established as $\underline{ca}$ 15.8 kcal mol^{-1} (Lipkowski, 1980). This is consistent with a clathration model which involves the host adopting a specific conformation in order to trap the guest molecules, thus preventing totally free diffusion of the guest molecules.

3.   [Ni(NCS)$_2$(4-EtPy)$_4$] STRUCTURES

The host complex with 4-ethylpyridine bases has thus far proved the most versatile. We characterised eight structures in all, two in its non porous α-phase (I and II) and six structures of its clathrates with $\underline{para}$ - (IV), $\underline{meta}$ (V) and $\underline{ortho}$ - (VI) xylenes, carbon disulphide (VII) and carbon tetrachloride (III,VIII) as guest molecules. The [Ni(NCS)$_2$(4-EtPy)$_4$] host molecules in all these structures have octahedral coordination and 'propeller' conformation.

Their diversity is summarised in Fig. 4, which is adapted from a similar scheme first proposed by Lipowski (1984). Compounds I and II are slightly different α-phases which crystallised in space group P$\bar{1}$ from different solvents. Their structures are essentially similar, differing only in their orientations of the ethyl groups, as shown in Fig. 5. The enclathration process requires the structure to change to a β_O phase, one in which the host molecules have realigned so as to form interstices. This β_O phase is in turn modified by the entry of guest molecules, into the various clathrate phases β, γ, δ, ... etc. These latter phases arise from the different host:guest ratios and from the varied

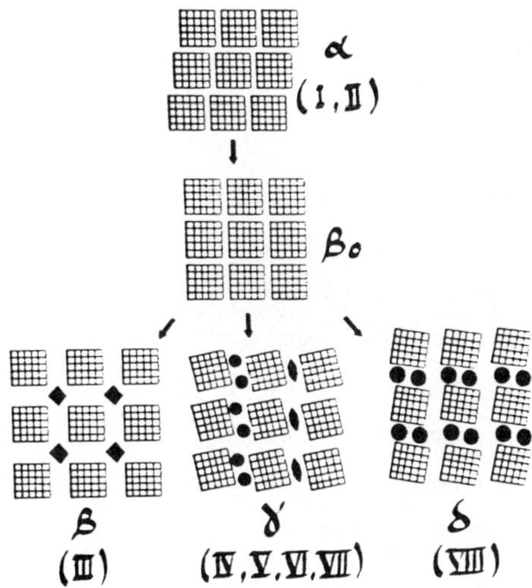

FIG. 4. Schematic diagram showing the different phases of [Ni(NCS)$_2$(4-EtPy)$_4$]·nG

molecular shapes of the guest molecules. In all
these compounds the nickel is coordinated to six
nitrogen donor ligands to form an octahedron with the
isothiocyanate moieties in the <u>trans</u> positions. The
observed bond lengths and angles are within the usual

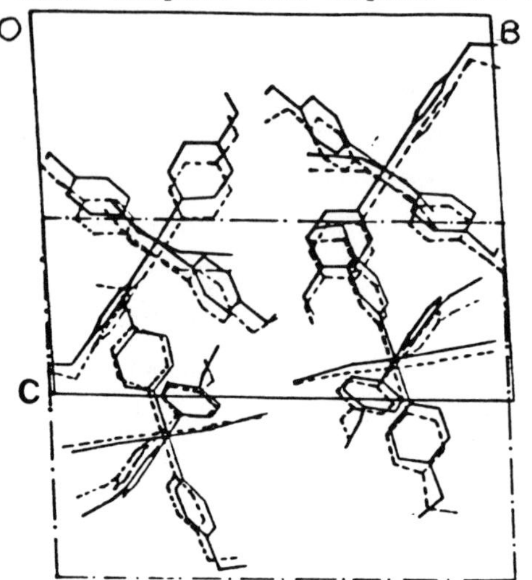

FIG. 5.   α-phases of [Ni(NCS)$_2$(4-EtPy)$_4$] viewed
along [100]

range for compounds of this type. A perspective view
of a host molecule of compound I, one of the α
phases, is shown in Fig. 6. This 'propeller'
conformation of the host molecule remains remarkably
constant throughout compounds I to VIII irrespective
of the crystallographic phase.

The enclathrated phases are stable only in the
presence of guest molecules so the closest one can
approximate a β_O phase, particularly as single
crystals, is a clathrate containing very small guest
molecules. Attempts to crystallise the host powder
from methanol always resulted in the formation of an

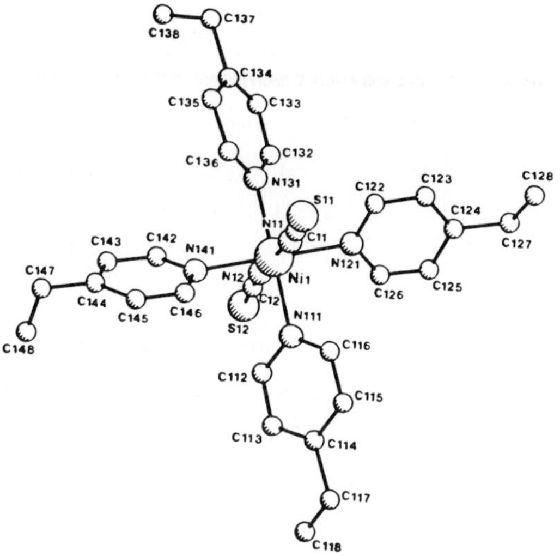

FIG. 6.   Perspective view of $[Ni(NCS)_2(4\text{-}EtPy)_4]$
host molecule

α phase. However the structure of $[Ni(NCS)_2(4\text{-}MePy)_4]\cdot 2CH_3OH$, which must be very close to the β_0 structure of the 4-Me compound, has been elucidated (Lipowski, Suwinska, Andreetti and Stadnicka, 1981).

Compound III, $[Ni(NCS)_2(4\text{-}EtPy)_4]\cdot\frac{1}{2} CCl_4$, crystallises in the well known β phase in the space group $I4_1/a$. However in all other similar structures the guest molecules occupy the $\bar{1}$ cavities, while in compound III the CCl_4 molecules fill the $\bar{4}$ cavities, thus matching the shapes of the guest molecules with those of the cavities.

Compounds IV (H.1$\underline{p}$ - xylene) V (H.1$\underline{m}$ - xylene), VI (H.1 $\underline{o}$ - xylene) and VII (H.2 carbon disulphide), all crystallise in the space group $P\bar{1}$ with similar unit cell parameters $a \approx b \approx c \approx 17{,}6\text{Å}$ and $\alpha \approx \beta \approx \gamma \approx 60°$ each with four molecules per unit

cell.   These triclinic cells can be related to the β
phase cell by the transformation

$$
\begin{bmatrix} a^1 \\ b^1 \\ c^1 \end{bmatrix} \begin{matrix} \\ \text{pseudo} \\ \text{tetragonal} \end{matrix} = \begin{bmatrix} 0 & 1 & -1 \\ 1 & 0 & 0 \\ 1 & -1 & -1 \end{bmatrix} \begin{bmatrix} a \\ b \\ c \end{bmatrix} \begin{matrix} \\ \\ \text{triclinic} \end{matrix}
$$

For example, applying the transformation matrix to
the triclinic cell of compound VII one obtains the
cell a^1 = 17.355, b^1 = 17.377, c^1 = 24.700 A
α = 89.96°, β = 89.97°, γ = 89.26°. This is similar
in dimension to the tetragonal cell of compound III
with a = 17.377 and c = 25.174 A. However the
similarities in the unit cell dimensions between the
β and pseudo -β phases are not reflected in the
relative configurations of the host molecules. This
is illustrated in Fig. 7 which schematically portrays
the difference between the β and pseudo -β phases.
This orientation difference in the host molecules
shows that structures IV, V, VI and VII can be
classified as a new phase, labelled γ.

Compound VIII, $[Ni(NCS)_2(4\text{-EtPy})_4] \cdot 2CCl_4$ has a
different stoichiometry from the chemically similar
compound III and the packing is totally different, as
may be expected from the four fold increase in
guest:host ratio. The monoclinic cell of compound
VIII cannot be related by any simple transformation
to any of the other clathrate structures, and thus
represents phase δ. Structural details of compounds
I to VIII have recently been published (Moore,
Nassimbeni and Niven, 1987).

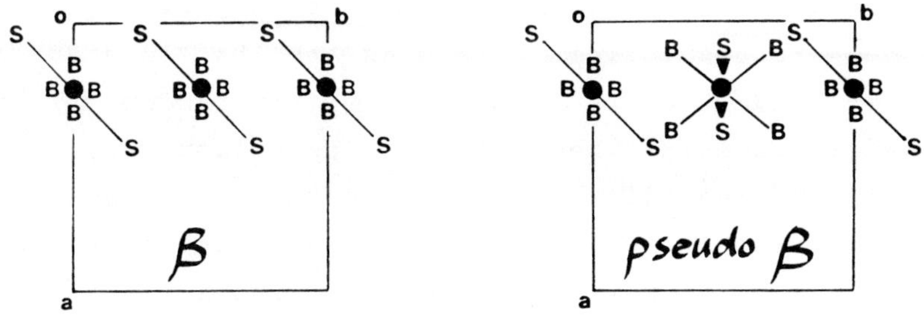

FIG. 7.   Schematic representation of β versus
pseudo-β structures, showing differing
host-molecule orientations.

Structures III and VIII, which contain carbon
tetrachloride as the guest, exhibit different modes
of enclathration. We used the program OPEC
(Gavezzotti, 1983) to map precisely the sizes and
shapes of the guest cavities. In compound III the
guest lies at a $\bar{4}$ site which is an approximately
spherical cage with a volume of 148 Å^3. According to
Kitaigorodskii (1973), an ordered carbon
tetrachloride molecule occupies 85Å^3. Final
refinement of this structure showed the guest
molecule to be partially disordered, so the
calculation of the cage volume was deemed
satisfactory. In contrast, compound VIII possesses
channels which zig-zag along a. The topology of a
channel is shown in Fig. 8, which displays two unit
cells sectioned at y = 0.5, where the channel is at
its widest. The channel has a distorted hour-glass
shape, similar to that found in Dianin's compounds.

The maximum width of the channel is 16Å, allowing two
carbon tetrachloride molecules to be located on
either side of a centre of inversion. The
constrictions in the channels are elliptical, with
cross sectional distances of approximately 2 and 3 Å.
This is too narrow to allow free movement of guest
molecules through the channel, but inspection of the
crystal structure shows that a simple rotation of a
pyridine moiety would ease this constriction.

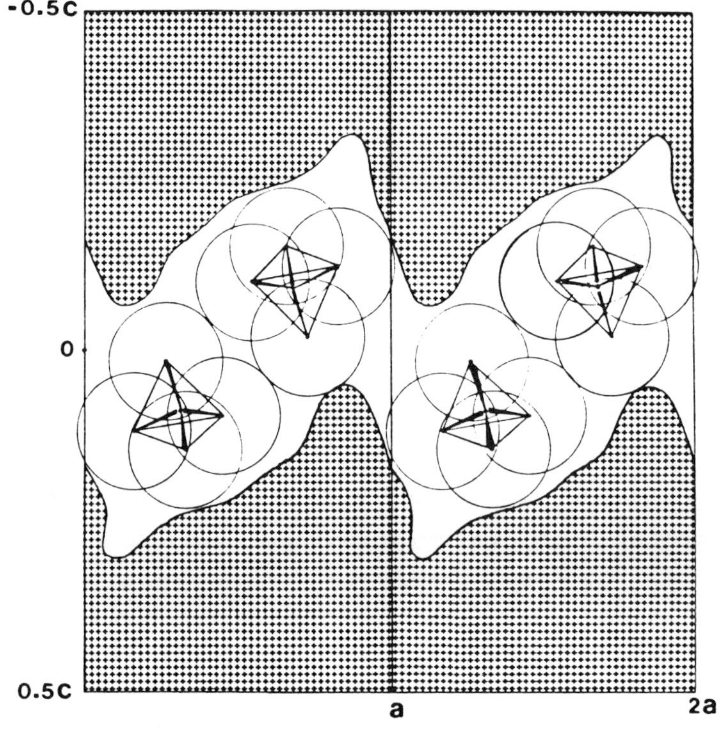

FIG. 8.   Two unit cells of
[Ni(NCS)$_2$(4-EtPy)$_4$]·2CCl$_4$
showing the topology of the channel
sectioned at y = ½.

4. $[Ni(NCS)_2(4\text{-}ViPy)_4]$ STRUCTURES

The clathrates formed with 4-vinylpyridine bases are more stable than their 4-ethylpyridine analogues. The structures of the α-phase complex and its β-phase clathrates with p-xylene, m- xylene and o- xylene have recently been reported (Moore, Nassimbeni, Niven and Taylor, 1986). The three inclusion compounds all crystallise in the space group $I4_1/a$, with the guest molecules located around centres of symmetry. In the case of m- and o- xylenes, this required the molecules to be disordered, resulting in a less efficient packing density than that of the p-xylene structure. Calculations employing the atom pair potentials previously outlined also established that the energies of the guests in their cavities are in the order p- xylene $<$ m- xylene $\approx$ o- xylene. This is in agreement with the view that coincidence of host cavity symmetry with guest molecular symmetry is energetically favoured (Arad-Yellin, Green, Knossow and Tsoucaris, 1983).

Differential Thermal Analysis (DTA) and Gravimetric Thermal Analysis (GTA) were carried out on a series of clathrates comprising the vinyl host and selected guest molecules. The GTA curves show that the mechanism of decomposition is invariably

$$NiX_2B_4 \cdot nG(s) \qquad \text{where} \quad X = NCS^-,$$
$$\downarrow \qquad\qquad\qquad\qquad B = 4ViPy$$
$$NiX_2B_4(s) + nG(g)$$
$$\downarrow$$
$$NiX_2B_2(s) + 2B(g)$$
$$\downarrow$$
$$NiX_2(s) \quad + 2B(g)$$

The DTA experiments were used to evaluate the enthalpy of the 'guest-release reaction'. This was established by measuring the area under the first endotherm in each DTA run, after calibration with suitable standards. The results for various guests are summarised below, where we list the temperature t_{g-r} at which the guests are released, as well as the enthalpy changes of the reactions.

nG	1.8 THF	2CHCl$_3$	0.25 ViPy	1-o-xylene	1-p-xylene
t_{g-r} /°C	47	62	90	102	125
ΔH_{g-r} kJmol^{-1}	87	102	44	81	74
thermogram	a	b	c	d	e

The thermograms are shown in Fig. 9.

These results correspond to our experiences in handling the single crystals of these clathrates for structural analyses. The tetrahydrofuran (THF) clathrate is so unstable that we never managed to isolate and mount a single crystal, and the chloroform clathrate is visibly more unstable than the corresponding xylene compounds.

In order to obtain a better understanding of the clathration process we measured the solubility of this host complex in the various guest solvents. The results are shown in Fig. 10 which depicts the mole fraction of the host complex which is soluble at various temperatures. The host complex is notably less soluble in the xylenes, and particularly p-xylene, which however are the guests that yield more stable crystalline clathrates.

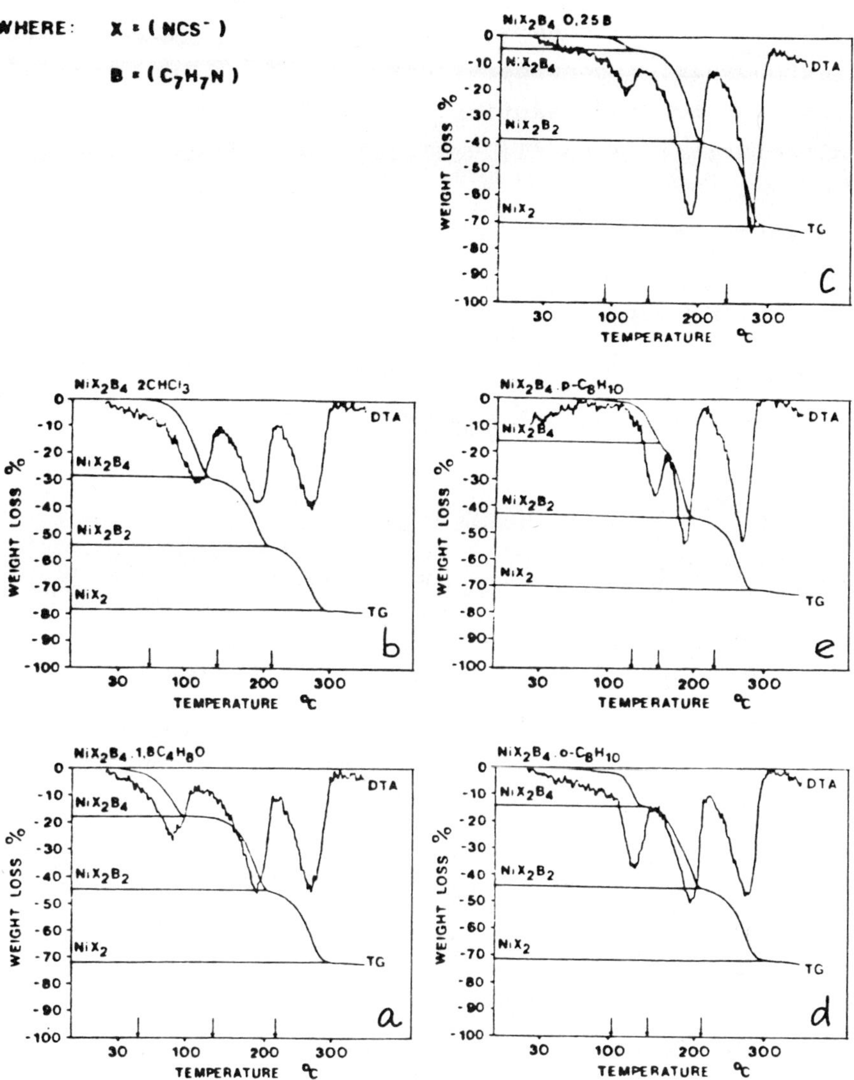

FIG. 9.   GTA and DTA curves obtained for the decomposition of [Ni(NCS)$_2$(4-ViPy)$_4$]·nG clathrates.

This can be understood in terms of simple thermodynamic considerations. The equilibrium constant for the reaction

H (solution) + nG (1) = HG_n(s) where H=host complex

G=guest solvent

is approximated by:

$$K = \frac{[HG_n]}{[H] \ [G]^n}$$

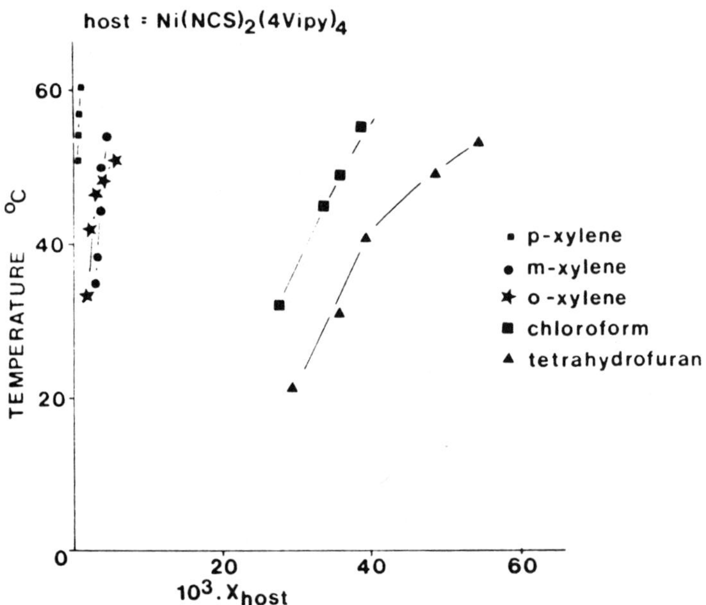

FIG. 10. Solubility curves for [Ni(NCS)$_2$(4-ViPy)$_4$]

Since the clathrate is solid and the guest concentration is in great excess and thus effectively constant, the equilibrium constant is inversely proportional to the host complex concentration in solution. K α 1/[H] (Lipkowski, 1984). Thus the less soluble the host complex is in a given guest solvent, the higher the value of K and the solid clathrate will be more easily formed.

5.   CHIRAL STRUCTURES

Another class of Werner complexes which have the ability to form inclusion compounds with aromatic species are complexes of the type $[Ni(NCS)_2 (1-arylalkylamine)_4]$. A recent review (Hanotier and de Radzitzky, 1984) reveals that little structural work had been carried out; as a result, the nature of the Ni(II) coordination in the host and clathrate complexes had not previously been understood.

Our initial interest in these compounds arises from the fact that certain 1-aryalkylamine ligands are chiral and the metal complexes they form may preferentially include one enantiomer from a racemic mixture of chiral guest molecules. Chiral discrimination by enclathration had previously been achieved with compounds of deoxycholic acid (Sabotka and Goldberg, 1932), cyclodextrin (Cramer and Dietsche, 1959) tri-o-thymotide (Arad-Yellin, Green, Knossow and Tsoucaris, 1983) and others. We elucidated the structures of host complexes of $[Ni(NCS)_2 (1-phenyl-1-ethylamine)_4]$ prepared from racemic and enantiomerically pure amine and clathrates of the racemic host with sec-butylbenzene and o- xylene (Nassimbeni, Niven and Zemke, 1985, 1986). As shown in Fig. 11, the coordinated -NCS groups are -cis to each other in the α-phase host complexes.

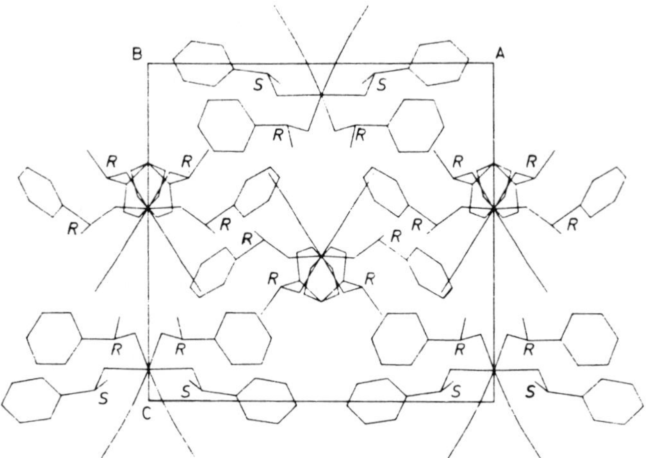

FIG. 11. Packing of the racemic structure
[Ni(NCS)$_2$(1-phenyl-1-ethylamine)$_4$]

However, in the clathrate structure shown in Fig. 12, the NCS groups are _trans_ to each other and the _sec-_butylbenzene is disordered about the two fold axes. This -_cis_ /_trans_ isomerism between the host and clathrate species is the first reported for inclusion compounds of Werner complexes and perhaps

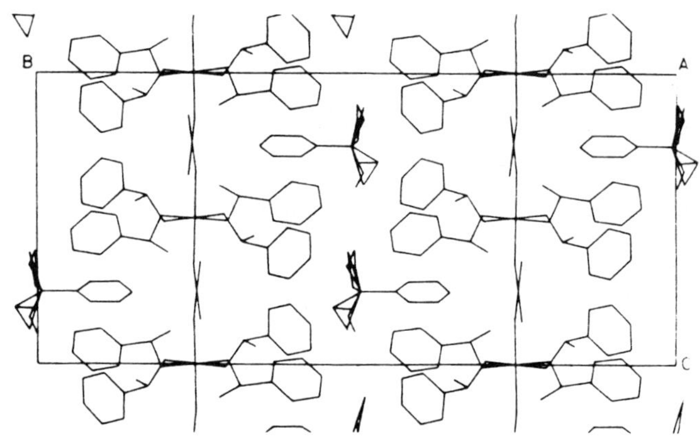

FIG. 12. Packing of the clathrate
[Ni(NCS)$_2$(1-phenyl-1-ethylamine)$_4$]
· _sec_ butyl benzene

explains why these clathrates cannot be formed by addition of guest to previously-prepared host complexes.

We have carried out an energy calculation of the -_trans_ to -_cis_ configurational transformation (without bond breaking) of the nickel (II) complex in a manner similar to the 'Berry pseudo rotation' which models configurational change in trigonal bipyramidal complexes (Berry, 1960). As shown in Fig. 13, we allowed the two opposite faces of the octahedron described by the nitrogen atoms of the NCS groups (A,A^1) and the nitrogens of the amines (B,B^1) to rotate. This was achieved as follows: the initial geometry of the nickel complex was taken from the crystal structure of $[Ni(NCS)_2(1$-phenyl-1-ethylamine$)_4]$. _sec_-butylbenzene. A conformation-governing torsion angle $\tau_0 = A$-C-C^1A^1 was defined with C,C^1 as the vector midpoints between the Ni atom and the midpoints of the N-N-N faces. τ_0 was varied in steps of 10° through 360°; after each 10° twist, two angles and fourteen torsion angles (which encompass all bonds capable of free rotation) were allowed to vary in order to achieve a minimum energy structure. We obtained energy barriers of 27-35 kcal mol^{-1} through a _gauche_ confirmation, and 19 kcal mol^{-1} for the _cis_- to-_cis_ transformation through an _anti_ eclipsed form. The energy difference between the _trans_ and _cis_ minima is 2-3 kcal mole^{-1}, the _cis_ configuration being of the lower energy. We appreciate that the real physical meaning of these values may be limited because Ni-ligand electronic

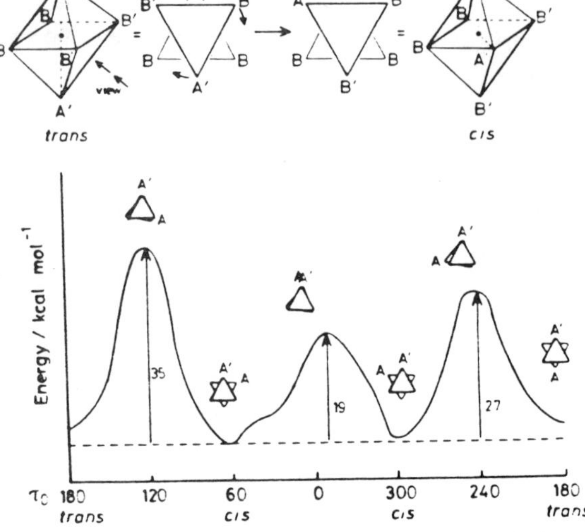

FIG. 13.   Schematic representation of the _trans_ →
           cis configurational change with accom
           -panying energy profile. τ_0 is a
           contrived torsion angle describing the
           angular separation between the NCS groups
           A. The coordinated amines are labelled B.

interactions have been ignored. However our results
concur with the fact that this Ni complex
crystallises in the -_cis_ rather than the -_trans_ form
unless in the presence of some other compound (guest)
with which it can cocrystallise.

     Considering our clathrate complexes with -_sec_-
butylbenzene and _o_-xylene: there is a 1 : 1 ratio of
host to guest molecules and the guests are
disordered. Our analysis of the disorder implies
that preferential enclathration of one guest
enantiomer over the other has not occurred. Our
previous work has shown that guest selectivity is
highest for compounds which conform to the guest site
symmetry.   We hence believe that for this type of

Werner clathrate, the selective enclathration of chiral molecules is likely to be most efficient in lattices of low symmetry; mixed ligand complexes may prove useful in this respect.

6. CONFORMATIONAL STUDIES

The rotational freedom of the substituted pyridines has been advanced as a possible reason for the host molecules to adopt a suitable conformation and accommodate a guest molecule. It is intriguing, therefore that the 3,5-dimethyl pyridine compound has thus far defied attempts at clathrate formation, while the 4-methylpyridine analogue does so readily. An intramolecular conformational energy study of both these compounds has been carried out (Nassimbeni, Papanicolaou and Moore, 1986). The host molecule was divided into five residues linked by four torsion angles as shown in Fig. 14.

Starting with the 3,5-dimethylpyridine compound, energy maps of the interaction between the linear S-C-N-Ni-N-C-S residue and the four lutidines were computed, allowing the four bases to twist about their Ni-N bonds. The resulting pseudo-five-dimensional layer diagram showing the variation of energy as a function of torsion angles is displayed in Fig. 15. The amount of rotational flexibility can be judged by the extent of the energy contours shown on the map. We repeated the study with 4-methylpyridine (R_1 = H, R_2 = Me in Fig. 14) and obtained a pseudo-five dimensional layer diagram which was almost identical to that shown in Fig. 15.

An analysis of the intramolecular close contacts
which occur upon torsion of the four substituted
pyridines shows that the energy is almost exclusively
governed by the two _ortho_-hydrogens on each base.

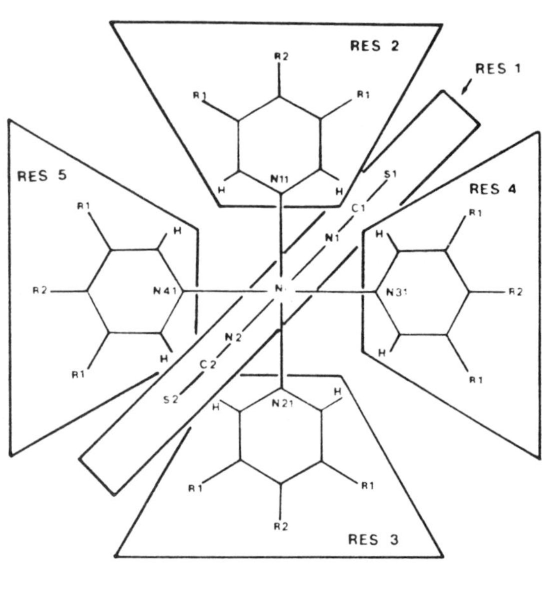

FIG. 14.   Division of host molecules into five
           residues linked by five torsion angles.

It is these which control the extent to which the
substituted pyridines may turn, while the variation
of substituents at the _meta_ and _para_ positions makes
only a secondary contribution.   This result is
similar to that reported by Lipkowski (1981) for
[Ni(NCS)$_2$(4-MePy)$_4$], except that in the latter only
the N-Ni-N subunit was used as the linear residue.

The extent of rotation allowed by the 3,5-dimethylpyridines may not, therefore, be the important factor governing clathrate formation.

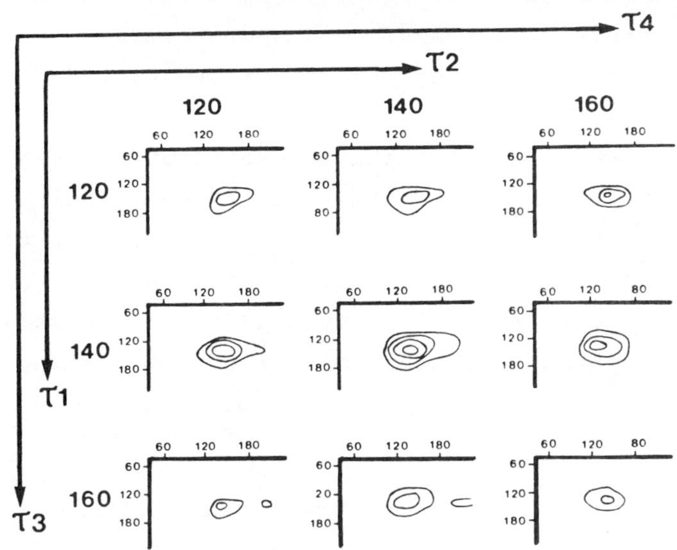

FIG. 15.   Pseudo-five-dimensional potential energy map.   Contours at 0, -2, -4 .....kcal.

However, inspection of the pseudo-five-dimensional potential energy map shows that the energy minimum is relatively broad, allowing rotations of approxiamtely 40 degrees of each substituted pyridine about its Ni-N bond.   We have therefore decided to explore this factor further, and have embarked upon an extended study of Werner clathrates in which the host is [Ni(NCS)$_2$ (4-PhPy)$_4$].   The first structure we characterised was [Ni(NCS)$_2$(4-PhPy)$_4$].4dmso (Nassimbeni, Papanicolaou and Moore, 1986).   For this compound we described the shapes of the channels by calculating the potential energy of a 'probe

molecule'.   We chose   a methyl   group as   a suitable
probe   and   allowed   this   to   move   systematically
throughout   the   unit   cell   while   calculating   its
intermolecular interaction with the host molecules at
suitable   intervals.   A   perspective   view   of   the
channels,   showing   the   surface   of   zero   potential
energy, is shown in Fig. 16.

   We simulated   the movement   of a   dmso   molecule
through   the   channels,   and   the   resulting   energy
profiles showed   no significant   energy   barriers   in
either channel,   suggesting an   unhindered motion   of
the guest   molecules through   the clathrate.   The 4-
phenylpyridine ligand may yield interesting

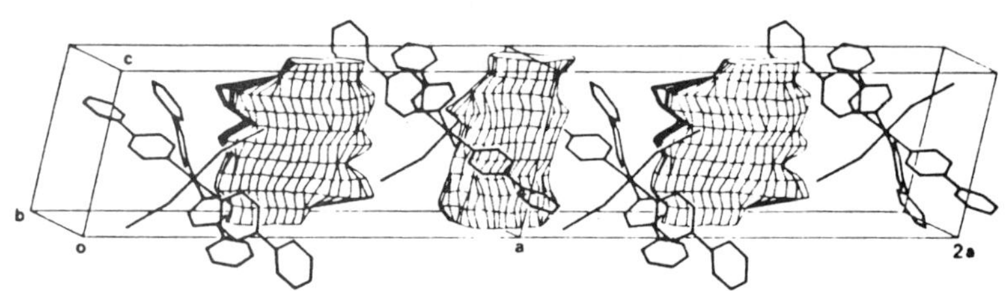

FIG. 16.   Two unit cells depicting the host
           molecules of [Ni(NCS)$_2$(4-PhPy)$_4$] and the
           zero energy surfaces of the channels.

inclusion compounds because of the added flexibility
of   the   phenyl   ring   with   respect   to   its   parent
pyridine.   Thus a   [Ni(NCS)$_2$(4-PhPy)$_4$] complex   will
have eight independent moieties, each having partial

rotational freedom.   We have obtained a number of clathrates with this host molecule, with promising results.

We thank the University of Cape Town, AECI (Modderfontein) and the CSIR Foundation for Research Development for research grants.

7.   REFERENCES

Allison, S.A. and Barrer, R.M. (1969). Journal of the Chemical Society, A, 1717.

Arad-Yellin, R., Green, B.S., Knossow, M. and Tsoucaris, G. (1983). Journal of the American Chemical Society, 105, 4561.

Berry, R.S.J. (1960). Journal of Chemistry & Physics, 32, 933.

Bond, D.R., Jackson, G.E. and Nassimbeni, L.R. (1983). South African Journal of Chemistry, 36, 19.

Borkowska, Z., Lipkowski, J., Moszynska,B. and Wolfram, W. (1972). Journal of Molecular Structure, 12, 265.

Cramer, F. and Dietsche, W. (1959). Chemische Berichte, 92, 378.

de Radzitzky, P. and Hanotier, J. (1962). Industrial Engineering Chemistry Process Design and Development, 1, 10.

Gavezzotti, A. (1983). OPEC, Organic Packing Energy Calculations Program. Journal of the American Chemical Society, 105, 5220.

Giglio, E. (1969). Nature, 222, 339.

Guarino, A., Occhiucci, G., Possagno, E. and
     Bassanelli, R. (1977). Spectrochimica Acta,
     33A, 199.

Hanotier, J. and de Radzitzky, P. (1984). In
     Inclusion Compounds (eds. J.L. Atwood, J.E.D.
     Davies and D.D. MacNicol) Vol. 1, pp.105-134.
     Academic Press, New York.

Kemula, W., Sybliska, D. and Lipkowski, J. (1981).
     Journal of Chromatography, 218, 465.

Kitaigorodskii, A.I. (1973). In Molecular Crystals
     and Molecules. Chapter 1. Academic Press, New
     York.

Lipkowski, J. (1980). Accademia Polacca delle
     Scienze, Biblioteca e Centro degli Studi a Roma,
     Conferenze 81.

Lipkowski, J. (1984). In Inclusion Compounds (eds.
     J.L. Atwood, J.E.D. Davies and D.D. MacNicol)
     Vol. 1, pp. 59-103. Academic Press, London.

Lipkowski, J., Starzewski, P. and Zielenkiewicz, W.
     (1982). Polish Journal of Chemistry, 56, 349.

Lipkowski, J., Suwinska, K., Andreetti, G.D. and
     Standicka, K. (1981). Journal of Molecular
     Structure, 75, 101.

Moore, M.H., Nassimbeni, L.R., Niven, M.L. and
     Taylor, M.W. (1986). Inorganica Chimica Acta,
     115, 211.

Moore,M.H., Nassimbeni, L.R. and Niven, M.L. (1987).
     Journal of the Chemical Society, Dalton
     Transactions, In Press.

Moszynksa, B., Lipkowski, J. and Janowski, A. (1973).
     Journal of Molecular Structure, 19, 347.

Motherwell, W.D.S. (1973). _EENY Potential Energy Program_. Cambridge University.

Nassimbeni, L.R., Niven, M.L. and Zemke, K.J. (1985). _Journal of the Chemical Society, Chemical Communications_, 1788.

Nassimbeni, L.R., Niven, M.L. and Zemke, K.J. (1986). _Acta Crystallographica_, _B42_, 453.

Nassimbeni, L.R., Papanicolaou, S. and Moore, M.H. (1986). _Journal of Inclusion Phenomena_, _4_, 31.

Schaeffer, W.D., Dorsey, W.S., Skinner, D.A. and Christian, J. (1957). _Journal of the American Chemical Society_, _79_, 5870.

Sobotka, H. and Goldberg, A. (1932). _Biochemistry Journal_, _26_, 905.

38. Non-bonded interactions in Werner clathrates

Janusz S. Lipkowski

Nonbonded interactions in clathrate-type molecular crystals discussed here include influence of the guest on the crystal and molecular structure of the host, and guest-guest interactions in orientationally disordered structures. The nomenclature Werner clathrates is being used for inclusion-type compounds in which Werner coordination complexes of the formula MX_2A_4 (M is a divalent metal ion, commonly $Ni^{2\oplus}$, $Co^{2\oplus}$ or $Fe^{2\oplus}$; X is a monovalent anionic ligand, mostly NCS-; A is a substituted pyridine or arylalkylamine neutral ligand) form the host framework structure. Suitable guests are usually organic compounds, inclusion of some inorganic species (e.g. noble gases) being also possible.

GUEST - HOST LATTICE INTERACTIONS

It is believed that some Werner complexes of the MX_2A_4 formula have molecular shape rather inappropriate for their close packing in a crystal structure. On the other hand, the molecules have some conformational freedom (cf. below). Thus, close molecular packing may be attained on crystal formation either through some deformation of the molecules, or through co-crystallisation with a suitable guest. In some cases, e.g. with the $Ni(NCS)_2$(4-Methylpyridine)$_4$ complex, it is difficult to obtain a crystalline, non-clathrated complex. The crystallization products often contain a mixture of the solvents used and 4-Methylpyridine (4-MePy), the presence of which is usually necessary in solution to avoid host-complex dissociation which may lead to precipitation of less soluble complexes containing 3 or 2 molecules of 4-MePy per $Ni(NCS)_2$ moiety (Lipkowski and Andreetti, 1978).

A variety of host crystal structures has been found and classified as organic zeolites (Fig. 1), layered inclusion compounds (Fig. 2), channel compounds (Dyadin et al, 1984) or real cage structures (Lewartowska et al, 1981). It seems that the primary effect of a guest compound is to determine the host polymorph most suitable for inclusion. The efficiency of packing of guest and host together plays an important role in this selection process, e.g. p-xylene and $Ni(NCS)_2$(4-MePy)$_4$ form a zeolite-type clathrate (commonly: beta form). Formation of the compound from liquid p-xylene and the solid non-clath-rated host is associated with an excess volume of -28 cm^3 (volume contraction). Substitution of 50 % guest p-xylene with o-xylene increases the molar volume of the clathrate by 6 cm^3. Higher o-xylene content in the guest mixture continues to ex-pand the host lattice, at o-xylene content greater than ca. 60 % a reconstruction of the host framework takes place and a layered clathrate forms.

Lattice expansion/contraction effects within the stability limits of one crystalline form, are described above. Any change in host lattice parameters should influence the energy of the guest-cavity nonbonded interactions. This has been con-firmed experimentally through calorimetric measurements of enthalpy of inclusion of xylene isomers in the beta form of

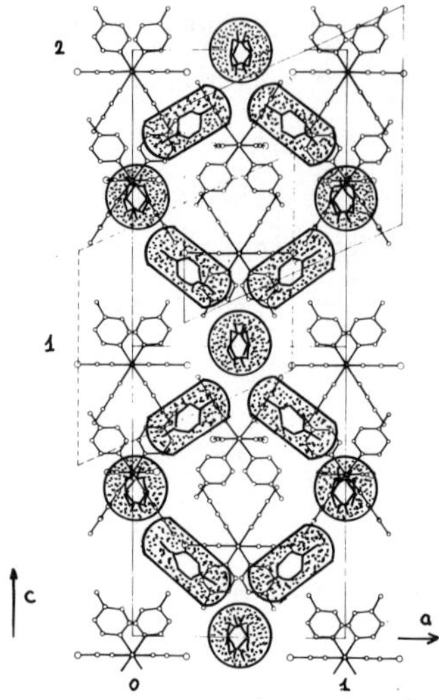

Fig. 1. Illustration of the guest (p-xylene, dotted areas)/host molecular arrangement in the "organic zeolite" Ni(NCS)$_2$(4-MePy)$_4$ structure, projected along 010 .

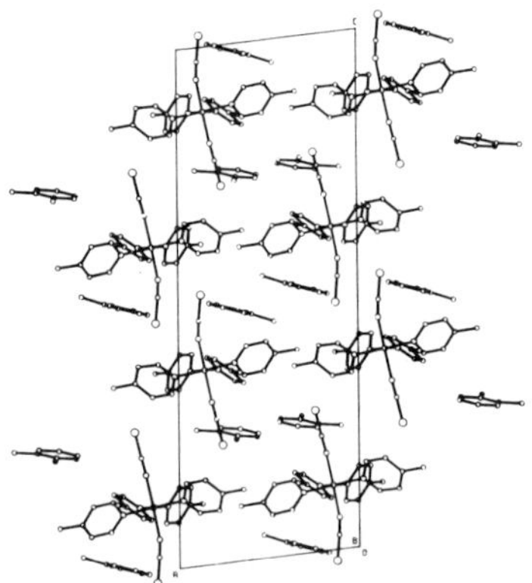

Fig. 2. Layered Ni(NCS)$_2$(4-MePy)$_4$, viewed along 010 .

Ni(NCS)$_2$(4-MePy)$_4$ (Lipkowski $\underline{et}$ $\underline{al}$, 1981). From equations:

$$(H_{clath})_o = 10.07 + 9.07*y_o$$
$$(H_{clath})_m = 13.38 + 5.64*y_m$$
$$(H_{clath})_p = 25.12 + 6.27*y_p$$

(where: $(H_{clath})_{o,m,p}$ is the partial molar enthalpy of clathration and $y_{o,m,p}$ the content of, resp., o-, m- and p-xylene; $y_o + y_m + y_p = 1$).

the attractive interaction energy between a given guest and
the host increases as the concentration of the guest enclathrated increases. This suggests that the cavities in the host
framework undergo continuous structural change on guest substitution. In this example the secondary effect has been attributed to guest-host interactions; EFF calculations suggest that
guest-guest interactions are negligible.

GUEST-HOST MOLECULE INTERACTION

There is another secondary effect of guest-host interactions
in the Werner clathrates, one which may often be observed visually as a colour change of the solids. Because of intramolecular steric interactions there are a limited number of stable
conformations for the host (Lipkowski 1981). In addition to
selecting a host polymorph, modifying it to an extent, the
guest can alter the host conformation _via_ guest-host nonbonded
interactions. As a result, minor changes can be introduced in
the coordination sphere of the central metal ion. These may be

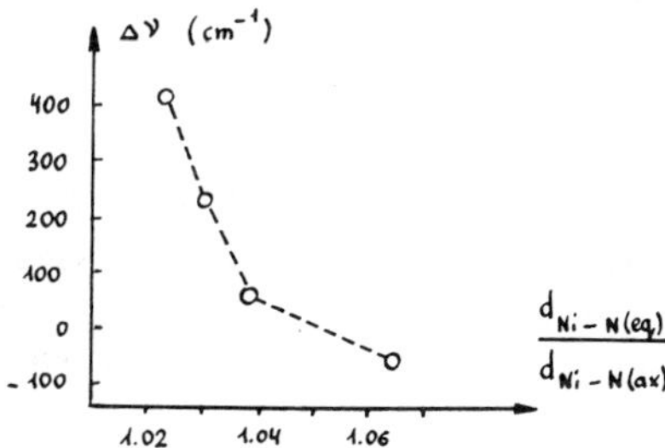

Fig. 3. Shift of the ^3A_2g-^3T_1g spectra of the $Ni(NCS)_2(4-MePy)_4$
(Guarino _et al_, 1977) vs equatorial/axial Ni-N distance ratio.

observed both as changes in bond lengths and angles and in the
visible spectra of the solids (Lipkowski 1984). Because of
axial/equatorial effects elongation of metal-pyridine nitrogen
equatorial bonds is accompanied by shortening of metal-isothio-
cyanate nitrogen bonds and _vice_ _versa_. Fig. 3 displays a correlation of spectral shift with bond length ratios.

GUEST-GUEST INTERACTIONS AND DISORDER

Guest orientational disorder is common in inclusion compounds. In the example given below the importance of EFF calculations in analysing guest disordering is suggested.

The $Ni(NCS)_2(4\text{-MePy})_4$ layered clathrate (Lipkowski <u>et al</u>, 1982) contains independent 1-Methylnaphthalene guest molecules with some differences in their geometry. The differences are small though apparently significant. A structure determination of the isomorphus 1-Bromonaphthalene complex demonstrated that there is orientational disorder of the guest. The less populated orientations have occupancy factors of about 0.2, consequently the structure could be determined only by using an EFF model to find the possible orientations of the guest (Fig. 4).

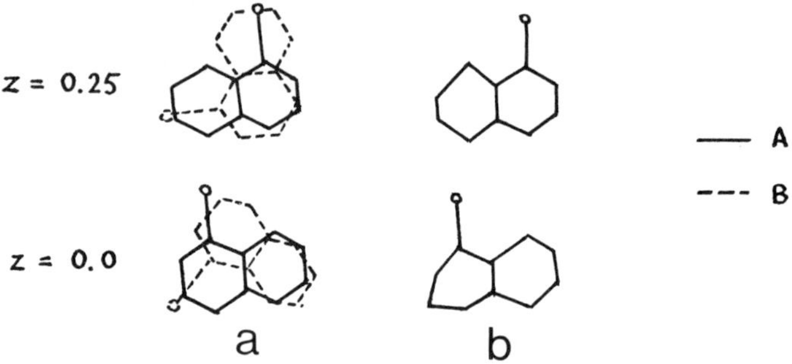

Fig. 4. The two independent 1-Bromonaphthalene molecules: (a) refined without any assumption of disorder and (b) refined as rigid bodies assuming two orientations for each molecule.

The calculations (summarized in the Table) indicate that in one of the two symmetry independent layers (at z = 0.0) random distribution of the guest is possible, in the other guest molecules having orientation B should have adjacent neighbors in the same orientation (Lipkowski <u>et al</u>, 1984).

Thus, there may be ordered layers of guests in orientation B (statistically every fifth) or two-dimensional ordered domains rather than random distribution of orientations A and B. Consequently further experimental study is of merit.

TABLE

Calculated energy (in kcal/mole) of non-bonded interaction for the optimized orientations of 1-Bromonaphthalene illustrated in Fig. 4 (CG is the "central guest" molecule within a layer; NG is the next neighbor guests; A and B correspond to Fig. 4).

	$z = 0$		$z = 0.25$	
CG:	A	B	A	B
NG				
A	-27.8	-26.1	-29.0	-24.7
B		-26.0		-27.6

ACKNOWLEDGEMENT: This work was performed within the 3.20 programme of the Polish Academy of Science (P.A.N.). Partial support from the Rigaku-Denki Corporation for participation in the symposium is gratefully acknowledged.

REFERENCES

DYADIN, YU.A., KISLYKH, N.N., CHEKHOVA, G.N., PODBEREZSKAYA, N.V., PERVUKHINA, N.V., LOGVINENKO, V.A. and OGELZENIVA, I.M. (1984) Journal of Inclusion Phenomena 2, 333.

GUARINO, A., OCCHIUCCI, G., POSSAGNO, E. and BASSANELLI, R. (1977). Spectrochimica Acta 33A, 199.

LEWARTOWSKA, A., BRZOZOWSKI, S. and KEMULA, W. (1981). Journal of Molecular Structure 75, 113.

LIPKOWSKI, J. (1981). Journal of Molecular Structure 75, 13.

LIPKOWSKI, J. (1984) in "Inclusion Compounds", (Atwood et al (Eds.)), Academic Press, Vol.1.

LIPKOWSKI, J.and ANDREETTI, G.D. (1978). Transition Metal Chemistry 3, 117.

LIPKOWSKI, J., STARZEWSKI, P. and ZIELENKIEWICZ, W. (1981). Thermochima Acta 49, 269.

LIPKOWSKI, J., GLUZINSKI, L., SUWINSKA, K. and ANDREETTI, G.D. (1984). Journal of Inclusion Phenomena 2, 327.

LIPKOWSKI, J., SGARABOTTO, P. and ANDREETTI, G.D. (1982). Acta Crystallographica B38, 416.

39. Structural studies of the inclusion properties of a novel coordinatoclathrate host

Ingeborg Csöregh, Mátyás Czugler and Edwin Weber

ABSTRACT

The present survey of seven crystal structures reveals the structural basis of the inclusion behaviour (stoichiometry, host/guest interactions, etc.) of the *trans*-dihydro-9,10-ethanoanthracene-11,12-dicarboxylic acid host molecule.

1.   INTRODUCTION

The novel host molecule, *trans-d*ihydro-9,10-*e*thano-anthracene-11,12-*d*icarboxylic *a*cid (DEADA [1]) like other coordinatoclathrate hosts (Czugler *et al*, 1983, Weber *et al*, 1984, Csöregh *et al*, 1986), has the following structural characteristics:

    I. a bulky molecular skeleton which gives rise to voids in the crystal lattice,

    II. appended carboxylic functions (sensor) groups which provide hydrogen bonding to guests or to other host molecules.

[1]

This roof-shaped molecule allows stoichiometric, selective inclusion of a great number of different polar organic guests, most of which are able to form hydrogen bonds. The structural principles of its inclusion properties have been demonstrated by x-ray structure analyses of the compounds listed in TABLE 1.

TABLE 1.

List of the studied compounds.

Host	Guest	Stoichiometry	Space group	Z	R-value for	N_{obs}
[1]	none	free host	$P\bar{1}$	4	0.048	3351
[1]	n-BuOH	(1:1)	$P\bar{1}$	2	0.065	2771
[1]	DMF	(1:1)	$P2_1/c$	4	0.048	2572
[1]	DMSO	(1:1)	$P2_1/n$	4	0.047	2489
[1]	HCOOH	(1:2)	$P\bar{1}$	2	0.048	2392
[1]	CH_3COOH	(1:1)	$P2_1/n$	4	0.046	3042
[1]	C_2H_5COOH	(1:1)	$P2_1/n$	4	0.056	2302

2.   DISCUSSION

In the structure of the free host the asymmetric unit consists of two molecules forming a hydrogen-bonded dimer. Each dimer is hydrogen bonded to two other centrosymmetrically related dimers. In this way molecules are linked to form infinite zig-zag chains (FIG. 1.). The chains are held together by

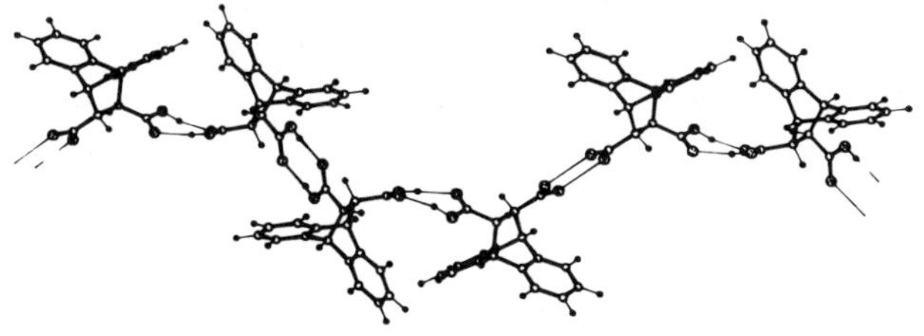

FIG. 1. H-bonded chain in the structure of the free host.

Van der Waals' forces and weak C-H..O type interactions.

The alcoholic oxygen in the butanol guest can function both as a proton donor and as an acceptor. By this virtue it can form a ring *via* H-bonds with the carboxyl group of the host. The twelve-membered (inclusive H atoms) loop formed with the present host (FIG. 2.) is very similar to those which occur in coordinatoclathrates of the 1,1'-binaphthyl-2,2'-dicarboxylic acid host with small alcoholic guests such as methanol, ethanol or *i*-propanol (Weber *et al*, 1984). These H-bonded rings link the hydrogen bonded host dimers together into infinite chains, which are held together by Van der Waals' interactions.

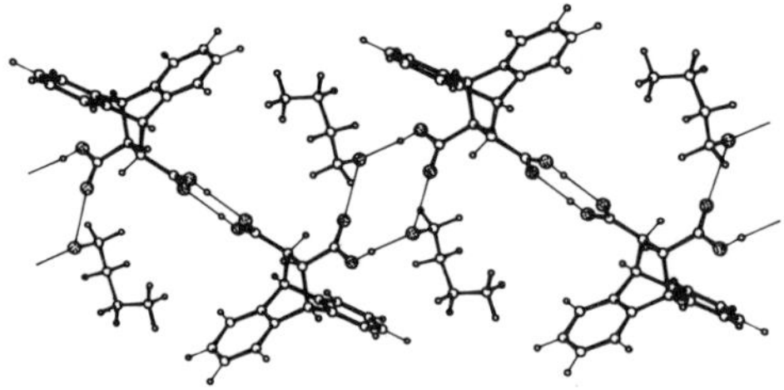

FIG. 2. Characteristic fragment of the *n*-BuOH inclusion of
    DEADA: host dimers linked together by loop of H-bonds.

Dimethylformamide (DMF), the next guest in our study, may be engaged in hydrogen bonding primarily as a proton acceptor, but also as a donor in C-H..O type interactions. Thus it can form a seven-membered ring with a carboxyl group, as we have seen in the crystals with other related coordinatoclathrate hosts (Czugler *et al*, 1983, Csöregh *et al*, 1986). DEADA, however, acts differently. In much the same way as in its butanol clathrate, a pair of DMF molecules are arranged around a center of symmetry and form a fourteen-membered loop with two carboxyl groups (FIG. 3.). The hosts still maintain their characteristic H-bondings to other hosts. In this way, infinite chains are formed just as in the clathrate with the *n*-butanol guest.

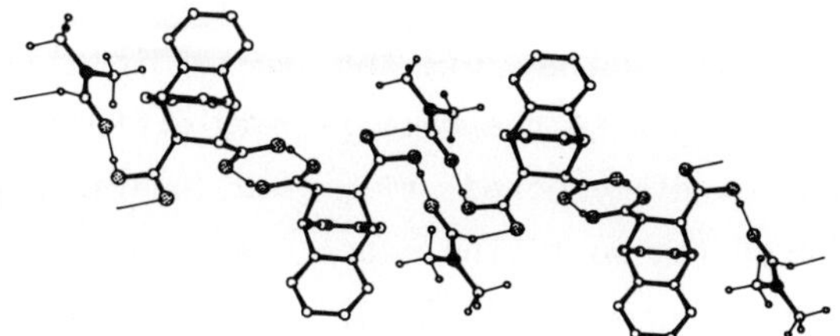

FIG. 3. H-bonded ring in the DMF inclusion compound of the
DEADA host links the host dimers to infinite chains.

Dimethyl sulphoxide (DMSO) has virtually no hydrogen donor
ability. This pyramide-shaped guest is, however, arranged so
that there is a short ($<$ 3.4 Å) distance from one of its methyl
carbon atom to a carboxyl oxygen of the host (FIG. 4.).

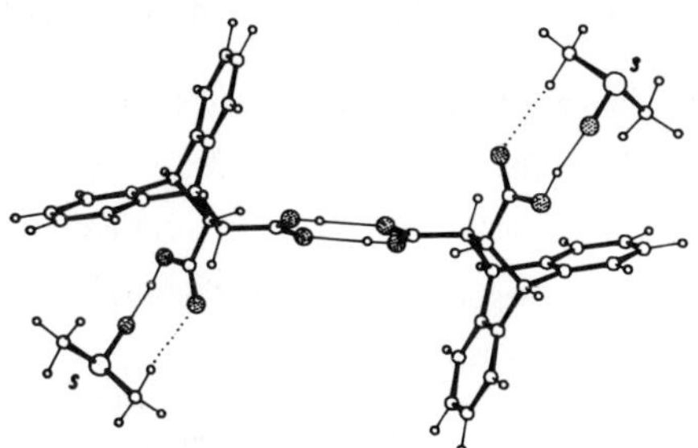

FIG. 4. View of H-bonded DEADA dimer with DMSO guests.

We have found similar host/guest contacts in three more DMSO
inclusion compounds (Csöregh *et al*, 1986, Csöregh and Czugler,
unpublished results). Because of the uncertainty in the
position of the methyl H atoms, it is not possible to make an
unequivocal conclusion. The pattern of the binding of the DMSO
guests suggests, however, the possibility of very weak C–H..O
type interactions between host and guest in these compounds.
The crystal structure of the present DMSO clathrate consists of
uniform entities of hydrogen bonded host dimers with their
respective guests held together by Van der Waals' forces.

The clathrate with formic acid guests is the only one in the present series that deviates from the 1:1 host/guest stoichiometry. The binding of the two formic acid molecules of the asymmetric unit to the host is very different. One of them closes an eight-membered ring *via* hydrogen bonds to a carboxyl group of the host. The other one forms H-bonded dimers around a center of symmetry (FIG. 5).

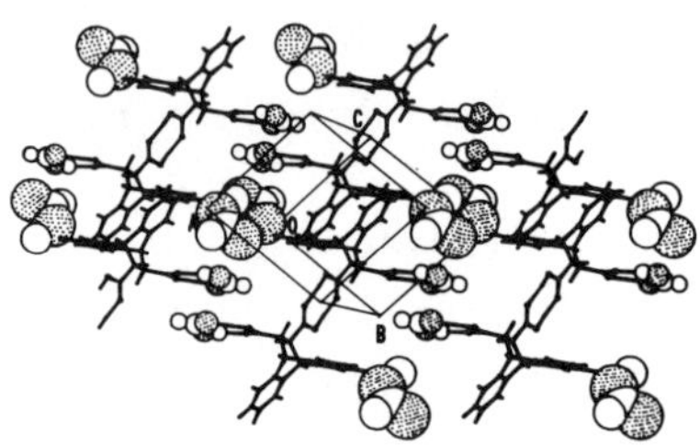

FIG. 5. Packing in the DEADA:formic acid (1:2) clathrate.

These guest dimers have exclusively Van der Waals' interactions with the host or with the host-bonded formic acid molecules. The guest dimers fill up the voids in the framework, formed by host-dimers with their respective formic acid guests.

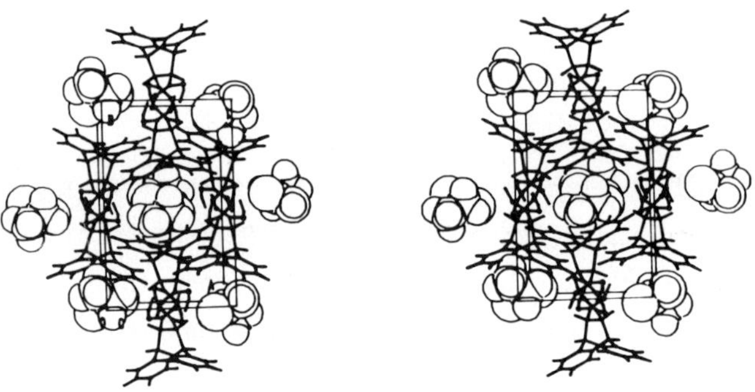

FIG. 6. Stereo packing illustration of the DEADA:propionic
acid (1:1) clathrate, viewed alon the *c* axis

The crystal structures of the two clathrates with acetic acid and propionic acid as guests are isomorphous. There is no

H-bond at all between host and guest in these two clathrates.
The framework is built up of hydrogen bonded zig-zag chains of
the host. These chains run approximately parallel to the *c* axis
and form tunnels, in which the hydrogen bonded guest dimers
are located (FIG. 6.). They are, in fact, real clathrates.

3.   CONCLUDING REMARKS.

As demonstrated by these structures, the *trans*-9,10-dihydro-
9,10-ethanoanthracene-11,12-dicarboxylic acid [1] host
   - forms a stable H-bonded framework even without a guest;
   - retains at least one of its sensor groups for binding
     to another host. Such a dimer, with two carboxyl groups
     free for binding of the guests, seems to be the basic
     motif of its structures, with the result that the most
     common host/guest stoichiometry is 1:1;
   - its functional groups display mostly those recognition
     features which are characteristic for carboxylic sensor
     groups.
These attributes, together with differing guest characteristics,
such as proton-donor/acceptor ability,size and shape, determine
the structural basis for the inclusion capacity of this host.

ACKNOWLEDGEMENT: We wish to thank prof. Peder Kierkegaard
(Univ. of Stockholm) for his support of this project.

REFERENCES.
CSÖREGH, I., SJÖGREN, A., CZUGLER, M., CSERZÖ, M. and WEBER, E.
(1986). Chem. Soc.,Perkin Trans,II, 507-513
CZUGLER, M., STEZOWSKI, J.J. and WEBER, E. (1983). J. Chem.
Soc.,Chem. Commun., 154-155
CZUGLER, M., WEBER, E. and AHRENDT, J. (1984). J. Chem.Soc.,
Chem. Commun., 1632-1634
WEBER, E., CSÖREGH, I., STENSLAND, B. and CZUGLER, M. (1984).
J. Am. Chem. Soc.,106, 3297-3306

40. Crystal structures of neutral host–guest associates: molecular recognition at a simple level

Mátyás Czugler

ABSTRACT.

Molecular recognition is described as a systematic structure pattern in congeneric series of multimolecular associates retrievable from the Cambridge Crystallographic Database (CCDB), too.

1.   INTRODUCTION

Recognition is an often used expression in the chemical literature and even more so in the biochemical one to characterize some kind of selective behaviour of different systems.

Clearly, inclusion chemists are also interested in the matter, too, for almost from its origin on this branch of the chemistry is dealing with phenomena like separation, selective encapsulation etc.

One is interested to learn whether there is a "behaviour", i.e. a consequent structural pattern towards a specific quest entity in a given family of compounds.

A few examples partly from our joint research work with Drs. E. Weber (Bonn) and I. Csöregh (Stockholm) will be cited to prove that even simple molecules do "behave" so. Then the more diverse and less systematic information hidden in the literature may be scrutinized. It is suspected that the nearly 3500 organic compounds enlisted as "solvates" in the CCDB (the

Cambridge Crystallographic Database) are in fact hetero-
molecular associates (inclusions). Looking at the number of
the most common solvents (628 methanol/ethanol, 86 dimethylformamide, 50 dimethylsulfoxide, 547 benzene, 42 acetic acid,
61 xylene etc.) one may realize that formation of hetero-
molecular associates is quite a common fact.

2. DISCUSSION

     Complementarity in molecular assemblies may be achieved by
the use of proper "sensor" groups and also by suitably tailored
basic skeletons of the would-be host molecules as well.
Now a link must be established between the host molecules wich
seem to differ principally on the first sight. Some properties
of their basic skeletons have been reviewed (Weber et al.,1983;
Czugler et al., 1983, "coordinato-clathrates"). The most
extensively studied carboxylic sensor types are shown in the
figure (FIG. 1.). This also accentuates their important feature
that they can be considered as homologous series of dicarboxylic acids of different chain length, centered around an
essentially rigid central portion.
The mean dihedral angles even for the mobile -COOH groups show
little variance (3-8°) suggesting that structure patterns in
such related inclusions may bear significance for other
related, though unknown, cases as well.

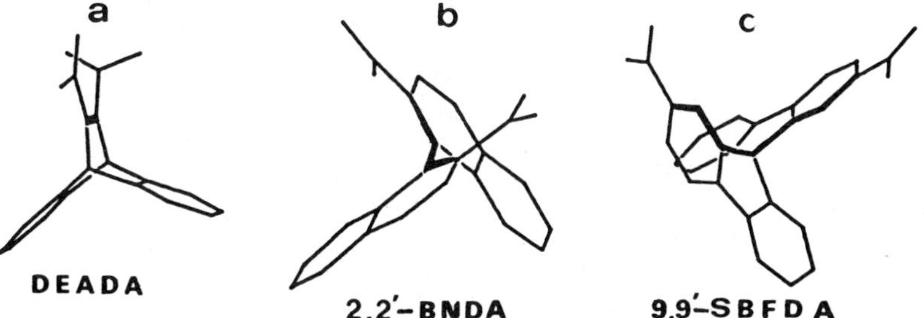

FIG. 1. Dicarboxylic acid portions emphasized by heavy
lines (a,b,c shows 1,2-; 1,4- and 1,7-diacids, respectively).

Recognition of alcohols by carboxylic groups is an example
showing how this function of the racemic 2,2' - BNDA recognizes
an alcoholic -OH. Depending on their stoichiometry which in
turn depends on the chain length of the alcohols we noticed
(WEBER et al., 1983) the following characteristic in the topology of their hydrogen bonding : those with 1:2 stoichiometry
tend to form 12-membered centrosymmetric rings (MeOH, EtOH,
i-PrOH guests, FIG. 2), while those with 1:1 stoichiometry (2-
BuOH and t-BuOH) appear to build up an asymmetric 10-membered
loop. The case of the bifunctional ethyleneglykol inclusion
merits also mention. Here apart from a closely associated
binding of the guest a huge centrosymmetric 24-membered closed
loop is formed. This pattern of forming loops is retained even
in the case of the n-propanol clathrate which has the unusual
2:1 stoichiometry. Similarity of this host's behaviour to those
of the inclusions of DEADA discussed preivously is obvious
(Csöregh et al., 1986a).
Accordingly, recognition of alcohols is characterized by the
basic structure pattern of forming extended closed loops of
H-bonding. Complementation appears in the form of the satura-
tion of the donor/acceptor sites available for hydrogen bonding, too. The grouping and distinction of hydrophyllic and
hydrophobic regions on both guest and host matrix sides is
also seen in the packing of these compounds.

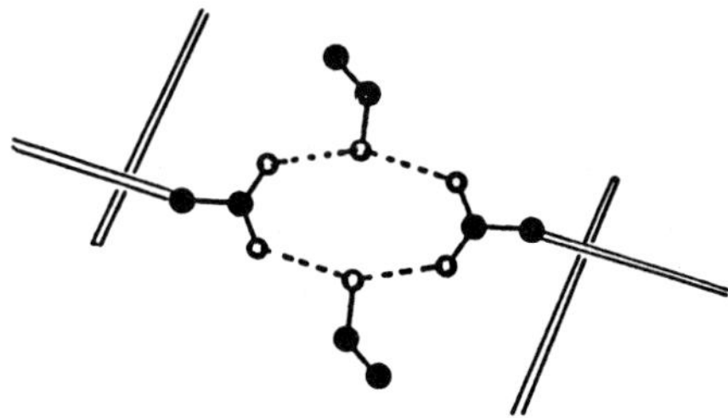

FIG. 2. Schematic representation of the hydrogen-bonded loop
in the ethanol inclusion of 2,2'-BNDA.

That formation of such pattern is a property not only of
these acids is illustrated now by three examples from the CCDB.
They also give an indication how a given structural pattern
may evolve from the interplay of many different effects (e.g.
sensor type, backbone shape etc.).

In the first example (CEXQZB, TOMASSINI et al., 1977; FIG.
3) an ethanol 'solvate' is shown, where the 'host' molecule
possesses many of those features attributed to coordinato-
clathrates: a corpulent X-shaped aromatic skeleton with appen-
ded functions. The keto - oxygen function of the host acts as
hydrogen - bond acceptor from the alcoholic - OH.

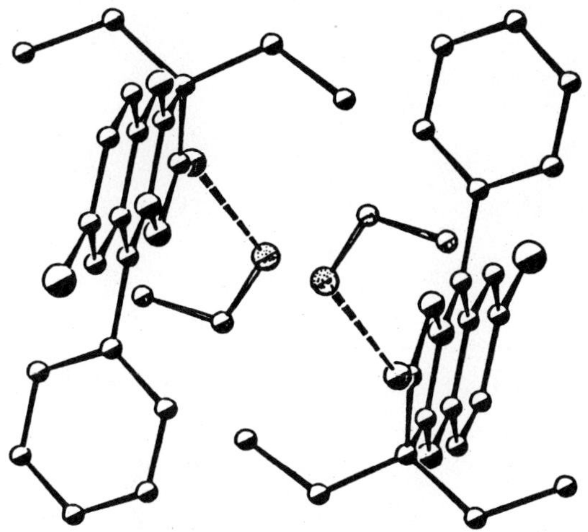

FIG. 3. Packing detail from the retrieved CEXQZB structure

The SULFZC example (KAMIYA et al., 1981; FIG. 4.) is another
step in the evolution of the loop - pattern. The - COOH groups
of the chiral host and the MeOH molecules form a helix which
is topologically related to the loop shown in the next example
(GARTLAND et al., 1974; FIG. 5.). Here, the obvious effect of
the   -COOH groups in the host molecule is coupled with additional steric advantage to ensure formation the same kind of H-loop
we have seen in the inclusions of 2,2'-BNDA.

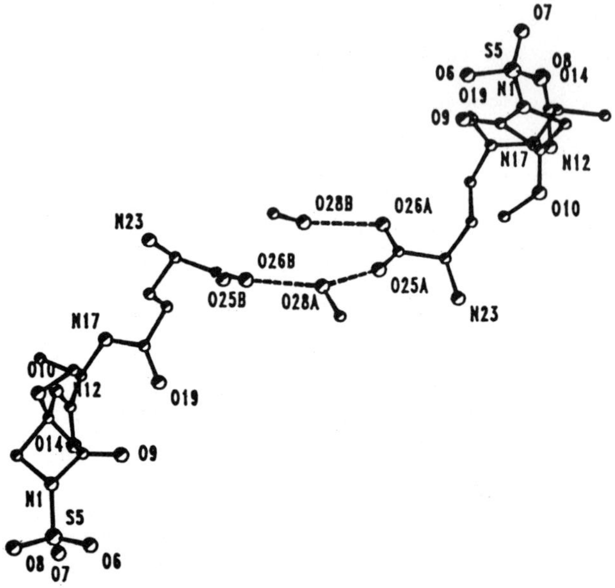

FIG. 4. Packing excerpt from the retrieved SULFZC structure

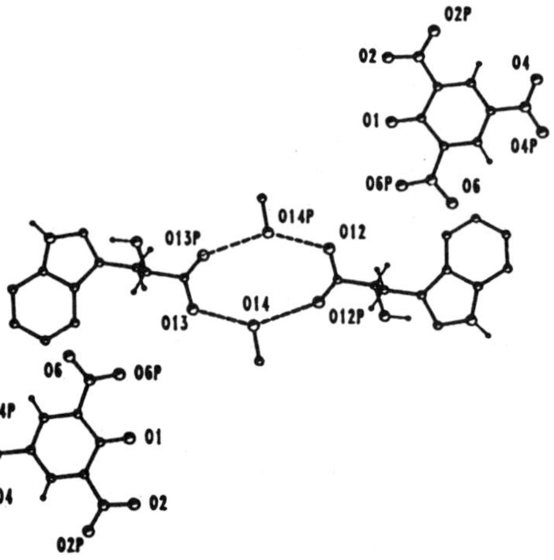

FIG. 5. Closed loop including methanol in the TPTPCM structure

3. SUMMARY

It has been demonstrated for alcohol classes of associates
that characteristic recognition patterns exist for small organic (solvent) molecules. This structural behaviour, albeit at
some variance, is tuned to fit shapes and electrostatic properties of host and guest molecules (matrices) in a complementary
manner.

ACKNOWLEDGEMENT

I am very much indebted to Prof. Dr. E. Weber (Universität
Bonn) and Dr. I. Csöregh (University of Stockholm) for inten-
sive collaboration. Grants of the Swedish Institute, the
Arrhenius Laboratory, the Hungarian Academy of Sciences and
of Rigaku Co. are gratefully acknowledged. Continuous interest
of professors P. Kierkegaard (Stockholm) and A. Kálmán
(Budapest) is also appreciated.

REFERENCES

CSÖREGH, I., CZUGLER, M. and WEBER, E., (1986a), previous paper
CSÖREGH, I., SJÖGREN, A., CZUGLER, M., CSERZÖ, M. and WEBER, E.,
 (1986b), J. Chem. Soc. Perkin II., 507.
CZUGLER, M., STEZOWSKI, J.J. and WEBER, E., (1983), J. Chem.
 Soc. Chem. Commun., 154.
GARTLAND, G.L., FREEMAN, G.R. and BUGG, C.E. (1974) Acta
 Crystallogr., Sect. B30, 1841.
KAMIYA, K., TAKAMOTO, M., WADA, Y. and ASAI, M., (1981) Acta
 Crystallogr., Sect. B37, 1626.
TOMASSINI, M., ZANAZZI, P.F. and ZANZARI, A.R., (1977), Acta
 Crystallogr., Sect. B33, 1197.
WEBER, E., CSÖREGH, I., STENSLAND, B. and CZUGLER, M. (1983),
 J. Amer. Chem. Soc., 106, 3297.

41. Chirality in two dimensions: asymmetric reactions on surfaces of single crystals

Mary Frances Richardson, Herbert L. Holland, P. Chinna Chenchaiah and Benito Munoz

1.   INTRODUCTION

Chiral synthesis and chiral recognition are important areas of chemistry. In general it is necessary to have some source of chirality among the reagents in order to generate chirality or to recognize it once created. Chiral information is clearly contained in chiral molecules, and the use of chiral reagents and catalysts in asymmetric synthesis is a well-developed area of synthetic chemistry.

Less well recognized by chemists, but well known to crystallographers and solid-state chemists, is the fact that crystals themselves can contain chiral information, even if they are composed of nonchiral entities. Both chiral and nonchiral molecules can pack in certain arrangements which are chiral in three dimensions. Solid-state reactions in such crystals can lead to the synthesis of chiral molecules, sometimes in quantitative enantiomeric yield. (Addadi, Van Mil, and Lahav, 1982; Green, Lahav, and Rabinovich, 1979; Thomas, 1974).

Almost unknown is the fact that there is such a thing as two-dimensional chirality, and that the chiral information

inherent in such a system can be utilized in the solution of
three-dimensional problems.   Some discussion of the
applications of two-dimensional chirality to molecular species
has been given by Mislow and Raban (1967), Prelog (1976), and
Wintner (1983). More recently, it has been recognized that
certain space groups, nonchiral in three dimensions, have
surfaces that are chiral (Holland and Richardson, 1980) and
divide space into enantiopolar regions (Addadi, Berkovitch-
Yellin, Weissbuch, Van Mil, Shimon, Lahav, and Leiserowitz,
1985).   Chiral surfaces can be used for absolute asymmetric
syntheses (Chenchaiah, Holland, and Richardson, 1982;
Chenchaiah, Holland, Munoz, and Richardson, 1986), for
determining the absolute configuration of chiral molecules
(Addadi, Berkovitch-Yellin, Weissbuch, Lahav, and Leiserowitz,
1982), and for generating and amplifying chirality by sorption
processes (Weissbuch, Addadi, Berkovitch-Yellin, Gati, Lahav,
and Leiserowitz, 1984). This paper describes our work on
absolute asymmetric synthesis by heterogeneous reactions on
chiral surfaces of single, nonchiral, crystals.

2.   TWO-DIMENSIONAL CHIRALITY

Consider a scalene triangle and its mirror image, as shown in
Fig. 1. The two triangles cannot be superimposed if one
requires that both triangles remain in the plane; they only
superimpose if one is lifted through the third dimension and
laid on the other. The scalene triangle (symmetry 1) is thus
chiral in two dimensions. A little thought shows that any
plane figure possessing only axial symmetry is chiral in two
dimensions (Figs. 1b, 1c). Figures possessing mirror lines of
symmetry have mirror images superimposable in two dimensions
and are thus not chiral (Fig. 1d).

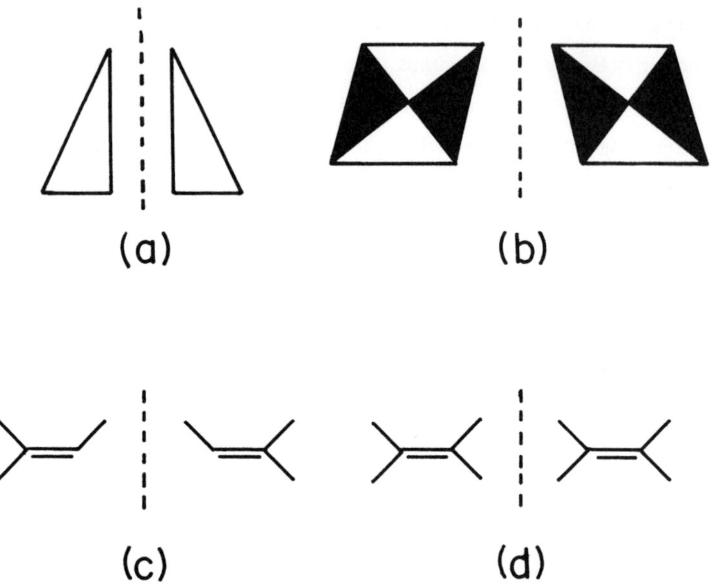

FIG 1. Some planar objects and their mirror images. The objects in (a)-(c) have nonsuperimposable mirror images in two dimensions and are thus chiral; the molecule in (d) is not chiral in two dimensions. The plane point symmetries are: (a) 1; (b) 2; (c) 1; (d) mm.

3.   PLANE GROUPS

Consider arranging an infinite number of identical objects in a plane so that the pattern repeats regularly in two directions. There are only 17 combinations of symmetry elements that are consistent with the requirement of translational symmetry; these define the 17 plane groups. The plane groups have been treated at length by Buerger (1956), Shubnikov and Koptsik (1974) and Stevens (1980), and their symmetries are described fully in International Tables for Crystallography

(Hahn, 1983). Concrete examples of the plane groups can be
seen in floor tilings, wallpapers, and fabrics, as well as in
some kinds of art: Chinese lattice work (Schattschneider,
1978), Escher's periodic drawings (MacGillavry, 1965), Islamic
art (Wade, 1976), Hungarian needlework (Hargittai and Lengyel,
1984), and general design (Heller, 1977; Hendler, 1976;
Wiliams, 1978). Patterns representing 13 of the plane groups
were designed by Brisse (1981) for the 12th International
Congress of Crystallography in Ottawa.

Of the 17 plane groups, five (p1, p2, p3, p4, p6) contain
only axes of rotation and are thus chiral in two dimensions.
The remaining twelve contain mirror or glide lines in the
pattern plane and are not chiral. Nonchiral figures may be
packed in either chiral or nonchiral arrangements; Fig. 2
shows a symmetrically substituted alkene in two different
plane groups, one chiral and one not. Chiral figures also
may be packed in either chiral or nonchiral plane groups, as
shown in Fig. 3. Thus the chirality of the individual units
is not necessarily a guide to the chirality of the packing
arrangement. The only rigid requirements are that a single
enantiomer must be packed in a chiral plane group; if a chiral
figure appears in a nonchiral plane group, the figure and its
enantiomer must be present in equal amounts.

4.   TRANSFER OF CHIRALITY FROM TWO DIMENSIONS TO THREE:
REACTIONS ON THE SURFACES OF SINGLE CRYSTALS
4.1.   Some Specific Examples
We initially investigated the cis-dihydroxylation of tiglic
acid by allowing single crystals of tiglic acid to react with
aqueous barium chlorate/osmium tetroxide (Chenchaiah, et al,
1982). The basic reaction is given in the Equation below.
Molecules of tiglic acid are not chiral, and tiglic acid
crystallizes in the nonchiral triclinic space group P$\bar{1}$

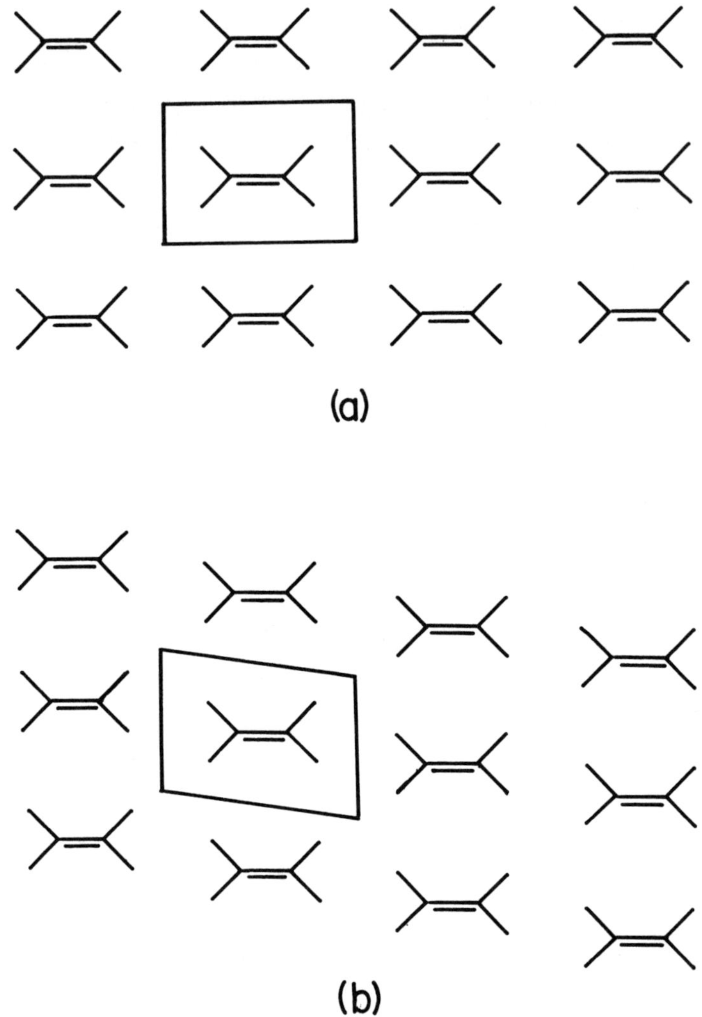

FIG 2. Nonchiral planar entities can pack in both nonchiral
and chiral arrangements. (a) A symmetrical alkene in
nonchiral plane group pmm; (b) The same alkene in chiral plane
group p2.

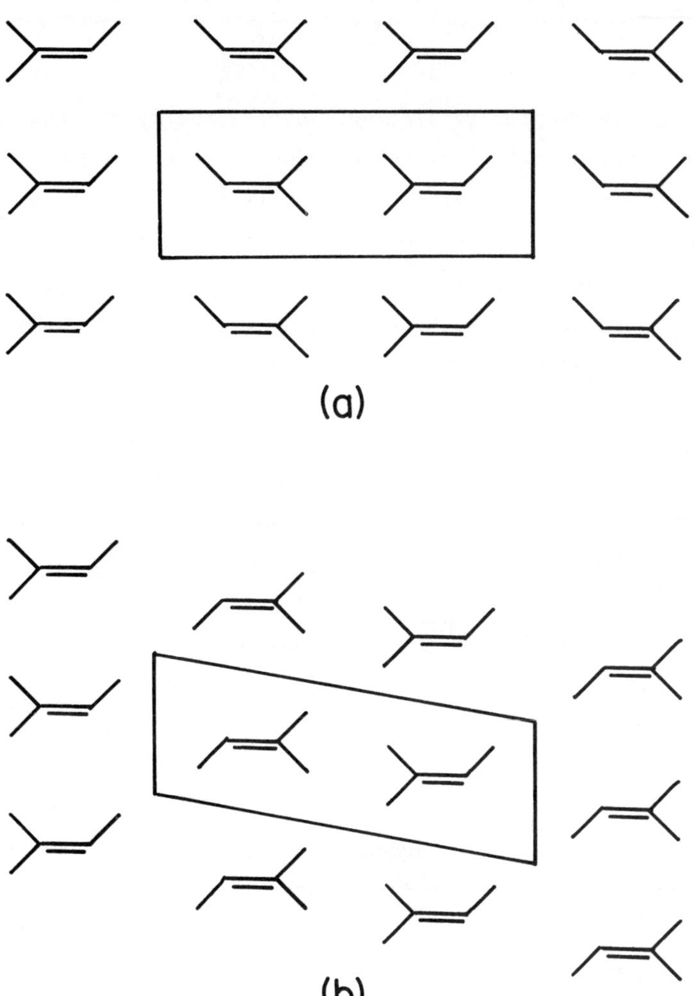

FIG 3. Chiral planar entities can also pack in both nonchiral
and chiral arrangements. (a) An unsymmetrical alkene in
nonchiral plane group pm. Note that the packing arrangement
contains both two-dimensional enantiomers. (b) The same
alkene in chiral plane group p2.

(Wyckoff, 1966). The molecules form layers whose plane
symmetry is p2; these layers are parallel to the ($\bar{2}$10) planes
of the crystal. Note that the only source of chirality is the
two-dimensional chirality of the layers. Successive layers
are related by simple translation. During the hydroxylation
reaction, layers in the crystal are converted to products. It
can be seen in Fig. 4 that if reaction is restricted to one
side of these layers, enantiomerically enriched products can
be obtained. That is, reaction on the ($\bar{2}$10) face generates an
excess of one enantiomer, whereas reaction on the opposite
(2$\bar{1}$0) face generates an excess of the other enantiomer
(Chenchaiah, _et al_, 1982).

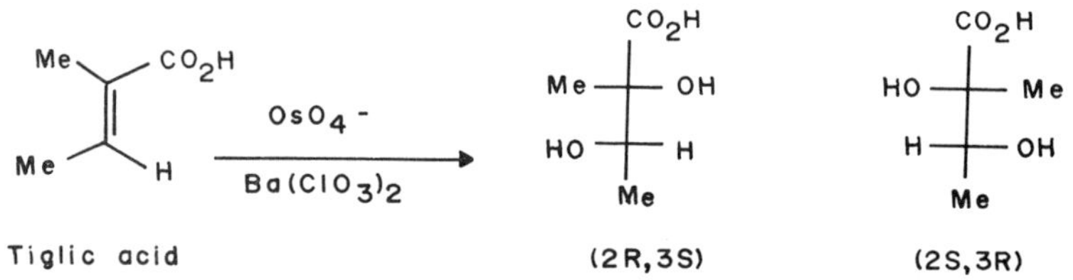

In subsequent work (Chenchaiah, _et al_, 1986) we carried
out reactions of single crystals of tiglic acid and crotonic
acid with various reagents including bromine, HCl, HBr, and
epoxidizing reagents. We forced reaction to occur in a given
orientation (on a particular crystal face) by coating all
surfaces except the desired one with glue or paraffin wax.
When layers were exposed the reactions proceeded with some
enantiomeric enrichment, the amount being dependent on the
crystal quality.   Enantiomeric excesses were determined from
nmr spectra of the products in the presence of chiral shift
reagents. Unfortunately, very high enantiomeric excesses were

not obtained for any of the reactions we tried, the maximum
being about 30%.

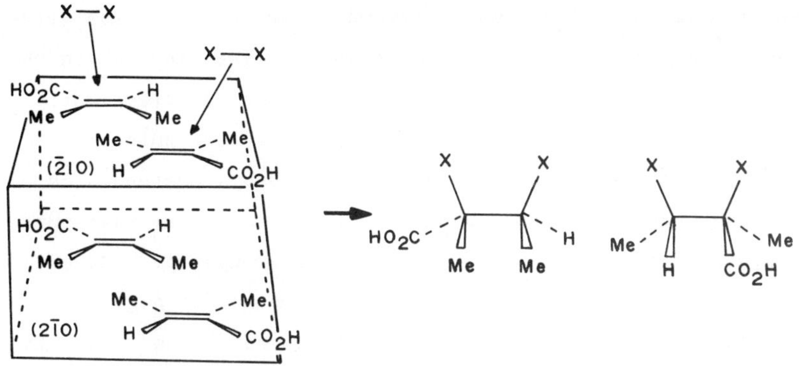

FIG 4.   Reactions of tiglic acid crystals. The X's represent
any reagent. The two molecules on the top plane $(\bar{2}10)$ are
related by a crystallographic centre of symmetry, yet generate
the same enantiomer upon reaction. The bounding planes for
this view of tiglic acid are $(\bar{2}10)$, $(2\bar{1}0)$, (001), $(00\bar{1})$,
(100), and $(\bar{1}00)$.

Bromine vapor reacts very rapidly with tiglic acid
crystals when a layer is exposed (Chenchaiah, _et al_, 1986);
the product is the trans-dibromide. The "sides" of tiglic acid
crystals react much more slowly than the layers and produce a
racemic product, as expected if the rate-determining step is
the diffusion of bromine into the crystal.

These results lead to a brief consideration of the factors
essential for chiral synthesis based on the transfer of
two-dimensional chirality from the surface.

4.2. Symmetry Requirements

Both the symmetries of the layers themselves and symmetry relationships between successive layers are important. The specific symmetry requirements depend on the nature of the reaction (Section 4.4); what follows refers to molecules with planar reaction centres, such as alkenes. The "layers" in the crystal are defined with respect to the reactive site, which may or may not lie in the best plane through the molecule.

For alkenes, carbonyls, or other such systems in which addition occurs to the double bond, it is essential that the layers belong to one of the chiral plane groups, p1, p2, p3, p4, or p6. Layers with symmetry p2 are common in organic crystals since close-packing requirements can be satisfied in this group (Kitaigorodskii, 1961). Layers belonging to one of the other twelve plane groups will, in general, produce a racemic mixture upon reaction (Holland and Richardson, 1980).

Since the entire crystal ultimately reacts, successive layers must produce the same enantiomer. This requirement determines what relationships are allowed between layers. In essence, the symmetry operation must not turn over the layer, since this allows the molecules to be attacked from the opposite side.

In three-dimensional terms, allowed symmetry operations both within and between the layers are: pure translation, inversion centres, rotational axes (including improper axes and screw axes) perpendicular to the layer plane, and mirror/glide planes parallel to the layer plane. Forbidden are rotational axes parallel to the layer plane or mirror/glide planes perpendicular to it. The common space groups $P\bar{1}$, $P2_1/c$, and $C2/c$ meet these conditions, provided in the last two cases that the layers are perpendicular to the

b-axis (Holland and Richardson, 1980). It is interesting that
the most frequent chiral space group, $P2_12_12_1$, is not
generally suitable for asymmetric syntheses by our method
because of the layer turnover problem (but see Section 4.3).

4.3.   Alignment of the Reaction Sites Within the Layers
It is intuitively desirable to have the reaction sites in the
layer aligned so that the incoming reagent has good access.
This is borne out by the bromination of tiglic acid crystals
(Chenchaiah, et al, 1986): reaction takes place much more
slowly at the "sides" of the crystal, i.e. parallel to the
layers, than at the top and bottom faces. The pi-electron
density is exposed on the $(\bar{2}10)$ and $(2\bar{1}0)$ faces, but is not
readily accessible along the other crystal faces. Directional
preferences for single-crystal reactions have been well
established for a number of systems (e.g. see Curtin and Paul,
1981). We expect that, in reactions of the type described for
tiglic acid, enantiomeric enrichment will become less and less
likely as the alkene plane becomes more and more inclined with
respect to the layer (Chenchaiah, Holland, and Richardson,
1985).

In space group $P2_12_12_1$ and similar groups, reaction
along one of the axial directions yields a racemic mixture no
matter what the alignment of the alkene groups with respect to
the crystal axes. However, it is possible in theory that
reaction along a polar direction , e.g. on the (111) face,
could yield chiral products provided that the alkene planes
were aligned parallel to this face. Predictions are hard to
make, but the enantiomeric enrichment is unlikely to be high
because only a small fraction of the molecules will be in
proper orientation for the reaction.

4.4.   Nature of the Reaction

4.4.1. The Reaction Mechanism

Fig. 4 shows clearly that we assumed <u>cis</u>-addition of the X_2 reagent to the alkene. In fact, the first reaction we tried (the reaction of tiglic acid with osmium tetraoxide) was chosen because the reaction was known to proceed in this way. Bromination, which normally proceeds by <u>trans</u>-addition in solution, also proceeds by <u>trans</u>-addition to solid tiglic acid. This means that one bromine is added from the "top side" and the other to the "bottom side," or in other words the molecule has to turn over before reaction is completed. Chiral induction must occur in the first step, presumably the formation of a bromonium cation. What happens next is unclear: possibilities include increased mobility of the cation as well as formation of a solution at the surface. If a solution forms the molecules in subsequent layers may no longer be held rigidly with a single face exposed to the attacking reagent, and this would reduce the enantiomeric enrichment.

4.4.2. The Reactant Molecules

Throughout this paper we have been wholly concerned with the reactions of prochiral alkenes. However, the principles developed above are readily applied to three-dimensional molecules; at least the required layer symmetries remain constant. The number of allowed symmetry relationships between the layers may be smaller than for planar molecules. Assume, e.g., that the reaction site in WX_2YZ is the W-X bond (W is any tetrahedral central atom, and X, Y, and Z are substituents). If one of the W-X bonds projects out of a given layer, a mirror or glide plane parallel to the layer will generate a new layer with the W-X bond pointing in towards the bulk of the crystal. Reaction might occur at the

other W-X bond, which will lie nearly in the layer plane; if
it does the same enantiomer will be generated (provided that
the reaction mechanism does not change). However, the
reaction might also stop because the orientation is not
suitable, or the mechanism might change, with unpredictable
results on the enantiomer distribution. This is clearly an
area in which much speculation is possible since the amount of
data is minimal.

4.5.   Crystal Perfection

The enantiomeric excesses obtained in the tiglic acid
reactions depend on crystal perfection; even crystals which
looked perfect to the naked eye only gave about 30%
enantiomeric enrichment, and crystals that looked poor gave
even lower yields. Most, if not all, macroscopic crystals
contain large numbers of imperfections (Thomas, 1974) which
facilitate reactions and diffusion of reagents into the
crystal. The necessity of growing crystals with a high degree
of perfection is a drawback for asymmetric synthesis by the
technique described here.

5.   SUMMARY

This work shows clearly that plane surfaces of the proper
symmetry can act as chiral templates, and that absolute
asymmetric synthesis can be achieved in the absence of any
source of three-dimensional chirality. Extensions of the
ideas herein to fields such as solid-state catalysis are
obvious. For example, chiral surfaces can be designed which,
in principle, could recognize the two-dimensional chirality of
a given prochiral alkene: the alkene would adsorb preferentially on either its _Re_ or _Si_ face, thus exposing only one
side of the alkene for reaction. Product desorption after
reaction would complete the catalytic cycle. We note that it

is not necessary to have a surface made up of prochiral
molecules (although all of our work so far has been done with
such systems); the packing itself, if chiral, is in theory
sufficient to guarantee chiral recognition. Of course there
are problems in finding a catalyst that both adsorbs and
activates the reactants while at the same time providing a
surface of well-defined chirality. However, work is
proceeding along these lines, and promises to yield exciting
results (M. Brook, McMaster University, personal
communication, 1985).

ACKNOWLEDGEMENTS

We thank the Natural Sciences and Engineering Research Council
of Canada and the Petroleum Research Fund of the American
Chemical Society for support of this research.

REFERENCES

ADDADI, L., VAN MIL, J., and LAHAV, M. (1982). _Journal of the
American Chemical Society_ 104, 3422.

ADDADI, L., BERKOVITCH-YELLIN, Z., WEISSBUCH, I., LAHAV, M.,
and LEISEROWITZ, L. (1982). _Journal of the American Chemical
Society_ 104, 2075.

ADDADI, L., BERKOVITCH-YELLIN, Z., WEISSBUCH, I. VAN MIL, J.,
SHIMON, L.J.W., LAHAV, M., and LEISEROWITZ, L. (1985).
Angewandte Chemie, International Edition 24, 466.

BUERGER, M. (1956). _Elementary Crystallography_. John Wiley
and Sons, New York.

BRISSE, F. (1981). _Canadian Mineralogist_ 19, 217.

CHENCHAIAH, P.C., HOLLAND, H.L., and RICHARDSON, M.F. (1982).
Journal of the Chemical Society. Chemical Communications,
436.

CHENCHAIAH, P.C., HOLLAND, H.L., AND RICHARDSON, M.F. (1985).
Acta Crystallographica Section C 41, 95.

CHENCHAIAH, P.C., HOLLAND, H.L., MUNOZ, B., and RICHARDSON, M.F. (1986). Journal of the Chemical Society. Perkin Transactions II: Physical Organic Chemistry, in the press.

CURTIN, D.Y., AND PAUL, I.D. (1981). Chemical Reviews 81, 525.

GREEN, B.S., LAHAV, M., and RABINOVICH, D. (1979). Accounts of Chemical Research 12, 191.

HAHN, T. (1983). International Tables for Crystallography. Vol. A. T. Hahn, Ed., D. Reidel Publishing Company, Dordrecht, Holland.

HARGITTAI, I. and LENGYEL, G. (1984). Journal of Chemical Education 62, 35.

HELLER, R. (1977). Designs for Coloring, Vol. 4. Grosset and Dunlap, New York.

HENDLER, M. (1976). Infinite Design Coloring Book. Dover Publications, New York.

HOLLAND, H.L. and RICHARDSON, M.F. (1980). Molecular Crystals and Liquid Crystals 58, 311.

KITAIGORODSKII, A.I. (1961). Organic Chemical Crystallography. Consultants Bureau, New York.

MACGILLAVRY, C. (1965). Symmetry Aspects of M. C. Escher's Periodic Drawings, A. Oosthoek's Uitgeversmaatschappij NV, Utrecht.

MISLOW, K., and RABAN, M. (1967). Topics in Stereochemistry 1, 1.

PRELOG, V. (1976). Science 193, 17.

SCHATTSCHNEIDER, D. (1978). American Mathematical Monthly 85, 439.

SHUBNIKOV, A.V. and KOPTSIK, V. A. (1974). Symmetry in Science and Art. Plenum Press, New York.

STEVENS, P.S. (1980). Handbook of Regular Patterns: An Introduction to Symmetry in Two Dimensions. The MIT Press, Cambridge, Massachusetts.

THOMAS, J.M. (1974). _Philosophical Transactions of the Royal Society of London. Series A: Mathematical and Physical Sciences_ 277, 251.

WADE, D. (1976). _Pattern in Islamic Art._ The Overlook Press, Woodstock, New York.

WILLIAMS, W. (1978). _Art-Form Designs to Color._ Grosset and Dunlap, New York.

WEISSBUCH, I., ADDADI, L., BERKOVITCH-YELLIN, Z., GATI, E., LAHAV, M., and LEISEROWITZ, L. (1984). _Nature_ 310, 16.

WINTNER, C.E. (1983). _Journal of Chemical Education_ 60, 550.

WYCKOFF, R.G. (1966). _Crystal Structures_, Vol. 5. Wiley Interscience, New York.

42. Chemical reaction in the crystalline state

Yuji Ohashi and Yoshio Sasada

A single crystal-to-single crystal transformation in the solid-state reaction is very attractive, since both structures of the reactant and product are obtained by X-ray analysis with only one crystal. Noteworthy examples reported include the photopolymerization of diacetylene derivatives (Enkelmann, leyrer, Schleier and Wegner 1980), the (2+2) photodimerization of 5-benzylidenecyclopentanone derivatives (Nakanishi, Jones, Thomas, Hursthouse and Motevalli 1981), the photoaddition of p-fluoroacetophenone to deoxycholic acid (Popovitz-Biro, Tang, Lahav and Leiserovitz 1985) and the thermal dimerization of o-benzenedithiol cobalt complex (Miller, Brill, Rheingold and Fultz 1983). In all these cases the reaction proceeds almost keeping the single crystal form and the movement of each atom can be presumed from the structures before and after the reaction.

Although the composite diagram of the reactant and product structures is very informative, the reaction does not always proceed smoothly, that is, obeying the first-order kinetics, from the initial stage to the final stage. The structures at several intermediate stages are essential for elucidating the reaction mechanism. We have found that the chiral cyanoethyl (cn) group in a series of cobaloxime complex crystals (Fig. 1) is racemized without degradation of the crystallinity (Ohashi and Sasada 1977).   The rate of the racemization is so slow that several intermediate structures are obtained from the three-dimensional intensity data.   Moreover, the quantitative relationship between the reaction rate and the crystal structure has been observed.   Since such a stepwise analysis is a very powerful method in elucidating the reaction mechanism, we call the analysis "dynamical structure analysis" and the reaction which enables the dynamical structure analysis is termed "crystalline-state reaction".   In this article we describe the crystalline-state racemization by the dynamical structure analysis.

Base:

(1) S-α-methylbenzylamine

(2) R-α-methylbenzylamine

(3) triphenylphosphine

(4) tributylphosphine

(5) diethylphenylphosphine

(6) diphenylethylphosphine

(7) pyridine

(8) 4-pyridinecarbonitrile

(9) 4-methylpyridine

(10) diphenylmethylphosphine

(11) 3-methylpyridine

(12) piperidine

(13) tri(p-chlorophenyl)phosphine

FIG. 1.   (R-1-cn)cobaloxime complexes with various axial base ligands.

1.RACEMIZATION IN THE CRYSTALLINE-STATE

1.1. DIRECT OBSERVATION OF THE STRUCTURAL CHANGE

The crystal of (R-1-cn)(S-α-methylbenzylamine)cobaloxime, (1),changes its unit-cell dimensions by X-ray exposure without destroying of a single crystal form (Ohashi, Yanagi, Kurihara Sasada and Ohgo 1981).   Fig. 2 shows the changes of a, b, c, β,

and V with exposure time to the conventional Mo Kα radiation.
The a and c axes contract slightly whereas the values of b and
β   increase to a considerable extent.   The changes seem to be
represented by the first-order kinetics.   The rate constants of
a, b, c, β and V were calculated to be 3.41 x 10^{-6}, 3.28 x 10^{-6},
2.95 x 10^{-6}, 2.82 x 10^{-6} and 2.86 x 10^{-6} s^{-1},
respectively, the average value being 3.06 x 10^{-6} s^{-1}.   The
space group, $P2_1$, remains unaltered.   If the irradiation was
interrupted,   the   change of the cell   dimensions   became
undetectable.   By resuming the irradiation, it continued as
before.

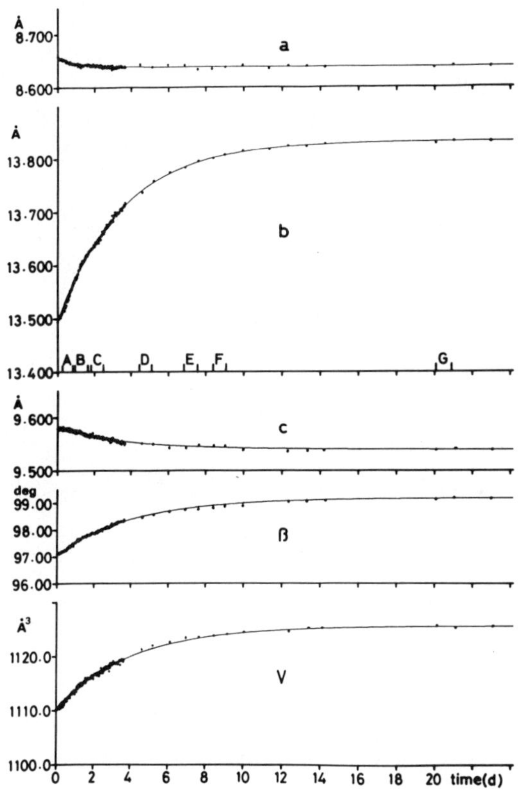

FIG. 2.   Change of the unit-cell dimensions of the crystal (1)
on exposure to X-rays.   First-order reaction curves are obtained
by the least-squares fitting using the observed data.

The three-dimensional intensity data were collected at the
stages A, B, C, D, E, F and G in Fig. 2.   Significant changes in
electron density were observed only in the vicinity of the cn
group.   Figs. 3a and b show the composite electron density maps
of the cn group viewed along the normal to the average plane of
the cobaloxime moity at the stages A and B, respectively.   Peaks
of the methyl group and the carbon atom bonded to the cobalt

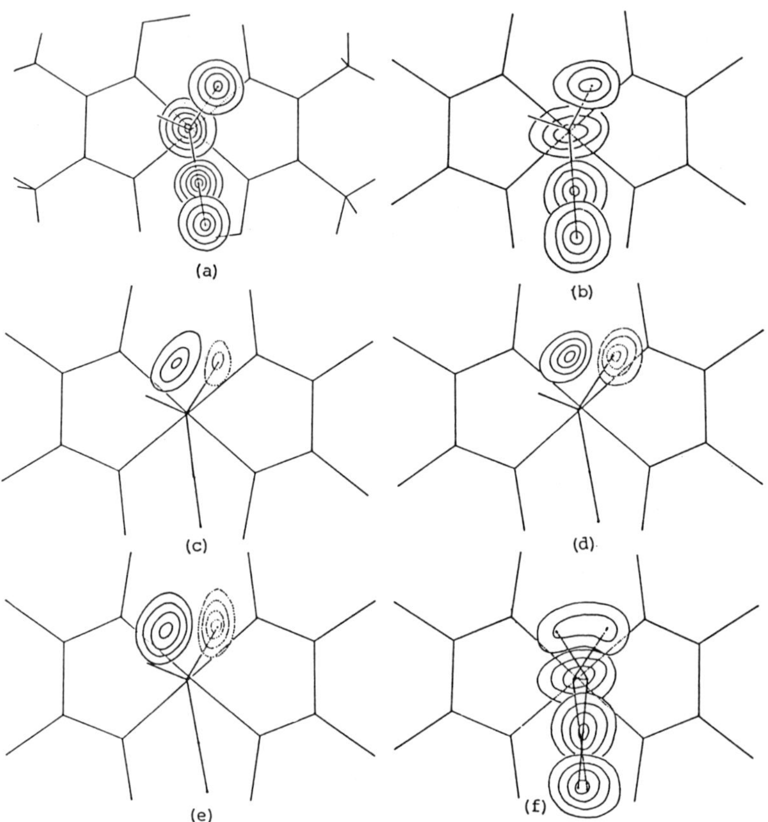

FIG. 3.   (a),   (b) and (f); Composite electron density maps of
the cn group viewed along the normal to the cobaloxime plane at
the stages A,   B and G, respectively, the contour interval being
1 e$\overset{\circ}{A}^{-3}$.   (c), (d) and (e); Composite difference electron density
maps of the cn group viewed along the same direction at the
stages B,   C and D respectively, contour interval being 0.2 e$\overset{\circ}{A}^{-3}$.

atom were lowered and expanded at the stage B.   Fig. 3c shows a composite difference electron density map for the stage B. There appear a trough at the position of the methyl group and a peak in the neighborhood of the methyl group. Figs. 3d and e are the corresponding composite difference electron density maps at the stages C and D, respectively. The new peak grows higher and the trough becomes deeper.   The composite electron density map at the final stage G is shown in Fig. 3f. The height of the new peak is the same as that of the original methyl carbon atom. The position of the new peak corresponds to that expected for the methyl group if the absolute configuration of the cn group is converted from R to S with the cyano group fixed.   This is the direct observation of the racemization process of the chiral cn group by X-ray irradiation.

1.2.   WHY DOES THE RACEMIZATION OCCUR ?

Conversion of the asymmetric carbon atom of R configuration to S implies that one or more of the four bonds, Co-C, NC-C, H_3C-C and H-C, are cleaved by X-rays in the transition state. The homolytic cleavage of the Co-C bond by visible light (λ > 420 nm) in alkyl(pyridine)cobaloxime complexes has been extensively studied from ESR measurement in solution (Giannotti and Bolton 1976).   Similar ESR spectra were observed after the crystal was exposed to X-rays for 30 min.   This suggests that the Co-C bond is cleaved homolytically by X-rays to produce the cn radical and the Co(II) complex in the crystal.   Recently Ohgo, Orisaku, Hasegawa and Takeuchi (1986) estimated the Co-C bond dissociation energy to be 117-122 kJ mol^{-1}, based on the rate of the racemization of chiral alkyl(pyridine)cobaloxime complexes at several temperatures.   The rate of racemization by visible light in solution is about 100 times that by X-rays in the crystalline-state.

On exposure to X-rays, the appearance of the crystals showed no decomposition or degradation.   Examination through a microscope with polarized light revealed no boundaries between the original and racemic phases in the crystal. The diffuse scattering due to long-range ordering was not found on Weissenberg photographs at intermediate stages. When the crystal was exposed to visible light, its surface was decomposed but its cell dimensions were the same as those of the fresh crystals. These facts indicate that X-rays introduce nucleation sites of racemic phase at random inside the crystal and that the nucleation is the rate-determining step of the crystalline-state racemization. Such a random nucleation may be essential for maintenance of the crystallinity.

Figure 4 shows the structure in the neighborhood of the cn group together with short interatomic distances in the crystal of (1). The cyano group makes close contacts with the

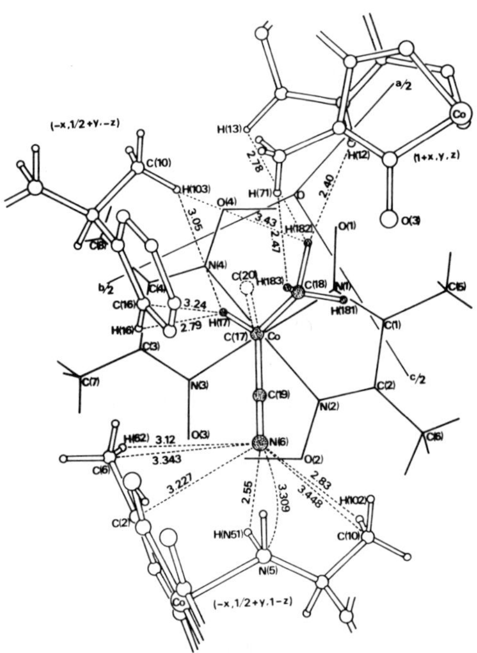

FIG. 4. The atoms in the neighborhood of the cn group viewed along the normal to the cobaloxime plane in the crystal of (1). The dotted atom, C(20), is the inverted methyl group.

neighboring complex.   A weak hydrogen bond of N...H-N is found between the cyano group and the amine of the neighboring complex.   The methyl group has usual van der Waals contacts with another complex.   However, there appears to be a void space around the H(17) atom.   When the crystal is irradiated by X-rays, the transferred methyl group of S configuration, C(20), fills up the void around H(17).   Since the void is not large enough to accomodate the methyl group, the crystal must expand along the b axis as the racemic cell grows.   If the crystal was cooled to 173 K, the crystal contracted anisotropically and the racemization was no longer observed. The void space disappeared in the structure at 173 K.   This brings about an idea that the cn radical produced by the Co-C bond cleavage may rotate around the C-C-N bond and would face the opposite side of the radical plane to the cobalt atom in the transition state.   The energy released by the Co-C recombination may reinforce the thermal motion of the neighboring complexes to bring about instantaneously a wide space enough for the rotation of the neighboring radicals.   Since it may be rare for the cn radical to have such a wide space, the rate of racemization in the crystalline-state is far smaller than that in solution.   Once a racemate is formed, the growth of the racemic structure would continue owing to the increase in entropy term.

2. STRUCTURAL REQUIREMENT FOR THE RACEMIZATION
2.1. REACTION CAVITY

     In the previous section, it has been revealed that the void space around the cn group is a requirement for the crystalline-state racemization.   In order to represent the packing of the cn group more quantitatively, we have defined the reaction cavity for the cn group as a space limited by a concave surface of the spheres of the surrounding atoms, as shown in Fig. 5.   The radius of each sphere is taken to be 1.2 Å greater than the van der Waals radius (Bondi, 1964) of the corresponding atom.   The

value of 1.2 Å is taken from the van der Waals radius of the
hydrogen atom which often occupies the surface of the reactive
molecule.   Any point in the cavity is considered to be accessed
by the centers of atoms of the cn group.

Fig. 6 shows a stereoscopic drawing of the reaction cavity
for the cn group in the crystal of (1).   Similar cavities are
obtained for the other crystals in Fig. 1. The volume of each
cavity was calculated. One of the principal advantages of the
reaction cavity is that it enables the comparison of the packing
around the cn group in one crystal with those in the other
crystals.   Another one is that the volume of the cavity is
regarded as a measure of the void space around the cn group.

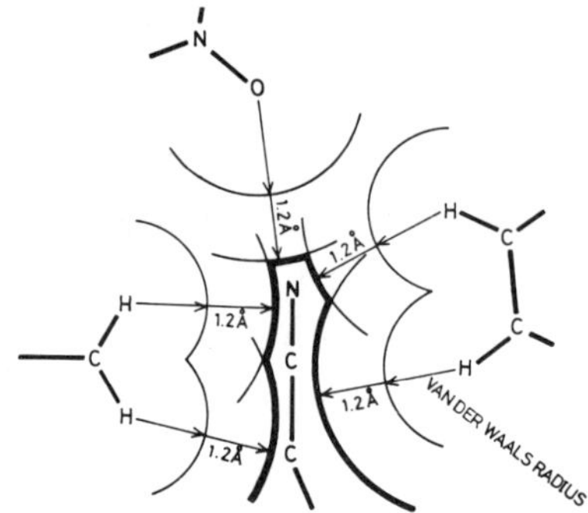

Fig. 5.   The definition of the cavity for the cn group.   One
section is drawn for clarity.

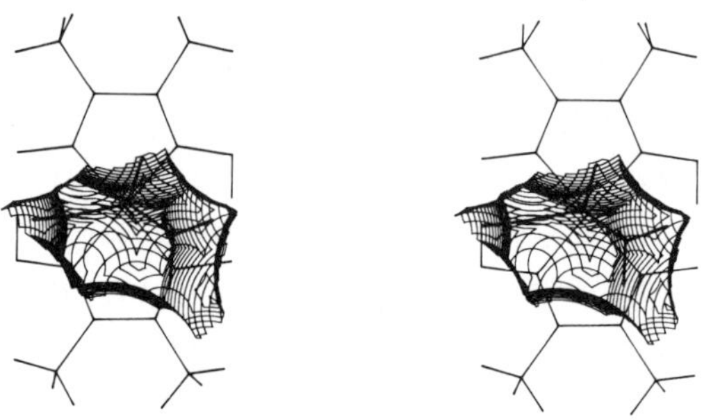

Fig. 6. A stereoscopic drawing of the cavity for the cn group.
The contours are drawn in sections separated by 0.1 Å.

TABLE 1

Volume of the cavity($Å^3$), rate constant(s^{-1}) and density for the crystals with one molecule per asymmetric unit at 293 K

Crystal	Volumne	Rate const.	Density	Reference
(1)	14.53	3.06×10^{-6}	1.388	see in text
(2)	12.23	2.10×10^{-6}	1.391	a
(3)	11.31	————	1.431	b
(4)	10.64	————	1.255	b
(5)	10.18	————	1.41	c
(6)	8.40	————	1.38	c

a) Ohashi, Sasada and Ohgo 1978; b) Kurihara, Uchida, Ohashi, Sasada, ohgo and Baba 1983; c) Tomotake, Uchida, Ohashi, Sasada, Ohgo and Baba 1984.

2.2. THE FIRST MODE (ORDER-TO-DISORDER) OF RACEMIZATION

In the first place the relation between the reaction cavity and the reactivity are examined for the crystals which have one molecule per asymmetric unit. If the crystalline-state racemization were observed for these crystals, the ordered cn group would be converted into the disordered racemates. Table 1 lists the volume of the reaction cavity and the rate constant of racemization for these crystals, although the crystals of (3), (4), (5) and (6) did not reveal the racemization at 293 K. It can be easily seen that the larger is the volume the greater is the rate. Moreover, The volume necessary for the racemization is greater than about 12 $Å^3$. The density of each crystal is also given in Table 1. It should be noted that the volume of the reaction cavity does not show any direct corrrespondence to the density of the crystal, that is, a measure of overall packing efficient.

3.   VARIOUS MODES OF RACEMIZATION CAUSED BY TWO MOLECULES
3.1.   THE SECOND MODE(ORDER-TO-ORDER) OF RACEMIZATION

The crystals containing two molecules per asymmetric unit reveals various types of racemization. Such a diversity is attributable to the cooperative motion of the two cn groups in the process of the racemization. The first example was found for the crystal of (7) (Ohashi, Yanagi, Kurihara, Sasada and Ohgo 1982). Fig. 7 shows the changes of a, b, c, β and V with the exposure time. The changes follow the first-order kinetics as observed in the first mode of racemization. The rate constant was calculated to be 2.83×10^{-6} s^{-1}. It is noteworthy that the volume of the unit cell gradually decreases.

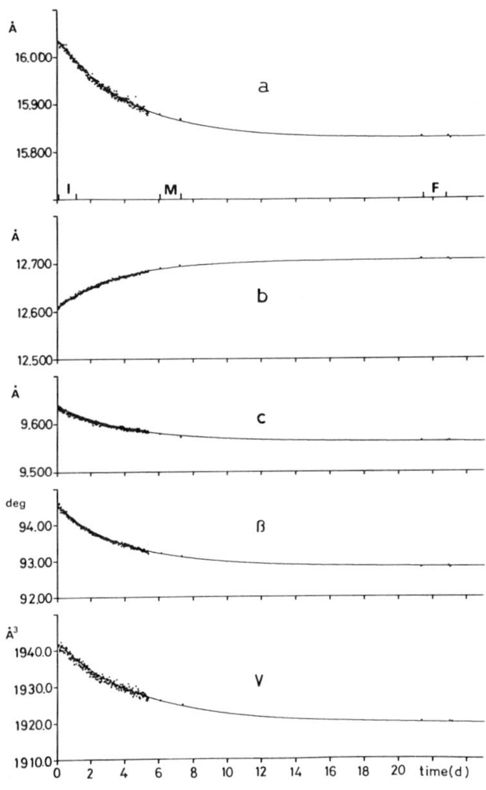

FIG. 7.   Change in the unit-cell dimensions of the (7) crystal

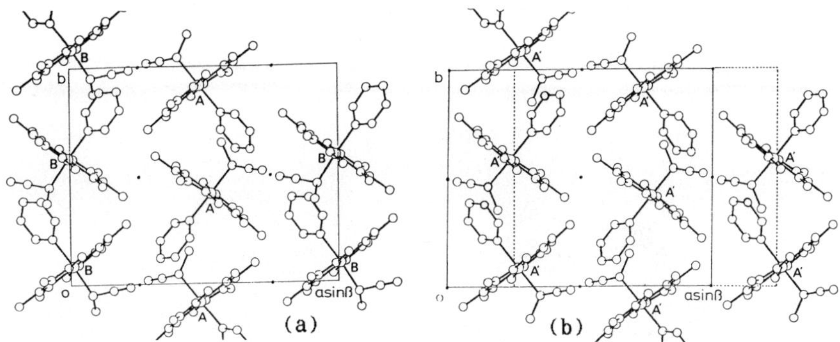

Fig. 8. (a) and (b); Crystal structures of (7) viewed along the c axis at the initial (I) and final (F) stages, respectively.

The three-dimensional intensity data were collected at the I, M and F stages. Figs. 8(a) and (b) show the structures of I and F, respectively. In the initial structure, the two crystallographically independent molecules, A and B, are related to each other by a pseudo inversion center. Only the cn groups of the two molecules destroy the true inversion symmetry. At the final stage, the cn group of the B molecule, B cn group, is inverted to the opposite configuration whereas the A cn group has the original configuration. The pseudo inversion center then becomes a crystallographic one. The space group is changed from $P2_1$ to $P2_1/n$. This indicates that the racemization accompanies an order-to-order conversion. Such a racemization process is classified as the second mode. The same racemic crystal as that transformed by X-ray irradiation was obtained from an aqueous methanol solution containing the racemic complex.

The reaction cavities for the A and B cn groups, A and B cavities, were drawn as shown in Fig. 9 and their volumes are calculated to be 8.89 and 11.34 $\overset{\circ}{A}{}^3$, respectively. Although the volume of the B cavity is slightly smaller than 12 $\overset{\circ}{A}{}^3$, the B cn group can be inverted using a part of A cavity which contacts with B cavity at the pseudo inversion center. On the other

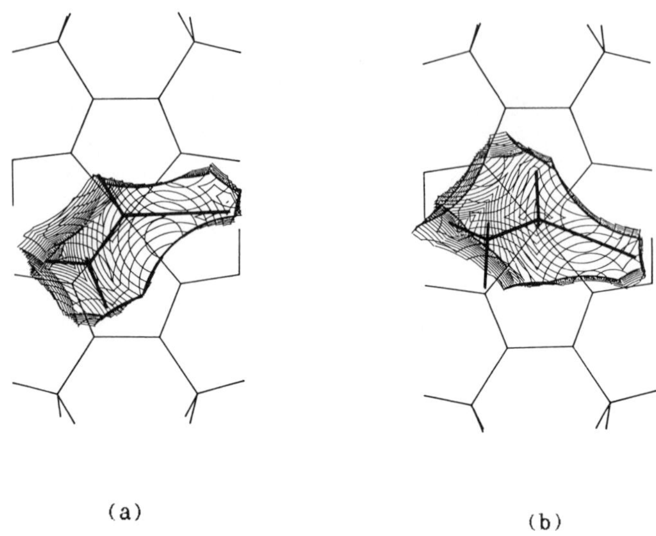

(a)   (b)

FIG.   9.   Cavities for the (A) (a) and B (b) cn groups   viewed
along the normal to the cobaloxime plane.   Contours are drawn in
sections separated by 0.1 Å.

hand, A cavity is too small for the inversion.   Since the
inverted B cn group is more closely packed in the crystal than
the original B cn group, the unit cell contracts after the
inversion of the B cn group.   Therefore, a driving force for the
racemization would be   decrease in enthalpy of the crystal.
Similar results were observed for the crystals of (8) and (9).
Table 2 gives the volume of the A and B cavities and the rate of
racemization for the crystals, (7), (8) and (9).   In either
crystal, the B cn group which has the greater cavity than the A
cn group is inverted to the opposite configuration and the unit-
cell volume decreases.   The rate constant has a positive
correlation with the volume of the B cavity except for (9), in
which the contact region between the A and B cavities is very
small because of the different packing mode.

TABLE 2

Volume of the cavity, $V(\overset{\circ}{A}^3)$, and rate constant, $k(\times 10^{-6}s^{-1})$, for the crystals with two molecules per asymmetric unit

Mode 2			Intermediate mode		
Crystal	V	k	Crystal	V(293K)	V'
(7) A B	8.89 11.34	2.83	(11) A B	10.24 14.29	10.84(343K) 12.79(343K)
(8) A B	7.97 10.37	1.56[a]	(12) A B	7.49 11.57	7.88(333K) 14.04(333K)
(9) A B	11.05 12.61	0.57[b]	(13) A B	9.54 10.04	

Mode 3		
Crystal	V	k
(10) A B	17.08 18.01	4.81

a) Ohashi, Uchida, Sasada and Ohgo 1983

b) Uchida, Ohashi, Sasada, Ohgo and Baba 1984

3.2. THE THIRD MODE(COMBINED ORDER-TO-DISORDER) OF RACEMIZATION

The crystal of (10) was at the first sight regarded as non-reactive on exposure to X-rays, since the change of the cell dimensions was negligibly small. However, the analyzed structure revealed the racemization of the cn group (Tomotake, Uchida, Ohashi, Sasada, Ohgo and Baba 1985). At the initial stage the two crystallographically independent molecules, A and B, are related by a pseudo inversion symmetry, which becomes a true crystallographic one at the final stage. The space group was changed from A2 to A2/a. Both of A and B cn groups are converted to the disordered racemates. The rate of racemization was measured by the intensity variation of the reflections which should disappear after the racemization according to the extinction rule. The variation follows the first-order kinetics. The rate constant is 4.81×10^{-6} s^{-1} and the volume

of the A and B cavities are 17.08 and 18.01 Å^3, which is much
larger than 12 Å^3. Such large cavities would cause the greatest
rate of racemization and the conversion of both cn groups to the
disordered racemates.   Fig. 10 shows the schematic drawing of
three modes of racemization.   The third mode observed in   the
present crystal is the combined one of the above two modes.

3.3   INTERMEDIATE MODES OF RACEMIZATION

The crystal of (11) is also racemized by X-ray exposure but
the racemization process does not belong to the above three
modes (Ohashi, Tomotake, Uchida and Sasada 1986). Fig. 11 shows
the variation of a, b, c and V in the $P2_12_12_1$ cell with exposure
time. Although the changes of a and c are well-explained by the
first-order kinetics, those of b and V have maxima at the
intermediate stage.   The stepwise structure analysis at six
stages I to VI were performed.

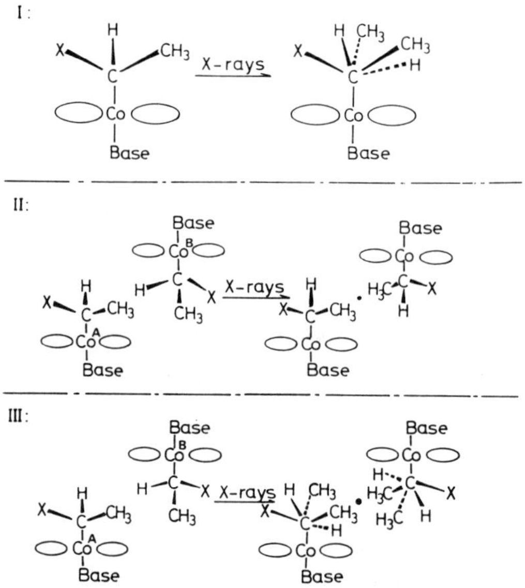

FIG. 10.   Schematic drawing of three modes of racemization. The
inversion center indicated by a point between the two molecules
and X is the cn group.

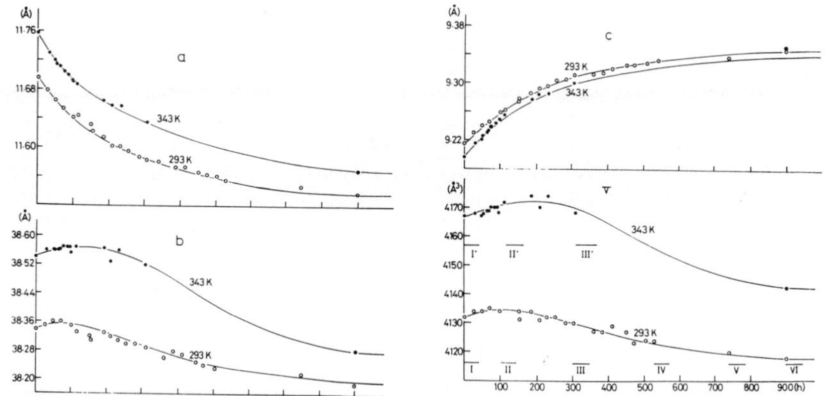

Fig. 11. Change in the cell dimensions of the (11) crystal at 293 K (circles) and at 343 K (black dots).

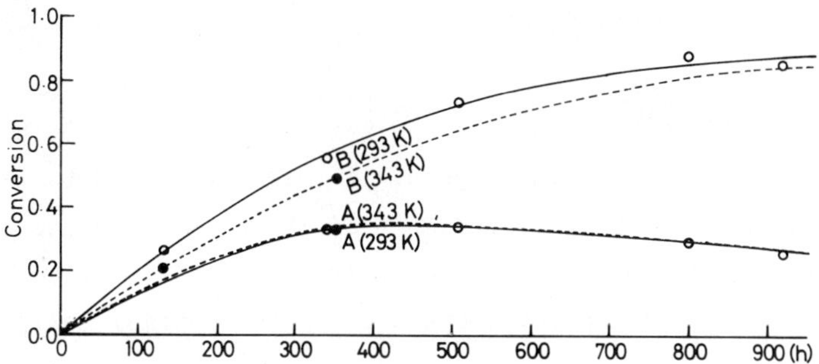

FIG. 12. Ratio of the cn groups with the inverted configuration to the total ones with the exposure time.

The crystal at the initial stage has two crystallographically independent molecules, A and B, which are related by a local inversion center except for the chiral cn groups. On exposure to X-rays, both of the A and B cn groups take the disordered structures. However, the local inversion center does not change to a crystallographic one and the non-centrosymmetric space group is conserved.

Fig. 12 shows the ratio of the cn groups with the inverted configuration to the total cn groups. The change of the B cn group follows the first-order kinetics. The conversion of the A

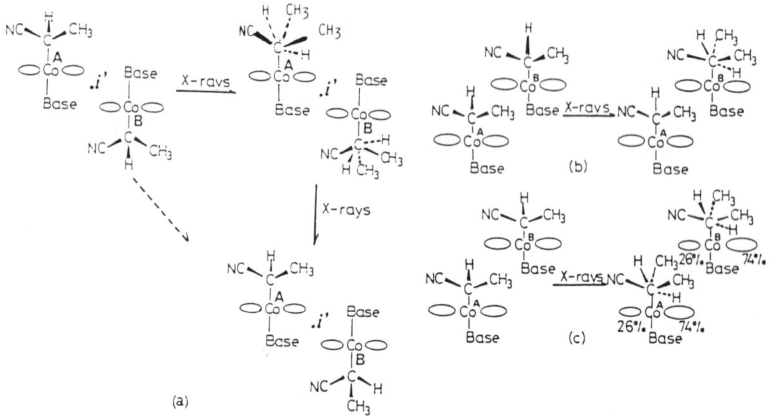

FIG. 13. Schematic drawing of three intermediate modes of racemization. (a), (b) and (c) are for the crystals of (11), (12) and (13), respectively.

cn group, on the other hand, has a maximum (0.35) at 400 h and becomes 0.25 at 900 h. The crystal is fully racemized by X-ray exposure for about 400 h, since 35 % of the A cn group and 65 % of the B cn group are inverted. However, the conversion of both groups still continues after 400 h. As shown schematically in Fig. 13(a), the B cn group appears to be completely inverted and the A cn group is probably restored to the original configuration after infinite exposure.

The volumes of the A and B cavities were calculated to be 10.24 and 14.29 $\mathring{A}^3$, respectively. These values interpret the different rate of inversion for the A and B cn groups. The B cavity has enough volume for the inversion of the cn group whereas the volume of the A cavity is only a little smaller than those of the B cavities in the second mode crystals as shown in Table 2. This may be a reason why not only B but also A cn groups can be inverted at the early stages. Either of the A and B cavities is too small to accomodate the disordered racemate and the molecule would be more favorably packed when the two cn groups have opposite configurations. This causes the reinversion of the A cn group and all the B cn groups will be converted to the opposite configuration. We are able to see the concerted change of the two reactive groups which would occur at the intermediate stages in the second mode racemization (indicated by the arrow with the dotted line in Fig. 13(a)).

Recently we have found further examples related to the above one. The crystal of (12) has also two crystallographically independent molecules, A and B, in the $P2_12_12_1$ cell (Danno, Uchida, Ohashi and Sasada 1986). The two molecules are related by none of pseudo symmetry. The A and B cavities have the volumes of 7.49 and 11.57 $\overset{o}{A}{}^3$, respectively, and the racemization was unobserved at 293 K. When the crystal was warmed at 333 K, the A and B cavities became 7.88 and 14.04 $\overset{o}{A}{}^3$, respectively. Only the B cn group was gradually converted to the disordered racemate (Fig. 13(b)). Further raising the temperature was impossible without degradation of the crystallinity.

The crystal of (13) has also A and B molecules in the P1 cell (Danno, Uchida, Ohashi, Sasada, Ohgo and Baba 1986). The volumes of the A and B cavities were calculated to be 9.54 and 10.04 $\overset{o}{A}{}^3$, respectively. On exposure to X-rays, only the B cn group was gradually inverted, though the rate was very slow. When the temperature was raised at 343 K, the unit-cell dimensions gradually changed by X-ray exposure. The pseudo glide plane in the initial structure became a crystallographic one. The space group was changed to Cc. The stepwise structure analysis of the crystal indicated that the change reached an equilibrium at which 26 % of the A cn group and 74 % of the B cn group are inverted after 700 h exposure (Fig. 13(c)). Although the volumes of the two cavities at 343 K were not obtained because of the poor intensity data, they may be larger by about 2 $\overset{o}{A}{}^3$ than the corresponding ones at 293 K, considered from the thermal expansion of the unit-cell.

Although the cn group is one of the simplest groups and the

racemization is the simplest chemical reaction, a wide variety of racemization modes have been observed in the crystalline-state reaction. This is due to the different environmental effect in the crystalline field. However, if the movement of each cn group is taken into account, it is easily interpreted by the reaction cavity for the cn group. The volume of the cavity is a very good guide for understanding the crystalline-state reaction.

When the cn group is replaced by the bulkier methoxycarbonylethyl (mce) group, more interesting but complicated crystalline-state racemization was observed, since the mce group has a rotational freedom around the C-C bond. The reactivity of the mce group can be also elucidated by the reaction cavity, although slight modification seems to be necessary (Kurihara, Uchida, Ohashi, Sasada and Ohgo 1984; Kurihara, Uchida, Ohashi and Sasada 1984).

4. APPLICATION OF THE REACTION CAVITY

It has been reported that the 2-cyanoethyl (2-cn) group in the powdered sample of the cobaloxime complexes is isomerized to the 1-cn group on exposure to visible light (Ohgo and Takeuchi

$$\begin{array}{ccc}
CH_2CH_2CN & & C^*H(CH_3)CN \\
| & \xrightarrow{h\nu} & | \\
Co(dmg)_2 & & Co(dmg)_2 \\
| & & | \\
base & & base
\end{array}$$

1985). The following two researches have revealed that the reaction cavity is applicable to the above solid-state reaction. In the first research (Uchida, Ohashi and Sasada 1986), we obtained a mixed crystal of (2-cn)(3-methylpyridine)cobaloxime and (rac-1-cn)(3-methylpyridine)cobaloxime which is the product of the above photoisomerization. The rate of the isomerization of the mixed crystal was observed to be about 12 times as large as that of the pure (2-cn)(3-methylpyridine)cobaloxime crystal.

The cavity for the 2-cn group in the mixed crystal becomes by 1.5 $\mathring{A}^3$ greater than that in the pure 2-cn crystal, since the racemic 1-cn group and the 2-cn group randomly occupy the same position in the mixed crystal.

The second research is related to the shape of the cavity. It has been found that the chirality of the produced 1-cn group depends on the asymmetry of the cavity for the 2-cn crystal in the initial reactant crystal, if the crystal has a chiral space group (Ohashi, Uchida, Sasada and Ohgo 1986).

All the above results suggest that the idea of the reaction cavity is valid not only for the crystalline-state reactions but also for the solid-state reactions. Further studies over a wide range of solid-state reaction are in progress.

The authors would like to express their thanks to Professor Y. Ohgo and Dr. S. Baba of Niigata College of Pharmacy and Dr. A. Uchida, Dr. T. Kurihara, Mr. K. Yanagi, Miss Y. Tomotake, Mr. K. Kato and Mr. M. Danno of Tokyo Institute of Technology for their advices and support throughout the works reported.

References

BONDI, A. (1964). J. Phys. Chem. 68, 441.

DANNO, M., UCHIDA, A., OHASHI, Y. and SASADA, Y. (1986). International Symposium on Molecular Structure, Beijing.

DANNO, M., UCHIDA, A., OHASHI, Y., SASADA, Y., OHGO, Y. and BABA, S. (1986). Acta Cryst. in press.

ENKELMANN, V., LEYRER, R. J., SCHLEIER, G. and WEGNER, G. (1980). J. Mat. Sci. 15, 168.

GIANNOTTI,C. and BOLTON,J.R.(1976). J. Organomet. Chem. 110, 383.

KURIHARA, T., UCHIDA, A., OHASHI, Y. and SASADA, Y. (1984). Acta Cryst. B40, 478.

KURIHARA, T., UCHIDA, A., OHASHI, Y., SASADA, Y. and OHGO, Y. (1984). J. Am. Chem. Soc. 106, 5718.

KURIHARA, T., UCHIDA, A., OHASHI, Y., SASADA, Y., OHGO, Y. and BABA, S. (1983). Acta Cryst. B39, 431.

MILLER, E., BRILL, T. B., RHEINGOLD, A. L. and FULTZ, W. C. (1983). J. Am. Chem. Soc. 105, 7580.

NAKANISHI, H., JONES, W., THOMAS, J. M., HURSTHOUSE, M. B. and MOTEVALLI, M. (1981). J. Phys. Chem. 85, 3636.

OHASHI, Y. and SASADA, Y. (1977). Nature(London) 267, 142.

OHASHI, Y., SASADA, Y. and OHGO, Y. (1978). Chem. Lett. 743.

OHASHI, Y., TOMOTAKE, Y., UCHIDA, A., and SASADA, Y. (1986). J. Am. Chem. Soc. 108, 1196.

OHASHI, Y., UCHIDA, A., SASADA, Y. and OHGO, Y. (1983). Acta Cryst. B39, 54.

OHASHI, Y., UCHIDA, A., SASADA, Y. and OHGO, Y. (1986). Eight IUPAC Conference on Physical Organic Chemistry, Tokyo.

OHASHI, Y., YANAGI, K., KURIHARA, T., SASADA, Y. and OHGO, Y. (1981). J. Am. Chem. Soc. 103, 5805.

OHASHI, Y., YANAGI, K., KURIHARA, T., SASADA, Y. and OHGO, Y. (1982). J. Am. Chem. Soc. 104, 6353.

OHGO, Y., ORISAKU, K., HASEGAWA, E. and TAKEUCHI, S. (1986). Chem. Lett. 27.

OHGO, Y. and TAKEUCHI, S.(1985). J. Chem. Soc. Chem. Commun. 21.

POPOVITZ-BIRO, R., TANG, C. P., LAHAV, M. and LEISEROWITZ, L. (1985). J. Am. Chem. Soc. 107, 4043

TOMOTAKE, Y., UCHIDA, A., OHASHI, Y., SASADA, Y., OHGO, Y. and BABA, S. (1984). Acta Cryst. C40, 1684.

TOMOTAKE, Y., UCHIDA, A., OHASHI, Y., SASADA, Y., OHGO, Y. and BABA, S. (1985). Israel J. Chem. 25, 327.

UCHIDA, A.,OHASHI, Y.and SASADA, Y.(1986) Nature(London) 320, 51.

UCHIDA, A., OHASHI, Y., SASADA, Y., OHGO, Y. and BABA, S. (1984). Acta Cryst. B40, 473.

43. Rate of *β*-isomerization of the cobaloxime complex in crystals containing the products and the composition of mixed crystals

A. Uchida, A. Yamano, Y. Sasada and Y. Ohashi

When powdered samples of β-cyanoethyl cobaloxime complexes with various base ligands (cobaloxime = bis(2,3-butanedione dioximato)cobalt(III)) are exposed to visible light, the cobalt-carbon bond is broken, and then cobalt combines with vicinal carbon. Thus, β-cyanoethyl group is converted into α-cyanoethyl group. The reaction can be monitored by the change of ir spectra. The reverse reaction, α-β isomerization, does not occur (Ohgo and Takeuchi 1985).

It was found that the cobaloxime complex containing 3-methylpyridine as a base ligand revealed abnormal behavior of the isomerization. Synthesis of this complex was made in two slightly different processes. Crude product of each synthetic batch was crystallized in aqueous methanol solution separately. The crystals obtained from one of the batches isomerized quickly, while ones from the other converted rather slowly (Uchida, Ohashi and Sasada, 1986).

To reveal such anomaly in the isomerization, structures of two kinds of crystals were examined by X-ray diffraction. Crystal **I** which isomerizes slowly is monoclinic, space group $P2_1/a$ with unit cell dimensions of a = 23.755(4), b = 9.509(1), c = 8.822(1) Å, β = 94.69(2) ° and V = 1986.0(5) Å^3. Crystal **II** which converts quickly is isomorphous to crystal **I** ; a = 23.764(5), b = 9.592(2), c = 8.843(2) Å, β = 93.90(2) ° and V = 2022.2(7) Å^3. Axial lengths of crystal **II** are slightly greater than those of crystal **I**, and unit cell volume of crystal **II** is larger by 36 Å^3.

Structure analysis indicated that crystal **II** contains small portion of α-isomer, while crystal **I** is ordinary crystal of β-isomer.

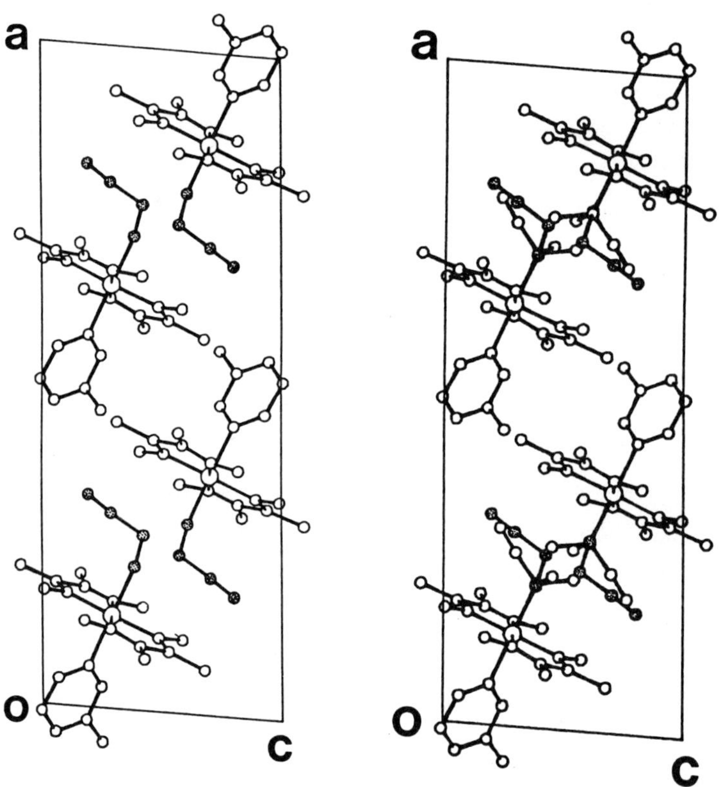

FIG. 1. Crystal structures of crystal **I** (left) and crystal **II** (right) viewed along the b axis.

That is to say, crystal **I** is pure crystal of reactant, and crystal **II** is mixed crystal of the reactant and product of the isomerization. As the isomerization does not occur in solution, α-isomer in mixed crystal is considered as a byproduct in synthesis. Fig. 1 shows crystal structures of pure crystal and mixed crystal. In both crystals the molecular arrangements are almost the same.

The molecular structures in pure and mixed crystal are shown in Fig. 2, where the atoms of the mixed α-cyanoethyl group are hatched. Both structures are very similar to each other except α-cyanoethyl group. The occupancy factors of α- and β-isomer are 0.34 and 0.66, respectively. α-isomer contains racemic α-cyanoethyl group. The disordered methyl groups correspond to that of R- or S-α-cyanoethyl group.

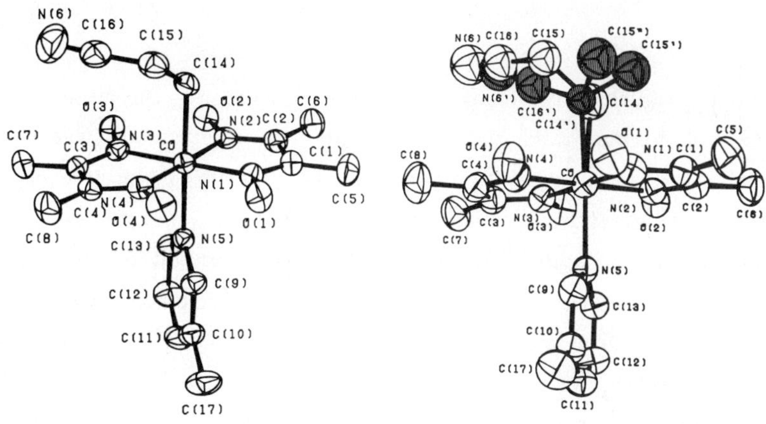

FIG. 2. Molecular structures in pure (left) and mixed crystal (right).

We have been interpreting the rate of the crystalline-state racemization in terms of the free space of the movement of reactive group, which is called cavity (Ohashi, Yanagi, Kurihara, Sasada and Ohgo 1981). The cavity is defined as a space in which the center of atom of the reactive group can be located freely (Ohashi, Uchida, Sasada & Ohgo, 1983).

The reaction cavities for the cyanoethyl group in the present pure and mixed crystals were drawn, as shown in Fig. 3.

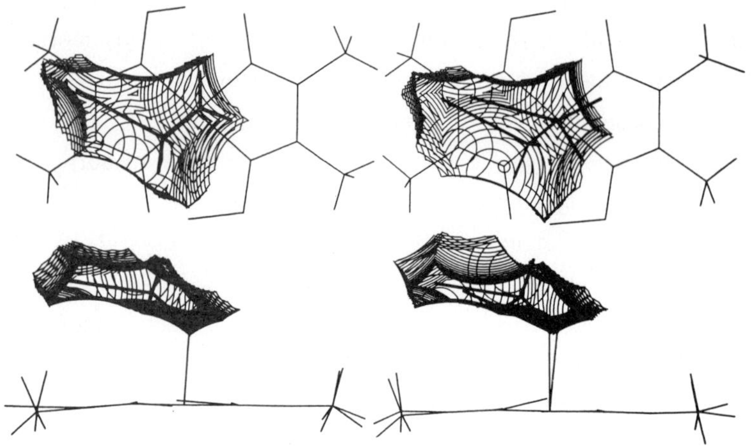

FIG. 3. Cavities for cyanoethyl groups in pure (left) and mixed crystal (right).

The upper figures are views along the normal to the mean plane of the cobaloxime, and the lower ones are side views, where the molecular skeletons are incorporated. The cavity in mixed crystal is greater than that in pure crystal; especially, the cavity in the former crystal is slightly thicker in the neighborhood of the methyl group of α-cyanoethyl group, as seen from the lower figures. The volumes of the cavities in pure and mixed crystal are 10.36 and 11.89 $\overset{\circ}{A}^3$, respectively. It means that the space for the mobility of β-cyanoethyl group and for the accomodation of α-cyanoethyl group formed is greater than that in pure crystal. This observation suggested that the reaction rate could be artificialy controlled by mixed crystal formation.

We have tried to prepare mixed crystals of different composition of α- and β-isomer from aqueous methanol solutions. Table 1 lists the crystal data of mixed crystals, which have good crystallinity. The composition of crystals was determined by means of ir spectra and structure determination. The ratios of α-isomer to β-isomer in crystalline state are lower than in solution by about 10%. The crystals are isomorphous to each other; monoclinic, space group $P2_1/a$ with $Z=4$. Crystals of pure α-isomer are not isomorphous to those of β-isomer. Thus, it is surprising that about 50% of α-isomer is contained in mixed crystal in spite of the remarkable difference in the molecular structures between α- and β-isomer.

TABLE 1

Crystal data of mixed crystals with various composition

a_s/%	a_c/%	a/Å	b/Å	c/Å	β/°	V/Å^3
0	0	23.755(4)	9.509(1)	8.822(1)	94.69(2)	1986.0(5)
10	8	23.760(4)	9.518(1)	8.829(1)	94.58(2)	1990.3(5)
20	15	23.761(4)	9.533(1)	8.835(1)	94.48(2)	1995.1(5)
30	23	23.770(6)	9.550(2)	8.835(2)	94.37(3)	1999.9(7)
40	32	23.81(2)	9.585(5)	8.839(4)	94.04(7)	2012(2)
50	40	23.724(5)	9.594(1)	8.843(2)	93.77(2)	2008.4(6)
60	48	23.713(4)	9.620(2)	8.751(3)	93.28(4)	2016(1)

a_s: composition of solution for crystallization

a_c: composition of mixed crystal

As seen from the Table, increase of α-isomer content results in gradual lengthening of the b axis, decrease of β, and expansion of the unit cell. It is very plausible that the cell expansion gives rise to the enlargement of the reaction cavity. Fig. 4 shows conversion curves of the β-α isomerization of these mixed crystals. With increase of α-isomer content, the reaction proceeded quickly. The initial rates of the reaction are 3.0×10^{-4}, 4.5×10^{-4} and 10.9×10^{-4} s^{-1} for pure and mixed crystals which contain 23% and 48% of α-isomer, respectively. The initial rate for mixed crystal containing 48% of α-isomer is about three times as large as that for pure crystal.

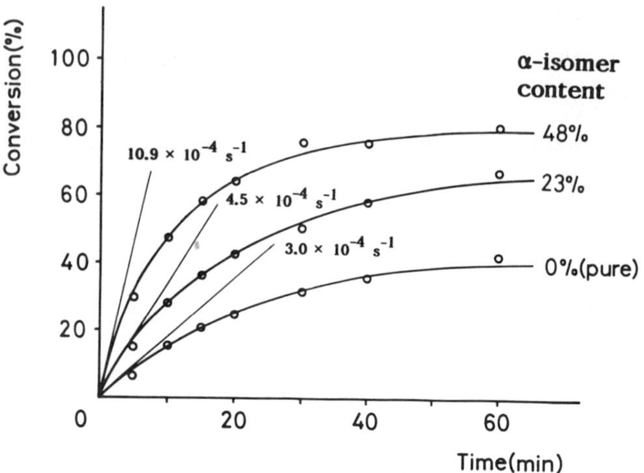

FIG. 4. Conversion curves of isomerization.

When β-isomer isomerizes to α-isomer, an asymmetric carbon appears. It is, therefore, expected that the product of the isomerization is optically active if the reaction proceeds in a chiral crystal. Instead of racemic α-isomer, optically active α-isomer is used for preparing this kind of mixed crystal, chiral mixed crystals could be obtained. The study of asymmetric synthesis by use of such chiral mixed crystal is in progress.

In conclusion, the present study indicated that artificial modification of the crystalline-field may control the reaction.

References

OHASHI, Y., YANAGI, K., KURIHARA, T., SASADA, Y. and OHGO, Y. (1981). J. Am. Chem. Soc. **103**, 5805.

OHASHI, Y. UCHIDA, A., SASADA, Y. and OHGO, Y. (1983). Acta Cryst. **B39**, 54.

OHGO, Y. and TAKEUCHI, S. (1985). J. Chem. Soc. Chem. Commun. 21.

UCHIDA, A. OHASHI, Y. and SASADA, Y. (1986). Nature **320**, 51.

44. Structural investigations of organic, organoelement and organometallic compounds and some aspects of their reactivity

Yu. T. Struchkov

ABSTRACT

Clearly it is impossible to cover even a part of such a vast field as relationships between molecular structure and chemical reactivity of organic, organoelement and organometallic compounds in a single paper. Thus the present report is only a brief review of some results on structures of a general chemical interest, recently obtained in our X-ray crystallography laboratory. Among transformations of organic compounds new addition and skeletal rearrangement reactions in the series of tricyclo[4.2.2.0^{2,5}]deca-3,7-diene and -3,7,9-triene have obtained structural interpretation as a result of strain relaxation in polycyclic skeletons. Structural studies of reaction pathways are illustrated by the novel 5-coordinated silicon derivatives as representatives of organoelement compounds covering, in fact, the entire reaction path of nucleophilic substitution at the non-carbon atom. In the field of organometallic compounds and metal clusters the analysis of M-C σ- and multiple bond lengths and the areas of applicability of magic number rules is carried out. Two modified magic number rules are introduced and a tendency to their fulfillment is an important driving force of reactions of σ- and π-complexes and small and medium clusters.

1. MOLECULAR STRUCTURE OF TRI- AND TETRACYCLODECANES. COMPARA-
TIVE ANALYSIS OF GEOMETRIC PARAMETERS, RING FORMS AND MOLECU-
LAR STRAIN

The chemistry of organic compounds with complicated polycyclic molecular skeletons and especially cage compounds is
currently a very promising and structurally attractive field.
In the studies of tricyclo[4.2.2.0^{2,5}]deca-3,7-diene, I, and
-3,7,9-triene, II, (Fig. 1) a number of new reactions were
found (Zefirov et al. 1982), proceeding stereoselectively and
involving either one double bond of the cyclobutene ring (electrophilic addition) or transannular participation of two spa-
tially close 3,4 and 7,8 double bonds with formation of cross-
cyclization or skeletal rearrangement products. Similar transformations of some diene I derivatives were also studied by
others (Kondo et al, 1978).

The reactions of diene I with electrophiles in non-polar
solvents proceed stereospecifically at the double cyclobutene
bond, yielding trans-adducts A. In polar solvents trans-annu-
lar cross-cyclization gives products B and skeletal rearrangement products C, while addition of electrolytes leads to exclu-
sive formation of rearrangement products C. Electrophilic addition to triene II proceeds similarly in the beginning, however, products D of multistage rearrangement are also formed.

The X-ray studies of four representative series A-D
provided the detection and correction of inaccuracies of IR
and NMR criteria for configuration assignement in the tricyclodecane series, proved the stereoselectivity of electrophilic
substitution and subsequent rearrangements in the systems of I
and II, proceeding via various carbocations, and finally, led
to the discovery of formation of covalent organic perchlorates
in the presence of LiClO$_4$ (Fig. 2, compounds 8-10 and
16-18). The covalent character of the C-O(perchlorate) bond is
demonstrated by its length of 1.48 Å, typical of organic

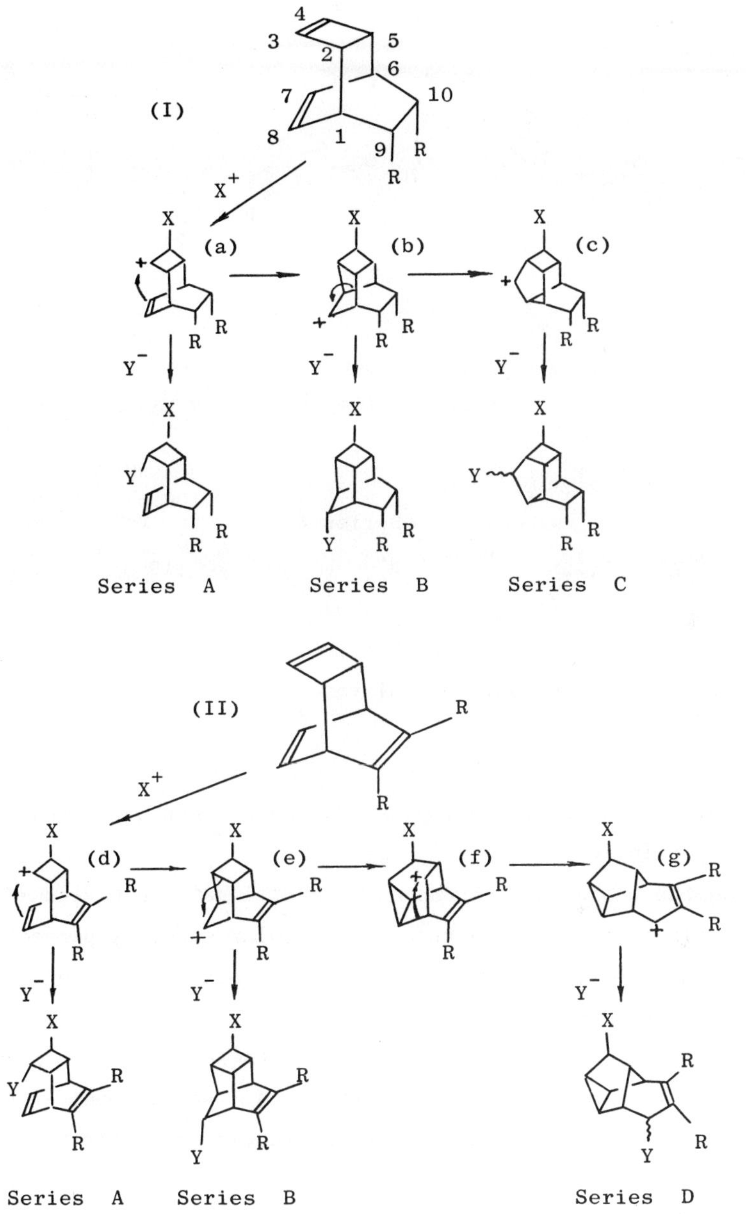

Fig. 1. Electrophilic addition and skeletal rearrangements of tricyclodecane diene I and triene II

	X	R	
8	I	COOMe	
9	OH	COOMe	Series B
10	I	COOMe	
16	$SC_6H_3(NO_2)_2$	Me	
17	I	Me	Series C
18	$SC_6H_3(NO_2)_2$	COOMe	

Fig. 2. Covalent organic perchlorates.

esters, and considerable (by ca. 0.2 Å) elongation of the corresponding Cl–O bond.

On the basis of numerous experimental data we have undertaken the systematic analysis of molecular geometry parameters in the series A, B, C and D to elucidate possible reasons for the novel skeletal rearrangements.

The bond lengths in all trans-adducts A (Fig. 3, compounds 3-7), as well as in the initial skeleton 1 and the only cis-adduct 2 studied, are close to the standard value for the $C(sp^3)$–$C(sp^3)$ bond. The average bending of cyclobutane ring is 13°, i.e. much less than 30° in unsubstituted cyclobutane. Valuable information on deformations of individual rings is provided by the puckering parameters (Cremer and Pople, 1975), modified by us (Palyulin et al, 1981) so that only angular (torsion angles) and not linear (bond distances) parameters are employed. In the molecules A all three 6-membered rings are close to the canonic boat conformation (ideal values of puckering parameters $\Theta=90°$ and $\varphi_2=270°$). In the molecules studied these angles are changed, depending, in particular, on the bulk of the substituent in the position 3. In the ClHg-substituted molecule 7 all 6-membered rings have the ideal boat conformation, whereas in the molecules 3-6 with the endo-orientation of bulkier Cl and Br atoms the rings are markedly (by <6°) twisted toward the canonic twist form (ideal values $\Theta=90°$ and $\varphi_2=240°$), Θ values deviate (by <4°) from 90° indicating ring flattening.

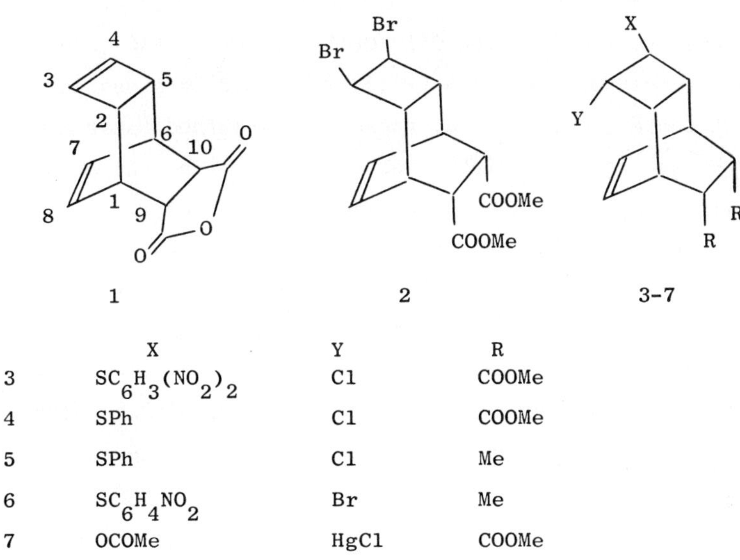

	X	Y	R
3	$SC_6H_3(NO_2)_2$	Cl	COOMe
4	SPh	Cl	COOMe
5	SPh	Cl	Me
6	$SC_6H_4NO_2$	Br	Me
7	OCOMe	HgCl	COOMe

Skeleton bond lengths

$d_{av.} = 1.550$ Å
$d_{max} = 1.566$ Å
$d_{min} = 1.531$ Å
$d_{stand} = 1.554$ Å

Average bond angles in n-membered rings and their scatter

$\Psi_4 = 89.8°$ $\Delta\Psi_4 = 1.9°$
$\Psi_6 = 108.2°$ $\Delta\Psi_6 = 6.9°$

Fig. 3. The initial diene I skeleton 1, the only cis-adduct 2 and the typical trans-adducts 3-7 (Series A).

Thus, relaxation of the strain in the skeleton A, induced by steric interaction of endo-substituents in position 3, results mainly in twisting of this polycyclic system.

The compounds 8-14 (Fig. 4), belonging to the tetracyclo-$[6.1.1.0^{2,7}.0^{5,10}]$decane series B, are formed from diene I and triene II derivatives by cross-cyclization, i.e. formation of the "cross" C(7)-C(8) and not "paralell" C(6)-C(8) bond. The cross-bonding in this system was substantiated by the analysis of frontier molecular orbitals (Inagaki <u>et al</u>, 1976) and force field calculations (Osawa, 1978). The specific structural feature of B skeletons is wide (ca. 0.1 Å) scatter in the $C(sp^3)$-$C(sp^3)$ bond lengths and in, the molecule 9, the C(2)-C(7) bond is elongated to 1.602 Å. The C(1)-C(10) and C(8)-C(10) bonds are also elongated to 1.567 Å (av.). It should be stressed that just these elongated and, therefore, weakened bonds are cleaved on subsequent skeletal rearrangements. The skeleton bond angles also vary in a rather wide range of 81-116°. All three 6-membered cycles have conformations close to the twist form, the torsion angle 1-2-7-6 being abnormaly increased to 95-98° (Fig. 5). In cages B the bicyclic norbornane system can also be considered as a building block. Due to its flexibility this system adopts a synchro-twist form as a result of relaxation of the strain induced by the additional 1-9-8 and 2-3-4-5 bridges (Fig. 6).

Thus, the polycyclic B cages are considerably strained, which is displayed in abnormally long bonds and unusually large bending of the cyclobutane ring by 45°. The strain results also in exotic forms of 6-membered rings with uncommonly high torsion angles (their algebraic sums over some rings are as large as 28°).

The group C compounds of the tetracyclo$[5.3.0^{2,5}.0^{3,8}]$dec-ane series (Fig. 1) are products of the Wagner-Meerwein rearrangement of the intermediate cation (b) due to migration of the C(2)-C(7) bond yielding the cation (c) and then the C pro-

duct. In the C skeletons (Fig. 7, compounds 15-20) the ordinary bond lengths are within the range of 1.516-1.566 Å and the
shortening of the C(6)-C(7) bond to 1.516 Å can be attributed
not only to skeletal strain but also to the influence of the
electron accepting substituent in the position 6. The bond
angles vary from 85.6 to 111.9° and some of them strongly
depend on this substituent orientation. So, in the syn-isomer
20, a steric repulsion between 6 and 4 substituents causes
variation of adjoining angles by 3-4° as compared with the
anti-isomers 15-19.

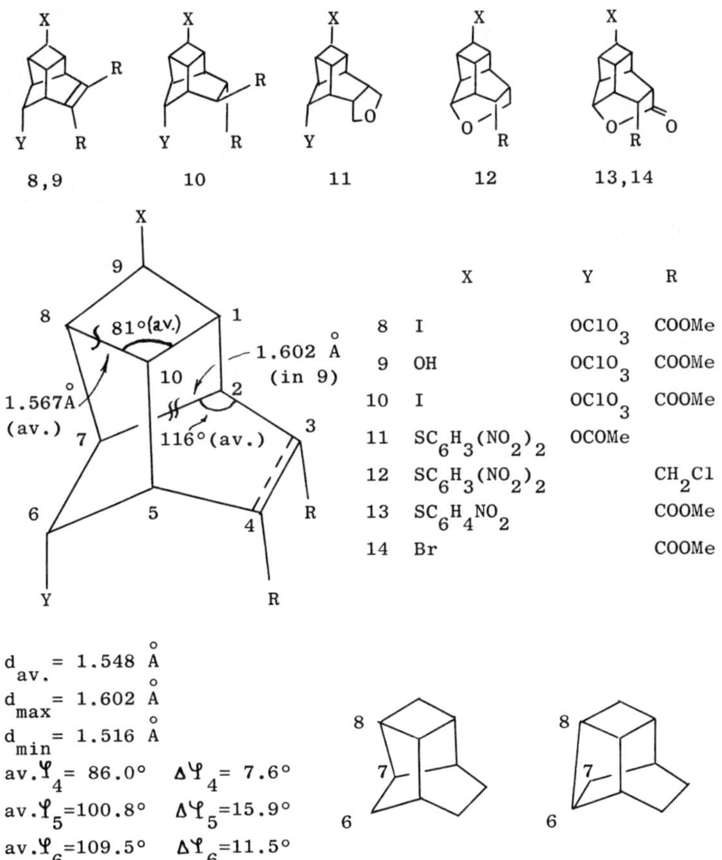

	X	Y	R
8	I	$OClO_3$	COOMe
9	OH	$OClO_3$	COOMe
10	I	$OClO_3$	COOMe
11	$SC_6H_3(NO_2)_2$	OCOMe	
12	$SC_6H_3(NO_2)_2$		CH_2Cl
13	$SC_6H_4NO_2$		COOMe
14	Br		COOMe

$d_{av.} = 1.548$ Å
$d_{max} = 1.602$ Å
$d_{min} = 1.516$ Å
av.$\Psi_4 = 86.0°$ $\Delta\Psi_4 = 7.6°$
av.$\Psi_5 = 100.8°$ $\Delta\Psi_5 = 15.9°$
av.$\Psi_6 = 109.5°$ $\Delta\Psi_6 = 11.5°$

Fig. 4. Cross-cyclization products (Series B). The cross and
parallel or cis cyclization is shown below (left and right).

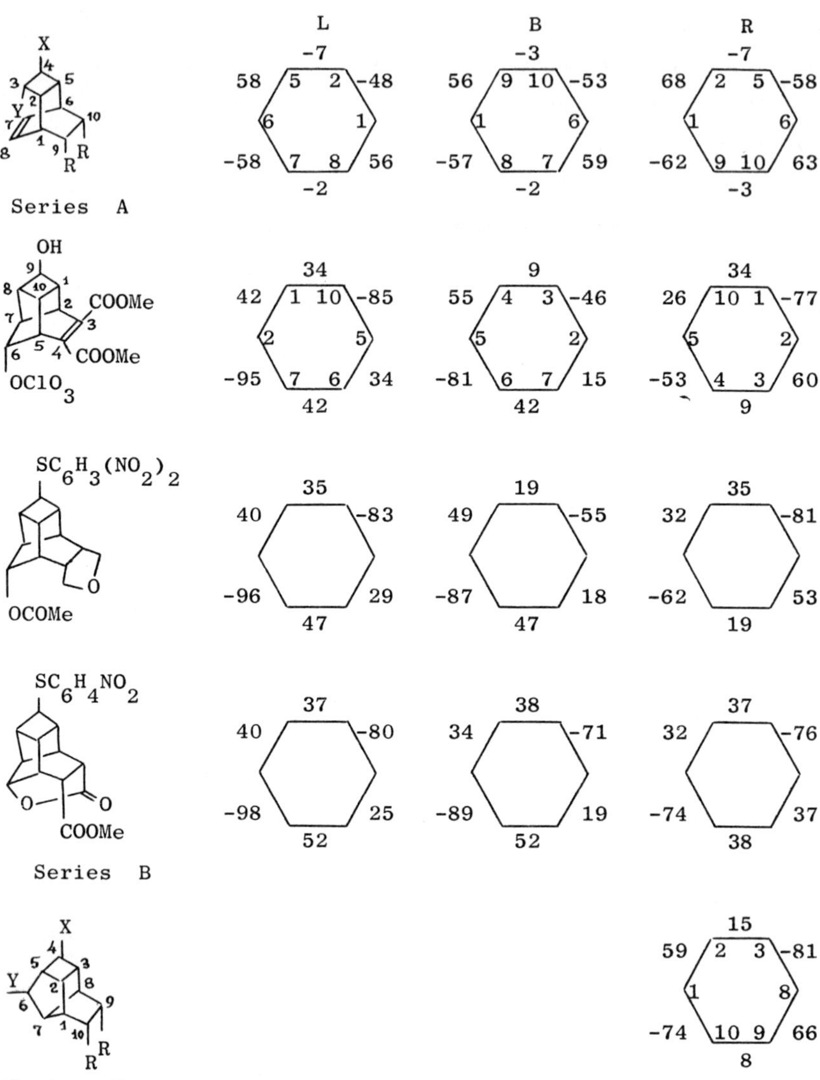

Fig. 5. Torsion angles in 6-membered rings.

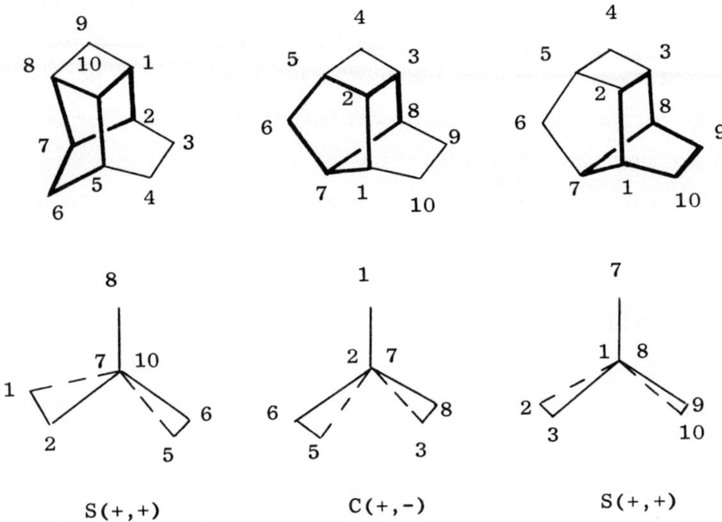

Fig. 6. Norbornane systems in the skeletons B (left) and C (centre and right).

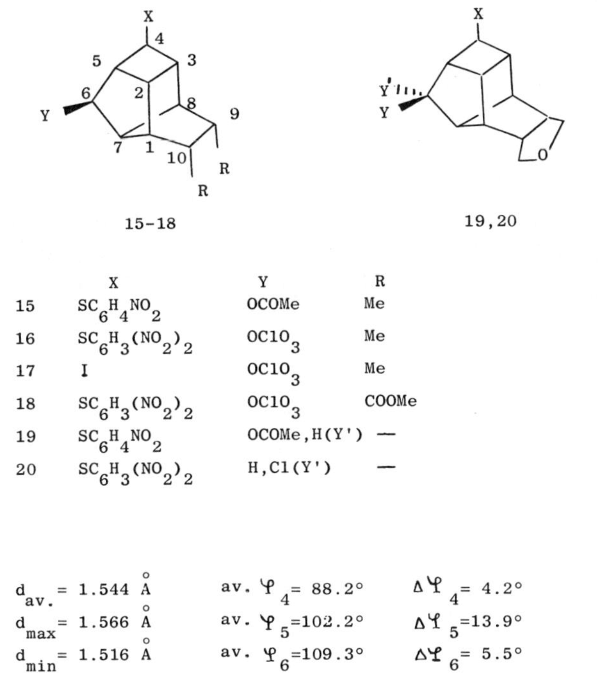

	X	Y	R
15	$SC_6H_4NO_2$	OCOMe	Me
16	$SC_6H_3(NO_2)_2$	$OClO_3$	Me
17	I	$OClO_3$	Me
18	$SC_6H_3(NO_2)_2$	$OClO_3$	COOMe
19	$SC_6H_4NO_2$	OCOMe,H(Y')	—
20	$SC_6H_3(NO_2)_2$	H,Cl(Y')	—

$$d_{av.} = 1.544\ \text{Å} \qquad \text{av.}\ \Psi_4 = 88.2° \qquad \Delta\Psi_4 = 4.2°$$
$$d_{max} = 1.566\ \text{Å} \qquad \text{av.}\ \Psi_5 = 102.2° \qquad \Delta\Psi_5 = 13.9°$$
$$d_{min} = 1.516\ \text{Å} \qquad \text{av.}\ \Psi_6 = 109.3° \qquad \Delta\Psi_6 = 5.5°$$

Fig. 7. Rearrangement products 15-20 of diene I (Series C).

As a whole the C skeletons have no significant strain, which is reflected in the closeness of bond lengths to the standard values and the "normal" cyclobutane ring bending of 30°. Nevertheless ring forms are significantly deformed relative to canonic conformations with considerable deviations of torsion angles from standard values. So in these molecules both norbornane systems 2-5-6-7-8-3-1 and 1-2-3-8-9-10-7 (Fig. 6) are forced to adopt contra- and synchro-twist forms respectively.

The last group of the tetracyclo[5.3.0^{2,10}.0^{3,8}]decane D series compounds represents products of the deeper skeletal rearrangement of triene II, involving four intermediate carbocations (d)-(f) and (g), the stable allyl cation (Fig. 1). These compounds 21 and 22 (Fig. 8) can be conveniently considered as the nortricyclene system 1-10-2-3-8-7-9 with an additional bridge 3-4-5-6-7.

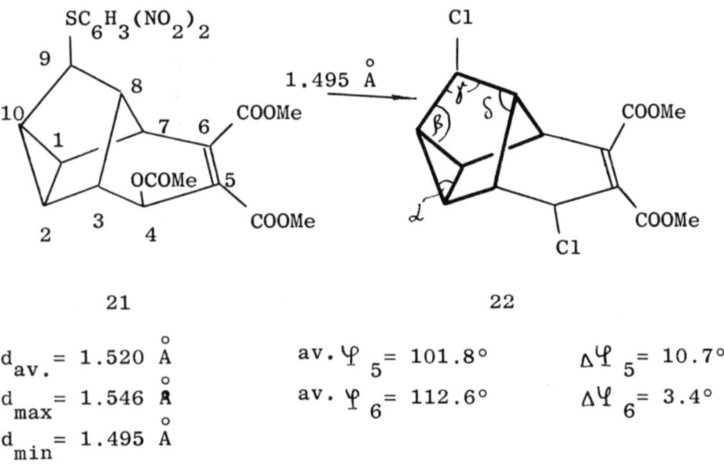

Fig. 8. Rearrangement products 21 and 22 of triene II (Series D)

Deviations from ideal three-fold symmetry of the nortricyclene moiety are indicative of its deformation and, therefore, molecular strain. The nortricyclene unit bond lengths in molecule 21 obey C_3 symmetry, while shortening of the C(9)- C(10) bond to 1.495 Å in molecule 22 may be attributed to the influence of the more electronegative Cl atom in position 9. In accordance with the ideal symmetry the nortricyclene angles are divided into four groups α, β, γ, δ. The α bond angles are equal to 60°, the scatter of β angles is as large as 10σ, for γ it is less, not exceeding 3σ, for δ it is again larger, being the sequence of the influence of additional bridge and the bulky substituent in the position 9. All three 5-membered rings have a very slightly twisted envelope conformation. Thus, according to rather small perturbations in the nortricyclene system C_3 symmetry and a closeness of the 3-4-5-6-7 bridge geometric parameters to standard values, the D polycyclic cage is almost unstrained.

Comparison of geometric parameters of the series A, B, C and D skeletons is carried out by means of the bond length histograms and bond angle spectra (Fig. 9). In skeletons of the trans-adducts A the distribution of the $C(sp^3)$-$C(sp^3)$ bond lengths corresponds to the normal law. On the contrary, in the skeletons of cross-cyclization products B real elongations and shortenings of some bonds are found, indicating more strain.

All bond angles are combined into three groups for 4-, 5- and 6-membered rings, angles belonging to two rings are assigned to the smaller ring. Comparison of the A and B series shows that formation of the cross C(7)-C(8) bond in the series B results in a marked increase of angle differences both in 4- and in 6-membered rings. In the B series the bond angle scatter in 5-membered rings is also high (15.5°). Such distinct anomalies are the result of a considerable strain, which is indicated also by the increase of the 4-membered cycle bending from 14° to 45° on going from the A to the B series.

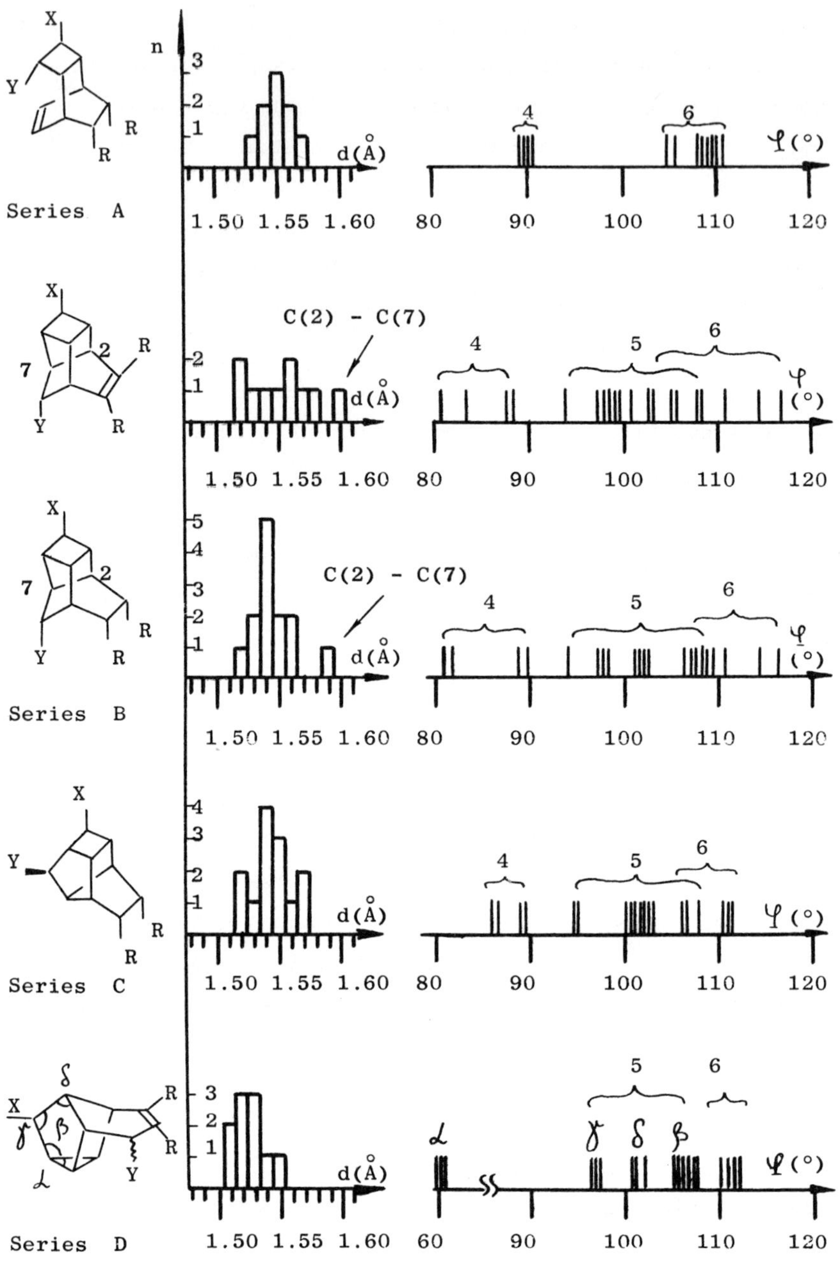

Fig. 9. Bond length histograms and bond angles spectra.

Thus, cross-cyclization of diene I and triene II gives
rise to formation of considerably more strained B systems with
the most obvious geometric anomalies, concentrated in the bicyclic moiety 1-2-7-8-9-10 (Fig. 4). The bond length histograms and bond angle spectra show that the further rearrangement B→C is accompanied by equalization of both bond lengths
and angles. The decrease of bond length scatter is effected by
decreasing of its upper limit, viz. instead of the abnormally
long C(2)-C(7) bond, which is cleaved on rearrangement, a
"new" C(7)-C(8) bond of ordinary length is formed (Figs. 1, 4
and 7). The scatter of bond angles in 4- and 6-membered cycles
of series C is also much smaller.

The rearrangement B→D also results in considerable equalization of bond lengths. The bond angle spectra are changed:
in the B skeletons the ranges of angles in 5- and 6-membered
rings overlap by almost 5°, whereas in the D cages this anomaly is not observed. Therefore rearrangements B→C and B→D are
accompanied by considerable decreasing of molecular strain.

The statistical analysis of experimental geometric parameters is in full agreement with force field calculations carried out with full optimization of molecular geometry. The
intermediate participation of carbocations has a vast chemical
substantiation (Koz'min, 1985). Thus, the following general
conclusions are fully justified (Fig. 1).

Flexibility of the polycyclic system A, i.e. the strain relaxation mainly by twisting, indicates a preparedness of this
skeleton to transannular cyclization with formation of not cis-
but cross-products B.

Geometric anomalies observed in the skeleton B demonstrate
its preparedness for rearrangements into cages C (in the case
of diene) and D (in the case of triene) with a cleavage of the
most elongated bonds. However, substituents in positions 3 and
4, interacting in some cases with intermediate reaction cen-

tres, can form additional O-bridges (Fig. 4, compounds 12-14), causing delocalization of anomalies over a larger part of the skeleton and, therefore, its stabilization toward further transformations.

Thus comparison of molecular geometries proves that the main driving force of skeletal rearrangements lies in a tendency to minimization of strain in polycyclic systems.

2. THE MODEL FOR THE PATHWAY OF NUCLEOPHILIC SUBSTITUTION OF SILICON

We also carried out studies of structure-reactivity relationships among various organoelement compounds. Here only penta-coordinated silicon derivatives are briefly considered. Of them the most well-known representatives are silatranes. In these molecules silicon-nitrogen secondary bonds (N.W. Alcock) are rather weak and correspond to the very beginning of the pathway of S_N2 substituted reaction at Si atom (Fig. 10).

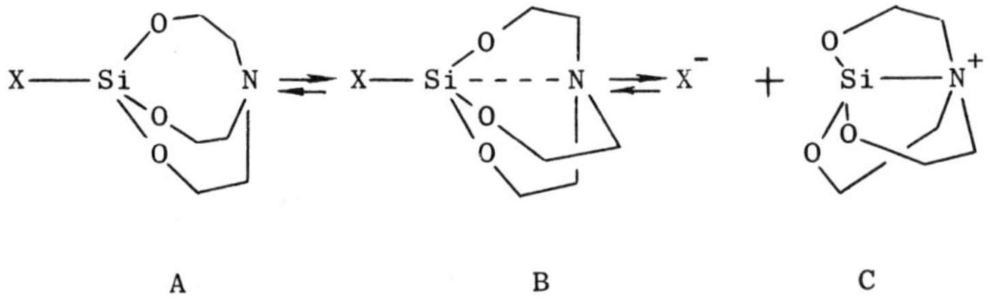

Fig. 10. The hypothetical transformation of exo-silatrane A into classical "intermediate" endo-silatrane B and then into silatranyl cation C.

A quite different situation is observed in molecules of lactam silicon derivatives, involving the 5-membered O, N and Si containing heterocycle (Fig. 11). The first compound of this type (III) was studied earlier (Onan et al, 1978). We studied a series of such compounds I, II and IV-VIII with various halogens on Si and various lactam ring sizes, which were prepared

by Yu.I. Baukov. In the iodo (I) and bromo (II) derivatives the
secondary Si$\cdots$X bonds are very weak and predominantly ionic
(both compounds are good electric conducters in polar solvents). Accordingly, the Pauling-Bürgi bond numbers, $\underline{n}$, correspond to the weak Si$\cdots$X and strong Si-O interactions. The
OSiC$_3$ tetrahedron is somewhat deformed, the exocyclic OSiC
angles being 106 and 103° respectively, though the endocyclic
angles vary from 90° (I) to 78° (VIII) due to the 5-membered
cycle closure. In contrast, in the Cl-derivatives III-VI the
"secondary" Si$\cdots$Cl interaction strength becomes comparable
with the "primary" Si-O bond, since the bond numbers of these
two hypervalent bonds are now practically equal. Moreover the
Si atom adopts an almost undistorted trigonal bipyramidal configuration (ClSiC and OSiC exocyclic angles being 92-94° and 91-
90°; the "axial" OSiX angles vary in the range of 162-173°). In
the fluoro-derivative VIII the Si-F bond becomes primary covalent and the Si$\cdots$O bond becomes secondary and the Si atom
adopts an inverted tetrahedral configuration, the FSiC angles
being 101 and 102° (the OSiC angles are 85 and 84°). Comparison
of the Cl-derivatives III-VI shows almost no influence of the
lactam ring size on the Si atom coordination. On the contrary,
in compound VII modification of this ring makes the Si-Cl interaction much stronger and this molecule approaches the limiting
case of the F-derivative VIII (possibly due to conjugation of
the peptide system with the exocyclic C=C bond).

3. MAGIC NUMBER RULES FOR ORGANOMETALLIC COMPOUNDS AND METAL CLUSTERS

Among organometallic compounds great theoretical interest
is devoted to molecules with σ- and multiple metal-carbon
bonds, which are formed, in particular, in many reactions of
π-complex transformations. On the basis of the Cambridge Structural Database (CSD) we proposed standard lengths for bonds of
various bond numbers between metals and carbon in various hyb-

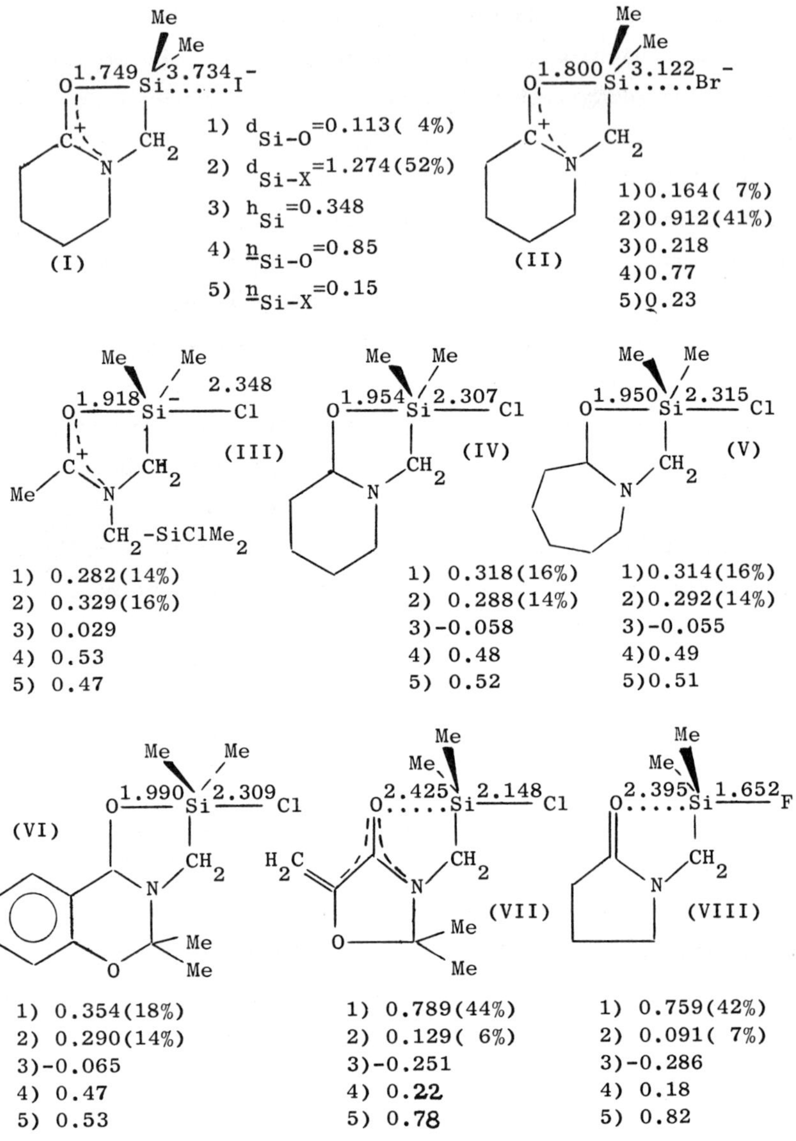

Fig. 11. The novel type of organosilicon molecules I-VIII with penta-coordinated silicon. $\sigma d = d(n)-d(1)$, h_{Si} is the Si atom distance from the plane of "equatorial" atoms.

ridization states (Batsanov <u>et al</u>, 1985). It is shown that shortening of ordinary M-C bonds, observed in the series $C_{sp}3 \rightarrow C_{sp}2 \rightarrow C_{sp}$ cannot be ascribed only to the associated decrease of the carbon atom covalent radius, but also requires consideration of the $d_\pi(M) \rightarrow p_\pi(C)$ dative interaction and, mainly, the M-C bond polarity.

Almost all σ-, π- and mixed σ,π-organometallic mono- and polynuclear complexes obey the modified effective atomic number (MEAN) rule (Slovokhotov <u>et al</u>, 1984)

$$N_{val} = 12 + 2d,$$

where N_{val} is a number of electrons in the outer metal atom shell, d is a dimension of its coordination. So for typical transition metals with a 3D coordination $N_{val}=18$, for metals in the end of transition rows (e.g. Pd, Pt) with a typical 2D square planar or trigonal coordination $N_{val}=16$ and for post-transition metals (Cu, Ag, Au, Hg) with one-dimensional usually linear coordination $N_{val}=14$ in most cases. A tendency to fulfillment of this rule is an important driving force of numerous reactions of σ- and π-complexes.

Among cluster compounds our main interest lies in structural studies of small triangular and tetrahedral Pd and Pt complexes, medium-sized (mainly octahedral) homo- (Fe) and heteronuclear (Fe, Rh, Co, Pt etc.) clusters. The CSD was again used to analyze in this case the areas of existence of so-called "magic" numbers of valent and skeletal electrons in 3-, 4-, 5- and 6-nuclear clusters. The first two cluster types with localized M-M bonds obey the MEAN rule and the 6-membered clusters with a superaromatic electron system obey the 2n+2 Wade rule, while in the 5-membered clusters both rules are often violated. For medium clusters M_n with n<12 and for multi-polyhedral or fused clusters with the same nuclearity of individual polyhedrons the modified rule of magic number of skeletal electrons is satisfied quite well (Slovokhotov <u>et al</u>, 1983):

$$N_{skeletal} = 2(n+f-c+1),$$

where n is the number of cluster metal atoms, f is the number of cluster "open" (non-triangular) faces and c is the number of common (shared) vertices in multi-polyhedral clusters.

However, both these magic number rules fail in the case of big clusters (n>12) and clusters of metals in the end of transition rows as is shown by structural studies of the medium (Pd_{10}) and big (Pd_{38} and two "isomers" of Pd_{23}) clusters prepared by dr. E.G. Mednikov. Here the cluster structures are controlled by the magic numbers not of electrons but of metal atoms in the cluster. However, the laws, governing these metal atom magic numbers, are still obscure. Moreover, big clusters, studied by us and other authors, do not obey the rules of metal atom close packing and, due to somewhat conflicting electronic and steric requirements of metal atoms and ligands, there seems to exist a nuclearity gap between big clusters and metallic microparticles, naked or surrounded by shells of chemisorbed small molecules.

ACKNOWLEDGEMENTS: I express my sincere gratitude to the chemists (Professors N.S. Zefirov and Yu.I. Baukov and Dr. E.G. Mednikov) and the X-ray crystallographers (Drs. K.A. Potekhin and Yu.L. Slovokhotov and Prof. V.E. Shklover) who have made this report possible.

REFERENCES

ZEFIROV, N.S., KOZ'MIN, A.S. AND ZHDANKIN, V.V. (1982). _Tetrahedron_ 38, 291.

KONDO, A., YAMANE, T., ASHIDA, T., SASAKI, T. and KANEMATSU, K. (1978). _Journal of Organic Chemistry_ 43, 1180 and references therein.

CREMER, D., and POPLE, J.A. (1975). _Journal of the American Chemical Society_ 97, 1354.

PALYULIN, V.A., ZEFIROV, N.S., SHKLOVER, V.E. and STRUCHKOV, Yu.T. (1981). _Journal of Molecular Structure_ 70, 65.

INAGAKI, S., FUJIMOTO, K. and FUKUI, K. (1976). _Journal of the American Chemical Society_ 98, 4054.

OSAWA, E. (1978), _Tetrahedron_ 34, 509.

KOZ'MIN, A.S. (1985). D. Sc. Dissertation. Moscow State Univ.

ONAN, K.D., McPHAIL, A.T., YODER, C.H. and HILLYARD, R.W. (1978). _Journal of the Chemical Society, Chemical Communications_ 209

BATSANOV, A.S. and STRUCHKOV, Yu.T. (1985). _Abstracts of the VI FECHEM Conference_, Riga, 149.

SLOVOKHOTOV, Yu.L. and STRUCHKOV, Yu.T. (1984), _Journal of Organometallic Chemistry_ 277, 143.

SLOVOKHOTOV, Yu.L. and STRUCHKOV, Yu.T. (1983), _Journal of Organometallic Chemistry_ 258, 47.

45. Research on the structure and reactivity of trinuclear molybdenum clusters

J. L. Huang, J. Q. Huang, M. Y. Shang, H. H. Zhuang, S. F. Lu,
X. T. Lin, Y. H. Lin, M. D. Huang and J. X. Lu

ABSTRACT

A series of trinuclear molybdenum clusters have been synthesized
by making use of the reactions of $MoCl_3 \cdot 3H_2O$ in an EtOH–HCl medium and characterized structurally by X-ray diffraction. Three
structural types (M_1, M_2, and B_2) are described and some structural regularities are discussed, including the dependence of
the Mo–Mo lengths on the numbers of cluster electrons as well as
on the different properties of the ligands, and the Jahn–Teller
effect on the B_2-type structures with 8 cluster electrons. The
IR absorption bands of the $Mo_3-(\mu_3-O)$ vibration are assigned. A
formation mechanism has been proposed for the B_2-type clusters
Several reactions are also investigated for the M_1-type clusters
with loose coordination sites.

1.   INTRODUCTION

The chemistry of the transition metal cluster compounds has become an active area in the past two decades on account of their
possible relevance to biological systems as well as to certain
catalytic reactions. The chemistry of Mo clusters has particularly aroused considerable interest as a result of the recognition of their being the active centers in several enzymes such
as nitrogenase. This has stimulated extensive investigation of
the Mo–Fe–S heteronuclear clusters in an attempt to obtain a
better understanding of the mechanism of nitrogen fixation in
the biological process. On the other hand, much attention has
been focused on the study of the basic chemistry of Mo clusters.

Among the Mo clusters obtained so far, the trinuclear ones exhibit rather wide varieties in both structure and chemistry. In 1980 Müller *et al.* reported the progress in research on trinuclear clusters of the early transition metals, and emphasized the importance of this area. A few years ago, our research group was able to find that there is an abundant cluster chemistry of medium-valenced molybdenum in an acidic ethanol medium. Since then a series of Mo clusters including binuclear, trinuclear, and tetranuclear clusters has been isolated from our reaction system and structurally characterized by X-ray diffraction work. In the meanwhile, several types of interesting reactions were discovered.

This paper will give a general account of our research work in the area of trinuclear Mo clusters along with some discussion based on the results from our research group together with those reported in the literature.

2.   CLUSTER CONFIGURATION AND STRUCTURAL PARAMETERS

All the trinuclear Mo clusters reported so far are either monocapped or bicapped species. For those obtained from our reaction system three basic types have been isolated, denoted by M_1, M_2, and B_2, respectively, according to their structural features ('M' and 'B' represent monocapped and bicapped clusters, respectively, while the subscripts are used to disignate difference in the mode of coordination of the Mo atoms).

2.1.   CRYSTAL STRUCTURE

2.1.1. STRUCTURE TYPE M_1

Fig. 1 shows the configuration of the M_1-type structure. In this structure type a triply bridging atom situated on one side of the plane of the three Mo atoms binds these Mo atoms together to form a trigonal pyramid. Besides, each pair of Mo atoms are attached to a doubly bridging atom located on the other side of the plane. Furthermore, each Mo atom is also coordinated to three terminal ligands to complete an octahedral coordination.

Tables 1 and 2 give a collection of important bond lengths for clusters of this type we have studied so far. All the clusters listed in Table 1 have bridging carboxylic ligands, while those listed in Table 2 have diethyldithiophosphate (dtp) or dithiocarbamate (dtc) ligands instead. The configurations of these clusters are shown in Fig. 2 and Fig. 3 respectively.

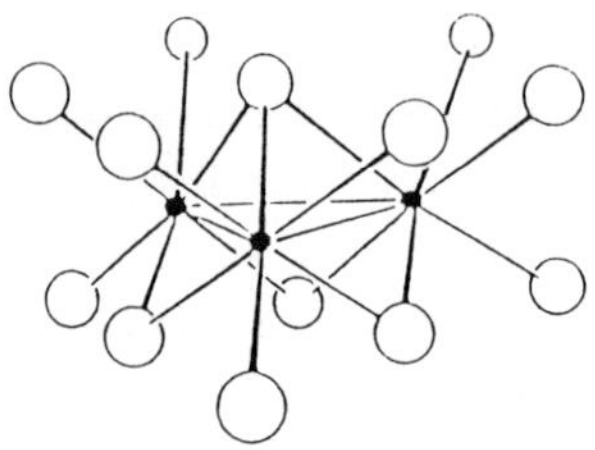

Fig. 1   Configuration of the M_1-type structure

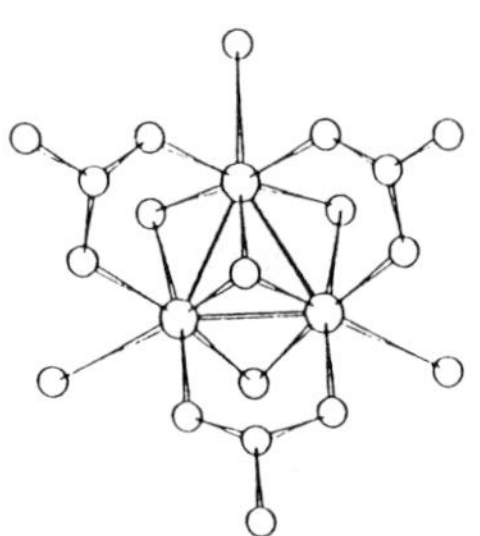

Fig. 2 M_1-type structure with
carboxylic ligands

Fig. 3 M_1-type structure
with dtp ligands

Table 1   Important bond lengths (in Å) of M_1-type clusters
contaning carboxylic ligands

No.	Cluster	Mo–Mo	Mo–μ_3	μ_2	Mo–μ_2	Mo–O	Mo–Y
1	$(Me_4N)[Mo_3OCl_3(O_2CH)_3Cl_3]$	2.577	1.982	Cl	2.414	2.084	2.412
2	$(Et_4N)[Mo_3OCl_3(O_2CH)_3Cl_3]$	2.573	1.978	Cl	2.410	2.086	2.412
3	$(C_5H_7S_2)^a[Mo_3OCl_3(OAc)_3Cl_3]$	2.577	1.992	Cl	2.419	2.075	2.419
4	$(Et_4N)[Mo_3OCl_3(OAc)_3(Cl,Br)_{1.5}]$	2.577	1.976	Cl	2.415	2.068	2.516
5	$(Et_4N)[Mo_3OCl_3(O_2CEt)_3Cl_3]$	2.576	1.973	Cl	2.448	2.064	2.430
6	$[(NH_2)_2CSH][Mo_3OCl_3(O_2CH)_3Cl_3]$	2.573	1.976	Cl	2.413	2.077	2.426
7	$(C_5H_7S_2)[Mo_3OBr_3(OAc)_3Cl_3]$	2.594	1.970	Br	2.549	2.076	2.442
8	$(Me_4N)[Mo_3OBr_3(O_2CH)_3Cl_3]$	2.596	1.976	Br	2.542	2.089	2.420
9	$(Et_4N)_2[Mo_3OCl_3(OAc)_2Cl_5]$	2.584[b]	1.996	Cl	2.407	2.086	2.451[c]
		2.594[b]					2.367[d]
		2.618					

[a] : dithioacetylacetonium.   [b] : bridged by acetate ligand.
[c] : cis to capping atom.   [d] : trans to capping atom.

Table 2   Important bond lengths (in Å) of M_1-type clusters
containing dtp or dtc ligands

No.[a]	Mo–Mo[b]		μ_3	Mo–μ_3	μ_2	Mo–μ_2	Mo–S_t[c]	Mo–S_b[d]	L	Mo–L
10	2.718	2.796,2.840(1)	S	2.340	S	2.295	2.532	2.551	Cl	2.430
					Cl	2.363				
11	2.647	2.704,2.730(2)	S	2.351	O	1.997	2.538	2.534	S	2.440
					Cl	2.393				
12	2.734	2.763,2.766(1)	S	2.346	S	2.283	2.571	2.586	O	2.361
13	2.731	2.748,2.753(1)	S	2.333	S	2.288	2.592	2.592	P	2.647
14	2.744	2.760,2.761(1)	S	2.341	S	2.286	2.582	2.608	S	2.618
15	2.742	2.760,2.762(1)	S	2.337	S	2.282	2.568	2.643	S	2.734
16	2.743	2.758,2.761(1)	S	2.339	S	2.282	2.573	2.623	N	2.360
17	2.752	2.760,2.767(1)	S	2.337	S	2.282	2.573	2.617	N	2.316
18	2.750	2.753,2.768(1)	S	2.341	S	2.284	2.577	2.654	N	2.250
	2.742	2.756,2.764(1)								
19	2.743	2.757,2.763(1)	S	2.344	S	2.281	2.572	2.633	N	2.306
20	2.722	2.740,2.744(1)	S	2.337	S	2.308	2.560	2.576	O	2.241
21	2.729	2.743,2.751(2)	S	2.338	S	2.315	2.553	2.551	N	2.271
22	2.669	2.684,2.690(1)	O	2.202	S	2.282	2.586	2.610	S	2.633
23	2.644	2.628,2.642(1)	O	2.027	S	2.282	2.590	2.584	N	2.283
24	2.658	2.646,2.649(2)	O	2.052	S	2.280	2.598	2.622	N	2.232
25	2.706	2.741,2.777(1)	S	2.343	S	2.289	2.523	2.593	N	2.365
26	2.703	2.743,2.778(1)	S	2.337	S	2.303	2.525	2.569	O	2.278
27		2.756,2.762,	S	2.339	S	2.287	2.571		N	2.266
		2.763(1)								

[a]: Structural formulas for the numbered clusters:

10. $Mo_3(\mu_3\text{–}S)(\mu\text{–}S)(\mu\text{–}Cl)_2(dtp)_4Cl$,   11. $Mo_3(\mu_3\text{–}S)(\mu\text{–}O)(\mu\text{–}Cl)_2(dtp)_4(SH)$,

12. $Mo_3S_4(dtp)_4(H_2O)$,   13. $Mo_3S_4(dtp)_4(PPh_3)\cdot0.86CH_2Cl_2$,

14. $Mo_3S_4(dtp)_4[SC(NH_2)(NHCH_2CHCH_2)]$,   15. $Mo_3S_4(dtp)_4(PhCH_2SH)$,

16. $Mo_3S_4(dtp)_4(Py)$,   17. $Mo_3S_4(dtp)_4(oxazole)$,

18. $Mo_3S_4(dtp)_4(PhCH_2CN)$,   19. $Mo_3S_4(dtp)_4(CH_3CH_2CN)$,

20. $[Mo_3S_4(dtp)_4(SbCl_3)(EtOH)]\cdot EtOH$,   21. $Mo_3S_4(dtp)_4(SbCl_3)(oxazole)$,

22. $Mo_3[\mu_3\text{–}(S,O)_{0.5}]S_3(dtp)_4[SC(NH_2)_2]$,   23. $Mo_3(\mu_3\text{–}O)S_3(dtp)_4(oxazole)$,

24. $Mo_3(\mu_3\text{–}O)S_3(dtp)_4(imidazole)$,   25. $[Mo_3S_4(dtc)_4Py](Py)_2(H_2O)$,

26. $Mo_3S_4(dtc)_4(dmf)$,   27. $[Mo_3S_4(dtp)_3(imidazole)_3](dtp)$.

[b]: The second column is the lengths of Mo–Mo bonds bridged by dtp or dtc.

[c]: S atom from terminal chelating group.

[d]: S atom from bridging chelating group.

2.1.2. STRUCTURE TYPE M_2

The configuration of the M_2-type structure is shown in Fig.4. In
this type of structure the three Mo atoms form an equilateral
triangle. On one side of this triangle plane a S or O atom binds
the three Mo atoms to form a trigonal pyramid. For each pair of
Mo atoms a S_2 group acts as a doubly bridging ligand in such a

manner that one S atom lies on the Mo_3 plane and the other is attached on the opposite side of the Mo_3 plane to the triply bridging atom. In addition, each Mo atom is coordinated to two Cl or S atoms, thus completing a distorted pentagonal bipyramidal coordination.

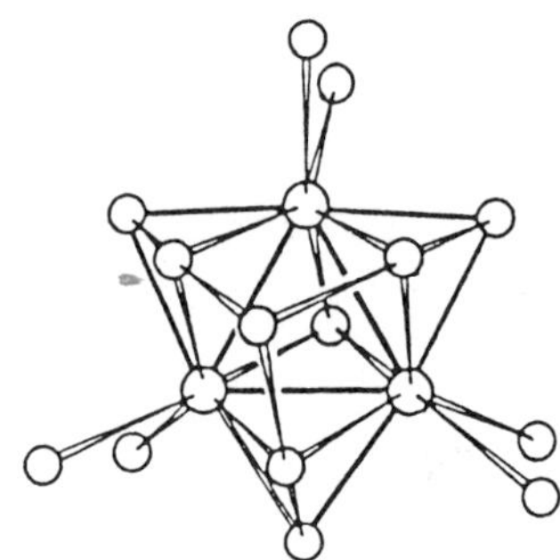

Fig. 4 Configuration of the M_2-type structure

Table 3   Important bond lengths (in Å) of M_2-type clusters

No.	Cluster	Mo–Mo	μ_3	Mo–μ_3	Mo–μ_2S		Mo–L_t		X	S–X
					a	b	c	d		
28	$(C_5H_7S_2)_3[Mo_3S_7Cl_7]$	2.751	S	2.348	2.469	2.396	2.485	2.449	Cl	3.028
29	$(PyH)_3[Mo_3S_7Cl_7]\cdot0.5H_2O$	2.750	S	2.356	2.448	2.387	2.499	2.446	Cl	2.969
30	$Mo_3S_7(dtp)_3Cl$	2.725	S	2.362	2.484	2.395	2.548	2.499	Cl	2.916
30	$Mo_3S_7(dtp)_3Cl^e$	2.715	S	2.369	2.478	2.396	2.549	2.510	Cl	2.893
31	$Mo_3S_7(dtp)_3I$	2.723	S	2.368	2.483	2.400	2.552	2.511	I	3.172
32	$[Mo_3S_7(dtp)_3]_2Cl\cdot FeCl_4$	2.726	S	2.371	2.478	2.405	2.546	2.503	Cl	3.003
33	$Mo_3OS_6(dtp)_3I$	2.626	O	2.036	2.460	2.404	2.540	2.510	I	3.202
34	$Mo_3OS_6(dtp)_3Cl$	2.631	O	2.046	2.465	2.402	2.554	2.519	Cl	2.862
35	$[Mo_3OS_6(dtp)_3]_2Cl\cdot FeCl_4$	2.627	O	2.028	2.461	2.404	2.533	2.490	Cl	3.010

[a] : S atom in the Mo_3 plane.   [b] : S atom out of the Mo_3 plane.
[c] : terminal ligand trans to μ_3 atom.   [d] : terminal ligand cis to μ_3 atom.
[e] : another crystal modification.

Table 3 summarizes all the clusters found in our research group together with some of the important bond lengths. All the cluster compounds with this M_2-type structure found so far have six cluster electrons without any exception. If the metal–metal interaction is taken into consideration, it is obvious that all the 9 valence orbitals of each Mo atom have participated in the formation of the molecular orbitals and that each Mo atom has 18 electrons in its outer shell. As a result, this kind of clusters are generally stable. It is, however, interesting to note that in this type of structure there is an additional Cl, I, or S atom, which is not at all bound directly with any of these Mo atoms, but approaches the three out-of-plane S atoms with distances of about 3.0 Å which are considerably shorter than the corresponding van der Waals contacts (ca. 3.7 Å). This gives an indication that the three out-of-plane S atoms have a higher electron affinity in character. Again, this might be the reason why clusters of this M_2 structure type can be converted to the M_1 structure type by an attack of anions. It is therefore not at all surprising why this kind of structure has been regarded as a model for the Mo-containing xanthine oxidase, which can rather easily lose its labile S atom upon attack by a nucleophile such as a CN^- anion (Müller and Reinsch, 1980).

2.1.3. STRUCTURE TYPE B_2

Two types of structures have been discovered for the bicapped clusters. The one with the formula $Mo_3(\mu_3-X)(\mu_3-Y)(O_2CR)_6(H_2O)_3$ reported by Bino *et al.* (1981) is designated as the B_1 structure type in our classification and the clusters synthesized by us are designated as the B_2 structure type simply for distinction. The configuration of this B_2-type structure is shown in Fig. 5. In this structure the Mo_3 plane is capped on both sides by two atoms (S, O, or Cl), while three doubly bridging Cl atoms are located essentially on the Mo_3 plane. Besides, there are two additional atoms bound terminally to each Mo atom. As a result of this arrangement each Mo atom is coordinated octahedrally to 6 ligand atoms, and the three distorted octahedra are connected to each other by sharing faces.

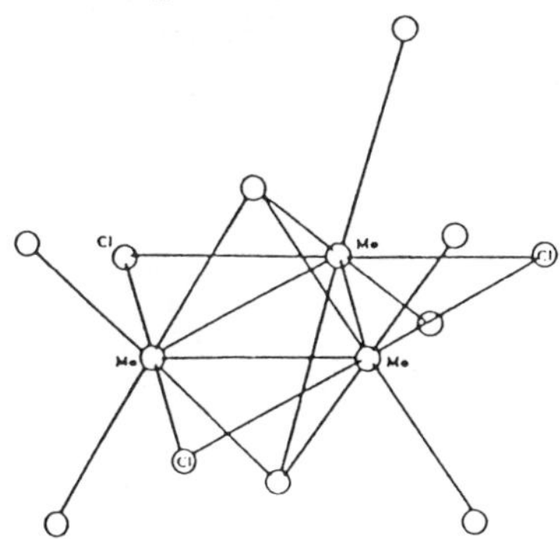

Fig. 5 Configuration of the B_2-type structure

Table 4 lists structural parameters for the three clusters obtained in our reaction system. It is interesting to note that the Mo_3 triangles in the first two clusters with Cl as terminal ligands are isosceles, while the third cluster with dtp as its terminal ligands, is equilateral. This structural difference will be discussed later in some detail.

Table 4   Important bond lengths (in Å) of the B_2-type clusters

No.	Cluster	Mo–Mo	μ_3 Mo–μ_3	Mo–Cl$_b$	L$_t$ Mo–L$_t$
36	$(C_5H_7S_2)_3[Mo_3S_2Cl_9]$	2.556,2.641,2.653(1)	S 2.375	2.467	Cl 2.447
37	$(Et_4N)_2[Mo_3OCl_{10}]\cdot EtOH$	2.601,2.608,2.642(1)	O 2.079	2.443	Cl 2.376
			Cl 2.496		
38	$Mo_3S_2Cl_3(dtp)_3$	2.606(1)	S 2.381	2.457	S 2.492

2.1.4. STRUCTURE OF M_1'-TYPE CLUSTER

In this structure type one Mo_2 pair of the Mo_3 triangle is not at all a Mo–Mo bond.

Cluster 39, $Mo_3S_3O_2(dtp)_3[SOP(OEt)_2]$

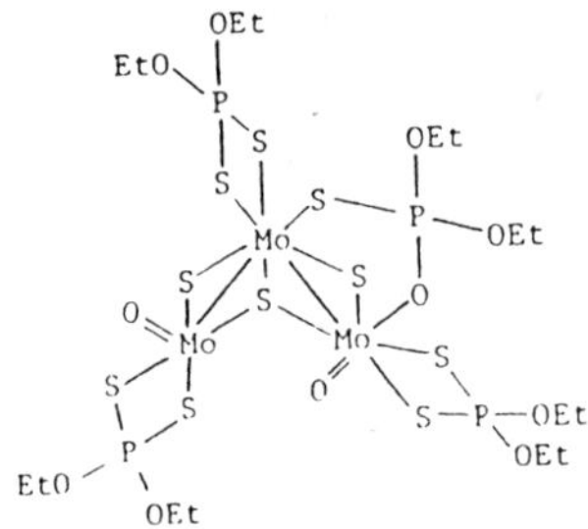

Mo–Mo: 2.808(1) Å;
2.839(1) Å
Mo–μ_3S: 2.391 Å
Mo^V–μ_2S: 2.333 Å
Mo^{IV}–μ_2S: 2.275 Å
Mo=O: 1.682 Å
Mo–O$_1$: 2.188 Å
Mo–S$_t$: 2.534 Å

Fig. 6 M_1'-type configuration

2.1.5. STRUCTURE OF (M_1+M_2)-TYPE CLUSTER

This type of Mo_3 cluster is a combination of the M_1 and M_2 types of cluster units bridged through S atoms, as shown in Fig. 7. The important bond lengths of Cluster 40 are listed in Table 5.

Cluster 40, $[Mo_3S_4(dtp)_4(NCS)][Mo_3S_7(dtp)_3]$

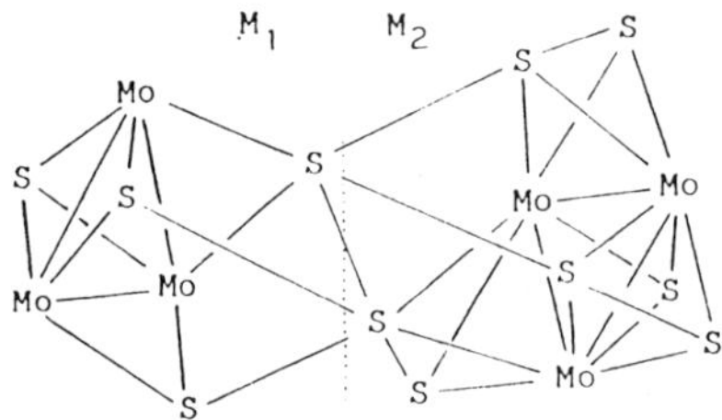

Fig. 7 Cluster core of the (M_1+M_2)-type structure

Table 5   Important bond lengths (in Å) of Cluster 40

	Mo–Mo	Mo–μ_3S	Mo–μ_2S		Mo–St		S_{M1}–S_{M2}	Mo–N
			a	b	c	d		
M_1	2.738,2.769,2.776(4)	2.339		2.288	2.576	2.563		2.22
M_2	2.722	2.358	2.490	2.396	2.530	2.481	3.26	

a : atom in the Mo_3 plane.　　b : atom out of the Mo_3 plane.
c : atom trans to μ_3 atom.　　d : atom cis to μ_3 atom.

2.2.　SOME STRUCTURAL REGULARITIES

2.2.1. RELATIONSHIP BETWEEN Mo–Mo BOND DISTANCE AND CLUSTER ELECTRONS

It is obvious that the metal–metal interaction is due first to the overlap of the valence shell orbitals of the metal atoms as well as to the cluster electron (c.e.) count occupying the molecular orbitals so formed. Therefore, for a given structure type the M–M distance depends on the cluster electrons. This will be made clear from a comparison of the M–M bond lengths for the M_1 type structures with different c.e. counts as shown in Table 6.

Table 6   Comparison of M–M bond lengths for the M_1-type clustres

Cluster	μ_3	μ_2	c.e.	Mo–Mo
$Zn_2Mo_3O_8$	O	O	6	2.524(2)
$LiZn_2Mo_3O_8$	O	O	7	2.578(1)
$Zn_3Mo_3O_8$	O	O	8	2.580(2)
$[Mo_3O(OCH_3)_3(NO)_3(CO)_6]^{2-}$	O	OCH_3	12	3.297(2)
$[Mo_3(OCH_3)_4(NO)_3(CO)_6]^-$	OCH_3	OCH_3	12	3.428
$Mo_3S_3O_2(dtp)_3[SOP(OEt)_2]$	O	O	4	2.808,2.839,3.337(1)

It can be seen from Table 6 that the compound $Zn_2Mo_3O_8$ with six c.e. has the shortest Mo–Mo distance (2.524 Å) corresponding to a Mo–Mo single bond. Increase in the c.e.c. leads to lengthening of the Mo–Mo distance (Torardi and McCarley, 1985). When the c.e.c. reaches 12 as found in $[Mo_3O(OMe)_3(NO)_3(CO)_6]^{2-}$ and $[Mo_3(OMe)_4(NO)_3(CO)_6]^-$ (Kirtley *et al*, 1980), the Mo–Mo distances are lengthened up to 3.3-3.4 Å, implying no Mo–Mo bonding interaction at all. In the case of the cluster $Mo_3S_3(dtp)_3[SOP(OEt)_2]O_2$ with 4 c.e., there are only two Mo–Mo bonds retained, and no M–M bond can at all be formed on the third side.

Birnbaum *et al.* (1985) have given a qualitative quantum

chemical treatment to this cluster skeleton, coming to the conclusion that the $1a_1$ and the $1e$ molecular orbitals are bonding, while the $2a_1$ lies at a slightly higher energy level than the $1e$ orbital. Upon examination of the role of the $2a_1$ molecular orbital in cluster compounds related to $Zn_2Mo_3O_8$ as shown in the first three rows of Table 6, Torardi *et al*. have come to the conclusion that the $2a_1$ is antibonding. This obviously agrees with the results from the two carbonyl clusters listed in Table 6, i.e. half of their 12 c.e. must occupy the three antibonding MO's to offset the bonding M–M interaction.

2.2.2. DEPENDENCE OF Mo–Mo BOND LENGTHS ON THE LIGAND ATOMS IN CLUSTER SKELETONS

<u>Triply bridging atoms</u>

In order to make a more reliable conclusion, it is desirable to carry out comparison between compounds belonging to the same structural type and having the same c.e. count and the same ligand atoms except the one to be compared. It is convenient to compare the Mo–Mo distances for M_2-type clusters with different μ_3-atoms as shown in Table 3. There are three clusters with the same formula $[Mo_3O(S_2)_3(dtp)_3]^+$ having an average Mo–Mo distance of 2.628 Å (2.631, 2.627, and 2.626 Å respectively), on the other hand there are four clusters with the formula $[Mo_3S(S_2)_3(dtp)_3]^+$ having an average Mo–Mo distance of 2.722 Å (2.725, 2.715, 2.723, and 2.726 Å, respectively). This shows clearly that the substitution of μ_3-S for μ_3-O causes the Mo–Mo bond to be lengthened by ca. 0.1 Å. Similarly, the increase of Mo–Mo bond distance caused by the replacement of μ_3-O by μ_3-S in the M_1-type clusters with six c.e. amounts to 0.11–0.12 Å, as can be seen in Table 2.

<u>Doubly bridging atoms</u>

There are two M_1 type clusters with the cluster core $Mo_3(\mu_3$-O$)$-$(\mu$-S$)_3$ in Table 2, whose average Mo–Mo distance is 2.645 Å. It follows from a comparison with those found for $Zn_2Mo_3O_8$ (2.524 Å) and $[Mo_3O_4(C_2O_4)_3(H_2O)_3]^{2-}$ (2.486 Å) (Bino *et al*, 1984) that the replacement of μ_2-O by μ_2-S leads to an increase of Mo–Mo bond distance by about 0.12–0.15 Å. A similar increase in the Mo–Mo distances has been observed in the change of the cluster core from Mo_3SO_3 (Shibahara *et al*, 1984) to Mo_3S_4. Sometimes, replacement of only one bridging O atom by a S atom can cause the Mo–Mo bond to increase by about 0.09 Å, as in the case of $\{Mo_3(\mu_3-S)(\mu-Cl)_2(\mu-X)\}$ (X=S,O) cluster core with 7 c.e. (Table 2). When bridging atoms become large, it is possible that there may be some stretching of the Mo–Mo bonds in order to accomodate them. However, the steric requirement of the bridging atoms can never be regarded as the sole contribution to such an increase as shown by a comparison of the Mo–Mo bonds in Mo_3OCl_3 and Mo_3OBr_3 cluster cores with the same c.e.. It can be seen from Table 1 that there are two clusters with the Mo_3OBr_3 core having a Mo–Mo bond length of 2.595 Å on the average. This is larger only by 0.018 Å than those with the Mo_3OCl_3 core. Therefore, we are led to the conclusion that an increase in the Mo–Mo distance arises not only from the steric effect of the bridging atoms, but also from their electronic effect. Taking into consideration that S atom is a good π donor, it may even donate its excess pπ electrons to fill in the Mo–Mo antibonding orbitals so as to weaken the Mo–Mo bonding interaction.

<u>Terminal atoms</u>

In the M_2-type clusters the average Mo–Mo distance is of 2.725 Å for the cluster anion $[Mo_3S(S_2)_3(dtp)_3]^+$, while the corresponding value for $[Mo_3S(S_2)_3Cl_6]^{2-}$ is 2.751 Å, as shown in Table 3. This means that the change of the terminal chelating ligands to Cl atoms leads to an increase in the Mo–Mo bond length by about 0.02 Å. A similar situation is also found in the B_2-type clusters, as shown in Table 4, where the average Mo–Mo bond length of $[Mo_3(\mu_3-S)_2(\mu-Cl)_3Cl_6]^{3-}$ (2.617 Å) is larger than that of the $Mo_3(\mu_3-S)_2(\mu-Cl)_3(dtp)_3$ (2.606 Å). In particular, when such chelating ligands (dtp, dtc, OAc, etc.) are in the bridging situation, the bridged Mo–Mo distances are shorter than the non-bridged, as shown in Tables 1 and 2, and the higher the number of such bridging ligands is, the shorter the average Mo–Mo bond length is. No doubt, chelating ligands may be expected to reduce the L–L repulsion and thus help the Mo atoms come to approach each other more closely.

2.2.3. JAHN–TELLER DISTORTION IN THE B_2–TYPE STRUCTURE

As mentioned above, the Mo_3 triangles in the two B_2-type structure with 8 c.e. are isosceles instead of being equilateral,

although the local environments of the three Mo atoms are very
much similar in these two clusters. In order to obtain a better
understanding of this distortion, an EHMO calculation was carried out on the cluster $[Mo_3S_2Cl_9]^{3-}$ in an idealized D_{3h} symmetry with the bond lengths and angles averaged from the experimental structure (Jiang *et al*, 1985). Fig. 8 gives a correlation diagram of the energy levels between the Mo_3 metal core and
the cluster anion. It is obvious that clusters with 6 c.e. occupying the three bonding MO's will possess three Mo–Mo sigma
bonds, while for an eight electron system, two electrons would
enter the two degenerate 7e" orbitals. In this case, the symmetry will be cut down so as to eliminate degeneracy, as required
by the Jahn–Teller effect. This is why the triangle becomes
isosceles. Nevertheless, the difference of the energy levels
between 7e" and $2a_1$" is not at all large. Thus the change of
terminal Cl atoms to chelating ligands would cause the change of
the sequence of these two obitals. As a result, the two excess
electrons would enter the $2a_1$" orbital. Therefore, no Jahn–
Teller distortion should be expected in this case, as found in
the cluster $Mo_3(\mu_3-S)_2(\mu-Cl)_3(dtp)_3$ (Table 4).

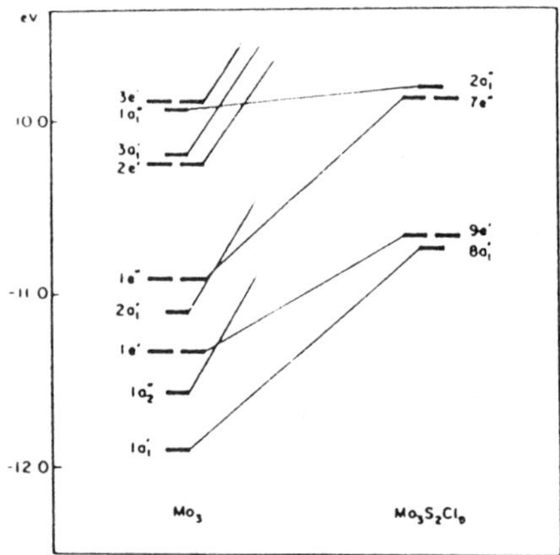

Fig. 8 Correlation diagram of the energy levels

3.   VIBRATIONAL BAND ASSIGNMENT OF IR SPECTRA OF TRINUCLEAR
       Mo CLUSTERS

3.1.   THE ABSORPTION BANDS OF THE $Mo_3-(\mu_3-S)$ MODE

Müller *et al.* (1978) have assigned $459cm^{-1}$ to the absorption of $Mo_3(\mu_3-S)$ vibration in $[Mo_3S_{13}]^{2-}$, using the isotopic substitution method. This assignment has been confirmed by our systematic IR work on the series of trinuclear Mo clusters with μ_3-S atoms, as shown in Table 7 (Zheng *et al*, 1983). In addition, a rather weak absorption band was also found at ca. $350cm^{-1}$ for the same group, although it is difficult to recognize in general on account of the overlap by other absorption spectra. Some typical spectra are shown in Fig. 9.

Table 7   IR absorption of $Mo_3-(\mu_3-S)$

Cluster	11*	12*	13*	14*	15*	16	17*	18*
ν (cm^{-1})	445.6	455.3	445.6	447.6	449.5	442	449.5	453.3

Cluster	19	20*	21*	28	29	30	36	38
ν (cm^{-1})	446	445.6	447.6	457	450	440	452	446

* FTS-20E/D-V, the rest on Perkin-Elmer 577.

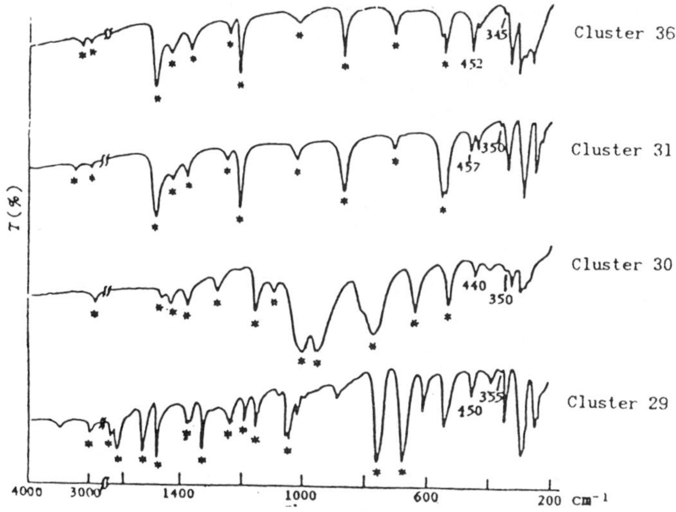

* IR absorption bands from the organic ligands

Fig. 9   IR absorption of the sulfido-capped clusters

3.2.   THE ABSORPTION BANDS OF THE $Mo_3-(\mu_3-O)$ MODE

The assignment of the IR spectra for this group has not been reported in literature so far as we are aware of. After a careful examination of the spectra for a series of oxo-capped clusters and comparison with those sulfido-capped ones, as shown in Fig. 10, we were able to find that there is a medium absorption band in the region of 700cm^{-1} and a weak band in the region of 500cm^{-1}. Table 8 lists the absorption data and the corresponding Mo$_3$-(μ-O) distances for several clusters. It is interesting to note that a longer bond distance corresponds to a lower vibrational frequency, giving a rather good correlation between these two sets of data, with only one exception observed in a cluster containing formic group. Some typical spectra are shown in Fig. 11.

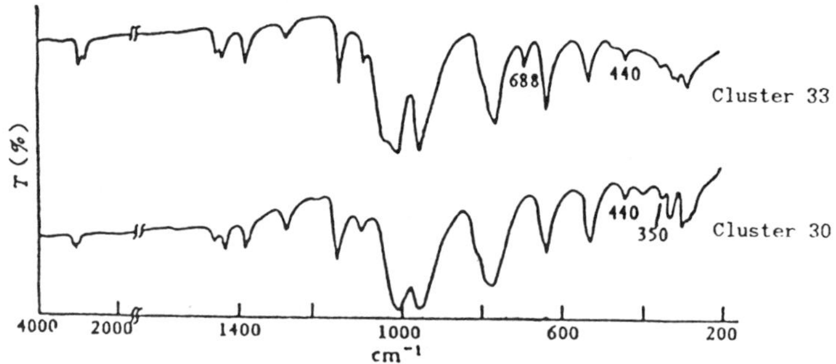

Fig. 10 Comparison of the IR absorption bands
for Mo$_3$OS$_6$(dtp)$_3$I and Mo$_3$S$_7$(dtp)$_3$Cl

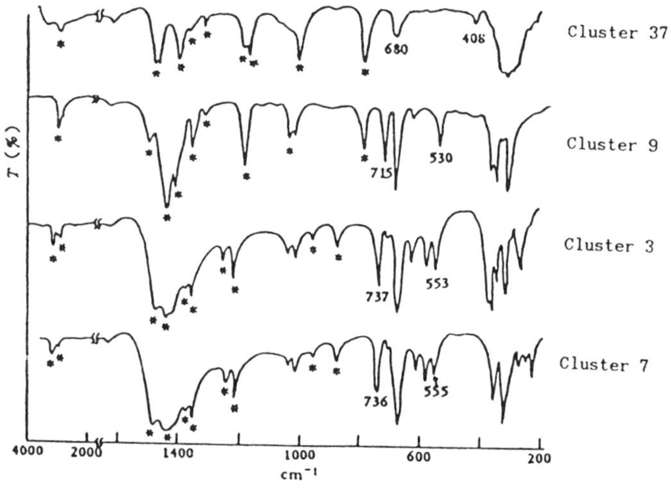

* absorption bands from the organic ligands
Fig. 11 IR absorption of the oxo-capped clusters

Table 8   IR absorption of $Mo_3-(\mu_3-O)$

Cluster	37	34	33	35	9	3	1
ν (cm^{-1})	680	675	688	680	715	737	775
	408		440		530	553	580
$Mo-(\mu_3-O)$	2.079	2.046	2.036	2.028	1.996	1.992	1.982
Cluster	2	6	8	4	5	7	
ν (cm^{-1})	724	688	730	732	732	736	
	571	426	579	553	575	555	
$Mo-(\mu_3-O)$	1.978	1.976	1.976	1.976	1.973	1.970	

It is worthwhile to point out that the frequency assignment of these IR absorption bands will be helpful to the prediction of the structual types for clusters as possible reaction products in our synthetic work and to the understanding of such chemical reactions as well.

4.   SYNTHESIS AND REACTIVITY

4.1.   SYNTHESIS

Our synthetic work was carried out under atmospheric conditions in an EtOH–HCl medium using $MoCl_3 \cdot 3H_2O$ as the starting material in most cases, although in some cases $MoBr_3 \cdot xH_2O$ and K_2MoOCl_5 were used. $MoCl_3 \cdot 3H_2O$ was prepared by dissolving MoO_3 in concentrated hydrochloric acid and then reducing the $Mo^{(VI)}$ in a diaphragm electrolytic cell with mercury cathode and graphite anode. The electrolytic process was carried out until the solution turned dark red. Then the solution was evaporated under reduced pressure at a temperature of 95–100°C. The product was dark brown instead of copper red, as described in the literature (Wardlaw and Wormell, 1924), in which the evaporation temperature was only 70–80°C. The latter product needed an appropriate oxidation in order to obtain a better yield of trinuclear clusters, as shown in our experiments. However, we are accustomed to using $MoCl_3 \cdot 3H_2O$ to denote the partially oxidized dark brown molybdenum trichloride. The synthetic procedures will be schematically presented in the sequence of the structure types.

4.1.1. SYNTHESIS OF THE M_1–TYPE CLUSTERS

Clusters containing carboxylic groups

$MoBr_3 \cdot 3H_2O$ ┬→ + HCOOH → + Me_4NBr, EtOH–HCl

└──→ $(Me_4N)[Mo_3OBr_3(O_2CH)_3Cl_3]$

└→ + HOAc ─→ + AcAcH, EtOH–HCl, H_2S

└──→ $(C_5H_7S_2)[Mo_3OBr_3(OAc)_2Cl_3]$

$MoCl_3 \cdot 3H_2O + Ac_2O, Et_4NCl, EtOH-HCl \longrightarrow (Et_4N)[Mo_3OCl_3(OAc)_2Cl_5]$
$\quad \llcorner + RCOOH \ \vdash + R'_4NI, EtOH-HCl \longrightarrow (R'_4N)[Mo_3OCl_3(O_2CR)_3Cl_3]$
$\quad\quad\quad \vdash + (NH_2)_2CS, EtOH \longrightarrow [(NH_2)_2CSH][Mo_3OCl_3(O_2CH)_3Cl_3]$
$\quad\quad\quad \vdash + AcAcH, EtOH-HCl, H_2S \longrightarrow (C_5H_7S_2)[Mo_3OCl_3(OAc)_3Cl_3]$
$\quad\quad\quad \llcorner + Et_4NBr, EtOH-HBr \longrightarrow (Et_4N)[Mo_3OCl_3(OAc)_3(Br,Cl)_{1.5}]$

It is interesting to note that in the synthesis of clusters containing carboxylic group, the doubly bridging halogen atoms always come from the molybdenum halides, although there may be other source of halogen atoms in the reaction mixture. Thus in the reaction system $MoBr_3 + HOAc + EtOH-HCl$, the product is $[Mo_3(\mu_3-O)(\mu-Br)_3(OAc)_3Cl_3]^-$, while in the system $MoCl_3 + HOAc + EtOH + HBr$, $[Mo_3(\mu_3-O)(\mu-Cl)_3(OAc)_3(Cl_3, Br_3)_{0.5}]^-$ is obtained instead.

<u>Clusters containing dtp groups</u>

$MoCl_3 \cdot 3H_2O \ \vdash \rightarrow + P_2S_5, EtOH/EtOH-HCl(1:4)$
$\quad\quad\quad\quad\quad \llcorner \rightarrow Mo_3(\mu_3-S)(\mu-S)(\mu-Cl)_2(dtp)_4Cl$
$\quad \vdash \rightarrow + SnCl_2, P_2S_5, EtOH/EtOH-HCl(1:1)$
$\quad\quad\quad\quad\quad \llcorner \rightarrow Mo_3(\mu_3-S)(\mu-O)(\mu-Cl)_2(dtp)_4(SH)$
$\quad \vdash \rightarrow + H_2S, P_2S_5, EtOH/EtOH-HCl(2:1) \rightarrow Mo_3S_4(dtp)_4(H_2O)$
$\quad \vdash \rightarrow + H_2S, P_2S_5, EtOH \longrightarrow Mo_3(\mu_3-O)(\mu-S)_3(dtp)_4(H_2O)$
$\quad\quad\quad\quad\quad\quad\quad\quad\quad\quad (\ Cluster \ 41 \ )$
$\quad \llcorner \rightarrow + PhCH_2SH, P_2S_5, EtOH/EtOH-HCl$
$\quad\quad\quad\quad \llcorner \rightarrow Mo_3(\mu_3-S)(\mu-S)_3(dtp)_4(PhCH_2SH)$

Cluster 41 has not been successfully characterized yet by X-ray diffraction owing to diffraction decay upon prolonged X-ray exposure. However, its IR spectrum is essentially similar to that of Cluster 12 and shows the characteristic of $Mo_3-(\mu_3-O)$ absorption ($680 cm^{-1}$). Moreover, upon replacement of H_2O by oxazole, the reaction product, has been fully confirmed by X-ray diffraction to te Cluster 23 with the above cluster structure.

4.1.2. SYNTHESIS OF THE M_2-TYPE CLUSTERS

$K_2MoOCl_5 \ \vdash \longrightarrow + P_2S_5, EtOH, Et4NI \longrightarrow Mo_3(\mu_3-O)(\mu-S_2)_3(dtp)_3I$
$\quad\quad \vdash \longrightarrow + P_2S_5, EtOH-HCl \longrightarrow Mo_3(\mu_3-S)(\mu-S_2)_3(dtp)_3Cl$
$\quad\quad \llcorner + AcAcH, EtOH-HCl, H_2S \rightarrow (C_5H_7S_2)_3[Mo_3(\mu_3-S)(\mu-S_2)_3Cl_7]$
$(PyH)_2MoOCl_5 + H_2S, EtOH/EtOH-HCl(1:1)$
$\quad\quad\quad\quad\quad \llcorner \longrightarrow (PyH)_3[Mo_3(\mu_3-S)(\mu-S_2)_3Cl_7]$

The M_2-type clusters are generally prepared from $Mo^{(V)}$. The reduction of $Mo^{(V)}$ to $Mo^{(IV)}$ by S^{2-} is accompanied inevitably by the oxidative dimerization $2S^{2-} \rightarrow (S_2)^{2-}$. An alternative synthetic route is from $MoCl_3 \cdot 3H_2O$. For example, in the preparation of Cluster 12, if the reaction system should be exposed too much to air the product would often turn out to be Cluster 30 instead.

4.1.3. SYNTHESIS OF THE ($M_1 + M_2$)-TYPE CLUSTER

$$MoCl_3 \cdot 3H_2O + P_2S_5, EtOH, H_2S \longrightarrow + (NH_4)SCN \longrightarrow$$
$$[Mo_3S_4(dtp)_4(NCS)][Mo_3S_7(dtp)_3]$$

It is evident that atmospheric oxygen takes part in the reaction.

4.1.4. SYNTHESIS OF THE B_2-TYPE CLUSTERS

When partially oxidized $MoCl_3 \cdot 3H_2O$ is dissolved in EtOH-HCl, and a tetraalkylammonium salt is then added to this reaction mixture, the Cluster 37 can be successfully seperated as black crystalline product, in which the Mo is present in the oxidation state $+3^1/_3$. It is perhaps reasonable to presume that the partially oxidized $MoCl_3 \cdot 3H_2O$ exists in EtOH-HCl in the form of $[Mo_3(\mu_3-O)(\mu_3-Cl)(\mu-Cl)_3Cl_6]^{2-}$, which can then be converted into Cluster 36 and Cluster 38 under appropriate conditions.

4.2. A PROPOSED FORMATION MECHANISM FOR THE B_2-TYPE CLUSTERS

In the synthetic work it is found that the formation of dark black crystals of the B_2-type clusters is usually accompanied by a few red crystals of $[Mo_2Cl_9]^{3-}$. In addition, the formation of the trinuclear cluster skeleton is also accompanied by a partial oxidation of the tervalent Mo atoms so that the Mo atoms attain an average oxidation number of $+3^1/_3$. These observations stimulate our speculation that the first step of building the cluster skeleton is the formation of a complex by the approach of a $MoCl_2^{2-}$ fragment towards a binuclear Mo anion $[Mo_2Cl_9]^{3-}$ as shown in Fig. 12. Then the triply bridging Cl atoms are substituted in some yet unspecified manner by either S or O atoms to produce the cluster anions with the corresponding capping atoms.

In order to explore the likelihood of this hypothetical mechanism, an EHMO calculation was carried out to see what would happen when these two fragments were brought together along the reaction path proposed (Huang *et al*, 1985). The results show that the whole system is stabilized to a maximum of ca. 4 e.v. as the two fragments are brought together to a distance where R is equal to 2.3-2.4 Å. The corresponding Mo—Mo distances are comparable to the bond lengths obtained.

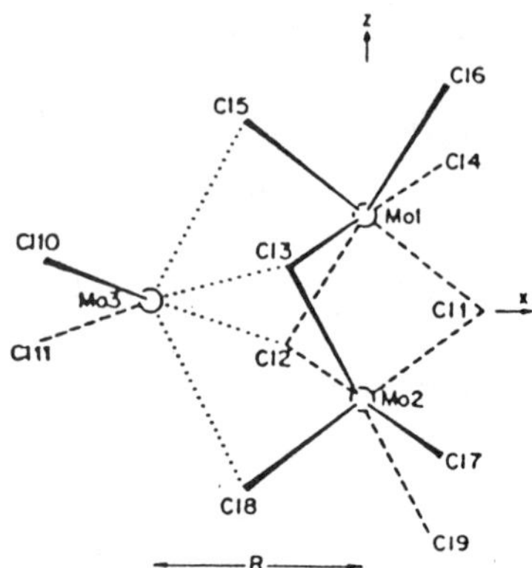

Fig. 12 Formation of the trinuclear cluster skeleton

Fig. 13 shows an interaction diagram for $[Mo_3Cl_{11}]^-$ formed from binuclear and mononuclear fragments to see where the stabilization comes from. It is evident that the strongest interaction between the fragments are through the $1a_1$ and $2a_1$ of the ML_2 fragment and the a_1 component of the M_2L_9 π bond, and through the $1b_1$ orbital of the ML_2 fragment and the corresponding b_1 component of the M_2L_9 π^* bond.

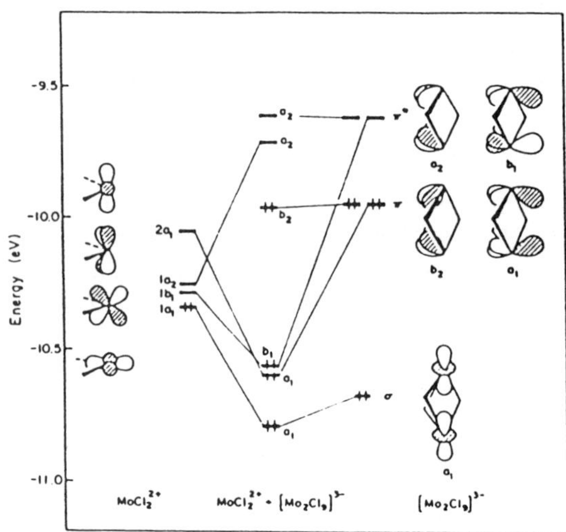

Fig. 13 Orbital interaction diagram for $[Mo_3Cl_{11}]^-$

4.3. REACTIVITY OF TRINUCLEAR Mo CLUSTERS WITH 'LOOSE COORDINATION SITE'

Fig. 3 shows the configuration of the trinuclear clusters with a 'loose coordination site' L.   The cluster skeleton belongs to the M_1-type structure with all the ligands being either bridging atoms or chelating groups except L. The ligand located in this coordination site is coordinated to the Mo atom with a rather weak bond, as compared with the corresponding dative bond, as shown in Table 2. As a result, the L ligand can be very easily substituted by another ligand.

4.3.1. SUBSTITUTION REACTION OF TERMINAL LIGANDS

$$Mo_3S_4(dtp)_4(H_2O) + L \longrightarrow Mo_3S_4(dtp)_4L + H_2O$$

(L = PPh_3, $SC(NH_2)NHCH_2CH{=}CH_2$, $PHCH_2SH$, Py, C_3H_3NO, $PhCH_2CN$.)

$$Mo_3S_4(dtp)_4(H_2O) + C_3H_4N_2 \longrightarrow [Mo_3S_4(dtp)_3(C_3H_4N_2)_3](dtp)$$

$$\quad\quad\quad + NaS_2CNEt_2, Py \longrightarrow [Mo_3S_4(dtc)_4(Py)] \cdot (Py)_2 \cdot (H_2O)$$

$$\quad\quad\quad + NaS_2CNEt_2, dmf \longrightarrow Mo_3S_4(dtc)_4(dmf)$$

4.3.2. SUBSTITUTION REACTION OF CAPPING ATOMS

$$Mo_3S_4(dtp)_4(H_2O) + SC(NH_2)_2 \longrightarrow Mo_3OS_3(dtp)_4[SC(NH_2)_2]$$

$$Mo_3OS_3(dtp)_4(H_2O) + H_2S, HCl\text{-}EtOH \longrightarrow Mo_3S_4(dtp)_4(H_2O)$$

4.3.3. ADDITION REACTION TO THE THREE BRIDGING S ATOMS

The three μ–S atoms in the cluster have a surplus of ability to bond with some other atoms so that the following reactions can be initiated.

$$Mo_3S_4(dtp)_4(H_2O) + H_2S, I_2 \longrightarrow Mo_3S(\mu\text{-}S_2)_3(dtp)_3I$$

$$\quad + H_2S, H_2O_2, EtOH\text{-}HCl \longrightarrow Mo_3S(\mu\text{-}S_2)_3(dtp)_3Cl$$

$$Mo_3S_4(dtp)_4L + SbCl_3, EtOH\text{-}MeCOMe \longrightarrow Mo_3S_4(dtp)_4(SbCl_3)L$$

(L = EtOH, C_3H_3NO.)

Metal atoms in the addition product $Mo_3S_4(dtp)_4(SbCl_3)L$ have a cubane–like arrangement, but no M–M bonding interaction is found between the Mo and Sb atoms. Nevertheless, it has been found recently by X. T. Wu *et al.* that the reaction with CuI leads to the formation of Mo–Cu bond (Mo–Cu: 2.807, 2.856, 2.885 Å):

$$Mo_3S_4(dtp)_4L + CuI, EtOAc\text{-}dmf \longrightarrow Mo_3S_4(CuI)(\mu\text{-}OAc)(dtp)_3(dmf).$$

4.3.4. OXIDATIVE REACTION

All the reactions mentioned above are characterized by keeping the cluster core basically unchanged. However, the following oxidative reaction may lead to the rupture of one Mo–Mo bond and the opening of one side of the Mo_3 triangle.

$$Mo_3S_4(dtp)_4(H_2O) + O_2 \longrightarrow Mo_3S_3O_2[SOP(OEt)_2](dtp)_3.$$

The reaction is accompanied by conversion of $[S_2P(OEt)_2]^-$ into $[SOP(OEt)_2]^-$ and the c.e. count is reduced from 6 to 4. In other words, two electrons have been removed from the bonding MO so that only two Mo–Mo bonds can be retained.

REFERENCES

Bino, A., Cotton, F. and Diri, Z. (1978). *J. Amer. Chem. Soc.* 100, 5252.

Bino, A., Cotton, F., Dori, Z. and Kolthammer, B. (1981). *J. Amer. Chem. Soc.* 103, 5779.

Huang, J., Lu, J., Jiang, Y. and Hoffmann, R. (1985). *Scientia Sinica* B10, 1035.

Jiang, Y., Tang, A., Hoffmann, R., Huang, J. and Lu, J. (1985), *Organomet.* 4, 27.

Kirtley, S., Chanton, J., Love, R., Tipton, D., Sorrell, T. and Bau, R. (1980), *J. Amer. Chem. Soc.* 102, 3451.

Müller, A., Jostes, R. and Cotton, F. (1980), *Angew. Chem. Int. Ed. Engl.* 19, 875.

Müller, A. and Reinsch, U. (1980), *Angew. Chem. Int. Ed. Engl.* 19, 72.

Torardi, C. and McCarley, R. (1985), *Inorg. Chem.* 24, 476.

Wardlaw, W. and Wormell, R. (1924), *J. Chem. Soc.* 195, 2370.

Zheng, Y., Shang, M., Zhang, L., Huang, J. and Lu, J. (1983), *Spect. Anal. (Chinese)* 2, 72.

46. Structural investigations of organometallic catalysts, catalyst precursors and important ligands

C. Krüger, K. Angermund and R. Goddard

ABSTRACT

Structural investigations of several catalytically active organometallic compounds were undertaken to study possible correlations between the molecular structure, the spatial electron distribution about the transition metal centers, NMR-shifts and the chemical reactivity of these compounds. The experimentally determined (X-X - method) electron density distributions (EDD) for several related isoelectronic compounds with various ligand arrangements are discussed.

Electronic reaction control of homogeneous catalysis may be supplemented by steric properties of ligands. It is found that small differences within the ligand framework of phosphine-modified nickel catalysts used for stereoselective dimerization reactions of olefins cause remarkable changes in product distributions as well as in reactivity of the catalysts. We used molecular modelling techniques to gain insight into possible sterical reasons for product differentiation of catalysts with optical active phosphines as ligands. Results of these investigations are described.

1. INTRODUCTION

Recent technical developments in diffraction methods as well as in NMR - spectroscopy showed impact onto the structural chemistry of organometallic compounds. Amongst these the results of high resolution X-ray (and neutron-) diffractometry give insight into the redistribution of electrons upon chemical bond formation within molecules (Angermund, Claus, Goddard and Krüger, 1985; Coppens, 1977; Coppens et al., 1982; 1984). The first part of this contribution is concerned with a critical description of the experimental methods, based on published material, and newly developed procedures for handling sensitive crystals and for improved methods of data collection at low temperatures. It is shown that frequent criticism of results obtained by X-X or X-N methods is based upon interpretation of erroneous measurments, over-interpreted experiments or neglect of basic methodical facts, e.g. the definition of the promolecule.

2. EDD DETERMINATIONS OF SELECTED ORGANOMETALLIC COMPOUNDS

Additional experimental difficulties arise during the course of an EDD determination of compounds containing heavier elements, typically transition metals. Techniques to overcome these difficulties are discussed using cyclopentadienyl (Cp) transition metal complexes as examples. In some of these compounds the metals are present in unusual electronic configurations. The molecular structure of the first example (Krüger et al., 1984), a 14e-metallacyclic chromium compound (Jonas, 1985), is shown in Fig. 1. The experimental EDD distribution in a plane passing through the metallacycle is shown in Fig. 2.

Furthergoing investigations include an estimation of errors, a direct integration of electron density about the chromium atom (charge + 0.3e), and refinement of valence shell populations. These results

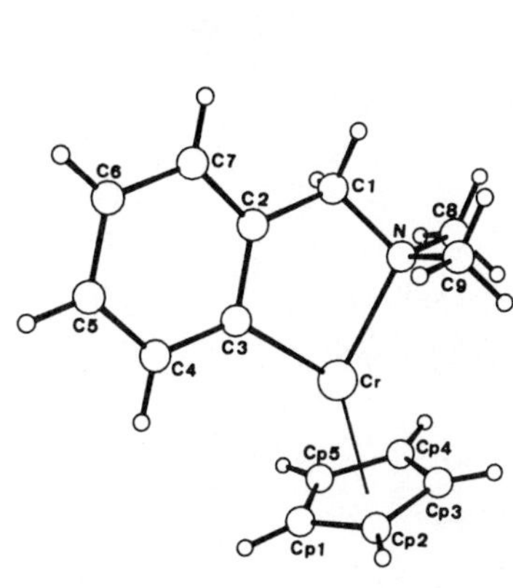

FIG.1.
Molecular structure of
$CpCrC_9H_{12}N$

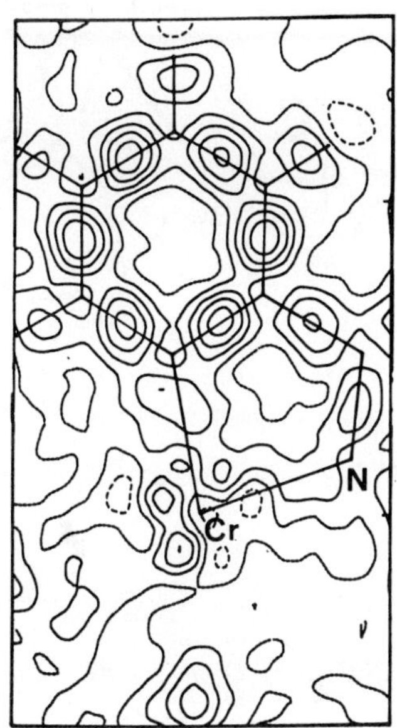

FIG.2.
EDD in the plane of
the metallacycle

are in accord with theoretical calculations on an <u>ab initio</u> level. EDD determinations of the 17-electron compound Cp-Fe-COD (COD = C_8H_{12}), (Fig. 3) and of related Cp-Co-COD compounds (Fig. 4) show an octahedral electron distribution about the transition metal.

Fig. 5 gives a schematic view of the spatial orientation of the EDD at the cobalt atom. Positive as well as negative regions of the deformation density distributions around the metal can be correlated with the ^{13}C-NMR-shifts of the olefinic carbon atoms and the [59]Co-NMR-shifts of the metal, with the chemical reactivity and in particular with the selectivity of these compounds in catalytic reactions (Angermund, 1986; Bönnemann, 1985). The program FRODO (Jones <u>et al.</u>, 1985) operating on an Evans & Sutherland Computer Graphics System PS300 in conjunction with a dedicated computer MicroVAX-II is used to predict the spatial arrangement of the EDD about the metal and thus possible reactive sites in a series of molecules of this type. It is to be noticed that isoelectronic and homologeous compounds show

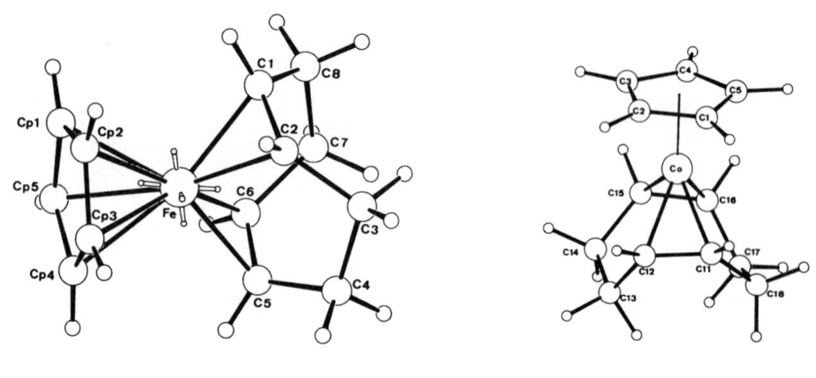

FIG.3.
Cp-Fe-COD

FIG.4.
Cp-Co-COD

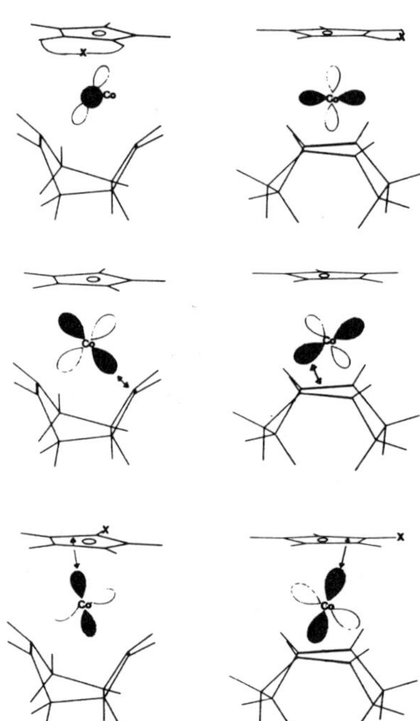

FIG.5.
Schematic view of the spatial orientation of the EDD at the cobalt atom (positive EDD is colored black).

similar EDD distributions about the transition metal.

3. MOLECULAR MODELLING OF ORGANOMETALLIC COMPOUNDS

The stereochemical course of a dimerization reaction of olefins at ligand modified Nickel(0) catalysts (Bogdanovič et al., 1970; Bogdanovič, 1979) based on Bis(η^3-allyl)nickel as one possible starting material is studied using X-ray data of model compounds (Fig. 6 and Fig. 7) and of several potential ligands.

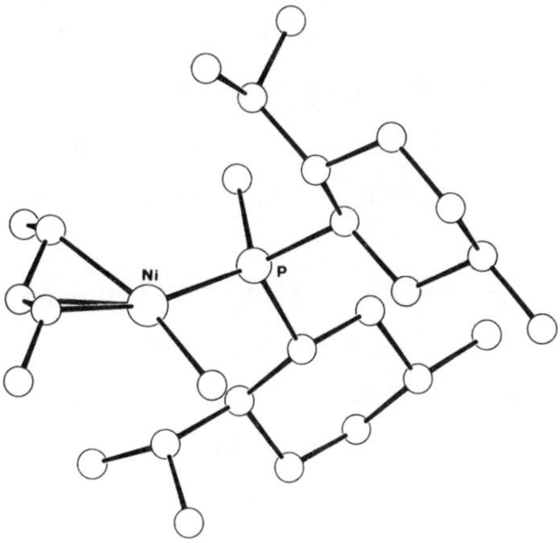

FIG.6.

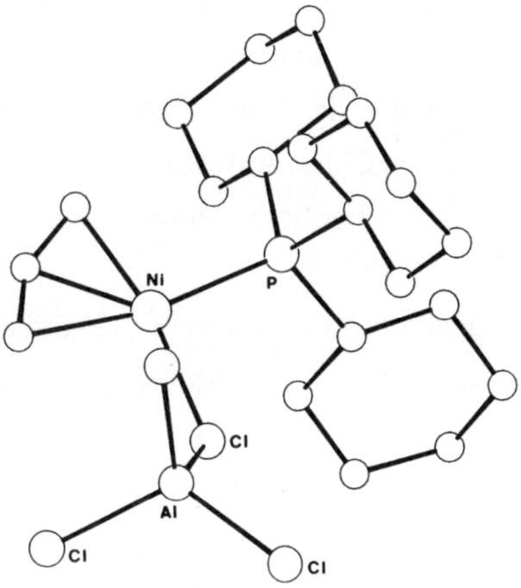

FIG.7.

In addition Molecular Modelling Techniques using the PS300 graphics system in conjunction with the program MOGLI are served to gain insight into the possible reaction mechanism, the product distribution, and the optical induction at the catalyst. In a series of graphical 'snapshots' the stereochemical course of a simulated catalytic reaction at a metal center is shown. These novel techniques may be used for 'tailoring of catalysts' (Wilke, 1965).

REFERENCES

Angermund. K. (1986). Dissertation Max-Planck-Institut für Kohlenforschung. Mülheim a. d. Ruhr. FRG.

Angermund. K.. Claus. K. H.. Goddard. R. and Krüger. C. (1985). Angew. Chem. Int. Ed. Eng. 24, 237.

Bönnemann. H. (1985). Angew. Chem. Int. Ed. Engl. 24. 248.

Bogdanovič. B. (1979). Adv. Organomet. Chem. 17. 105.

Bogdanovič. B., Henc, B.. Karmann. H. G., Nüssel. H. G.. Walter, D., and Wilke ,G. (1970). Ind. Eng. Chem. 62. 34.

Coppens. P. (1977). Angew. Chem. Int. Ed. Engl. 16. 32.

Coppens. P. and Hall. M. M. (1982). Electron Distribution and the Chemical Bond. Plenum Press. New York. and lit. cited therein.

Coppens. P.. Dam. J.. Harkema. S.. Feil. D.. Feld. R.. Lehmann. M. S., Goddard, R., Krüger. C., Hellner, E., Johansen. H.. Larsen. F. K.. Koetzle, T. F.. McMullan. R. K.. Maslen. E. N., Stevens. E. D. (1984). Acta Crystallogr. A40, 184.

Jonas, K. (1985). Angew. Chem. Int. Ed. Engl. 24, 295.

Jones, A., Bush, B. L.. Pflugrath. J. W.. Sack, J. S. and Saper, M. A. (1985). Program FRODO version 6.1.

Krüger. C., Goddard, R., Allibon, J. (1984). Acta Crystallogr. A40, (Suppl.) C167.

Wilke, G. (1965). Proc. R. A. Welch Found. Conf. Chem. Res. 9, 179.

47. Building chains and clusters for mixed metal carbynes by designed synthesis

Judith A. K. Howard, G. P. Elliot, C. M. Nunn and F. G. A. Stone

Major advances in organoplatinum and mixed metal cluster chemistry have been made since the discovery of $[Pt(C_2H_4)_3]$ (Green et al, 1975; Howard et al, 1983) and complexes of the type $[Pt(PR_3)(C_2H_4)_2]$, (Harrison et al, 1978), following an improved synthetic route to $[Pt(cod)_2]$ (Spencer 1979, Howard 1982) [cod = cyclo-octa-1,5-diene].   These reagents are extremely reactive, the former functioning as a source of 'ligand free' platinum.

Serendipitous discovery that $[Pt(cod)_2]$ reacted with hexa-fluoropropene to give a dimetallic carbene-bridged species (Green et al, 1977) suggested that these new Pt(0) complexes might react more generally with mononuclear metal carbene and carbyne compounds to afford species containing heterometal-metal bonds and bridging alkylidene or alkylidyne ligands. Many successful syntheses of this form were undertaken and new species isolated and structurally characterised (Stone 1981(a), (b), and refs. therein).   We drew the analogy at this time, between the metal alkene and alkyne complexes [I(a) and I(b)] and the dimetallic species [II(a) and II(b)] and attempts were made to investigate analogous chemistry.   The elegant isolobal model (Hoffmann, 1982) gave a supporting and illuminating rationalisation of these early observations and new synthetic routes were immediately explored.

$$I(a) \qquad I(b) \qquad\qquad II(a) \qquad II(b)$$

Preparation and characterisation of $[Pt(C_2R_2)_2]$ (Boag et al, 1980(a)), together with a new appreciation for the isolobal nature of $W(CO)_2Cp$ and CR and prompted the synthesis of the Group VIII metal-ditungsten compounds (III).

$$(III)$$

(III) M = Ni, Pd, Pt.   R = C_6H_4Me-4, Me, Ph.

An interesting feature of the $Pt(alkyne)_2$ compounds is their ability to add further $Pt(O)$ groups, to form di- and tri-platinum species with bridging alkyne ligands (Boag et al, 1980(b), 1981). These results suggested that compounds II and III might coordinate suitable ML_n fragments _via_ the C=W bonds. Depending on the bonding requirements of the added 'ML_n' group, the products would be expected to show either open, chain-like structures or alternatively closed structures in which the carbon atom of the alkylidyne ligand caps a metal triangle. Thus we can envisage for example, adding a Pt(cod) fragment or other ML_2 carbene-like groups to one or both ends of III in order to increase the chain length systematically. The synthetic route to V(a) or V(b), each a chain of 5 metal atoms, dictates the product, but not the isomer. The existence of

$W \equiv W(CO)_2 Cp$; $C \equiv CR$; $R = C_6H_4Me-4$, Me, Ph

isomers can be detected by n.m.r. spectroscopy and the formation of such can be understood in terms of the direction of
addition of Pt(cod) or $W(Cp)(CO)_2$ to the precursors.   Both
V(a) and V(b) have been characterised fully by single crystal
X-ray diffraction studies (Elliott et al, 1986(a)).   V(a) is
shown to be the symmetrical isomer with two 'butterfly' μ_3-CPt$_2$W
moieties sharing a wing tip vertex on a crystallographic two-
fold axis.   [Pt-W 2.748(1), 2.751(1) Å, Pt....Pt 3.089(1) Å,
W-Pt-W 176°]. V(b) in contrast, contains one triply bridging and

two edge-brdging carbyne ligands in an essentially planar metal
framework. W-Pt distances range from 2.713(2) $\overset{\circ}{A}$ (μ_2-C) to
2.773(2) $\overset{\circ}{A}$ (μ_3-C) and a Pt....Pt separation of 2.949(2) $\overset{\circ}{A}$ again
implies little or no direct bonding.

Systematic continuation of these synthetic routes has led
us to bind together 6, 7, and 8 metal atoms (Elliott et al,
1986(b),(c), Davies et al, 1986).   The fully characterised
products are shown in Figure 1.   Several isomers of each
species have been identified in solution, but only those shown
in Figure 1 were successfully investigated by diffraction
methods.   The ring closure achieved by the 8 metal "chain" can
be understood again in terms of the orientation of approach by
the additional fragment to a given isomer of the precursor and
we can now extend an open chain of 8, to 9 and 10 metal atoms.

Also, since the discovery and extended chemical study of
these molecules, their syntheses have been achieved by routes
other than simple one-fragment additions, viz. 8 can be grown
from 7 + 1 or 4 + 4, such that dimerization of the 4 metal atom
species also produces the 'star' cluster of 8, and addition of
an excess of $Ni(cod)_2$ to certain PtW_2 compounds affords clusters
containing a $\overline{NiWPtWNiWPtW}$ ring (Elliott et al, 1986(b)).

Our studies have been extended to the syntheses of
compounds containing molybdenum in conjunction with W, Pt and Ni,
e.g. $[Pt_3Mo_3W_2(\mu\text{-CMe})_2(\mu_3\text{-CC}_6H_4Me\text{-}4)_2(CO)_8(\eta\text{-C}_5H_5)_4]$ and $[Pt_3Mo_2\text{-}$
$(\mu_3\text{-CC}_6H_4Me\text{-}4)_2(CO)_4(cod)_2(\eta\text{-C}_5H_5)_2]$ (Davies et al, 1986), and
we envisage that with fine tuning of substituent groups and
metal centres, clusters with unique reactivities will be discovered.   There is also the interesting possibility of 'trapping'
small molecules in the cavities of the star clusters, where the
transannular separations are between 3.6 and 4.1 $\overset{\circ}{A}$.   Single
crystal X-ray diffraction studies have established the conformations of all diastereoisomers discussed, thereby placing these

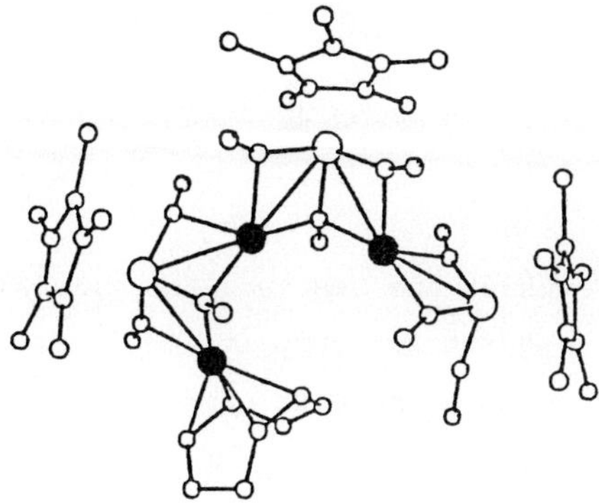

(VI) $Pt_3W_3(\mu\text{-}CMe)(\mu\text{-}CMe)_3(CO)_6(cod)(\eta\text{-}C_5Me_5)_3$

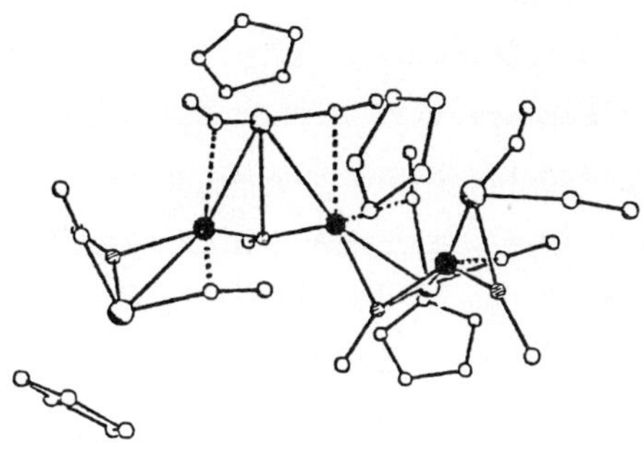

(VII) $[Pt_3W_4(\mu_2\text{-}CR)_2(\mu_3\text{-}CR)_2(CO)_8(\eta\text{-}C_5H_5)_4]$

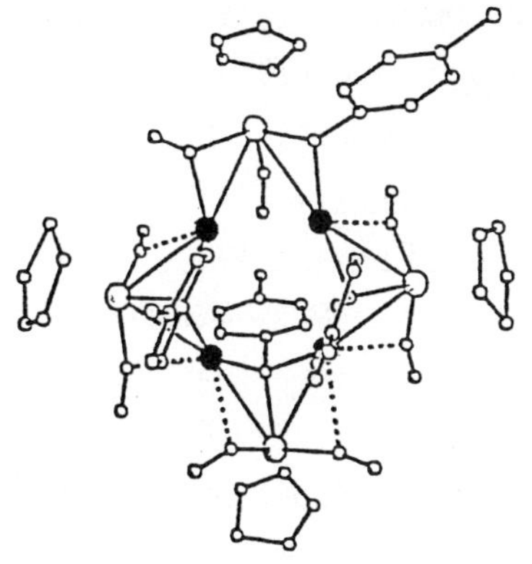

(VIII) $[Pt_4W_4(\mu_2\text{-}CR)(\mu_3\text{-}CR)_3(CO)_8(\eta\text{-}C_5H_5)_4]$

Figure 1.

stepwise cluster building reactions on a firm structural foundation.

Acknowledgements. We thank the SERC and the U.S. Air Force Office of Scientific Research for support, together with our many co-workers referenced in the text papers.

REFERENCES

BOAG, N.M., GREEN, M., GROVE, D.M., HOWARD, J.A.K., SPENCER, J.L., and STONE, F.G.A. (1980(a)). J.Chem.Soc., Dalton Trans., 2170; (1980(b)). ibid, 2181; (1981). ibid, 1051.

DAVIES, S.J., ELLIOTT, G.P., HOWARD, J.A.K., NUNN, C.M., and STONE, F.G.A. (1986). Submitted to J.Chem.Soc., Dalton Trans.

ELLIOTT, G.P., HOWARD, J.A.K., MISE, T., MOORE, I., NUNN, C.M., and STONE, F.G.A. (1986)(a)). J.Chem.Soc., Dalton Trans., 2091.

ELLIOTT, G.P., HOWARD, J.A.K., MISE, T., NUNN, C.M., and STONE, F.G.A. (1986(b)). Angew.Chem., Int.Ed.Engl., 25, 191-192.

ELLIOTT, G.P., HOWARD, J.A.K., NUNN, C.M., and STONE, F.G.A. (1986(c)). J.Chem.Soc., Chem.Commun., 431-433.

GREEN, M., HOWARD, J.A.K., SPENCER, J.L., and STONE, F.G.A. (1975). J.Chem.Soc., Chem.Commun., 449-451.

GREEN, M., LAGUNA, A., SPENCER, J.L., and STONE, F.G.A. (1977). J.Chem.Soc., Dalton Trans., 1010-1016.

HARRISON, N.C., MURRAY, M., SPENCER, J.L., and STONE, F.G.A. (1978). J.Chem.Soc., Dalton Trans., 1337-1342.

HOWARD, J.A.K., (1981). Acta Crystallogr., B38, 2896-2898.

HOWARD, J.A.K., SPENCER, J.L., AND MASON, S.A. (1983). Proc.R. Soc.Lond.A. 386, 145-161.

SPENCER, J.L. (1979). Inorg.Syntheses (XIX), 213-218.

STONE, F.G.A. (1981(a)). Acc. of Chem.Research. 14, 318-325.

STONE, F.G.A. (1981(b)). Inorg.Chimica Acta, 50, 33-42.

48. Diphosphinomethane-bridged diplatinum complexes: steric effects, molecular structure and reactivity

Ljubica Manojlović-Muir

1. INTRODUCTION

The synthesis and characterization of the platinum(I) complex $[Pt_2Cl_2(\mu\text{-}Ph_2PCH_2PPh_2)_2]$ (Glockling and Pollock, 1974; Brown et al, 1977) has been followed by a rapid expansion of our knowledge of dinuclear transition metal complexes stabilized by diphosphinomethane ligands, $R_2PCH_2PR_2$. Important factors in this development are the search for new catalysts and the realization that dinuclear complexes can offer useful models for homogeneous catalytic processes and coordination chemistry at metal surfaces. It is hoped that new catalytic reactions can be induced by cooperative interactions of the metal centres.

$R_2PCH_2PR_2$-bridged complexes are now known for many transition metals but they are particularly numerous for Rh, Pd and Pt (Puddephatt, 1983). In most cases the bridging ligand is $Ph_2PCH_2PPh_2$, but other ligands, notably $Me_2PCH_2PMe_2$ and $(EtO)_2PCH_2P(OEt)_2$, are increasingly used.

2. DIPLATINUM COMPLEXES

Our own work has been concerned mostly with $R_2PCH_2PR_2$-diplatinum complexes, which have proved particularly rich in novel types of molecular structure and reactivity. So far we have characterized 12 different structural types, which show great

diversity of structural properties, such as different conformations of the $R_2PCH_2PR_2$ ligands and of various dimetallahetero-
cycles.   The metal atoms display a variety of oxidation states
(0, 1, 2, 4) and coordination numbers (3 - 6), and the Pt-Pt
distances range from 2.6 to 4.7 Å.   In most complexes the metal
atoms are not bonded to one another. However, σ-covalent Pt-Pt
bonds, 3-centre-2-electron Pt-Pt interactions and donor-acceptor Pt→Pt single bonds are all represented. Space precludes a
detailed discussion of these properties.

Accordingly, we shall concentrate further on complexes of
a single structural type only, (1), and show that their structural properties and chemical behaviour can be profoundly influenced by the substituents R of the bridging ligands.

3.   STRUCTURES OF cis,cis-[$Pt_2Me_4(\mu-R_2PCH_2PR_2)_2$] COMPLEXES
We have characterized crystallographically three tetramethyl-
diplatinum complexes which differ only in the nature of R: (1a)
R = OEt (Manojlović-Muir, unpublished work), (1b) R = Me, and
(1c) R = Ph (Manojlović-Muir _et al_, 1984a). An example of their
molecular structures is shown in Figure 1. In each dimer the

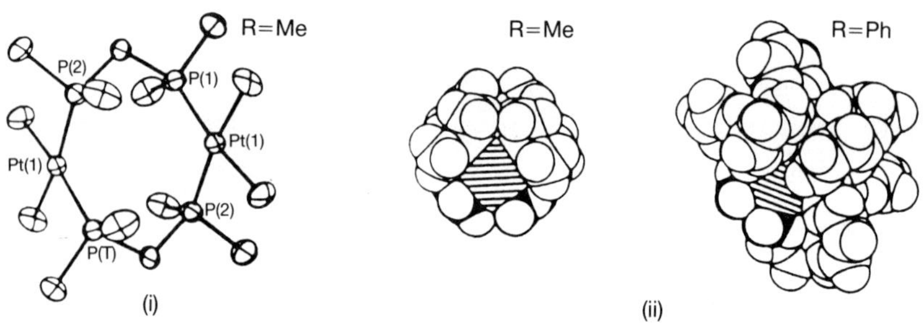

Fig. 1. (i) The molecular structure of (1b). (ii) Space-filling
models of (1b) and (1c); Pt atoms are cross-hatched and methyl
C atoms bonded to platinum are shown in black.

$R_2PCH_2PR_2$ ligands span two cis-$PtMe_2$ fragments to form a 8-membered $Pt_2P_4C_2$ ring, in which both Pt(II) centres lie in square-planar environments and the transannular Pt...Pt distance is too large to permit direct metal-metal bonding.

TABLE 1

Geometry of the $Pt_2P_4C_2$ rings (Å and °) in cis,cis-
[$Pt_2Me_4(\mu-R_2PCH_2PR_2)_2$] complexes

	R	Pt..Pt	Pt–P–C	P–C–P	P...P	Pt–P–C–P		Conformation
(1a)	OEt	3.46	115	114	3.06	-57	57	C_s boat-chair
(1b)	Me	4.24	115	115	3.12	-44	-44	C_{2h} twist-chair
(1c)	Ph	4.36	121	119	3.19	-22	73	C_1 twist-boat

The Pt...Pt separation varies substantially in the three complexes (Table 1), increasing along the series which reflects the steric requirements of R: OEt ⩽ Me < Ph. This variation can be ascribed partly to the tendency of the ring angles subtended at P and C atoms, and of the ligand bite, to increase with the steric bulk of R.   A much more important factor, however, is the change in conformations of the 8-membered rings, which arises from different conformations of the bridging ligands evident from the Pt-P-C-P torsion angles. The boat-chair

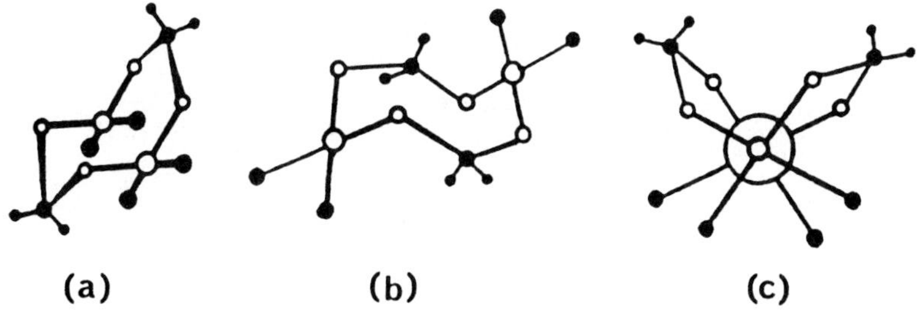

(a)   (b)   (c)

conformation (a) of the $Pt_2P_4C_2$ ring in the sterically least crowded (1a), R = OEt, is similar to that of the cyclooctane conformer of lowest energy. The twist-chair conformation (b) of (1b), R = Me, and the twist-boat conformation (c) of (1c), R = Ph, lead to longer Pt...Pt distances. They are also reminiscent of low energy forms of cyclooctane.

A related structural feature is illustrated in Figure 1: the metal centres in the R = Me complex are relatively more exposed to the environment, and thus could be more easily accessible to molecules of reagents, than those in the R = Ph complex. One might therefore conclude that (1b) should be more reactive than (1c), and that oxidation of the Pt(II) centres to the +4 state and the concurrent increase of the metal coordination number from 4 to 6 would occur more easily in (1b).

4.   REACTIVITY OF cis,cis-$[Pt_2Me_4(\mu-R_2PCH_2PR_2)_2]$ COMPLEXES

The chemical behaviour of (1b) and (1c) when treated with the oxidizing agents I_2 and MeI bears out this prediction.

I_2 reacts with (1c) to give the unusual cation $[Pt_2Me_3(\mu-Ph_2PCH_2PPh_2)_2]^+$ (2) (Brown et al, 1981). The reaction occurs with electrophilic cleavage of a methyl group and formation of a donor-acceptor Pt→Pt bond; both Pt atoms retain oxidation state +2, but one increases its coordination number from 4 to 5. The complex (1b) undergoes a different reaction with I_2, to yield a novel mixed-valence complex, $[Me_3Pt(\mu-I)(\mu-Me_2PCH_2-PMe_2)_2PtMe]^+$ (3) (Ling et al, 1983 ; Manojlović-Muir and Muir,

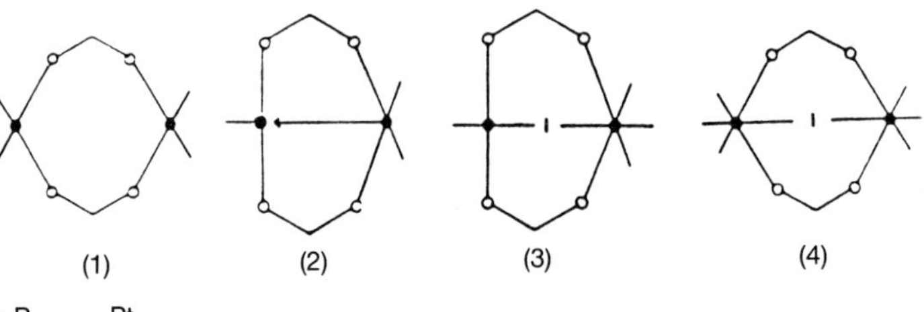

(1)   (2)   (3)   (4)

o P   • Pt

1984b). In this reaction one Pt(II) centre is oxidized to Pt(IV), expanding its coordination geometry to octahedral, and a methyl group is transferred from one metal centre to the other. Transannular oxidative addition is a novel process in the chemistry of platinum.

With MeI (1c) fails to react, while (1b) undergoes oxidative addition at both metal centres to form a highly strained Pt(IV)-Pt(IV) species, $[Me_3Pt(\mu-I)(\mu-Me_2PCH_2PMe_2)_2PtMe_3]^+$ (4), in which both metal centres are octahedrally coordinated (Ling <u>et al</u>, 1985).

In all reactions described here the $Pt_2(\mu-R_2PCH_2PR_2)_2$ nucleus is retained without fragmentation. This is achieved at the expense of conformational changes and substantial angular distortions of the diphosphinomethane ligands. Thus in (4), for example, the $Pt-P-CH_2$ and $P-CH_2-P$ angles (122 and 126°) are grossly distorted.

5.   SUMMARY

The work presented here shows that the steric requirements of the phosphine substituents, R, can have a major influence on both the molecular conformation and reactivity of $R_2PCH_2PR_2$-bridged dinuclear complexes. This influence appears even more striking than the steric effects in mononuclear phosphine complexes previously considered by Tolman (1977).

The role of $\mu-R_2PCH_2PR_2$ ligands in dinuclear species is to prevent fragmentation. Integrity of the $Pt_2(\mu-R_2PCH_2PR_2)_2$ nucleus is maintained in diverse reactions, even when stereochemical demands on both metal centres are drastically altered by oxidation state changes. We attribute this to the strength and kinetic stability of the Pt-P bonds and to the stereochemical flexibility of the $R_2PCH_2PR_2$ ligands - their ability to twist internally about P-C bonds and to permit large distortions of the bond angles subtended at P and C atoms.

ACKNOWLEDGEMENTS

It is a pleasure to thank Dr. M.P. Brown, University of
Liverpool, Dr. K.W. Muir, University of Glasgow, and
Professor R.J. Puddephatt, University of Western Ontario,
for fruitful collaboration.

REFERENCES

BROWN, M.P., PUDDEPHATT, R.J., RASHIDI, M., MANOJLOVIĆ-MUIR,
LJ., MUIR, K.W., SOLOMUN, T. and SEDDON, K.R. (1977).
Inorganica Chimica Acta, 23, L33.

BROWN, M.P., COOPER, S.J., FREW, A.A., MANOJLOVIĆ-MUIR, LJ.,
MUIR, K.W., PUDDEPHATT, R.J., SEDDON, K.R. and THOMSON, M.A.
(1981). Inorganic Chemistry, 20, 1500.

GLOCKLING, F. and POLLOCK, R.J.I. (1974). J. Chem. Soc., Dalton
Transactions, 2259.

LING, S.S.M., PUDDEPHATT, R.J., MANOJLOVIĆ-MUIR, LJ. and MUIR
K.W. (1983). Journal of Organometallic Chemistry, 255, C11.

LING, S.S.M., JOBE, I.R., MANOJLOVIĆ-MUIR, LJ., MUIR, K.W. and
PUDDEPHATT, R.J. (1985). Organometallics, 4, 1198.

MANOJLOVIĆ-MUIR, LJ., MUIR, K.W., FREW, A.A., LING, S.S.M.,
THOMSON, M.A. and PUDDEPHATT, R.J. (1984a). Organometallics,
3, 1637.

MANOJLOVIĆ-MUIR, LJ. and MUIR, K.W. (1984b). Croatica Chemica
Acta, 57, 587.

PUDDEPHATT, R.J. (1983). Chemical Society Reviews, 99.
TOLMAN, C.A. (1977). Chemical Reviews, 77, 313.

49. Phase transitions in λ-Co(sepulchrate)(NO$_3$)$_3$

Finn Krebs Larsen, Palle Jørgensen, Rita Grønbæk Hazell, Bente Lebech, Robert Thomas, Rodney J. Geue and Alan M. Sargeson

INTRODUCTION

The cobalt sepulchrate complex is a cage compound with a capped type sexidentate ligand involving nitrogen atom donors. Sargeson (1984) has reviewed the chemistry of these capped hexaazamacropolycycles, which could be interesting electrode and ion exchange materials. λ-Co(sepulchrate)(NO$_3$)$_3$ exhibits a strongly temperature dependent circular dichroisme (C.D.) (Dubicki et al., 1980), indicating a stronger dependence on the outer sphere configuration than anticipated. The cobalt sepulchrate nitrate structure at room temperature (FIG. 1) is

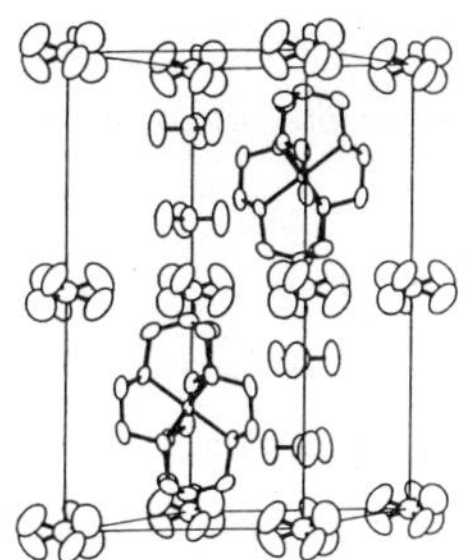

FIG. 1. Unit cell contents for λ-Co(sepulchrate)(NO$_3$)$_3$ at room temperature with 50% displacement ellipsoids at the atomic positions. Hydrogen atoms have been omitted.

hexagonal, space group $P6_3 22$ ($Z = 2$). The cations lie on D_3 sites and are surrounded by six nitrate ions which occupy C_3 sites and form hydrogen bonds with the amino groups. The next coordination sphere consists of six nitrate ions which are disordered at D_3 sites. The pronounced temperature dependence of the C.D. was explained (Dubicki et al., 1980) to be due to the six NO_3^- ions which are hydrogen bonded to the sepulchrate cage. Further examinations using optical crystallography and single-crystal Raman spectroscopy were reported by Dubicki et al. (1984).

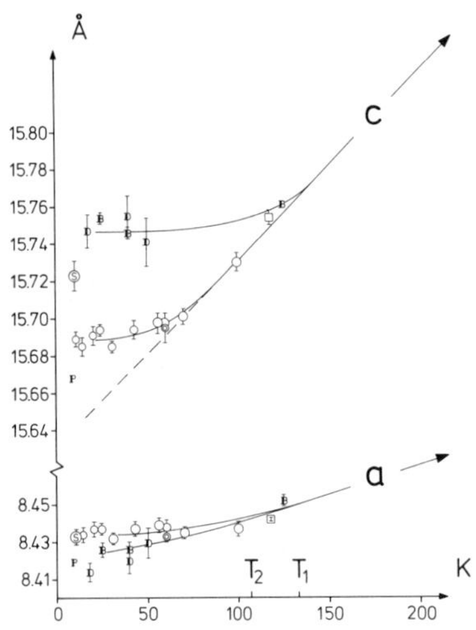

FIG. 2. Cosep cell parameters as function of temperature.

X-ray and neutron single crystal studies were performed at four different diffractometers and using several different crystals. At 295K both the normal P- and the amino group deuterated D-Co(sepulchrate)$(NO_3)_3$ belong to the hexagonal space group $P6_3 22$ and have cell dimensions a = 8.500(2) Å, c = 15.925(6) Å. The lengths of a and c are found to have the temperature dependence shown in FIG. 2.

In the 120 to 300K temperature range P and D have identical linear thermal coefficients of expansion. The cell

dimensions for P and D develop differently below 120K. P at
10.5K has the values a = 8.420(5) Å and c = 15.669(7) Å,
while a range of values have been observed for the deuterated
crystals. The **B** entries are results from a neutron diffraction
study conducted at Brookhaven National Laboratories. This
crystal was found to have a deuterium occupancy of 95.6%
for the amino hydrogen atoms. The **S** entry is from an X-ray
experiment on a spherical crystal of 0.40(2) mm diameter
prepared from a fragment of the **B** crystal which partly shatter-
ed by renewed cooling long after the Brookhaven data collection. Yet a few months later, as shown with open circles,

FIG. 3. P-cosep structural
satellites measured with neutron
diffraction. Satellites are
observed close to ($\bar{1}$20) in the
[110]-direction. The satellite
positions shift with temperature. Characteristic satellite
positions $(\bar{1},2,0)+\vec{q}$, where
$\vec{q} = \vec{q}_1$, $\vec{q}_2$, or $\vec{q}_3$ are indicated.
Similar satellites are observed
at $(\bar{1},2,0)-\vec{q}$. The appearance of
the $\vec{q}_1$ satellites agrees with
the only phase transition temperature T_1 = 133K deduced from
Raman data. The neutron diffraction experiment shows that
P-cosep undergoes further phase
transitions at 106K and
about 98K.

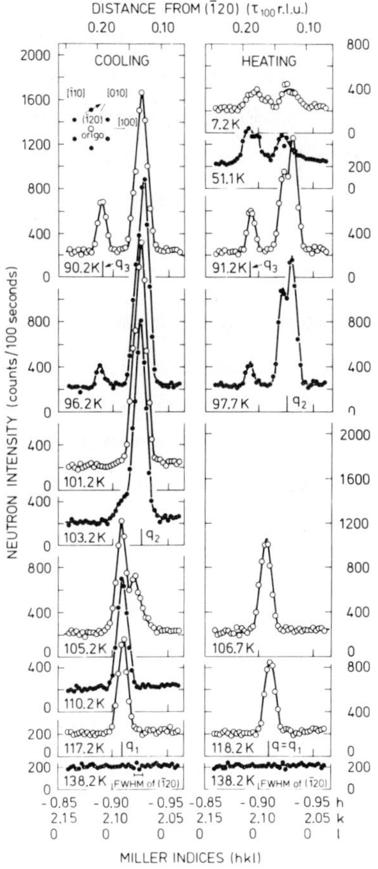

the same 0.4 mm diameter crystal gave c = 15.669(4) Å at
10.5K. The entries marked with bold face **D** letters are from
an X-ray determination of cell constants carried out on a
"fresh" deuterated crystal. Experimental values for the
c axis length at 10K for deuterated cosep crystals seem to
cluster around either of two values c = 15.75 Å or c = 15.69 A.

P-cosep phase transition temperatures.

A high resolution neutron diffraction study on P carried out
at Research Establishment Risø revealed a complicated set of
phase transitions, more involved than deduced from spectroscopic data (Dubicki et al., 1984). General scans were
performed along a number of directions in reciprocal space
through the strong $(\bar{1}20)$ reflection for many different temperatures. Results from scans along the [110]-direction are summarized in FIG. 3 (note that brackets are used to indicate a direction parallel to the reciprocal vector). Satellite peaks appear at positions $(\bar{1},2,0) \pm (q_i,q_i,0)$. The satellite positions
shift with varying temperature and at least three positions
$q_1 = 0.09$, $q_2 = 0.075$ and $q_3 = 0.11$ may easily be identified.
All phases appear to be incommensurate structures.

D-cosep phase transitions.

None of the satellites found for P have been observed for D,
but instead superstructure peaks corresponding to a commensurate structure appeared at low temperature. The structure of
D-cosep was investigated at Brookhaven National Laboratory by
single crystal neutron diffraction at 125K, 40K, and 25K.
Reciprocal space was searched for features which might not be
explained by the space group $P6_322$. Nothing unusual was
observed at 125K, but at both 40K and 25K a general scan along
the [110]-direction showed six extra peaks equidistantly
spaced between neighbouring main peaks (FIG. 4).

The hexagonal lattice can be described by a centered
orthohexagonal cell which has $a_{ORT} = a_{HEX}$, $b_{ORT} = \sqrt{3}b_{HEX}$ and

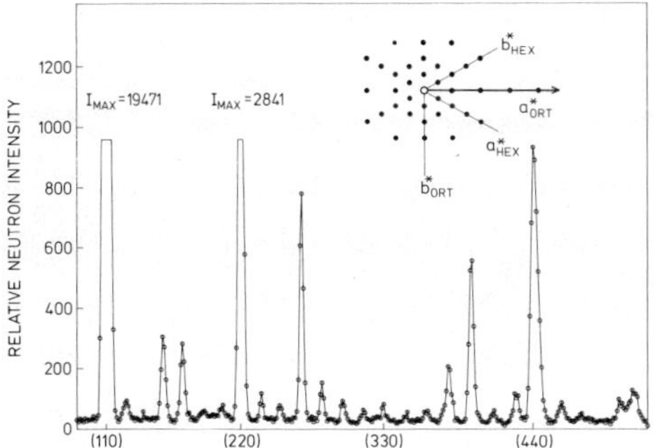

FIG. 4. D-cosep satellites at 40K measured with neutron diffraction. The scan direction is [110] through reflection (110).

$c_{ORT} = c_{HEX}$. For D in the low temperature phase a_{ORT} increased by a factor of seven giving a centered cell of volume fourteen times that of the primitive hexagonal cell. The phase transition is accompanied by a formation of orthorhombic domains, and since three equivalent sets of hexagonal a_{HEX}, b_{HEX} axes can be chosen with the same probability, three sets of orthorhombic domains may be set up, consistent with the optical crystallographic findings (Dubicki et al., 1984). The relative values of superreflection intensities along the $[110]$, $[\bar{2}10]$, and $[1\bar{2}0]$ directions can be used to deduce relative populations of the triplet domains. Main reflections from the different domains cannot be resolved, but their intensities do not obey the hexagonal symmetry at 40K, the internal agreement being 0.10 instead of 0.04 as at higher temperatures.

The commensurate structure has been observed for a number of D crystals at temperatures below 125K, but the phase transition does not always take place even after prolonged cooling to very low temperatures. Two complete and quite different sets of MoKα X-ray diffraction data sets have been collected at the same very low temperature of 10.5 ± 0.5K using the same spherical crystal of 0.40(2) mm diameter, one where the crystal had undergone the phase transition and

another where it had not. During the first experiment the
crystal had cell dimensions (listed as hexagonal axes)
a = 8.433(4) Å, c = 15.723(8) Å, and showed satellite reflections through the entire data collection period. Some months
later, after a renewed cool-down, the crystal no longer showed
satellite reflections. The cell dimensions now were
a = 8.434(2) Å, c = 15.689(4) Å. Another full set of data was
collected and both sets of data were averaged according to
hexagonal symmetry. $R_{INT}(I) = 0.127$; $R_{INT}(II) = 0.036$. Set II
refined reasonably to an agreement factor $R(F) = 0.077$ in the
same model as the room temperature phase. Displacement parameters were, however, found to be too big for a 10K structure
indicating a frozen-in statistical disorder of the NO_3^- group
on the D_3 position identical to the room temperature phase.
A trial refinement in the $P6_3 22$ model using Set I of the X-ray
data, which showed satellite reflections, gave a poor agreement factor $R(F) = 0.132$, but this average structure for the
big orthorhombic cell gave an indication that the phase transformation is accompanied by an ordering of the D_3 NO_3^- ions.

Intensities of the satellites are relatively bigger for
the neutron diffraction data than for the X-ray data which
means that also the cations with their shroud of hydrogen and
deuterium atoms in the commensurate phase are systematically
rearranged relative to the mean structure. A constrained rigid
body refinement of the commensurate phase of D-cosep nitrate
is in progress.

REFERENCES

DUBICKI, L., FERGUSON, J., GEUE, R.J., and SARGESON, A.M.
(1980). Chem. Phys. Lett. **74**, 393.
DUBICKI, L., FERGUSON, J., and WILLIAMSON, B. (1984).
J. Phys. Chem. **88**, 4254.
SARGESON, A.M. (1984). Pure and Appl. Chem. **56**, 1603.

50. Gas-phase UV-photoelectron spectra and DV-Xα calculations of μ_2-XO (X=C, N) bridged binuclear complexes

Maurizio Casarin, Andrea Vittadini, David Ajò, Gaetano Granozzi
and Renzo Bertoncello

1. INTRODUCTION

The description of the multi-centered metal-metal and metal-
ligand interactions in binuclear bridged organometallic complexes has been the object of several experimental and theoretical investigations (Trogler, 1980; Templeton, 1980; Cotton,
1975; Mitscher et al, 1978; Bénard, 1978; Bénard, 1979;
Granozzi et al, 1980; Granozzi et al, 1982; Bottomley, 1983).
The basic problem to be solved is the existence of direct
metal-metal bonding. In fact, qualitative electron counting
methods suggest the presence of single or multiple metal-metal
bonds even in those cases where experimental and/or theoretical evidence excludes them. The reason of the failure of such
qualitative theories is to be traced back to the neglect of
back-bonding interactions which, on the other hand, may play a
relevant role in such multi-centered bonds when low-valent
electron-rich metal atoms are involved. As a matter of fact,
it is now well established that, in several carbonyl bridged
dimers of VIII group transition metals, the electron pairing
between the metal atoms occurs $\underline{via}$ a concerted back-donation
into carbonyl $2\pi^*$ orbitals (Granozzi $\underline{et}$ $\underline{al}$, 1982). Such an
interaction, however, may be influenced by several factors,
e.g. the π acidity of the bridge, the $\underline{d}$ occupancy of the metal

atomic orbitals and the nature of the other terminal ligands. One interesting consequence of the subtle balance between all these factors is the high conformational flexibility of the $(\mu\text{-L})_2\text{-M}_2$ core, which can assume either a planar or a puckered form.

In order to pick up those electronic factors which could play a crucial role in determining the structural arrangement of this kind of molecules, the electronic structure of $[(\mu\text{-CO})\text{CpNi}]_2$ (I) and $[(\mu\text{-NO})\text{CpCO}]_2$ (II) $\{\text{Cp}=\eta^5\text{-C}_5\text{H}_5\}$ complexes has been accurately investigated carrying out several DV-Xα numerical experiments assuming both planar and bent geometries for the $[(\mu\text{-XO})\text{-M}]_2$ (X=C,M=Ni;X=N,M=Co) moiety (Fig.1).

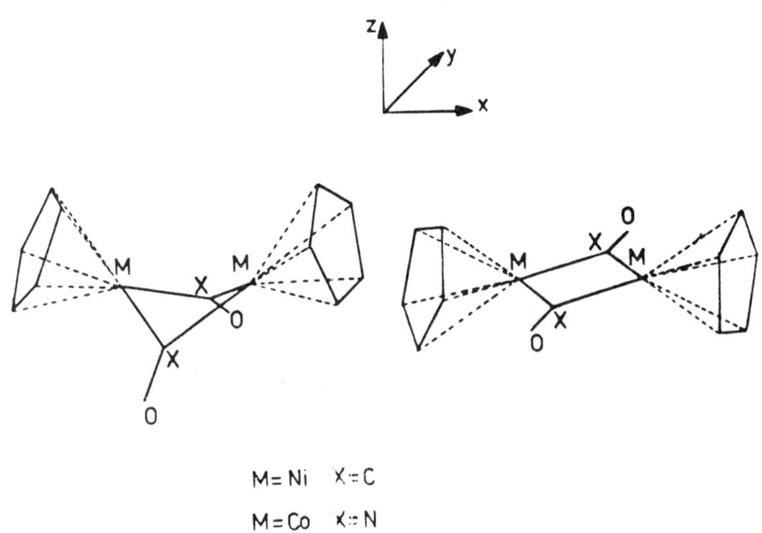

Fig. 1. Schematic view of the bent and planar geometries of $[(\mu\text{-CO})\text{CpNi}]_2$ and $[(\mu\text{-NO})\text{CpCO}]_2$

In the solid state the core of I is bent (Byers and Dahl, 1980) while it is planar when the Cp rings are monomethylated (Byers and Dahl, 1980). On the contrary, the reported X-ray structural determination of II shows a planar arrangement (Bernal et al, 1977). On the other hand, vapour-phase and solution experimental evidence (Byers and Dahl, 1980; Pilloni et al, 1987) indicates a planar and a bent structure for $[(\mu\text{-CO})Ni]_2$ and $[(\mu\text{-NO})CO]_2$, respectively.

2. THEORETICAL SECTION

Hartree-Fock-Slater (HFS) discrete variational (DV) Xα calculations (Averill and Ellis, 1973; Trogler et al; 1979, Rosen et al, 1976) of I and II were performed on a VAX-11/750 (Digital Equipment Corporation) computer.

Numerical atomic orbitals (through 4p on Co and Ni, 2p on C, N, O and 1s on H) obtained for the neutral atoms were used as basis functions. Due to the size of the systems, orbitals 1s-3p(Co,Ni) and 1s on both carbon, nitrogen and oxygen were treated as part of a frozen core in the molecular calculations. Atomic orbital and bond overlap populations (OPs) were computed using the Mulliken's scheme (Mulliken, 1955). The molecular geometry for the planar form (C_{2h}) of I was taken from the crystal structure of the monomethylated Cp derivative (Byers and Dahl, 1980), while the geometry of the bent form (C_{2v}) refers to the average of the two symmetry independent units of I in the crystal (Byers and Dahl,1980). Geometrical parameters of planar form of II were derived from its crystal structure (Bernal et al,1977) while the bent version was set with the same Co-Co, Co-Cp and Co-N distances as the planar molecule assuming an angle between the two nitrosyls and the two Cp rings of 120° and 30°, respectively.

3. RESULTS

Complexes I and II show a very different behavior on going from the planar to the bent form. This is very well summarized by making reference to the total electron density plots reported in figure 3.1 and 3.2. In the planar form (Figure 3.1) there is not a definite charge accumulation between the metal centres, while, quite amazingly, an unambigous charge accumulation between the metal atoms is present in the bent Co dimer (at variance with the Ni analogue) (Figure 3.2).

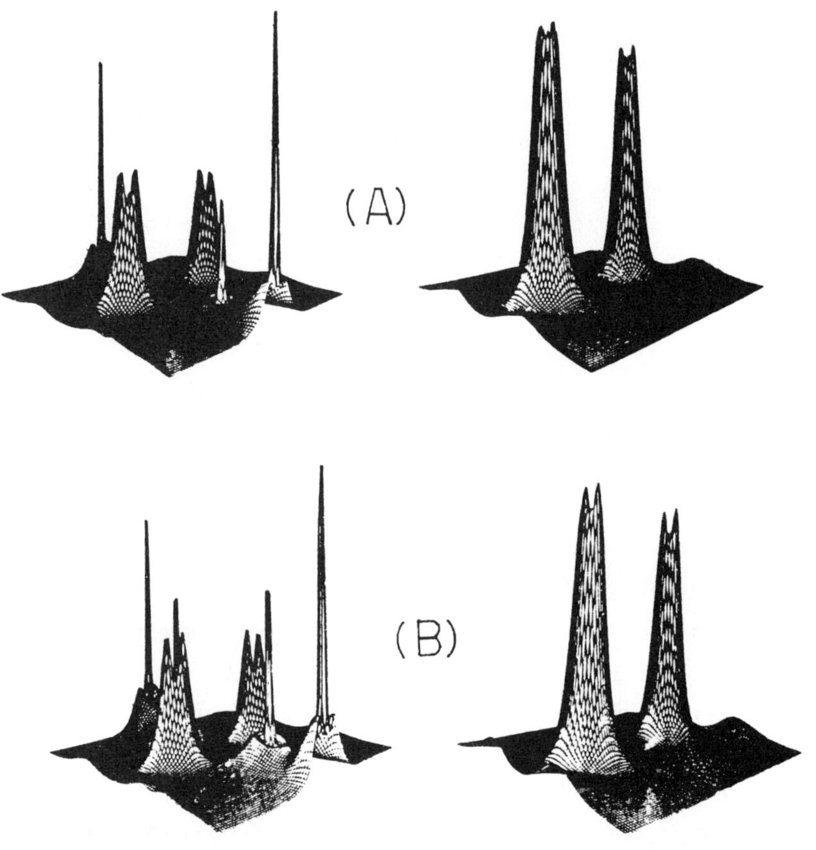

Figure 3.1 DV-Xα total electron density map in the xy plane (left) and xz plane (right) of planar I (A) and II (B)

As already pointed out (Pilloni et al, 1987), such a feature
supports the existence of a direct Co-Co bond in the bent
form. With respect to its source, it is not possible to select
a single MO responsible for that. At the same time it is hard
to explain such a feature on the basis of a simple balance
between the number of bonding and antibonding M-M levels. Here
it is unavoidable to invoke the actual percentage localization
of each MO on the metal centres. The application of such a criterion to the present cases gives us the basis to understand
the presence of a net M-M bond in the bent form of II. The
Co-Co antibonding MOs are significantly more delocalized over
the $(\mu\text{-L})_2\text{-M}_2$ core (i.e. π^* bridge NO are heavily involved in
the relative eigenvectors) than the corresponding Ni-Ni ones.
In other words, the higher π acceptor capability of the NO vs.
CO gives rise to a charge removal from an antibonding Co-Co
region.

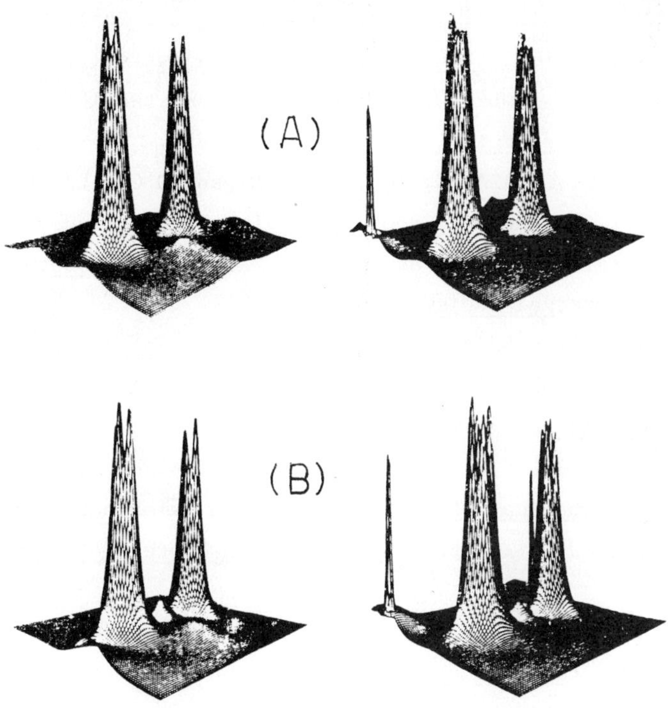

Figure 3.2. DV-Xα total electron density map in the xy plane
(left) and xz plane (right) of bent I (A) and II (B).

ACKNOWLEDGEMENT

We thank Mr. Franco De Zuane for invaluable technical assistance.

REFERENCES

AVERILL, F.W. and ELLIS, D.E., (1973). Journal of Chemical Physics 59, 6411.

BERNAL, I., KORP, J.D., REISNER, G.M., HERMANN, W.A. (1977), Journal of Organometalic Chemistry 139, 321.

BENARD, M. (1978). Journal of the American Chemical Society 100, 7740.

BENARD, M. (1979). Inorganic Chemistry 18, 2782.

BYERS, C.R. and DAHL, C.F. (1980), Inorganic Chemistry 19, 680.

BOTTOMLEY, F. (1983), Inorganic Chemistry 22, 2656.

COTTON, F.A. (1975), Chemical Society Reviews 4, 27.

GRANOZZI, G., TONDELLO, E., BENARD, M. and FRAGALA, I. (1980), Journal of Organometalic Chemistry 194, 83.

GRANOZZI, G., CASARIN, M., AJO, D. and D'OSELLA, D. (1982), Journal of the Chemical Society, Dalton Transactions 2047.

MITSHER, A., REES, B. and LEHMANN, M.S. (1978), Journal of the American Chemical Society 100, 3390.

MULLIKEN, R.S. (1955), Journal of Chemical Physics 23, 1833.

PILLONI, G., ZECCHIN, S., CASARIN, M. and GRANOZZI, G. (1987), Organometallics, in press.

ROSEN, A., ELLIS, D.E., ADACHI, H. and AVERILL, F.W. (1976), Journal of Chemical Physics 65, 3629 and references therein.

TEMPLETON, J.L. (1980), Progress in Inorganic Chemistry 26, 211.

TROGLER, W.C., ELLIS, D.E. and BERKOWITZ, J. (1979), Journal of the American Chemical Society 101, 5896.

TROGLER, W.C. (1980), Journal of Chemical Education 57, 424.

51. Cluster dynamics

G. Huttner

The clusters RP Fe$_3$(CO)$_9$ R'CCR" (Huttner et al. 1984, 1986, 1987), occur
in three isomeric forms (A–C Fig.1).

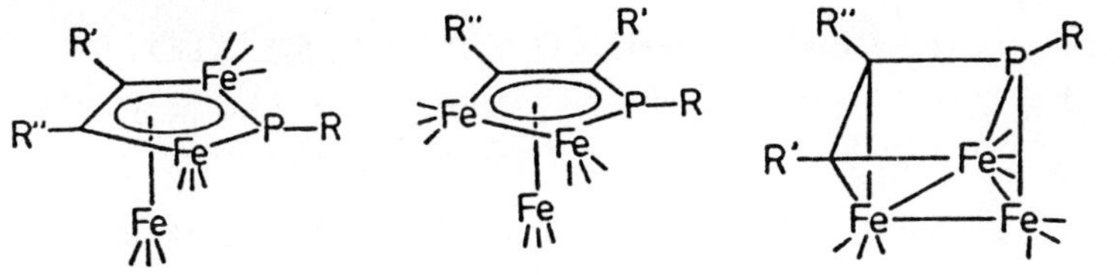

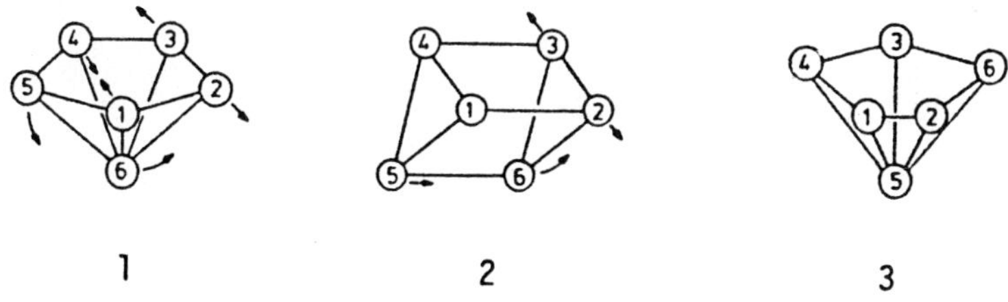

FIG.1. Three isomeric forms of RP Fe$_3$(CO)$_9$ R'CCR".

Experiments with these species show that pentagonally pyramidal and
distorted trigonally prismatic frameworks may be close in energy for six
vertex polyhedra.

FIG.2. "Tandem" isomerisation for six vertex polyhedra.

A "tandem" mechnism according to fig.2 is proposed for the isomerisation
of A into B.

EHT calculations for $B_6H_6^{4-}$, which is isolobal to A-C indicate that the tandem process (Fig.2.) may be a general pathway for the isomerisation of six vertex polyhedra.

Literature

Knoll,K., Orama,O. and Huttner,G. (1984) Angew. Chem.96, 989, Angew.Chem.Int.Ed. 23, 976.

Knoll,K., Huttner,G., Zsolnai,L., Orama,O. and Wasiucionek,M. (1986) J. Organmetal. Chem. 310, 225.

Knoll,K., Huttner,G. and Zsolnai,L.(1986), J. Organmetal. Chem. 312, C57.

Huttner,G. and Knoll,K. (1987) Angew.Chem. in press (Review).

Knoll,K., Huttner,G., Zsolnai,L. and Orama,O. (1987). Angew.Chem. 98, 1099; Angew.Chem.Int.Ed. 25, 1119.

Knoll,K., Fässler,Th. and Huttner,G. (1987)J. Organmetal.Chem. in press.

Knoll,K., Huttner,G., Fässler,Th. and Zsolnai,L. (1987).J.Organmetal.Chem. in press

Knoll,K., Huttner,G., Zsolnai,L. and Orama,O. (1987).J. Organmetal.Chem. in press.

Knoll,K., Huttner,G. and Zsolnai,L.(1987) J. Organmetal.Chem. in press.

52. Using the inorganic crystal structure database for the systematic examination of inorganic crystal structures

I. D. Brown, S. M. Bradley and D. Altermatt

The Inorganic Crystal Structure Database (ICSD) (Bergerhoff, Hundt, Sievers and Brown, 1983) contains full structural data on each of the 24,000 inorganic crystal structures (structures that are not metals or alloys and do not contain organic carbon) which have so far been published. Information stored includes the unit cell, the space group, atomic positional and mean atomic displacement parameters, as well as the literature citation and flags describing the structure determination. The database is currently available with a program that allows it to be searched interactively for information that is directly stored in the database (the elements present, the year of publication, mineral name etc).

We have been developing programs to permit a systematic analysis of the data in the ICSD. The first stage was to write a program (SINDBAD, Altermatt and Brown (1985)) that determines which of the interatomic distances are chemical bonds. This defines a bond as any distance between an anion and a cation whose bond valence (Brown, 1978) is less than the anion valence and greater than 0.034 times the cation valence. The use of bond valences permits the program to check on the correctness of its assignment by comparing the valence of each atom with the sum of the bond valences of the bonds that it forms. The program is able to assign bonds in 90% of the structures stored in the ICSD and it recognises that it has made a correct assignment in two thirds of these cases. In the remaining cases, disorder or lack of accuracy in the structure determination results in the calculation of bond lengths that are implausably long or short but which in any case are unsuitable for systematic study. The output of SINDBAD is a file which lists the chemical bonds in the asymmetric unit. In addition to the bond lengths and their experimental error this file contains the bond valence, the components of the bond vectors and special position labels for the atoms (Altermatt and Brown, 1987). With this information it is possible to reconstruct the bonding network in real space over any desired volume of the crystal.

The bond file is a valuable resource for systematic studies of chemical bonding. It has been used to produce a bond index to the entries in ICSD similar to that used in the publication BIDICS. The index can be searched by bond type (defined by the two elements forming the bond and their oxidation numbers) as well as by cation coordination number, average bond length and flags that indicate the accuracy of the structure, the presence of disorder and the type of radiation used (x-ray, neutron). The bond index points to the location of the data in the ICSD and in the original literature.

A search through the bond file has also been used to list all the cation coordination spheres which contain a single type of ligand. When ordered by cation this brings together a wealth of information on the coordination numbers and bond lengths of individual cations (e.g. 988 environments around Si, 634 around Na). We have already used this listing to calculate improved bond valence parameters (Brown and Altermatt, 1985) and other systematic studies are currently under way. The purpose of these studies is to determine empirical relationships between chemical properties and structure which can be exploited by the programs described below.

The bond file can also be used to explore bonding network topologies by expanding the asymmetric bond set to fill one (or more) unit cells. Since the file contains bond valences as well as bond vectors, it is possible to restrict the expansion of the topology to bonds of a certain strength. If one is interested, for example, in the alumino-silicate framework of a mineral but not the bonds to the alkali metal and alkaline earth atoms, the network can be expanded using only bonds with a bond valence greater than 0.6 valence units (v.u.). Alternatively the complete bond network can be found, including

Table: STRUMO network analysis of β-Na$_2$Si$_2$O$_5$ (Pant (1968).

```
------------------------------------------------------------------
FRAGMENT NUMBER    1
Spanning tree forms sheets perpendicular to  1  0  0
Formula of fragment is: Si4 O10
   Connection table
SI1        1  0   0   0   0 1 1 1 1   0 0 0 0 0   0 0 0 0 0   0 0 0 0 0
O 5        1  0   0   0   1 0 0 0 0   0 0 0 0 0   0 0 0 0 0   0 0 0 0 0
O 1        2  0   0   0   1 0 0 0 0   1 0 0 0 0   0 0 0 0 0   0 0 0 0 0
O 1        1  0   0   0   1 0 0 0 0   0 1 0 0 0   0 0 0 0 0   0 0 0 0 0
O 3        1  0   0   0   1 0 0 0 0   0 0 1 0 0   0 0 0 0 0   0 0 0 0 0
SI1        2  0   0   0   0 0 1 0 0   0 0 0 1 1   1 0 0 0 0   0 0 0 0 0
SI1        2  0-12   0   0 0 0 1 0   0 0 0 0 0   0 0 0 0 0   0 0 0 0 0
SI2        1  0   0   0   0 0 0 0 1   0 0 0 0 0   0 1 1 1 0   0 0 0 0 0
O 5        2  0   0   0   0 0 0 0 0   1 0 0 0 0   0 0 0 0 0   0 0 0 0 0
O 1        1  0  12   0   0 0 0 0 0   1 0 0 0 0   0 0 0 0 0   0 0 0 0 0
O 3        2  0   0   0   0 0 0 0 0   1 0 0 0 0   0 0 0 0 1   0 0 0 0 0
O 4        1  0   0   0   0 0 0 0 0   0 0 1 0 0   0 0 0 0 0   0 0 0 0 0
O 2        2  0-12  12   0 0 0 0 0   0 0 1 0 0   0 0 0 0 0   1 0 0 0 0
O 2        1  0   0   0   0 0 0 0 0   0 0 1 0 0   0 0 0 0 0   0 1 0 0 0
SI2        2  0   0   0   0 0 0 0 0   0 0 0 0 0   1 0 0 0 0   0 0 1 1 1
SI2        2  0-12  12   0 0 0 0 0   0 0 0 0 0   0 0 1 0 0   0 0 0 0 0
SI2        2  0   0  12   0 0 0 0 0   0 0 0 0 0   0 0 0 1 0   0 0 0 0 0
O 4        2  0   0   0   0 0 0 0 0   0 0 0 0 0   0 0 0 0 1   0 0 0 0 0
O 2        1  0 0-12   0 0 0 0 0   0 0 0 0 0   0 0 0 0 1   0 0 0 0 0
O 2        2  0   0   0   0 0 0 0 0   0 0 0 0 0   0 0 0 0 1   0 0 0 0 0
The charge on the fragment is   -4.0
The base strength of the fragment is about  0.17
Excess base valence on the fragment is  3.97 v.u.
Excess acid valence on the fragment is -0.03 v.u.
Atom SI1       has  4 bonds and probably forms  0 more bonds of strength  0.00
Atom O 5       has  1 bonds and probably forms  5 more bonds of strength  0.17
Atom O 1       has  2 bonds and probably forms  0 more bonds of strength  0.00
Atom O 1       has  2 bonds and probably forms  0 more bonds of strength  0.00
Atom O 3       has  2 bonds and probably forms  1 more bonds of strength  0.09
Atom SI1       has  4 bonds and probably forms  0 more bonds of strength  0.00
Atom SI2       has  4 bonds and probably forms  0 more bonds of strength  0.00
Atom O 5       has  1 bonds and probably forms  5 more bonds of strength  0.17
Atom O 3       has  2 bonds and probably forms  1 more bonds of strength  0.09
Atom O 4       has  1 bonds and probably forms  5 more bonds of strength  0.18
Atom O 2       has  2 bonds and probably forms  1 more bonds of strength  0.08
Atom O 2       has  2 bonds and probably forms  1 more bonds of strength  0.08
Atom SI2       has  4 bonds and probably forms  0 more bonds of strength  0.00
Atom O 4       has  1 bonds and probably forms  5 more bonds of strength  0.18
```

The network is expanded around Si(1) using only bonds of valence greater than 0.6. The columns in the connection table give the element identifier, symmetry operator and three components of the translation vector in units of 1/12th unit cell. The remaining columns constitute the connection matrix. The terms "acid strength" and "base strength" are defined by Brown (1981) The predictions can be compared with the actual bonds formed to Na by O(1) (0.19 v.u.), O(2) (0.16 v.u.), O(3) (0.15 v.u.), O(4) (0.12, 0.18, 0.21, 0.24 v.u.) and O(5) (0.12, 0.19, 0.21, 0.26 v.u.).

the bonds formed to the low charge cations, by setting the threshold to 0.0 v.u. Our program carries out this expansion and explores the topological and chemical properties of the network. It determines whether the network is a chain, sheet, framework or isolated cluster, calculates its net charge, calculates the acid and base strengths of each of the atoms in the network and estimates the number of additional bonds (other than those in the network) that would be needed to satisfy their chemical requirements. The computer output of a simple example of the network expansion around Si in clino-enstatite is shown in the table. We intend to exploit this program to provide methods of studying bonding topologies along the lines proposed by Klee (1980), Hawthorne (1985) and Bosch (1986).

We have also started to explore the use of ICSD in modelling defect structures and surfaces. Here we use the checked data stored in ICSD as a starting model to avoid the tedious job of retyping the atomic coordinates and checking the space group settings. We have, for example, searched ICSD for all the structures of olivine and downloaded them to our computer where we have selected one to use as input to the structure modelling package STRUMO. In STRUMO it is possible to explore diffusion paths using a valence related function that determines which positions in the crystal the diffusing cation will find favourable for bonding (Waltersson, 1978) (see the

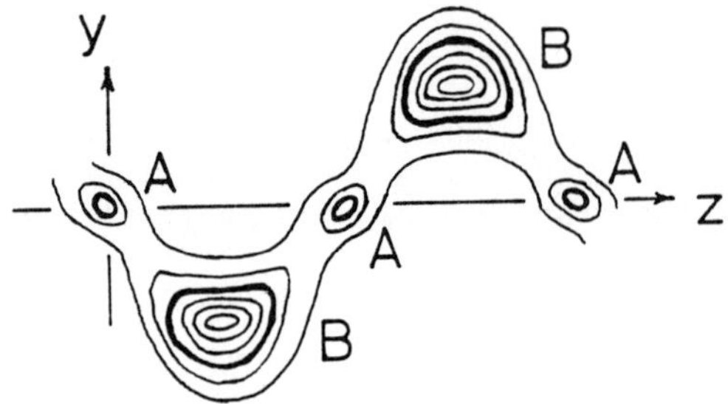

Figure: The c axis diffusion path for Mg in Forsterite.   The function plotted is $(V/2)^{-8}$ where V is the sum of the valences of the bonds that a Mg atom would form if placed at that point in the crystal.   The thick contour corresponds to the function having a value of 1.0, i.e. it represents ideal locations for a Mg atom.   The Mg atoms normally occupy the A sites but diffuse via the large cavity B whose maximum has a value of 2.5.

figure).   The effects of introducing defects or occupational disorder into the structure can be examined by performing a distance least squares refinement using target distances generated automatically by STRUMO.   This procedure, which is very rapid, currently predicts atomic positions in Forsterite that lie within 0.06Å of those found by x-ray diffraction.   With refinements under development we hope to achieve a better agreement with experiment.

ACKNOWLEDGEMENTS

This work has been funded by an operating grant from the Natural Science and Engineering Research Council of Canada.

REFERENCES

ALTERMATT, D. and BROWN, I.D. (1985). Acta Crystallographica B41, 240.

ALTERMATT, D. and BROWN, I.D. (1987). Acta Crystallographica in press.

BERGERHOFF, G., HUNDT, R., SIEVERS, R. AND BROWN, I.D. (1983). Journal of Chemical Information and Computer Science 23, 69.

BOSCH, W. (1986). Diplomarbeit, Univ. of Bonn, F.R.G., 137 pp.

BROWN, I.D. (1978). Chemical Society Reviews 7, 359.

BROWN, I.D. (1981). in Structure and Bonding in Crystals Vol. 2. (Eds. O'Keefe and Navrotsky), Academic Press, New York, pp. 1-30.

HAWTHORNE, F.C. (1985). American Minerologist 70, 455.

KLEE, W.E. (1980). MATCH, Communications in Mathematical Chemistry 9, 105.

PANT, A.K. (1968) Acta Crystallographica B24, 1077.

WALTERSSON, K. (1978). Acta Crystallographica A34, 901.

53. A general structural hierarchy for oxy-salt minerals

Frank C. Hawthorne

ABSTRACT

Recently, it has been proposed that structures may be ordered
or classified according to the polymerization of those coordination polyhedra with higher bond-valences (Hawthorne, 1983).
Here I consider oxysalt minerals with triangular, tetrahedral
and octahedral coordination of small highly-charged (2+)
cations. Structures are described and ordered according to
their basic heteropolyhedral cluster, or fundamental building
block, and the way in which this cluster polymerizes to form
the structure module, a complex anionic polyhedral array whose
excess charge is balanced by the presence of large low-valence
cations. These descriptions are particularly useful for loosely
packed and highly hydrated minerals, the structures of which
often resist adequate description in terms of such traditional
concepts as close-packing.

1. INTRODUCTION

As increasing amounts of crystal structure information have
become available, structural correlations among quite large
chemical groups of minerals have been set up: silicates

(Bragg, 1927; Liebau, 1985), aluminofluorides (Pabst, 1954), phosphates (Moore, 1982). These schemes focus on the polymerization of a specific complex anionic group, and are essentially geometrical in nature. The next step in the development of general structural systematics of minerals is to broaden the class of minerals considered to include all oxysalt minerals.

2. GENERAL APPROACH

The principal thrust of this work is <u>not</u> to produce a structural classification of minerals, but to develop a method of description of a structure that incorporates both graphical and chemical information. That this is feasible is suggested by simple molecular orbital theory, which indicates that there is considerable chemical information inherent in the graphical aspects (often called 'topology') of bond connectivity. In dealing with the structural complexity of hydroxy-hydrated oxysalts, it is advantageous to deal with ideas that are simple and intuitive, and what I will use to approach this problem is Pauling's (1960) second rule, together with its generalizations that Brown (1981) has systematized into the coherent framework of bond-valence theory. In this regard, it should be noted that this approach is <u>not</u> based on an ionic model; use of the terms 'cation' and 'anion' indicates merely that some atoms are more electropositive than others, and carries no connotation as to chemical bonding.

Formal bond-strength may be defined as cation charge divided by cation coordination number, and Pauling's second rule states that the sum of the bond-strengths around an anion is approximately equal to the magnitude of the anion valence. There is a very useful way in which we can think about this law with regard to the satisfaction of bond-strength requirements around each atom in a structure. For the cations, by definition the bond-strength requirements are satisfied by the formation of anion coordination polyhedra around them. Thus, we can think of the structure as an array of complex anions, which polymerize in order to satisfy their anionic bond-strength requirements. The most important polymerizations are those involving the coordination polyhedra of higher bond-strengths, as these contribute most to the satisfaction of the anion bond-strength requirements.

If the major imperative of a structure is the satisfaction of its cation and anion bond-strength requirements, the most important features of the structure are the type of cation co-ordination polyhedra, and the way in which the higher bond-strength polyhedra polymerize. This suggests the following hypothesis: structures may be profitably described and ordered according to the polymerization of those coordination polyhedra with higher bond-valences (Hawthorne, 1983).

In order to exploit the chemical content of the graphical aspects of structure, the following definitions are useful. The most strongly bonded cluster of coordination polyhedra in a structure is the fundamental building block of that structure. This is repeated (often polymerized) by the translational symmetry operators to form the module of the structure, an anionic polyhedral array whose excess charge is balanced by the presence of large low-valence cations.

3. MAJOR CATEGORIES OF STRUCTURES

Structures may be ordered into classes according to the dimensionality of their structure modules: (i) isolated polyhedra; (ii) finite clusters; (iii) infinite chains; (iv) infinite sheets; (v) infinite frameworks. From a graphical viewpoint, classes (i) and (ii) could be combined; however, paragenesis suggests that it will be more convenient to have them separate.

4.   STRUCTURAL TRENDS IN OXYSALTS

The idea of fundamental building blocks and structure modules
can be used in conjunction with the structure–based scale of
Lewis acid and base strengths developed by Brown (1981) to provide considerable insight into structure type as a function of
chemical composition. The Lewis acid strength of a cation is
the average valence of a bond formed by that cation. The Lewis
base strength of an anion is the average valence of a bond
formed by that anion. Simple anions often show a great range of
bond–valences, but this is generally greatly reduced for
complex anions (Hawthorne, 1985).

The hypothesis introduced above considers the polymerization
of the most tightly–bonded oxyanions in a structure. Increasing
polymerization will decrease the resultant Lewis base strength,
suggesting that the degree of polymerization in such structures
should be related to the average Lewis base strength of their
component oxyanions; that this is the case has been shown by
Hawthorne (1985, 1986).

5.   LARGE CATION CHEMISTRY IN OXYSALTS

The structure module may be considered as a very complex
oxyanion, and a Lewis basicity may be calculated for it. The
valence–matching principle (Brown, 1981) states that the most
stable structures will form when the Lewis acid strength of the
cation most closely matches the Lewis base strength of the
anion. This indicates that the Lewis base strength of the
structure module should match the Lewis acidity of the extra-
module cation(s). Hawthorne (1985, 1986) has shown this to be
the case for a large and diverse group of oxysalt minerals.
Thus the characteristics of the structure module exert a strong
control over the chemical identity of the extra–module cations.

                    OXY–SALT MINERALS

ACKNOWLEDGEMENTS: This work was supported by the Natural
Sciences and Engineering Research Council of Canada.

REFERENCES

BRAGG, W.L. (1930). Zeitschrift für Kristallographie 74, 237.

BROWN, I.D. (1981). Structure and Bonding in Crystals II, 1.

HAWTHORNE, F.C. (1983). Acta Crystallographica A39, 724.

HAWTHORNE, F.C. (1985). American Mineralogist 70, 455.

HAWTHORNE, F.C. (1986). Canadian Mineralogist 24 (in press).

LIEBAU, F. (1985). Structural Chemistry of the Silicates.
    Springer–Verlag, New York.

MOORE, P.B. (1980). International Mondial du Phosphate. 2nd
    International Congress (Boston), 105.

PABST, A. (1950). American Mineralogist 35, 149.

PAULING, L. (1960). The Nature of the Chemical Bond (3rd
    Edition). Cornell University Press, Ithaca, New York.

54. The direct method of X-ray crystallography: an overview*

Herbert Hauptman

ABSTRACT

The electron density function, $\rho(\mathbf{r})$, in a crystal determines
its diffraction pattern, i.e. both the magnitudes and phases
of its X-ray diffraction maxima, and conversely. If however,
as is always the case, only magnitudes are available from the
diffraction experiment, then the density function $\rho(\mathbf{r})$ cannot
be recovered. If one invokes prior structural knowledge,
usually that the crystal is composed of discrete atoms of
known atomic numbers, then the observed magnitudes are, in
general, sufficient to determine the positions of the atoms,
i.e. the crystal structure.

The intensities of a sufficient number of X-ray
diffraction maxima determine a crystal structure. The
available intensities usually exceed the number of parameters
needed to describe the structure. From these intensities a
set of numbers $|E_{\mathbf{H}}|$ can be derived, one corresponding to each

intensity. However the elucidation of the crystal structure
requires also a knowledge of the complex numbers $E_H = |E_H| \times$
$\exp(i\phi_H)$, the normalized structure factors, of which only the
magnitudes $|E_H|$ can be determined from experiment. Thus, a
"phase" ϕ_H, unobtainable from the diffraction experiment, must
be assigned to each $|E_H|$, and the problem of determining the
phases when only the magnitudes $|E_H|$ are known is called "the
phase problem". Owing to the known atomicity of crystal
structures and the redundancy of observed magnitudes $|E_H|$, the
phase problem is solvable in principle.

1.   THE PHASE PROBLEM

Denote by ϕ_H the phase of the structure factor F_H:

$$F_H = |F_H| \exp(i\phi_H), \tag{1}$$

where H is a reciprocal lattice vector (having three integer
components) which labels the corresponding diffraction
maximum. Then the relationship between the structure factors
F_H and the electron density function $\rho(r)$ is given by

$$F_H = \int_V \rho(r) \exp(2\pi iH\bullet r)dV \tag{2}$$

and

$$\rho(r) = \frac{1}{V} \sum_H F_H \exp(-2\pi iH\bullet r) = \frac{1}{V} \sum_H |F_H| \exp i(\phi_H - 2\pi H\bullet r)$$

$$\tag{3}$$

in which V represents the unit cell or its volume. Thus the
structure factors F_H determine $\rho(r)$. The x-ray diffraction
experiment yields only the magnitudes $|F_H|$ of a finite number
of structure factors, but the values of the phases ϕ_H, which
are also needed if one is to determine $\rho(r)$ from (3), cannot
be determined experimentally. If observed values for the
magnitudes $|F_H|$, but arbitrary values for the phases ϕ_H are
specified in Eq. (3), then density functions $\rho(r)$ are defined

which, when substituted into (2) yield structure factors F_H
the magnitudes of which agree with the observed magnitudes
$|F_H|$. Thus diffraction intensities alone do not determine a
unique density function $\rho(r)$. Even if the known non-
negativity of $\rho(r)$ is assumed, thus greatly restricting the
values of the phases (Harker <u>et al</u>, 1948; Karle <u>et al</u>, 1950),
the observed diffraction intensities are, in general, still
not sufficient to determine $\rho(r)$ uniquely. It follows that
the phase problem, to determine the values of the phases ϕ_H of
the structure factors F_H when only the magnitudes $|F_H|$ are
given, is, in principle, unsolvable when formulated in these
terms. It was this argument which led the crystallographic
community, prior to 1950, to believe also that crystal
structures could not, even in principle, be determined from
the diffraction intensities alone. However, by invoking the
prior structural knowledge that crystals consist of discrete
atoms, one readily refutes this argument, as shown next.

If one replaces the real crystal, with continuous
electron density $\rho(r)$, by an idealized one, the unit cell of
which consists of N discrete, non-vibrating, point atoms, then
the structure factor F_H is replaced by the normalized
structure factor E_H and (1), (2) and (3) are replaced by

$$E_H = |E_H| \exp(i\phi_H) \tag{4}$$

$$E_H = \frac{1}{\sigma_2^{1/2}} \sum_{j=1}^{N} Z_j \exp(2\pi i H \cdot r_j), \tag{5}$$

$$\left\langle E_H \exp(-2\pi i H \cdot r) \right\rangle_H = \frac{1}{\sigma_2^{1/2}} \left\langle \sum_{j=1}^{N} Z_j \exp[2\pi i H \cdot (r_j - r)] \right\rangle_H,$$

$$= \frac{Z_j}{\sigma_2^{1/2}} \quad \text{if } r = r_j,$$

$$= 0 \text{ if } \mathbf{r} \neq \mathbf{r}_j, \qquad (6)$$

respectively, where Z_j is the atomic number and $\mathbf{r}_j$ is the position vector of the atom labeled j, and

$$\sigma_n = \sum_{j=1}^{N} Z_j^n, \quad n = 1,2,3,\ldots . \qquad (7)$$

In practice the magnitudes $|E_{\mathbf{H}}|$ of the normalized structure factors $E_{\mathbf{H}}$ are obtainable (at least approximately) from the observed magnitudes $|F_{\mathbf{H}}|$ while the phases $\phi_{\mathbf{H}}$, as defined by (4) and (5), cannot be determined experimentally. Since one now requires only the 3N components of the N position vectors $\mathbf{r}_j$, rather than the much more complicated electron density function $\rho(\mathbf{r})$, it turns out that, in general, the known magnitudes are more than sufficient. This is most readily seen by equating the magnitudes of both sides of (5), thus eliminating the unknown phases $\phi_{\mathbf{H}}$, in order to obtain

$$|E_{\mathbf{H}}| = 1/\sigma_2^{1/2} \left| \sum_{j=1}^{N} Z_j \exp(2\pi i \mathbf{H} \cdot \mathbf{r}_j) \right|, \qquad (8)$$

a system of equations in which the only unknowns are the 3N components of the position vectors $\mathbf{r}_j$. Since the number of equations (8), equal to the number of reciprocal lattice vectors $\mathbf{H}$ for which the magnitudes $|E_{\mathbf{H}}|$ are observed, usually exceeds the number of unknowns, 3N, by far, the system (8) is redundant. Thus the phase problem is, in principle, solvable when reformulated in terms of fixed, point atoms, as reference to Eq. (5) then shows. In summary then, the lost phases ϕ are to be found among the measurable magnitudes $|E|$.

It should be observed in passing that the system of

equations (5) implies the existence of relationships among the normalized structure factors E_H since the (relatively few) unknown position vectors r_j may, at least in principle, be eliminated. By the term "direct methods" is meant that class of methods which exploits relationships among the normalized structure factors in order to go directly from the observed magnitudes $|E|$ to the needed phases ϕ.

2. THE STRUCTURE INVARIANTS.

Equation (6) implies that the normalized structure factors E_H determine the crystal structure. However (5) does not imply that, conversely, the crystal structure determines the values of the normalized structure factors E_H since the position vectors r_j depend not only on the structure but on the choice of origin as well. It turns out nevertheless that the magnitudes $|E_H|$ of the normalized structure factors are in fact uniquely determined by the crystal structure and are independent of the choice of origin, but that the values of the phases ϕ_H depend also on the choice of origin. Although the values of the individual phases depend on the structure and the choice of origin, there exist certain linear combinations of the phases, the so-called structure invariants, whose values are determined by the structure alone and are independent of the choice of origin.

If the origin of coordinates is shifted to a new point having position vector r_0 with respect to the old origin, then, from the definition, (5), of E_H, it follows readily that the phase ϕ_H of the normalized structure factor E_H with respect to the old origin is replaced by the new phase ϕ'_H with respect to the new origin given by

$$\phi'_H = \phi_H - 2\pi H \cdot r_0 \ . \tag{9}$$

Equation (9) implies that the linear combination of three phases,

$$\psi_3 = \phi_H + \phi_K + \phi_L,\qquad (10)$$

is a structure invariant (triplet) provided that

$$H + K + L = 0 ;\qquad (11)$$

the linear combination of four phases,

$$\psi_4 = \phi_H + \phi_K + \phi_L + \phi_M\qquad (12)$$

is a structure invariant (quartet) provided that

$$H + K + L + M = 0 ;\qquad (13)$$

and so on.

Since the values of the individual phases depend not only on the structure but on the choice of origin, it follows that magnitudes $|E|$ alone cannot determine unique values of the individual phases. It is clear that magnitudes $|E|$ alone determine only the values of the structure invariants (and not even uniquely at that, because of the enantiomorph problem, as clarified below) and only then, after suitable specification of the origin (and enantiomorph when necessary), may the individual phases be determined.

It should be noted finally that the theory of the structure invariants leads directly to recipes for origin specification called for by the techniques of direct methods. For example, when no crystallographic element of symmetry is present (space group P1) the rule states simply that the values of any three phases

$$\phi_{h_1 k_1 \ell_1},\ \phi_{h_2 k_2 \ell_2},\ \phi_{h_3 k_3 \ell_3},$$

where the determinant

$$\begin{vmatrix} h_1 k_1 \ell_1 \\ h_2 k_2 \ell_2 \\ h_3 k_3 \ell_3 \end{vmatrix} = \pm 1,$$

are to be specified arbitrarily, thus fixing the origin
uniquely.

3. THE STRUCTURE SEMINVARIANTS

If a crystal possesses elements of symmetry then the origin
may not be chosen arbitrarily if the simplifications permitted
by the space group symmetries are to be realized. For
example, if a crystal has a centre of symmetry it is natural
to place the origin at such a centre while if a two-fold screw
axis, but no other symmetry element is present, the origin
would normally be situated on this symmetry axis. In such
cases the permissible origins are greatly restricted and it is
therefore plausible to assume that many linear combinations of
the phases (not only the structure invariants) will remain
unchanged in value when the origin is shifted only in the
restricted ways allowed by the space group symmetries. One is
thus led to the notion of the structure seminvariant, those
linear combinations of the phases whose values are independent
of the choice of permissible origin.

If the only symmetry element is a centre of symmetry, for
example (space group $P\bar{1}$), then it turns out (from Eq. (5) and
(9)) that a single phase ϕ_H is a structure seminvariant
provided that the three components of the reciprocal lattice
vector H are even integers; the linear combination of two
phases $\phi_H + \phi_K$ is a structure seminvariant provided that the
three components of $H + K$ are even integers; etc.

If the only symmetry element is a two-fold rotation axis
(or two-fold screw axis) then one finds (again from (5) and
(9)) that the single phase $\phi_{hk\ell}$ is a structure seminvariant

provided that h and ℓ are even integers and $k = 0$; the linear combination of two phases

$$\Phi_{h_1 k_1 \ell_1} + \Phi_{h_2 k_2 \ell_2}$$

is a structure seminvariant provided that $h_1 + h_2$ and $\ell_1 + \ell_2$ are even and $k_1 + k_2 = 0$; etc.

The structure invariants and seminvariants have been tabulated for all the space groups (Hauptman _et al_, 1953, 1956, 1959; Karle _et al_, 1961; Lessinger _et al_, 1975). In general the collection of structure invariants is a subset of the collection of structure seminvariants. If no element of symmetry is present, that is the space group is P1, then the two classes coincide.

4.   THE FUNDAMENTAL PRINCIPLE OF DIRECT METHODS

It is known that the values of a sufficiently extensive set of cosine seminvariants (the cosines of the structure seminvariants) lead unambiguously to the values of the individual phases (Hauptman, 1972). Magnitudes $|E|$ are capable of yielding estimates of the cosine seminvariants only or, equivalently, the magnitudes of the structure seminvariants; the signs of the structure seminvariants are ambiguous because the two enantiomorphous structures (related to each other by reflection through a point) which are permitted by the observed magnitudes $|E|$ correspond to two values of each structure seminvariant differing only in sign. However, once the enantiomorph has been selected by specifying arbitrarily the sign of a particular enantiomorph sensitive structure seminvariant (i.e. one different from 0 or π), then the magnitudes $|E|$ determine both signs and magnitudes of the structure seminvariants consistent with the chosen enantiomorph. Thus, for fixed enantiomorph, the observed magnitudes $|E|$ determine unique values for the structure

seminvariants; the latter, in turn, as certain well defined
linear combinations of the phases, lead to unique values of
the individual phases. In short, the structure seminvariants
serve to link the observed magnitudes $|E|$ with the desired
phases Φ (the fundamental principle of direct methods). It is
this property of the structure seminvariants which accounts
for their importance and which justifies the stress placed on
them here.

5. THE NEIGHBORHOOD PRINCIPLE

It has long been known that, for fixed enantiomorph, the value
of any structure seminvariant ψ is, in general, uniquely
determined by the magnitudes $|E|$ of the normalized structure
factors. Recently it has become clear that, for fixed
enantiomorph, there corresponds to ψ one or more small sets of
magnitudes $|E|$, the neighborhoods of ψ, on which, in favorable
cases, the value of ψ most sensitively depends; that is to say
that, in favorable cases, ψ is primarily determined by the
values of $|E|$ in any of its neighborhoods and is relatively
independent of the values of the great bulk of remaining
magnitudes. The conditional probability distribution of ψ,
assuming as known the magnitudes $|E|$ in any of its
neighborhoods, yields an estimate for ψ which is particularly
good in the favorable case that the variance of the
distribution happens to be small (Hauptman, 1975a,b) (the
neighborhood principle).

The first neighborhood of the triplet ψ_3, (Eq. (10)),
consists of the three magnitudes

$$|E_H|, \ |E_K|, \ |E_L| \ . \tag{14}$$

The first neighborhood of the quartet ψ_4, (Eq. (12)) consists
of the four magnitudes

$$|E_{\mathbf{H}}|, \ |E_{\mathbf{K}}|, \ |E_{\mathbf{L}}|, \ |E_{\mathbf{M}}| \ . \tag{15}$$

The second neighborhood of the quartet consists of the four magnitudes (15) plus the three additional magnitudes

$$|E_{\mathbf{H+K}}|, \ |E_{\mathbf{K+L}}|, \ |E_{\mathbf{L+H}}| \ , \tag{16}$$

i.e. seven magnitudes $|E|$ in all (Hauptman, 1975b). The neighborhoods of all the structure invariants are now known.

6. THE EXTENSION CONCEPT.

By embedding the structure seminvariant T and its symmetry related variants in suitable structure invariants Q one obtains the extensions Q of the seminvariant T. Owing to the space group dependent relations among the phases, T is related in a known way to its extensions. In this way the theory of the structure seminvariants is reduced to that of the structure invariants. In particular, the neighborhoods of T are defined in terms of the neighborhoods of its extensions. The procedure will be illustrated in some detail only for the two-phase structure seminvariant in the space group $P\bar{1}$ which serves as the prototype for the structure seminvariants in general, in all space groups, noncentrosymmetric as well as centrosymmetric.

6.1. The two-phase structure seminvariant in $P\bar{1}$.
It has already been seen (§ 3) that the linear combination of two phases

$$T = \phi_{\mathbf{H}} + \phi_{\mathbf{K}} \tag{17}$$

is a structure seminvariant in $P\bar{1}$ if and only if the three components of the reciprocal lattice vector $\mathbf{H} + \mathbf{K}$ are all

even. Then the components of each of the four reciprocal
lattice vectors

$$\frac{1}{2} \, (\pm\mathbf{H} \pm \mathbf{K})$$

are all integers. Note also that in this space group the
structure factors are real and all phases are 0 or π.

6.2 The extensions of T.

One embeds the two-phase structure seminvariant T (17) and its
symmetry related variant

$$T_1 = \phi_{-\mathbf{H}} + \phi_{\mathbf{K}} \tag{18}$$

in the respective quartets

$$Q = T + \phi_{-\frac{1}{2}(\mathbf{H}+\mathbf{K})} + \phi_{-\frac{1}{2}(\mathbf{H}+\mathbf{K})} \, , \tag{19}$$

$$Q_1 = T_1 + \phi_{-\frac{1}{2}(-\mathbf{H}+\mathbf{K})} + \phi_{-\frac{1}{2}(-\mathbf{H}+\mathbf{K})} \, . \tag{20}$$

In view of (17) and (18) and the space group-dependent
relationships among the phases, it is readily verified that Q
and Q_1 are in fact (special) four-phase structure invariants
(quartets) and

$$T = T_1 = Q = Q_1 . \tag{21}$$

The quartets Q and Q_1 are said to be the extensions of the
seminvariant T. In this way the theory of the two-phase
structure seminvariant T is reduced to that of the quartets.
In particular, the neighborhoods of T are defined in terms of
the neighborhoods of the quartet.

6.3.   The first neighborhoods of the extensions.

Since two of the phases of the quartet Q (19) are identical, only three of the four main terms are distinct. The first neighborhood of Q is accordingly defined to consist of the three magnitudes

$$\left[\text{since } \left| E_{-\frac{1}{2}(H + K)} \right| = \left| E_{\frac{1}{2}(H + K)} \right| \right] :$$

$$|E_H|, \; |E_K|, \; \left| E_{\frac{1}{2}(H + K)} \right| . \qquad (22)$$

In a similar way the first neighborhood of the extension Q_1, (20), is defined to consist of the three magnitudes

$$|E_H|, \; |E_K|, \; \left| E_{\frac{1}{2}(H - K)} \right| . \qquad (23)$$

6.4.   The first neighborhood of T.

The first neighborhood of the two-phase structure seminvariant T is defined to consist of the set-theoretic union of the first neighborhoods of its extensions, i.e., in view of (22) and (23), of the four magnitudes

$$|E_H|, \; |E_K|, \; \left| E_{\frac{1}{2}(H + K)} \right|, \; \left| E_{\frac{1}{2}(H - K)} \right| . \qquad (24)$$

7.   THE SOLUTION STRATEGY.

One starts with the system of equations (5). By equating real and imaginary parts of (5) one obtains two equations for each reciprocal lattice vector H. The magnitudes $|E_H|$ and the atomic numbers Z_j are presumed to be known. The unknowns are the atomic postion vectors r_j and the phases ϕ_H. Owing to the redundancy of the system (5), one naturally invokes probabilistic techniques in order to eliminate the unknown position vectors r_j, and in this way to obtain relationships among the unknown phases Φ_H having probabilistic validity.

Choose a finite number of reciprocal lattice vectors $\mathbf{H}$, $\mathbf{K}$,... in such a way that the linear combination of phases

$$\psi = \Phi_H + \Phi_K + \ldots \tag{25}$$

is a structure invariant or seminvariant whose value we wish to estimate. Choose satellite reciprocal lattice vectors $\mathbf{H}'$, $\mathbf{K}'$, ... in such a way that the collection of magnitudes

$$\left|E_H\right|, \ \left|E_K\right|, \ldots; \quad \left|E_{H'}\right|, \ \left|E_{K'}\right|, \ldots \tag{26}$$

constitutes a neighborhood of ψ. The atomic position vectors $\mathbf{r}_j$ are assumed to be the primitive random variables which are uniformly and independently distributed. Then the magnitudes $|E_H|$, $|E_K|$,...; $|E_{H'}|$, $|E_{K'}|$,...; and phases Φ_H, Φ_K,...; $\Phi_{H'}$, $\Phi_{K'}$,... of the complex, normalized structure factors E_H, E_K,...; $E_{H'}$, $E_{K'}$,..., as functions [Eq. (5)] of the position vectors $\mathbf{r}_j$, are themselves random variables, and their joint probability distribution P may be obtained. From the distribution P one derives the conditional joint probability distribution

$$P(\Phi_H, \ \Phi_K, \ldots \mid |E_H|, \ |E_K|, \ldots; \ |E_{H'}|, \ |E_{K'}|, \ldots), \tag{27}$$

of the phases Φ_H, Φ_K,..., given the magnitudes $|E_H|$, $|E_K|$,...; $|E_{H'}|$, $|E_{K'}|$,..., by fixing the known magnitudes, integrating with respect to the unknown phases $\Phi_{H'}$, $\Phi_{K'}$,... from 0 to 2π, and multiplying by a suitable normalizing parameter. The distribution (27) in turn leads directly to the conditional probability distribution

$$P(\Psi \mid |E_H|, \ |E_K|, \ldots; \ |E_{H'}|, \ |E_{K'}|, \ldots) \tag{28}$$

of the structure invariant or seminvariant ψ, assuming as

known the magnitudes (26) constituting a neighborhood of ψ.
Finally, the distribution (28) yields an estimate for ψ (for
example, the mode) which is particularly good in the favorable
case that the variance of (28) happens to be small

8. ESTIMATING THE TRIPLET IN P1

Let the three reciprocal lattice vectors **H**, **K**, and **L** satisfy
(11). Refer to § 5 for the first neighborhood of the triplet
ψ_3 [Eq. (10)] and to § 7 for the probabilistic background.

Suppose that R_1, R_2, and R_3 are three specified non-
negative numbers. Denote by

$$P_{1/3} = P(\Psi|R_1, R_2, R_3)$$

the conditional probability distribution of the triplet ψ_3,
given the three magnitudes in its first neighborhood:

$$|E_{\mathbf{H}}| = R_1, \quad |E_{\mathbf{K}}| = R_2, \quad |E_{\mathbf{L}}| = R_3. \tag{29}$$

Then, carrying out the program described in § 7, one finds
(Cochran, 1955):

$$P_{1/3} = P(\Psi|R_1, R_2, R_3) \approx \frac{1}{2\pi I_0(A)} \exp (A \cos\Psi) \tag{30}$$

where

$$A = \frac{2\sigma_3}{\sigma_2^{3/2}} R_1 R_2 R_3 , \tag{31}$$

I_0 is the modified Bessel function, and σ_n is defined by (7). Since
$A>0, P_{1/3}$ has a unique maximum at $\Psi = 0$, and it is clear that the larger
the value of A the smaller is the variance of the distribution. See
Figure 1, where A = 2.316, Figure 2, where A = 0.731. Hence in the
favorable case that A is large, say, for example, A>3, the distribution

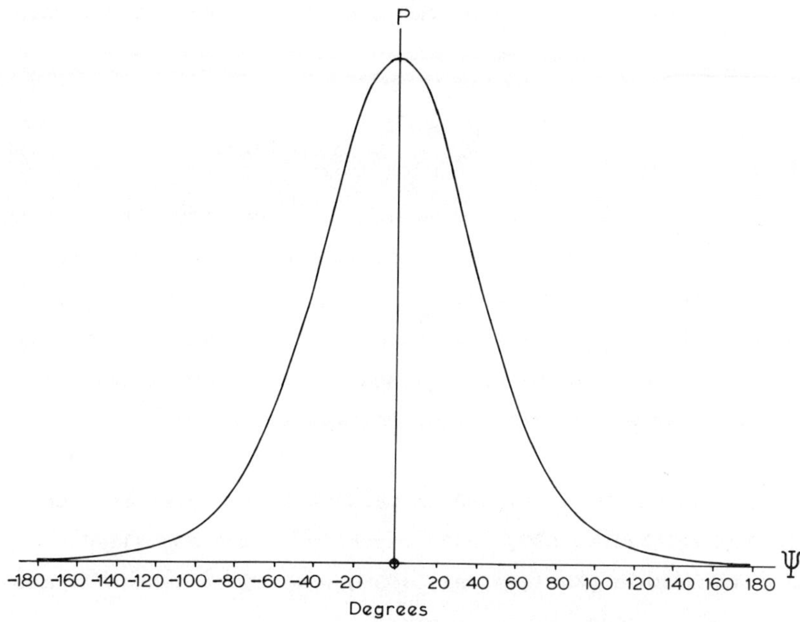

FIG. 1. The distribution $P_{1/3}$, equation (30), for A = 2.316.

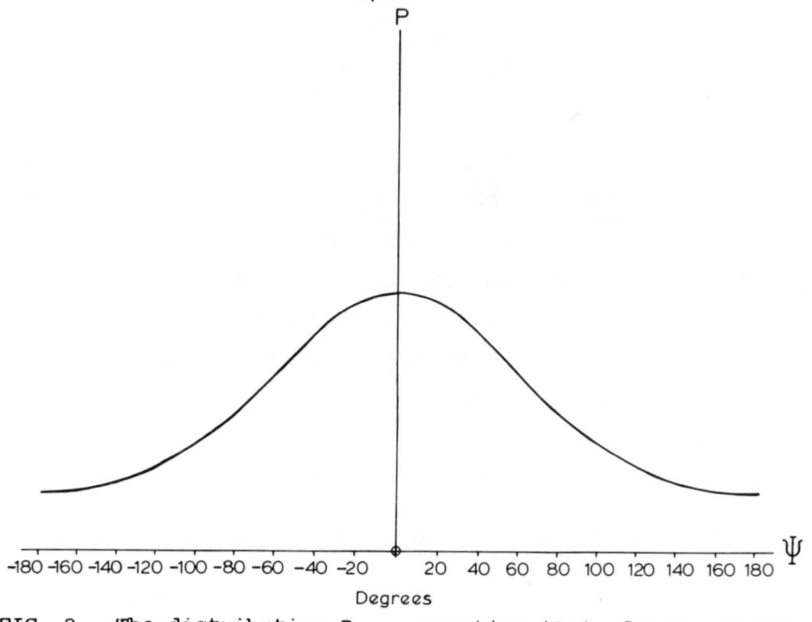

FIG. 2. The distribution $P_{1/3}$, equation (30), for A = 0.731

leads to a reliable estimate of the structure invariant ψ_3, zero in this case:

$$\psi_3 \approx 0 \text{ if A is large.} \qquad (32)$$

Furthermore, the larger the value of A, the more likely is the probabilistic statement (32). It is remarkable how useful this relationship has proven to be in the applications; and yet (32) is severely limited because it is capable of yielding only the zero estimate for ψ_3, and only those estimates are reliable for which A is large, the favorable cases.

It should be mentioned in passing that a distribution closely related to (30) leads directly to the so-called tangent formula (Karle <u>et al</u>, 1956) which is universally used by direct methods practitioners:

$$\tan \phi_h = \frac{\left\langle |E_K E_{h-K}| \sin(\phi_K + \phi_{h-K}) \right\rangle_K}{\left\langle |E_K E_{h-K}| \cos(\phi_K + \phi_{h-K}) \right\rangle_K} \quad , \quad (33)$$

in which h is a fixed reciprocal lattice vector, the averages are taken over the same set of vectors K in reciprocal space, usually restricted to those vectors K for which $|E_K|$ and $|E_{h-K}|$ are both large, and the sign of $\sin \phi_h$ ($\cos \phi_h$) is the same as the sign of the numerator (denominator) on the right hand side. The tangent formula is usually used to refine and extend a basis set of phases, presumed to be known.

9. ESTIMATING THE QUARTET

Two conditional probability distributions are described, one assuming as known the four magnitudes $|E|$ in the first neighborhood of the quartet, the second assuming as known the seven magnitudes |E| in its second neighborhood.

Suppose that **H**, **K**, **L**, and **M** are four reciprocal lattice vectors which satisfy (13). Refer to (15) for the first neighborhood of the quartet ψ_4 (12). Suppose that R_1, R_2, R_3, and R_4 are four specified non-negative numbers. Denote by

$$P_{1/4} = P(\Psi | R_1, R_2, R_3, R_4)$$

the conditional probability distribution of the quartet ψ_4, given the four magnitudes in its first neighborhood:

$$|E_{\mathbf{H}}| = R_1, \quad |E_{\mathbf{K}}| = R_2, \quad |E_{\mathbf{L}}| = R_3, \quad |E_{\mathbf{M}}| = R_4 \qquad (34)$$

Then (Hauptman, 1975a,b, 1976)

$$P_{1/4} = P(\Psi | R_1, R_2, R_3, R_4,) \sim \frac{1}{K} \exp(B \cos \Psi) \qquad (35)$$

where

$$B = \frac{2\sigma_4}{\sigma_2^2} R_1 R_2 R_3 R_4, \qquad (36)$$

K is a normalizing parameter not relevant here, and σ_n is defined by (7). Thus $P_{1/4}$ is identical with $P_{1/3}$, but B replaces A. Hence similar remarks apply to $P_{1/4}$. In particular, (35) always has a unique maximum at $\Psi = 0$ so that the most probable value of ψ_4, given the four magnitudes (34) in its first neighborhood, is zero, and the larger the value

of B the more likely it is that $\psi_4 \approx 0$. Since B values, of
order $1/N$, tend to be less than A values, of order $1/\sqrt{N}$, at
least for large values of N, the estimate (zero) of ψ_4 is in
general less reliable than the estimate (zero) of ψ_3. Hence
the goal of obtaining a reliable non-zero estimate for a
structure invariant is not realized by (35). The decisive
step in this direction is made next.

Employ the same notation as in the previous paragraph but
refer now to (15) and (16) for the second neighborhood of the
quartet ψ_4. Suppose that R_1, R_2, R_3, R_4, R_{12}, R_{23}, and R_{31},
are seven non-negative numbers. Denote by

$$P_{1/7} = P(\Psi|R_1, R_2, R_3, R_4; R_{12}, R_{23}, R_{31}) \qquad (37)$$

the conditional probability distribution of the quartet ψ_4,
given the seven magnitudes in its second neighborhood:

$$|E_H| = R_1, \quad |E_K| = R_2, \quad |E_L| = R_3, \quad |E_M| = R_4; \qquad (38)$$

$$|E_{H+K}| = R_{12}, \quad |E_{K+L}| = R_{23}, \quad |E_{L+H}| = R_{31}. \qquad (39)$$

The explicit form for $P_{1/7}$ has been found (Hauptman, 1975a,b,
1976) but is too long to be given explicitly here. Instead,
Figures 3-5 show the distribution (37) (solid line ———) for
typical values of the seven parameters (38) and (39). For
comparison the distribution (35) (broken line – – – –) is also
shown. Since the magnitudes $|E|$ have been obtained from a
real structure with $N = 29$, comparison with the true value of
the quartet is also possible. As already emphasized, the
distribution (35) always has a unique maximum at $\Psi = 0$. The
distribution (37), on the other hand, may have a maximum at Ψ
$= 0$, or π, or any value between these extremes, as shown by
Figures 3-5. Roughly speaking, the maximum of (37) occurs at

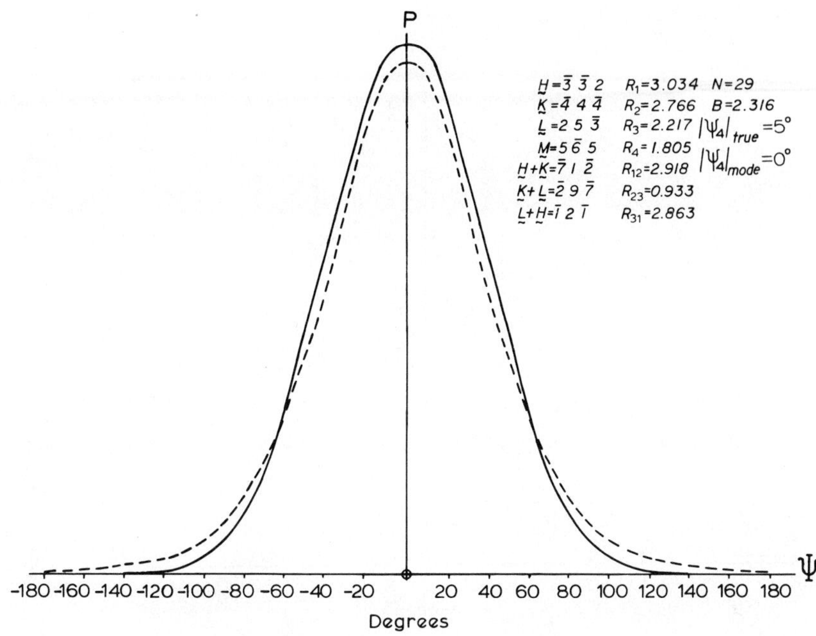

FIG. 3. The distribution (38) (——) and (35) (– – –) for the values of the seven parameters (36) and (37) shown. The mode of (38) is 0, of (35) always 0.

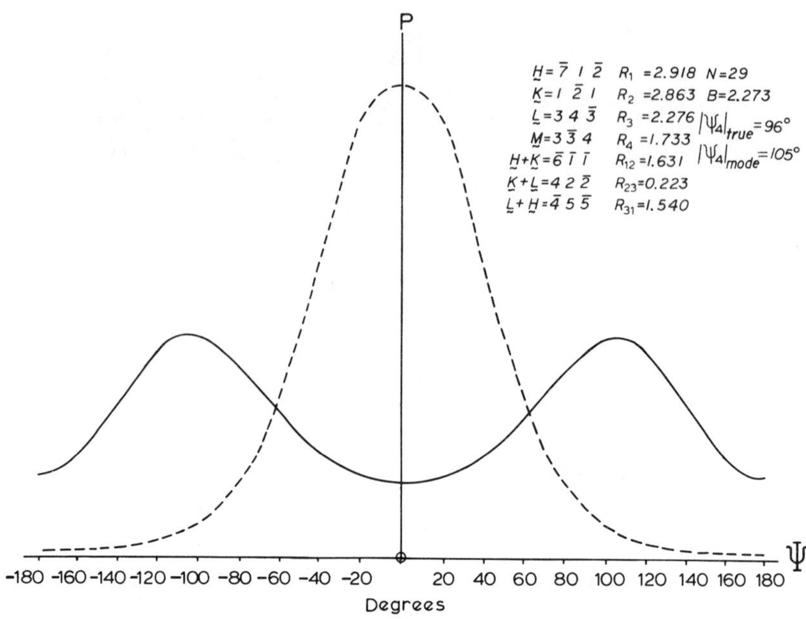

FIG. 4. The distribution (38) (———) and (35) (– – –) for the
values of the seven parameters (36) and (37) shown. The mode
of (38) is 105°, of (35) always 0.

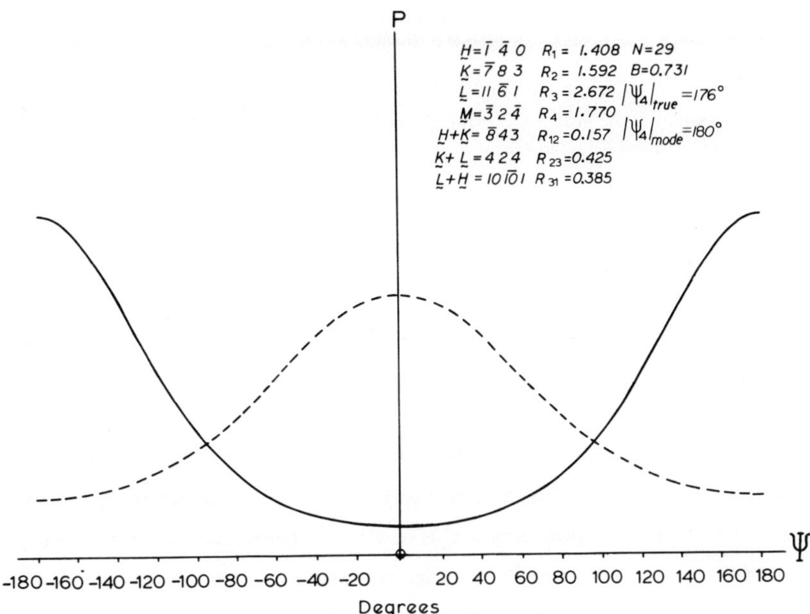

FIG. 5. The distribution (38) (———) and (35) (— — —) for he
values of the seven parameters (36) and (37) shown. The mode
of (38) is 180°, of (35) always 0.

0 or π according as the three parameters R_{12}, R_{23}, R_{31} are all
large or all small, respectively. These figures also clearly
show the improvement which may result when, in addition to the
four magnitudes (38), the three magnitudes (39) are also
assumed to be known. Not unexpectedly, when more information
is available, i.e. seven magnitudes |E| rather than only four,
the potential for determining more reliable estimates of the
structure invariants is increased. Finally, in the special
case that

$$R_{12} \sim R_{23} \sim R_{31} \sim 0, \tag{40}$$

the distribution (37) reduces simply to

$$P_{1/7} \sim \frac{1}{L} \exp(-2B'\cos\Psi), \tag{41}$$

where

$$B' = -\frac{1}{\sigma_2^3} (3\sigma_3^2 - \sigma_2\sigma_4)R_1R_2R_3R_4 , \tag{42}$$

and L is a normalizing parameter not relevant here, which has a unique maximum at $\Psi = \pi$ (Fig. 5).

10.   ESTIMATING THE TWO-PHASE STRUCTURE SEMINVARIANT IN $P\bar{1}$

Suppose that **H** and **K** are two reciprocal lattice vectors such that the three components of **H** + **K** are even integers. Then the linear combination T of two phases (17) is a structure seminvariant. Refer to § 6.4 for the four magnitudes (24) in the first neighborhood of T and to § 7 for the probabilistic background. Suppose that R_1, R_2, r_{12}, and $r_{1\bar{2}}$ are four non-negative numbers. In this space group every phase is 0 or π so that T = 0 or π and the conditional probability distribution of T, assuming as known the four magnitudes in its first neighborhood, is discrete. Dentote by $P_+(P_-)$ the conditional probability that T = 0 (π), given the four magnitudes in its first neighborhood:

$$|E_H| = R_1, \ |E_K| = R_2, \ |E_{\frac{1}{2}(H+K)}| = r_{12}, \ |E_{\frac{1}{2}(H-K)}| = r_{1\bar{2}} . \tag{43}$$

In the special case that all N atoms in the unit cell are identical, the solution strategy described in § 7 leads to (Green et al, 1976)

$$P_{\pm} \sim \frac{1}{M} \exp \left\{ \mp \frac{R_1 R_2 (r_{12}^2 + r_{1\bar{2}}^2)}{2N} \right\} \cosh \left\{ \frac{r_{12} r_{1\bar{2}} (R_1 \pm R_2)}{N^{1/2}} \right\} \tag{44}$$

where upper (lower) signs go together and

$$M = \exp \left\{ -\frac{R_1 R_2 (r_{12}^2 + r_{1\bar{2}}^2)}{2N} \right\} \cosh \left\{ \frac{r_{12} r_{1\bar{2}} (R_1 + R_2)}{N^{1/2}} \right\} +$$

$$\exp \left\{ +\frac{R_1 R_2 (r_{12}^2 + r_{1\bar{2}}^2)}{2N} \right\} \cosh \left\{ \frac{r_{12} r_{1\bar{2}} (R_1 - R_2)}{N^{1/2}} \right\} \tag{45}$$

It is easily verified that, under the assumption that R_1 and R_2 are both large, $P_+ \gg 1/2$ if r_{12} and $r_{1\bar{2}}$ are both large, but $P_+ \ll 1/2$ if one of r_{12}, $r_{1\bar{2}}$ is large and the other is small. Hence $T = 0$ or π respectively (the favorable cases of the neighborhood principle for T).

11. CONCLUDING REMARKS

The estimate (zero) of the triplet ψ_3 (32) and the related tangent formula (33) are the corner stones of most computer programs (Main et al, 1980; Karle et al, 1963) used for the direct solution of crystal structures. However, estimates of the quartets, in particular (41), used to identify those quartets whose values are close to π (the so-called negative quartets because their cosines are negative), as well as of the higher order structure invariants and seminvariants, often play an important (sometimes indispensable) role (DeTitta et al, 1975; Gilmore, 1983; Langs et al, 1975) particularly for complex structures when diffraction data may be limited in number and quality (Karle et al, 1958; Blank et al, 1976;

Bonnett et al, 1978; Busetta, 1976; Freer et al, 1977; Gilmore, 1977; Glidewell et al, 1979; Sax et al, 1976; Silverton et al, 1978a,b).

Major emphasis has been placed on the neighborhood principle and the important role played by the structure seminvariants. The conditional probability distribution of a structure seminvariant T, given the magnitudes $|E|$ in any of its neighborhoods, yields a reliable estimate for T in the favorable case that the variance of the distribution happens to be small. Since the structure seminvariants are the essential link between magnitudes $|E|$ and phases Φ, probabilistic methods play the central role in the solution of the phase problem.

ACKNOWLEDGEMENTS

This research was support in part by the National Science Foundation Grant No. CHE 8508724.

REFERENCES

BLANK, G. RODRIGUES, M., PLETCHER, J. and SAX, M. (1976). Acta Crystallogr. B32, 2970.

BONNETT, R., DAVIES, J.E., HURSTHOUSE, M.B. and SHELDRICK, G.M. (1978). Proc. R. Soc. London Ser. B 202, 249.

BUSETTA, B. (1976). Acta Crystallogr. A32, 139.

COCHRAN, W. (1955). Acta Crystallogr. 8, 473.

DE TITTA, G.T., EDMONDS, J.W., LANGS, D.A., and HAUPTMAN, H. (1975). Acta Crystallogr. **A31**, 472.

FREER, A.A. and GILMORE, C.J. (1977). J. Chem. Soc. Chem. Commun. **1977**, 296.

GILMORE, C.J. (1977). Acta Crystallogr. **A33**, 712.

GILMORE, C.J. (1983). A Computer Program for the Automatic Solution of Crystal Structures from X-Ray Data (University of Glasgow, Glasgow).

GILMORE, C.J., HARDY, A.D.W., MAC NICOL, D.D., and WILSON, D.R. (1977). J. Chem. Soc. Perkin Trans. 2, 1427.

GLIDEWELL, C. LILES, D.C., WALTON, D.J., and SHELDRICK. G.M. (1979). Acta Crystallogr. **B25**, 500.

GREEN. E. and HAUPTMAN, H. (1976). Acta Crystallogr. **A32**, 940.

HARKER, D. and KASPER, J.S. (1948). Acta Crystallogr. **1**, 70.

HAUPTMAN, H. (1972). Crystal Structure Determination: The Role of the Cosine Semininvariants. New York:Plenum Press.

HAUPTMAN, H. (1975a). Acta Crystallogr. **A31**, 671.

HAUPTMAN, H. (1975b). Acta Crystallogr. **A31**, 680.

HAUPTMAN, H. (1976). Acta Crystallogr. **A32**, 877.

HAUPTMAN, H. AND KARLE, J. (1953). Solution of the Phase Problem. I. The Centrosymmetric Crystal. ACA Monograph No. 3. Polycrystal Book Service.

HAUPTMAN, H. AND KARLE, J. (1956). Acta Crystallogr. **9**, 45.

HAUPTMAN, H. AND KARLE, J. (1959). Acta Crystallogr. **12**, 92.

KARLE, I.L., HAUPTMAN, H., KARLE, J., and WING, A.R. (1958). _Acta Crystallogr._ **11**, 257.

KARLE, I.L. and KARLE, J. (1963). _Acta Crystallogr._ **16**, 1963.

KARLE, J. and HAUPTMAN, H. (1950). _Acta Crystallogr._ **3**, 181.

KARLE, J. and HAUPTMAN, H. (1956). _Acta Crystallogr._ **9**, 635.

KARLE, J. and HAUPTMAN, H. (1961). _Acta Crystallogr._ **14**, 217.

LANGS, D.A. and DE TITTA, G.T. (1975). _Acta Crystallogr._ **A31**, 516. (Abstr. 02.2–14)

LESSINGER, L. and WONDRATSCHEK, H. (1975). _Acta Crystallogr._ **A31**, 521.

MAIN, P. _et al._ (1980). MULTAN 80: _A System of Computer Programs for the Automatic Solution of Crystal Structures from X-Ray Diffraction Data_ (University of York, York, England, and University of Louvain, Louvain, Belgium).

SAX, M., RODRIGUES, M., BLANK, G., WOOD, M.K. and PLETCHER, J. (1976). _Acta Crystallogr._ **B32**, 1953.

SILVERTON, J.V. and AKIYAMA, T. (1978a). _Am. Crystallogr. Assoc. Program_ Abstr. Ser. 2, (Abstr. PB15).

SILVERTON, J.V. and KABUTO, C. (1978b). _Acta Crystallogr._ **B34**, 588.

55. New developments in direct methods

Hai-fu Fan

1. INTRODUCTION

Even in the early days of the development of direct methods, the
investigators had to answer constantly the question: "Is the method now going to the end?". In reply to this, some people said
"Yes" and withdrew from the competition, others said "No" and
made further development of the method. Now direct methods have
become routine works in the determination of small molecular
structures. However it does not mean the development of direct
methods should be finished. Contradictorily it signals that the
time is now coming for direct methods to explore new fields of
applications. In fact, new achievements have already been made
in recent years. In this paper, three topics are to be discussed
: (1) Solving structures with pseudo-translational symmetry;
   (2) Tackling the phase problem of macromolecular structures by
       direct phasing of the SIR or OAS data;
   (3) Image processing in high resolution electron microscopy.

2. SOLVING STRUCTURES WITH PSEUDO-TRANSLATIONAL SYMMETRY

Structures with pseudo-translational symmetry are important in
mineralogy, structural chemistry and solid state physics, However they are difficult to solve owing to the existence of certain kind of systematically weak reflections. Accordingly a direct method has been developed in order to get rid of this dif-

ficulty. A structure $\rho(\underline{r})$ possesses the pseudo-translational symmetry $\underline{t}$ of the order p, if there exists $\underline{t} = \underline{T}/p$ so as

$$\rho(\underline{r}) \sim \rho(\underline{r} + \underline{t}) \qquad . \qquad (2.1)$$

(2.1) means that

$$\rho(\underline{r}) \neq \rho(\underline{r} + \underline{t}) \quad \text{and} \quad \int_V |\rho(\underline{r}) - \rho(\underline{r} + \underline{t})| d\underline{r} \ll \int_V \rho(\underline{r}) d\underline{r} \quad .$$

Where $\underline{t}$ is a pseudo-translation vector; $\underline{T}$ is the shortest lattice vector parrallel to $\underline{t}$; p is an integer greater than 1; V is the volume of unit cell. In the above case, the structure factor can be written approaximately as:

$$F_{\underline{H}} \sim \sum_{j=1}^{N/p} f_j \exp(i2\pi \underline{H} \cdot \underline{r}_j) + f_j \exp\{i2\pi \underline{H} \cdot (\underline{r}_j + \underline{t})\} +$$

$$+ f_j \exp\{i2\pi \underline{H} \cdot (\underline{r}_j + 2\underline{t})\} + \cdots + f_j \exp\{i2\pi \underline{H} \cdot (\underline{r}_j + (p-1)\underline{t})\}$$

$$= \sum_{j=1}^{N/p} f_j \exp(i2\pi \underline{H} \cdot \underline{r}_j) \times$$

$$\left[1 + \exp(i2\pi \underline{H} \cdot \underline{t}) + \exp(i2\pi \underline{H} \cdot 2\underline{t}) + \cdots + \exp\{i2\pi \underline{H} \cdot (p-1)\underline{t}\}\right]$$

$$(2.2)$$

The sum of the series in the bracket of (2.2) is given by

$$S = \{\exp(i2\pi \underline{H} \cdot p\underline{t}) - 1\} / \{\exp(i2\pi \underline{H} \cdot \underline{t}) - 1\}$$

$$= \begin{cases} p, & \text{if } \underline{H} \cdot \underline{t} = \text{integer} \quad ; \\ 0, & \text{if } \underline{H} \cdot \underline{t} \neq \text{integer} \quad . \end{cases}$$

Notice that $\underline{H} \cdot p\underline{t} = \underline{H} \cdot \underline{T}$ = integer. Hence all reflections with $\underline{H} \cdot \underline{t} \neq$ integer will be systematically weak leading to an effect of pseudo systematic extinction. In other words, all the strong reflections will satisfy the condition $\underline{H} \cdot \underline{t}$ = integer. Usually in solving a structure with pseudo-translational symmetry the phases of the systematically 'strong' reflections are relatively easy to determine. However the derivation of phases for the systematically 'weak' reflections will be extremely difficult. In order to solve the phase problem for the 'weak' reflections by making use of the phases previously obtained for the 'strong' ones, a modified Sayre equation was derived (Fan, 1975; Fan, He, Qian & Liu, 1978; see also Fan, Yao, Main & Woolfson, 1983):

$$F_{H,wk} = 2\frac{\Theta}{V} \sum_{H'} F_{H',str} F_{H-H',wk} \qquad . \qquad (2.3)$$

An automatic procedure to use formula (2.3) was established under the collaboration between the direct methods group in York and that in Beijing. The main points of the procedure are as follows:

(1) Automatic search of the pseudo systematic extinction rule is first performed and the reflections are grouped accordingly;

(2) Each reflections group is normalized independently;

(3) The $\sum_2$ relationships involving three 'weak' reflections are eliminated and the phase development process is devided into two steps. In the first step only the phases of the 'strong' reflections are developed. Then in the second step the phases of the 'weak' reflections are derived by making use of the phases of the 'strong' ones.

This procedure has been incorporated in the structure analysis program system SAPI-85 (Yao, Zheng, Qian, Han, Gu & Fan, 1985; Fan, 1986). An example of automatically solving an unknown structure with pseudo-translational symmetry by SAPI-85 is given below:

RLD6, $C_{27}H_{31}NO_9$, belongs to space group $P2_1/c$ with a=24.420, b=8.363, c=24.747 Å, β=90.22°. There are 74 non-hydrogen atoms in the asymmetric unit which includes two independent molecules related by a pseudo-translation vector $\underline{t}$ = ($\underline{a}$ + $\underline{c}$)/2. RANTAN-81 (Yao,1981) failed to solve the structure from 100 starting sets with the default control parameters. On the other hand SAPI-85 with default control yielded almost the complete structure from the 70th random starting set (see Fig.2.1). The diffraction data and refined atomic parameters of RLD6 were kindly provided by Professor T. C. W. Mak.

Studies on the application of direct methods to structures with pseudo-translational symmetry have also been reported by other authors (Gramlich, 1975, 1978, 1984; Böhme, 1982; Prick, Beurskens & Gould, 1983; Giacovazzo, 1984).

Among the structures having pseudo-translational symmetry, there is a typical class known as superstructure. Superstructure and incommensurate structure are both modulated structures. Up to now, there is no straightforward way to solve incommensurate structures, in spite of their significance in pure and applied sciences. Further development of the direct method described above may be of use in dealing with incommensurate structures. Recently an interesting type of condensed materials called quasicrystal has been discovered and extensively investigated. The structure of quasicrystals, though not containing translational symmetry in 3-dimensional space, can be described in reciprocal space by a way similar to that used for incommensurate structures. It can be expected that direct methods can also find their use in solving the structures of quasicrystals.

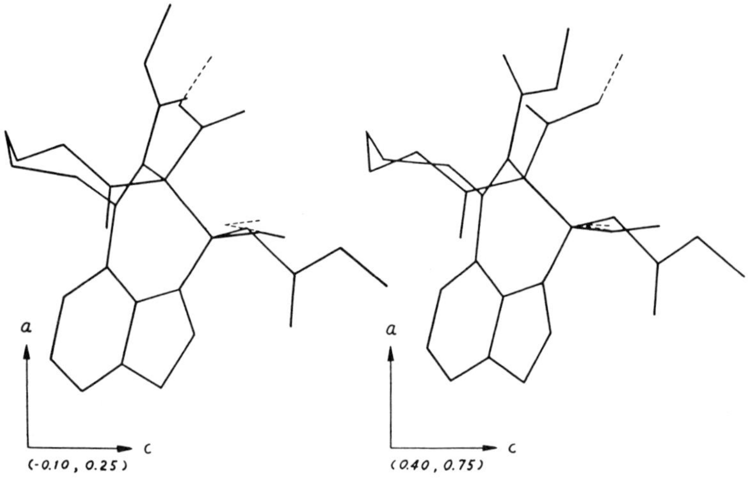

Figure 2.1

Two independent molecules of RLD6 projected along the b axis. The fragments shown with solid lines were obtained from the E-map output by SAPI-85 directly.

3. TACKLING THE PHASE PROBLEM OF MACROMOLECULAR STRUCTURES BY DIRECT PHASING OF THE SIR OR OAS DATA

Since the 1960's attempts have been made to combine direct methods with the single isomorphous replacement (SIR) and the one-wavelength anomalous scattering (OAS) methods (Coulter, 1965; Fan, 1965; Karle, 1966). Such a combination would be important for the structure analysis of proteins due to the possibilities of reducing the number of heavy-atom derivatives needed for solving a protein structure, saving the time in data collection thus enhancing the effective lifetime of the protein crystal and simplifying the process of the structure determination. However during the last decade, procedures proposed were not as successful as expected. Recently some progresses have been achieved. Hauptman (1982a,b) integrated the probabilistic theory of the triplet structure invariants with the SIR and OAS techniques. Giacovazzo (1983) reported a similar theory. Karle (1983; 1984a, b) proposed simple rules for the evaluation of triplet structure invariants from SIR or OAS data. All these methods have been successful in deriving large number of reliable triplet structure invariants using error free data. An alternative procedure has been proposed (Fan, Han, Qian & Yao, 1984; Fan, Han & Qian, 1984; Fan & Gu, 1985; Yao & Fan, 1985; Qian, Fan & Gu, 1985), which, by the test with experimental data, has been proved to be efficient in breaking the enantiomorphous phase ambiguities yielding large number of reliable individual phases.

3.1. Enantiomorphous phase ambiguities in SIR and OAS methods

In the SIR case, for a given reciprocal vector $\underline{H}$, we have

$$F_{H,N} = F_{H,D} - F_{H,R} \quad . \tag{3.1.1}$$

Where $F_{H,N}$, $F_{H,D}$ and $F_{H,R}$ are the structure factors of the native protein, the derivative and the replacing atoms respectively. The moduli of $F_{H,N}$ and $F_{H,D}$ can be obtained experimentally. Accordingly the parameters of the replacing atoms can be found

and $F_{H,R}$ be calculated. Consequently, we have two ways for drawing the triangle of (3.1.1) leading to an phase doublet for both
$F_{H,N}$ and $F_{H,D}$ in the phase vector diagram, Fig. 3.1.
In the case of OAS, we have

$$F_H^+ = F_H + F_{H,A}'' \quad \text{and} \quad F_H^{-*} = F_H - F_{H,A}'' \quad . \quad (3.1.2)$$

Here F_H is the contribution of both the normal and the real part
anomalous scattering from the whole unit cell. F_H^{-*} denotes the
conjugate of F_H^- . $F_{H,A}''$ is the contribution from the imaginary-
part scattering of the anomalous scatterrers, i.e.

$$F_{H,A}'' = \sum_{A=1}^{N_A} i\Delta f_A'' \exp(i2\pi \underline{H}\cdot\underline{r}_A) \quad .$$

It follows from (3.1.2) that

$$F_H^+ = F_H^{-*} + 2F_{H,A}'' \quad . \quad (3.1.3)$$

The moduli of F_H^+ and F_H^{-*} can be obtained experimentally and
then $F_{H,A}''$ can be derived. Hence we have also two ways to draw
the triangle of (3.1.3) leading to an enantiomorphous phase doublet for F_H as shown in Fig. 3.2. Both the phase doublets in SIR
and OAS cases can be expressed in the generalized form

$$\phi_H = \phi_H' \pm |\Delta\phi_H| \quad , \quad (3.1.4)$$

In the case of SIR:

$$\phi_H' = \phi_{H,R} \quad ,$$

$$\Delta\phi_{H,N} = \pm \cos^{-1}\{(F_{H,D}^2 - F_{H,R}^2 - F_{H,N}^2)/2F_{H,R}F_{H,N}\}$$

$$\Delta\phi_{H,D} = \pm \cos^{-1}\{(F_{H,D}^2 + F_{H,R}^2 - F_{H,N}^2)/2F_{H,R}F_{H,D}\}$$

$$(3.1.5)$$

Where $\Delta\phi_{H,N}$ and $\Delta\phi_{H,D}$ are the phase differences of the native
and the derivative respectively.
In the case of OAS:

$$\phi_H' = \phi_{H,A}'' \qquad \text{or} \qquad \phi_H' = \phi_{H,A} + \omega_H$$

where $\phi_{H,A}''$ is the phase of $F_{H,A}''$, $\phi_{H,A}$ is the phase of $F_{H,A}$, ω_H is the phase difference between $F_{H,A}''$ and $F_{H,A}$. If there is only one kind of anomalous scatterer in the unit cell, then $\omega_H = \pi/2$. We have for the OAS case (see Blundell & Johnson, 1976)

$$\Delta\phi_H = \pm \cos^{-1}\{(F_H^+ - F_H^-)/2F_{H,A}''\} \qquad . \qquad (3.1.6)$$

Notice that ϕ_H now is the phase of $F_H = (F_H^+ + F_H^-)/2$.

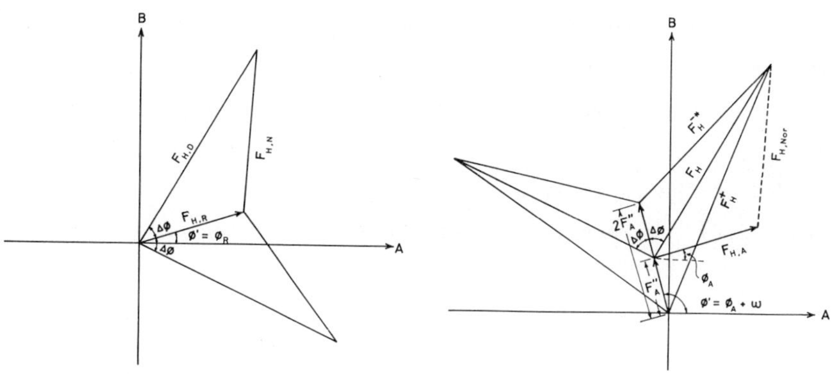

Figure 3.1
Enantiomorphous phase doublet
in SIR case

Figure 3.2
Enantiomorphous phase doublet
in OAS case

3.2. Treatment of errors and the best starting phases in the presence of enantiomorphous phase ambiguity

The 'best phase relationship'

Following Blow & Crick (1959), we can introduce into direct methods the concepts of 'best phase' and 'figure of merit' for individual reflections. A best normalized structure factor is defined as

$$E_{H,best} = m_H|E_H|\exp(i\phi_{H,best}) \qquad , \qquad (3.2.1)$$

where

$$\vec{m}_H = m_H \exp(i\phi_{H,best}) = \int \exp(i\phi_H)P(\phi_H)d\phi_H \qquad .$$

A triplet phase relationship which consists of the best normalized structure factors is called the 'best' phase relationship'.

Expressions of the best phase and the figure of merit for a

single reflection with enantiomorphous phase ambiguity

Defining $\Delta\phi_{H,best} = \phi_{H,best} - \phi'_H$, the $\Delta\phi_{H,best}$ and m_H can be
expressed as follows (Fan, Han & Qian, 1984)

$$\tan(\Delta\phi_{H,best}) = 2\{P_+(\Delta\phi_H) - \tfrac{1}{2}\}\sin|\Delta\phi_H| / \cos\Delta\phi_H \quad (3.2.2)$$

$$m_H = \exp(-\sigma_H^2/2)\left[\{2(P_+ - \tfrac{1}{2})^2 + \tfrac{1}{2}\}(1-\cos2\Delta\phi_H)+\cos2\Delta\phi_H\right]^{\tfrac{1}{2}}$$
$$(3.2.3)$$

where $P_+(\Delta\phi_H)$ is the probability that $\Delta\phi_H$ has a positive sign,
while σ_H is related to the experimental error and can be derived
from the standard deviation D of the 'lack of closure error'
(Blow & Crick, 1959). m_H may be regarded as a measure of reliability of $\Delta\phi_{H,best}$. As can be seen, there are three factors
included in the expression of m_H :

$\exp(-\sigma_H^2/2)$ a measure of the sharpness of the experimental
                 distribution of ϕ_H

$(P_+ - \tfrac{1}{2})^2$ a measure of the bias of $\Delta\phi_H$ towards positive or
                 negative. It reaches the maximum when $P_+ = 0$ or 1

$\cos2\Delta\phi_H$ a measure of the closeness of the two possible
                 phases, $\phi_H^+ = \phi'_H + |\Delta\phi_H|$ and $\phi_H^- = \phi'_H - |\Delta\phi_H|$. It
                 reaches the maximum when $\Delta\phi_H = 0$ or π .

Either of the last two factors will have no effect on m_H when
the other one reaches the maximum.

The best start

Substitute $P_+ = P_- = \tfrac{1}{2}$ into (3.2.2). It follows that

$$\Delta\phi_{H,best} = \begin{cases} 0 , & \text{if } SIGN(\cos\Delta\phi_H) = 1 \\ \pi , & \text{if } SIGN(\cos\Delta\phi_H) = -1 \end{cases}$$

or $\qquad \exp(i\phi_{H,best}) = SIGN(\cos\Delta\phi_H)\exp(i\phi'_H)$ (3.2.4)

Meanwhile, (3.2.3) reduces to

$$m_H = \exp(-\sigma_H^2/2)|\cos\Delta\phi_H| \qquad . \qquad (3.2.5)$$

Substituting (3.2.4) and (3.2.5) into (3.2.1), one obtains

$$E_{H,best} = \exp(-\sigma_H^2/2) \; \cos\Delta\phi_H |E_H| \exp(i\phi_H') \qquad (3.2.6)$$

This is the 'best' normalized structure factor which could be obtained at the begining from the corresponding doublet. In other words, when enantiomorphous ambiguities are present, the best way to start a direct method process is to use the averaged value of the phase doublet as the starting phase and use the weight $\exp(-\sigma_H^2/2)|\cos\Delta\phi_H|$ for the corresponding E_H.

3.3. Incorporating the phase doublet information into direct method formulas

Use of modified Sayre equation

Denote the heavy atoms with known positions in the unit cell by p and the atoms of the unknown part by u. Then according to Fan (1965;1975) we have the so-called modified Sayre equation as

$$F_H = \frac{\Theta_{H,u}}{V} \sum_{H'} F_{H'} F_{H-H'} - \sum_p \left(\frac{\Theta_{H,u}}{\Theta_{H,p}} - 1\right) F_{H,p} \quad , \qquad (3.3.1)$$

where Θ is an atomic form factor. Replacing F_H by $|F_H|\exp(i\phi_H)$ and replacing ϕ_H by $\phi_H' + \Delta\phi_H$, (3.3.1) becomes

$$|F_H|\exp(i\Delta\phi_H) = \frac{\Theta_{H,u}}{V} \sum_{H'} |F_{H'} F_{H-H'}| \exp\{i(-\phi_H'+\phi_{H'}'+\phi_{H-H'}'+$$
$$+\Delta\phi_{H'}+\Delta\phi_{H-H'})\} - \sum_p \left(\frac{\Theta_{H,u}}{\Theta_{H,p}} - 1\right) |F_{H,p}| \exp\{i(\phi_{H,p}-\phi_H')\} \; . \qquad (3.3.2)$$

Taking the imaginary part of (3.3.2) and denoting $-\phi_H'+\phi_{H'}'+\phi_{H-H'}'$ by Φ_3' , we have

$$|F_H|\sin\Delta\phi_H = \frac{\Theta_{H,u}}{V} \left\{ \sum_{H'} |F_{H'} F_{H-H'}| \sin(\Phi_3'+\Delta\phi_{H'}+\Delta\phi_{H-H'}) \right\}$$
$$- \sum_p \left(\frac{\Theta_{H,u}}{\Theta_{H,p}} - 1\right) |F_{H,p}| \sin(\phi_{H,p}-\phi_H') \quad . \qquad (3.3.3)$$

Equation (3.3.3) can be used to refine the sign as well as the magnitude of $\Delta\phi_H$, once a large starting set of $\Delta\phi_H$ is available.

Use of the combination of Cochran's distribution and Sim's distribution

With the expression $\phi_H = \phi_H' + \Delta\phi_H$, the Cochran distribution (Cochran, 1955) can be modified to give (Fan et al., 1984):

$$P_{Cochran}(\Delta\phi_H) = \{2\pi I_o(\alpha')\}^{-1}\exp\{\alpha'\cos(\Delta\phi_H - \beta')\} \qquad (3.3.4)$$

where $I_o(\alpha')$ is the modified Bessel function,

$$\alpha' = \left[\{\sum_{H'}\kappa_{HH'}\sin(\Phi_3'+\Delta\phi_{H'}+\Delta\phi_{H-H'})\}^2 + \{\sum_{H'}\kappa_{HH'}\cos(\Phi_3'+\Delta\phi_{H'}+\Delta\phi_{H-H'})\}^2\right]^{\frac{1}{2}}$$

$$\tan\beta' = \sum_{H'}\kappa_{HH'}\sin(\Phi_3'+\Delta\phi_{H'}+\Delta\phi_{H-H'}) \,/\, \sum_{H'}\kappa_{HH'}\cos(\Phi_3'+\Delta\phi_{H'}+\Delta\phi_{H-H'})$$

$$\Phi_3' = -\phi_H'+\phi_{H'}'+\phi_{H-H'}' \qquad , \qquad \kappa_{HH'} = 2\sigma_3\sigma_2^{-3/2}|E_H E_{H'} E_{H-H'}| \qquad .$$

In the same way, Sim's distribution (Sim, 1959) can be modified as

$$P_{Sim}(\Delta\phi_H) = \{2\pi I_o(x)\}^{-1}\exp\{x\cos(\Delta\phi_H-\delta_H)\} \qquad , \qquad (3.3.5)$$

where $\qquad x = 2|E_H E_{H,p}|/\{\sum_u Z_u^2/\sigma_2\} \qquad , \qquad \delta_H = \phi_{H,p} - \phi_H' \qquad .$

Combination of (3.3.4) and (3.3.5) gives the total probability distribution of $\Delta\phi_H$ (Fan & Gu, 1985):

$$P(\Delta\phi_H) = \{2\pi I_o(\alpha)\}^{-1}\exp\{\alpha\cos(\Delta\phi_H - \beta)\} \qquad , \qquad (3.3.6)$$

where $\qquad \alpha = \left[\{\sum_{H'}\kappa_{HH'}\sin(\Phi_3'+\Delta\phi_{H'}+\Delta\phi_{H-H'})+x\sin\delta_H\}^2 + \right.$

$$\left. \{\sum_{H'}\kappa_{HH'}\cos(\Phi_3'+\Delta\phi_{H'}+\Delta\phi_{H-H'})+x\cos\delta_H\}^2\right]^{\frac{1}{2}} \qquad , \qquad (3.3.7)$$

$$\tan\beta = \frac{\sum_{H'}\kappa_{HH'}\sin(\Phi_3'+\Delta\phi_{H'}+\Delta\phi_{H-H'})+x\sin\delta_H}{\sum_{H'}\kappa_{HH'}\cos(\Phi_3'+\Delta\phi_{H'}+\Delta\phi_{H-H'})+x\cos\delta_H} \qquad . \qquad (3.3.8)$$

Since $|\Delta\phi_H|$ is a known quantity when phase doublet information is available, the probability that $\Delta\phi_H$ has a positive sign can be derived from (3.3.6):

$$P_+(\Delta\phi_H) = \frac{1}{2} + \frac{1}{2}\tanh\left[\sin|\Delta\phi_H|\{\sum_{H'}\kappa_{HH'}\sin(\Phi_3'+\Delta\phi_{H'}+\Delta\phi_{H-H'})+x\sin\delta_H\}\right].$$

$$(3.3.9)$$

With (3.3.9) the phase problem is now reduces to a sign problem.

On the other hand, by maximizing (3.3.6) we have $\Delta\phi_H = \beta$. Hence we can calculate the 'most probable' value of $\Delta\phi_H$ from (3.3.8). By replacing the $E_{H'}$ and $E_{H-H'}$ with their 'best' values, we can modify (3.3.8) and (3.3.9) to give

$$P_+(\Delta\phi_H) = \frac{1}{2} + \frac{1}{2}\tanh\left[\sin|\Delta\phi_H|\{\sum_{H'} m_{H'}m_{H-H'}\kappa_{HH'} \times\right.$$

$$\left.\sin(\Phi_3'+\Delta\phi_{H'best}+\Delta\phi_{H-H'best})+x\sin\delta_H\right] \quad (3.3.10)$$

and

$$\tan(\Delta\phi_H) = \frac{\sum_{H'} m_{H'}m_{H-H'}\kappa_{HH'}\sin(\Phi_3'+\Delta\phi_{H'best}+\Delta\phi_{H-H'best})+x\sin\delta_H}{\sum_{H'} m_{H'}m_{H-H'}\kappa_{HH'}\cos(\Phi_3'+\Delta\phi_{H'best}+\Delta\phi_{H-H'best})+x\cos\delta_H}$$

$$(3.3.11)$$

Formulas (3.2.2), (3.2.3), (3.3.10) and (3.3.11) can be used for ab initio phasing of the SIR or OAS data leading to unique estimates of individual phases.

Use of Hauptman's distribution

Hauptman (1982a,b) integrated the probabilistic theory of the three-phase structure invariants with SIR and OAS techniques leading to a series of formulas, which give unique estimates of the three-phase structure invariants. However, initial applications (Z.-B. Xu et al., 1984; Furey et al., 1986) showes that Hauptman's distribution tended to produce protein phases which were overly biased toward those of the heavy-atom substructure. Fortier (1985) incorporated the heavy-atom information into Hauptman's formulas resulting in the increase of accuracy of phase estimates. An alternative method was proposed recently by Hao (1986). He incorporated the phase doublet information into Hauptman's distribution leading to unique estimates of individual phases. Unlike Fortier's method, Hao makes use not only the magnitudes but also the phases of the heavy-atom substructure. On the other hand Hao's method is related to that of Fan et al. (1984) by replacing Cochran's distribution with Hauptman's. It is expected that Hauptman's distribution will give better result.

3.4. Direct phasing experimental OAS data of two known proteins

In a test calculation, the experimental OAS data of insulin and
APP were directly phased by the procedure of Fan et al. (1984,
1985). Satisfactory results have been obtained.

Data

Insulin crystallizes in space proup R3 with a=82.5, c=34.0 $\overset{o}{A}$,
γ=120° and Z=9. There are $\sim$6400 independent reflections at 1.9 $\overset{o}{A}$
resolution. The data was kindly provided by Drs G. Dodson and E.
Dodson. APP crystallizes in space group C2 with a=34.18, b=32.92
, c=28.44 $\overset{o}{A}$, β=105.30^o and Z=4. There are $\sim$2100 independent reflections at 2.1 $\overset{o}{A}$ resolution. The data was kindly provided by
Professor T. Blundell. 1000 largest E's and 60000 strongest $\sum_2$
relationships from each structure were used in the test calculation.

Test and results

Starting with $P_+ = \frac{1}{2}$, values of $\Delta\phi_{H,best}$ and m_H were calculated
using (3.2.2) and (3.2.3) respectively and then substituted into
(3.3.10) to calculate new values of P_+ . Most of them differ
from $\frac{1}{2}$ considerably. With the new set of P_+ , one more cycle of
iteration led to further improvement on the reliability. In order to examine the result statistically, the reflections were
arranged in descending order of $\left| P_+ - \frac{1}{2} \right|$ and then cummulated
into 5 groups, which contain the top 200, 400, 600, 800 and all
reflections respectively. The average error in phase estimates
and the percentage of reflections with the signs of $\Delta\phi_H$ correct-
ly determined were calculated for each group. The results are
listed in Table 3.1a and 3.2a for insulin and APP respectively.
Multiple isomorphous replacement (MIR) phases of insulin and
SIR-OAS phases of APP were also cummulated according to their
figures of merit. The results are listed respectively in Tables
3.1b and 3.2b for comparision. In the case of insulin, the isomorphism of the sample crystals was not very good. The direct
method phases are obviously better than those from MIR method,

at least for a thousand of reflections with large E values. On
the other hand the isomorphism in APP was nearly perfect. The
direct method phases are still comparable in quality with those
of SIR-OAS method, so far as one half of reflections at 2.1 Å
resolution are concerned. It can be concluded that the direct
method under examine is very efficient in breaking the enantiomorphous phase ambiguity of protein OAS data. It can provide a
very large number of reliable starting phases, which in turn can
be the base of further phase development and refinement.

Table 3.1

Results on phasing the experimental data of insulin

(a)   (b)
Direct phased OAS method   MIR method

Group	%	Error	Group	%	Error
1	93.0	33°	1	70.0	41°
2	94.5	35°	2	69.5	41°
3	94.0	37°	3	63.5	44°
4	93.5	38°	4	60.9	48°
5	92.3	41°	5	58.8	54°

Table 3.2

Results on phasing the experimental data of APP

(a)   (b)
Direct phased OAS method   SIR-OAS method

Group	%	Error	Group	%	Error
1	95.0	24°	1	92.0	16°
2	90.8	30°	2	88.8	20°
3	91.3	29°	3	87.5	24°
4	89.0	32°	4	84.1	30°
5	86.3	36°	5	79.4	39°

4. IMAGE PROCESSING IN HIGH RESOLUTION ELECTRON MICROSCOPY

High resolution electron microscopy (HREM) is a powerful tool in
the investigation of crystal structures, especially when the
sample is not suitable for X-ray analysis. However in many cases
,high resolution electron micrographs without special processing
can not reflect directly the true structure. On the other hand,
direct methods are actually some kind of image processing technique, which may be of use in the processing of high resolution
electron micrographs. Under the collaboration of the research
group on crystal structure analysis and that on HREM in the Institute of Physics in Beijing, a new technique of image processing in HREM using direct methods has been proposed, which is a
new junction of X-ray crystallography and electron microscopy.
The procedure is devided into two parts, i.e. the image deconvolution and the resolution enhancement.

4.1. Image deconvolution

Let T(H) denotes the Fourier transform of the intensities of an
electron micrograph (EM). Under the weak-phase-object approximation[*], in which the dynamic diffraction effect is neglected, the
structure factor F(H) is related to T(H) by the following formula:

$$F(H) = T(H) \; / \; 2\sigma \sin\chi_1(H) \cdot \exp\{-\chi_2(H)\} \qquad . \qquad (4.1.1)$$

Here $\sigma = \pi/\lambda U$, λ is the electron wavelength and U the accelerating voltage. $\sin\chi_1(H) \cdot \exp\{-\chi_2(H)\}$ is the contrast transfer function. in which

$$\chi_1(H) = \pi \Delta f \lambda H^2 + \frac{1}{2}\pi C_s \lambda^3 H^4 \qquad ,$$

$$\chi_2(H) = \frac{1}{2}\pi^2 \lambda^2 H^4 D^2 \qquad .$$

---

* The applicability of the weak-phase-object approximation has
  been demonstrated by Unwin & Henderson (1975) for biological
  specimens and by Klug (1978/1979) for an inorganic compound.

Here Δf is the defocus value, C_s is the spherical aberration co-efficient and D is the standard deviation of the Guassian dis-tribution of defocus due to the chromatic aberration (Fijes, 1977). The task of image deconvolution is to find out F(H) from the corresponding T(H). For this purpose we have to know Δf, C_s and D in advance. Among these three factors, C_s and D can be de-termined experimentally without much difficulties. However it is usually very difficult to measure the value of Δf accurately. Hence in order to carry out image deconvolution, the main pro-blem is to find out Δf. With the estimated values of C_s and D, we can calculate a set of F(H) from (4.1.1) for a given value of Δf. If this value is correct, the corresponding set of F(H) should obey the Sayre equation (Sayre, 1952)

$$F(H) = \frac{\Theta}{V} \sum_{H'} F(H') \, F(H-H') \qquad . \qquad (4.1.2)$$

Hence the true Δf can be found by a systematic change of the trial Δf. The practical procedure should be as follows:

(1) Calculate a set of T(H) from an EM.

(2) Assign trial values of Δf in a wide range with a small interval, say 10 Å. For each trial Δf, a set of F(H) is calcu-lated from T(H) using equation (4.1.1). Reflections with $|\sin\chi_1(H) \cdot \exp\{-\chi_2(H)\}| < 0.2$ will be neglected.

(3) Calculate the figure of merit S for each set of F(H) using the following formula (Debaerdemaeker, Tate & Woolfson, 1985)

$$S = \frac{\{\sum_H E^*(H) \sum_{H'} E(H')E(H-H')\}^2}{\sum_H |E(H)|^2 \sum_H |\sum_{H'} E(H')E(H-H')|^2} \qquad .$$

S has a value between 0 and 1. The greater the value of S, the better the set of F(H) fits the Sayre equation.

(4) Find out the greatest S and then Fourier Transform the corresponding set of F(H) to deconvolute the image.

The procedure has been test by simulating calculations with a series of theoretical EM of chlorinated phthalocyanine at 2 Å resolution taken under different Δf values. The other conditions are: Accelerating voltage = 500 kV; C_s = 1 mm; D = 150 Å . Some of the results are shown in Fig. 4.1. It can be seen that the deconvolution was quite successful. For further detail the reader is referred to Han, Fan, Zheng & Li (1986).

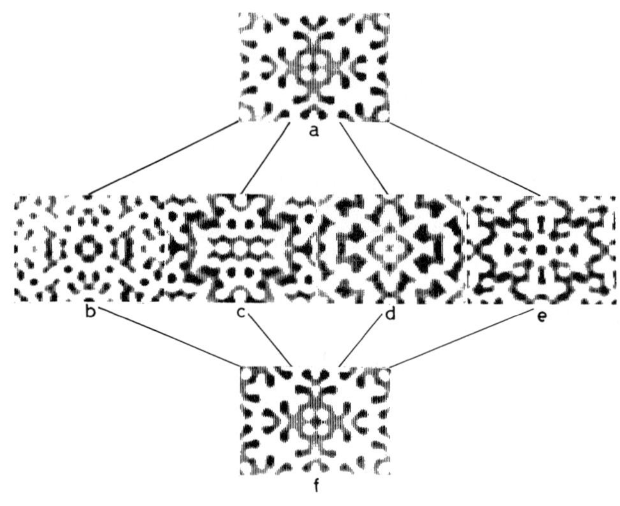

Figure 4.1

**Deconvolution of theoretical EM's
of copper chlorinated phthalocaynine**

a — the expected image; b,c,d and e — theoretical EM's taken with Δf equal to -1000, +1000, -600 and +600 Å respectively; f — the deconvolution result.

4.2. Resolution enhancement

Phase extension

An electron diffraction (ED) pattern usually contains information up to 1 Å resolution, which is considerably higher than that which can be reached by an EM. In addition, the intensities of ED are independent of defocus and spherical aberration of the objective lens. Accordingly, under the weak-phase-object approximation a set of high resolution structure amplitudes of good quality can be obtained from an ED pattern. However, the struc-

ture analysis by ED alone is subject to the well known difficulty of the 'phase problem'. On the other hand, an EM after suitable deconvolution can provide phase information corresponding to
about 2 Å resolution. This can greatly reduce the complexity of
the solution of the phase problem. Hence an improved high-resolution image may be obtained by a phase extension procedure using
the amplitudes of the structure factors from ED and starting
phases from EM. Test of the procedure has been done using also
the model structure of copper chlorinated phthalocaynine. The
results are shown in Fig. 4.2, from which it can be seen that,
while direct methods failed to solve the structure from the ED
data alone, phase extension gave satisfactory result. For further
detail see Fan, Zhong, Zheng & Li (1985).

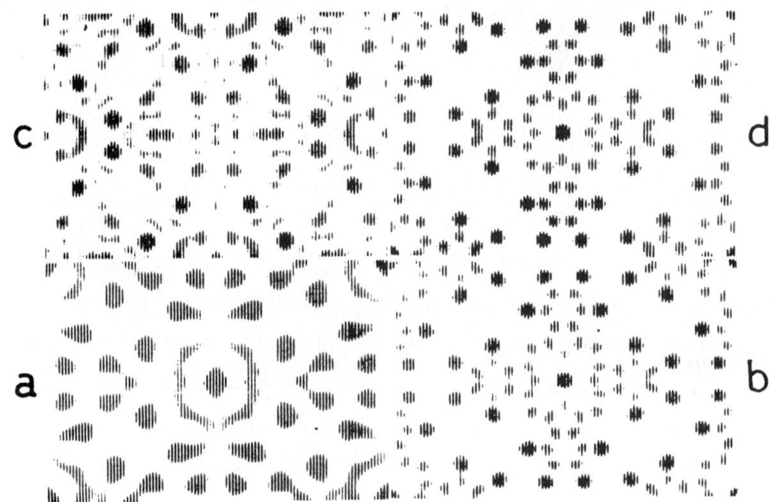

Figure 4.2

Phase extension from 2 to 1 Å resolution

a — 2 Å EM after deconvolution; b — image resulted
from phase extension using the 2 Å EM and the corresponding ED data; c — E-map from the direct method
solution of ED data; d — the expected 1 Å image.

Structure factor extrapolation

While the technique of phase extension has been known for a long time, the extrapolation of structure factors on both magnitudes and phases is seldom used. Fan & Zheng (1975) pointed out that, even at very low resolution, there will be often enough reflections to set up simultaneous equations solving, at least in theory, the complete structure. Hence it is possible to extrapolate a set of low resolution structure factors to obtain that of high resolution ones. Sayre equation can be used for this purpose. Substituting a low resolution set of structure factors in the right hand side of Sayre equation (4.1.2), structure factors beyond the resolution limit can be obtained from the left hand side. This can be made iteratively to improve the result. For the indication of discrepancy, an R factor was defined as

$$R = \frac{\sum_{H} \left| F(H) - \frac{\Theta}{V} \sum_{H'} F(H')F(H-H') \right|}{\sum_{H} |F(H)|} \quad , \quad (4.2.1)$$

where $\sum_{H}$ includes only the low resolution structure factors, while $\sum_{H'}$ includes also those structure factors obtained from extrapolation. Normally during the initial cycles of iteration the R factor will decrease. The process should stop when R reaches a minimum. The extrapolation can also be performed using least squares method which minimizes the R factor. The above procedure was verified in 1975 with a one-dimensional model structure. Recently Liu, Fan & Zheng (1986) used a similar procedure to enhance successfully the resolution of a theoretical EM. The result is shown in Fig. 4.3.

In conclusion, direct methods have entered the field of HREM. The preliminary results were encouraging. We can expect that, direct methods will be as successful in HREM as they were in X-ray crystallography.

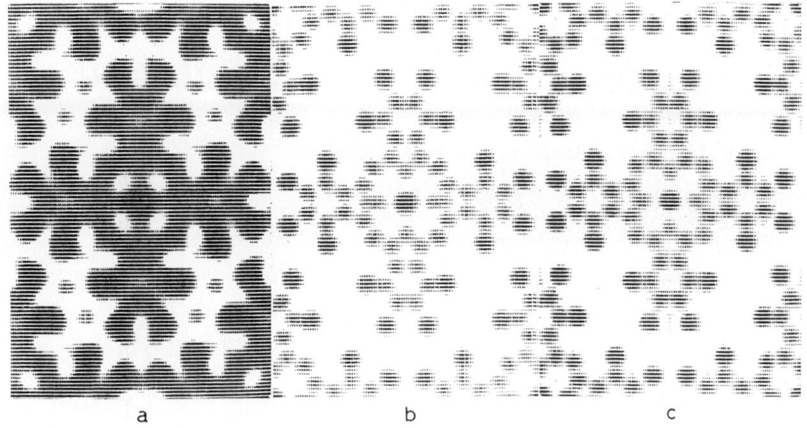

Figure 4.3

Structure factor extrapolation
of copper chlorinated phthalocaynine

a — 2 Å EM after deconvolution; b — resulted image of
extrapolation from 2 - 1 Å resolution; c — expected
image at 1 Å resolution.

REFERENCES

Blow,D.M. & Crick,F.H.C.(1959).Acta Cryst.12,794-802

Blundell,T.L. & Johnson,L.N.(1976)."Protein Crystallography"
    p177.London:Academic Press

Boehme,R.(1982).Acta Cryst.A38,318-326

Cochran,W.(1955).Acta Cryst.8,473-478

Coulter,C.L.(1965).J.Mol.Biol.12,292-295

Debaerdemaeker,T.,Tate,C. & Woolfson,M.M.(1985).Acta Cryst.
    A41,286-290

Fan,H.F.(1965).Acta Phys.Sin.21,1105-1113; 1114-1118

Fan,H.F.(1975).Acta Phys.Sin.24,57-60

Fan,H.F.(1986).Rigaku Journal,3,25-30

Fan,H.F. & GU,Y.X.(1985).Acta Cryst.A41,280-284

Fan,H.F.,Han,F.S. & Qian,J.Z.(1984).Acta Cryst.A40,495-498

Fan,H.F.,Han,F.S.,Qian,J.Z. & Yao,J.X.(1984).Acta Cryst.
    A40,489-495

Fan,H.F.,He,L.,Qian,J.Z.,Liu,S.X.(1978).Acta Phys.Sin.27,554-558

Fan ,H.F.,Yao,J.X.,Main,P. & Woolfson,M.M.(1983)
    Acta Cryst.A39,566-569

Fan,H.F. & Zheng,Q.T.(1975).Acta Phys.Sin.24,97-104

Fan,H.F.,Zhong,Z.Y.,Zheng,C.D. & Li,F.H.(1985).Acta Cryst.
     A41,163-165

Fijes,P.L.(1977).Acta Cryst.A33,109-113

Fortier,S.,Moore,N.J. & Fraser,M.E.(1985).Acta Cryst.A41,571-577

Furey,W.F.,Robbins,A.H.,Clancy,L.L.,Winge,D.R.,Wang,B.C.
     & Stout,C.D.(1986). Science,231,704-710

Giacovazzo,C.(1983).Acta Cryst.A39,585-592

Giacovazzo,C.(1984).private communication

Gramlich,V.(1975)Acta Cryst.A31,S90;(1978).ibid.A34,S43

Gramlich,V.(1984).private communication

Han,F.S.,Fan,H.F.,Zheng,C.D. & Li,F.H.(1986).Acta Cryst.in press

Hao,Q.(1986).in this meeting

Hauptman,H.(1982a).Acta Cryst.A38,289-294;(1982b).ibid.632-641

Karle,J.(1966).Acta Cryst.21,273-276

Karle,J.(1983).Acta Cryst.A39,800-805

Karle,J.(1984a).Acta Cryst.A40,4-11;(1984b).ibid.374-379

Klug,A.(1978/1979).Chem.Scr.14,245-256

Liu,Y.W.,Fan,H.F. & Zheng,C.D.(1986).in this meeting

Prick,P.A.J.,Beurskens,P.T. & Gould,R.O.(1983).Acta Cryst.
     A39,570-576

Qian,J.Z.,Fan,H.F. & Gu,Y.X.(1985).Acta Cryst.A41,476-478

Sayre,D.(1952).Acta Cryst.5,60-65

Sim,G.A.(1959).Acta Cryst.12,813-815

Unwin,P.N.T. & Henderson,R.(1975).J.Mol.Biol.94,425-440

Xu,Z.B. et al.(1984).in XIIIth IUCr Congress

Yao,J.X.(1981).Acta Cryst.A37,642-644

Yao,J.X. & Fan,H.F.(1985).Acta Cryst.A41,284-285

Yao,J.X.,Zheng,C.D.,Qian,J.Z.,Han,F.S.,Gu,Y.X. & Fan,H.F.(1985)
     "SAPI-85: A Computer Program for Automatic Solution of
       Crystal Structures from X-ray Data" Institute of Physics,
     Academia Sinica, Beijing, China.

56. XOOM crystallography in relation to other problems

L. Massa

Introduction

Structure is a driving idea in science. X-ray crystallography provides an experimental image of structure at the atomic level. An x-ray orthornormal orbital model (XOOM) of crystallography yields an enhanced image of structure at the electronic level. Thus experimental orbitals are obtained, and the x-ray diffraction experiment is formalized with greater surety within the framework of quantum mechanics.

The problem of obtaining valid experimental orbitals is one in a confluence of important physical problems now seen in relation to one another. These include:

1) Evaluation of the universal HK functional of the density.

2) Obtaining a correlation potential for the Kohn-Sham equations

3) Calculation of the correlation energy, and

4) finding an effective solution to the mathematical problem of electron N-representability.

Of great interest to crystallography is the assertion of Hohenberg and Kohn (1964) to the effect that all electronic properties are knowable through the electron density, ie,

$$\rho(r) \rightarrow v[\rho] \rightarrow H[\rho] \rightarrow \Psi[\rho] \rightarrow O[\rho]. \tag{1}$$

The density fixes the external potential, the Hamiltonian, its eigenfunction, and every electronic operator property. The HK theorem clearly lends a certain basic character to the x-ray experiment. The HK theorem asserts the fundamental carrier of information in an electronic system is the density, and XOOM crystallography provides an image of that density in a formalism having immediate and comprehensive quantum mechanical meaning.

XOOM Crystallography

Experimental structure factors determine the density and a set of orbitals which deliver that density exactly, ie,

$$F(k) \rightarrow \rho(r) \rightarrow \phi(r). \tag{2}$$

Now expression (2) hinges on the Fourier transform relation connecting structure factors and density, ie,

$$F(k) = \int e^{ik \cdot r} \rho(r) dr \tag{3}$$

being cast in an orbital formalism. Hence expand the orbitals $\underline{\phi}$ in an orthonormal basis $\underline{\psi}$ according to

$$\underline{\phi} = \mathbf{C}\underline{\psi}. \tag{4}$$

With the definition

$$\mathbf{P} = \mathbf{C}^\dagger \mathbf{C}. \tag{5}$$

The density matrix

$$\rho_1(r\,r') = tr\underline{\phi}(r)\underline{\phi}^\dagger(r') \tag{6}$$

becomes

$$\rho_1(r\,r') = tr\mathbf{P}\underline{\psi}(r)\underline{\psi}^\dagger(r'). \tag{7}$$

The density, i.e., the diagonal elements of ρ_1, inserted into the Fourier transform relation gives

$$F(k) = tr\mathbf{P}\,\mathbf{f}(k) \tag{8}$$

where the elements of $\mathbf{f}$ are Fourier transforms of basis orbital products. Now the quantum orbitals of crystallography are obtained by treating the elements of Lowdin's charge-bond order matrix $\mathbf{P}$ as experimental parameters varied to fit the measured x-ray structure factors $F(k)$. However, a crucial requirement for quantum mechanical validity is that variation of $\mathbf{P}$ must conform to the restrictions of N-representability (Clinton and Massa 1982). If

$$\rho_1(1\,1') = N \int \Psi^*(1\dots N)\Psi(1'\dots N)d2\dots dN\,dS \tag{9}$$

where Ψ is an antisymmetric N-body wavefunction then ρ_1 is called N-representable. Slater determinant N-representability is completely characterized by normalized $\mathbf{P}$ matrices which are projectors (idempotent), i.e.,

$$\mathbf{P}^2 = \mathbf{P}. \tag{10}$$

All such projectors capable of delivering x-ray orbitals may be obtained as solutions of Clinton's equations, in the iterative form,

$$\mathbf{P}_{n+1} = 3\mathbf{P}_n^2 - 2\mathbf{P}_n^3 + \lambda_k \mathbf{f}(k) + \lambda_N \mathbf{1} \tag{11}$$

where the Lagrangian multipliers λ_k and λ_N are determined by constraints satisfying structure factors and normalization. As these constraints are explicit functionals of the density so too is $\mathbf{P}$. Considerable numerical experience shows the formalism of XOOM crystallography delivers physically meaningful experimental orbitals. (Frishberg and Massa 1981)

Correlated Determinant Wave Function (CDWF) N-Representability

A set of exact density orbitals, i.e., x-ray orbitals, may be organized into a Slater determinant. This will deliver the exact density but in general is not an exact wavefunction. However a CDWF in the form

$$\Psi(1\dots N) = \Psi_{\underline{\det}}(1\dots N)\prod_{i<j}(1 - \phi(i\,j)) \tag{12}$$

is capable of very high accuracy. We indicate by subscript 'det' any quantity obtained from a Slater determinant, and if underlined one based on exact density orbitals. The spinless two-particle correlation function $\phi(1\,2)$ is symmetric in the coordinates of a particle pair, and as a Slater determinant is antisymmetric, so too is the overall CDWF. Notice the product

$$\Psi^{*}\Psi = \Psi^{*}_{\underline{\det}}\Psi_{\underline{\det}}(1 + \Sigma b + \Sigma bb + \ldots) \tag{13}$$

where b is defined to be

$$b(1\,2\,1'\,2') \equiv -\phi(1\,2) - \phi(1'\,2') + \phi(1\,2)\phi(1'\,2'). \tag{14}$$

The summations indicated in equation (13) are over the various possible pairs of particle coordinates. We may work with equation (13) to various levels of approximation. If we neglect correlation altogether, ie, $b \equiv 0$, we obtain, of course, for density matrices just pure determinental expressions.

In the next approximation, the only one dealt with here, we retain terms involving a single b, but neglect all products of b's. We also Taylor series expand the interparticle coordinate dependence of $\Psi^{*}_{\underline{\det}}\Psi_{\underline{\det}}$ retaining only the leading term, a reasonable approximation if the correlation function $\phi(1\,2)$ decreases sharply enough with increasing distances between particles. The density matrices resulting are

$$\rho(1\,2\,1\,2) = \rho_{2\underline{\det}}(1 + b(1\,2\,1\,2)) \tag{15}$$

and

$$\rho(1\,1') = \rho_{1\underline{\det}}(1\,1') + \int \rho_{2\underline{\det}}(1\,2\,1'\,2)b(1\,2\,1'\,2)d2. \tag{16}$$

Within the approximations noted, a condition necessary and sufficient for CDWF N-representability of these is

$$\int \rho_{2\underline{\det}}(1\,2\,1\,2)b(1\,2\,1\,2) = 0. \tag{17}$$

This may be solved to completely determine b by adopting a density functional model for the correlation function $\phi(1\,2)$.

The Correlation Energy

Given any Ψ_{det}, we adopt a generalized definition of correlation energy E_c according to

$$E_c \equiv E_{\text{exact}} - E_{\text{det}}. \tag{18}$$

Using the density matrices (15) and (16) we obtain a CDWF correlation energy

$$E_c = -\frac{1}{2}\int \nabla^2 \{\rho_{2\underline{\text{det}}}b(1\,2\,1'\,2)\}_{1'\to 1}d2d1 + \frac{1}{2}\int \rho_{2\underline{\text{det}}}\frac{b(1\,2\,1\,2)}{|1-2|}d2d1. \tag{19}$$

This correlation energy is a functional of the density, because $\rho_{2\underline{\text{det}}}$ and b are such functionals.

Density Functional Theory

The HK theorem is implemented in the main via the Kohn-Sham (1965) (KS) equations

$$\{-\frac{1}{2}\nabla^2 + v(1) + \int \frac{\rho(2)}{|1-2|}d2 + v_c(1)\}\phi_i(1) - \int \frac{\rho_1(1\,2)}{|1-2|}\phi_i(2)d2 = \epsilon_i\phi_i(1). \tag{20}$$

The orbitals and their energies are ϕ_i and ϵ_i respectively. The successive terms on the left of equation (20) represent kinetic energy, external potential, Coulomb repulsion, correlation potential and the exchange potential, respectively. We point out that using (19) it is possible to obtain an expression for the KS correlation potential, viz.,

$$v_c(1) \equiv \frac{\delta E_c[\rho]}{\delta \rho}. \tag{21}$$

since E_c is a functional of the density (Massa 1986). Such a correlation potential is:

(i) directly related to the Hamiltonian of the problem under consideration, (and is not taken from a model Hamiltonian such as that of a free electron gas) and

(ii) broken explicitly into its kinetic and potential energy contributions.

Finally the HK universal functional

$$F[\rho] = \langle \Psi[\rho]|T + U|\Psi[\rho]\rangle, \tag{22}$$

where T and U are kinetic and electron interaction energies respectively, can be approximately evaluated using density matrices (15) and (16), as they are N-representable and functionals of the density. The result is

$$F[\rho] = -\frac{1}{2}\int \nabla^2\{\rho_{1\underline{det}}(1\,1') + \int \rho_{2\underline{det}}b(1\,2\,1'\,2)d2\}_{1'\to 1}d1$$

$$+\frac{1}{2}\int \rho_{2\underline{det}}\frac{1 + b(1\,2\,1\,2)}{|1 - 2|}d2d1. \tag{23}$$

Summary

It is now possible to obtain physically meaningful x-ray orbitals, and these may be used to form the Slater determinant reference state for a CDWF. In this way crystallography is brought into contact with and contributes to the solution of other important problems such as those of correlation energy, density functional theory, and wavefunction N-representability.

References

W. L. Clinton, L. Massa, Phys. Rev. Lett., **29**, 1363 (1972)

C. A. Frishberg, L. Massa, Phys. Rev., **B24**, 7018 (1981)

P. Hohenberg, W. Kohn, Phys. Rev., **136**, B864 (1964)

W. Kohn, L. J. Sham, Phys. Rev. **140**, A1133, (1965)

L. Massa, Chemica Scripta **26**, 469 (1986)

57. Correlation between spectroscopic and crystallographic methods for molecular crystals

Carlo M. Gramaccioli

In the last years,crystallographic measurements have become more accurate -on the average- than before: hence,there is a possibility of an extended com parison with other physical-chemical results,such as for instance spectroscopic or thermodynamic data.

An important example of such possibility is pro vided by temperature factors: whereas many years ago the anisotropic B's were considered by most crystallo graphers as essentially nothing else than a vague set of additional parameters for improving the final fit, we realize nowadays that in many cases there is a re- markably good agreement with the corresponding lat- tice-dynamical estimates (see Filippini et al, 1976; Gramaccioli & Filippini,1983,1985; Filippini & Gramac cioli,1986, and Table 1).

The limited extent of such comparison is mainly due to the difficulties of calculation,and for this reason the procedure has been carried on so far only for molecules which are free of charge,such as hydrocar- bons or sulfur (Gramaccioli & Filippini,1984). There is however no reason for doubting the validity of crystallographic B's also in other cases: for instan ce,the physical meaning of the B's is particularly evident in showing anomalous Jahn-Teller effect in some complexes (Ammeter et al.,1979).

If there is an essential agreement between crystallographic and 'spectroscopic' temperature factors, then it is possible either to implement information concerning bond-length correction ,or to check the validity of atom-atom potentials not only with respect to molecular conformation in the crystal,but also by reproducing the B's. At the same time,application of atom-atom potentials to the crystallographic structural model should reproduce IR or Raman-active vibrational frequencies,or/and phonon dispersion curves. This frequency comparison may be very useful for improving potentials,both inter- and intra-molecular: for instance, Rinaldi & Pawley (1975) followed this way for final calibration of their van der Waals parameters concerning sulfur; Filippini et al.(1984) derived a general force field for 'out-of-plane' vibrations in aromatic hydrocarbons; the first derivations of frequencies for aromatic hydrocarbons,as well as dispersion curves,are mainly due to Pawley (1967,1968):see also Pawley et al.,1980,Dorner et al.,1982).

For rigid molecules, a 'classic' interpretation of B's in terms of molecular vibration tensors T,L and S is given by Schomaker & Trueblood (1968). A comparison between these tensors and their lattice-dynamical correspondents shows substantial agreement for T and L,whereas for S the agreement is not outstanding.In some cases,due to instability of the normal equation matrix,the agreement is not good also for some components of L: a review of such an argument is given by Gramaccioli & Filippini (1983).

For a crystallographer,the possibility of carrying on a thermal-libration correction for bond length even for non-rigid molecules is particularly

Table 1.-   Temperature factors $(\times 10^4)$ for two non-rigid polyphenyls;calculated values from Gramaccioli & Filippini(1985),or Filippini & Gramaccioli (1986); observed values from Brown & Levy (1979),or Robbins et al. (1975).

Atom		B_{11}	B_{22}	B_{33}	B_{12}	B_{13}	B_{23}
				o-terphenyl			
C1	(cal)	33	290	65	-11	-1	-1
	(obs)	30	277	55	-10	-2	4
C2	(cal)	32	350	84	2	6	-20
	(obs)	28	324	78	-1	2	-20
C3'	(cal)	26	387	80	13	3	-10
	(obs)	30	398	73	28	7	14
C4'	(cal)	25	513	92	24	4	6
	(obs)	28	566	93	36	9	15
				tetraphenylmethane			
C1	(cal)	55	58	122	-2	0	-5
	(obs)	56(1)	60(1)	111(3)	-5(1)	10(2)	0(2)
C2	(cal)	62	73	134	-6	-8	-7
	(obs)	71(1)	83(2)	129(4)	-5(1)	-4(2)	-6(2)
C3	(cal)	81	86	155	-19	-8	-23
	(obs)	83(2)	111(2)	134(4)	-25(2)	1(2)	-20(2)
C4	(cal)	104	77	194	-15	1	-40
	(obs)	111(2)	86(2)	175(4)	-26(2)	23(1)	-42(3)
C5	(cal)	103	63	206	5	1	-31
	(obs)	107(2)	67(2)	195(4)	3(2)	23(3)	-21(2)

Table 2.- Bond distances (Å) in tetraphenylmethane. For each bond, the columns show, in sequence: (1) the uncorrected bond length; (2) the corrected bond length according to the rigid body ; (3) the corrected bond length for 'riding' motion; (4) the corrected bond length for a general non-rigid motion (Filippini & Gramaccioli, 1986; data from Robbins et al., 1975)

	(1)	(2)	(3)	(4)
CO-C1	1.551	1.554	1.553	1.557
C1-C2	1.403	1.406	1.412	1.411
C1-C6	1.389	1.392	1.397	1.399
C2-C3	1.385	1.387	1.393	1.394
C3-C4	1.381	1.384	1.386	1.392
C4-C5	1.384	1.387	1.397	1.394
C5-C6	1.392	1.395	1.403	1.401
Average in				
the ring	1.3890	1.3918	1.3963	1.3985

Table 3.- Thermodynamic functions for orthorhombic sulfur (S_8) at 298K (from Gramaccioli & Filippini, 1984; experimental data from Guthrie et al.(1954) and Eastman & Gavock (1937)).

		calc	obs
Entropy	(cal/mol.K)	60.86	61.0(0.4)
c_v	(cal/mol.K)	41.17	40.8(0.4)
c_p	(cal/mol.K)	43.27	43.2(0.2)

useful:for instance,in tetraphenylmethane the correct
ion for general 'non-rigid' motion gives almost identical results for C-C bonds as from other sources,
such as theoretical or spectroscopic (see Filippini &
Gramaccioli,1986,and table 2).For such a correction,
not only temperature factors,but also the 'displacement tensors' between the movement of different atoms
are necessary: such tensors cannot be derived from
'usual' Bragg diffraction.

From crystal frequencies,$\underline{via}$ the partition function,also thermodynamic functions such as entropy,c_v,
etc. can be derived. An example is given in table 3
for sulfur crystals:here the internal force field has
been readjusted in order to give a reasonable agreement with the measured frequencies. The agreement
with the experimental data is excellent (Gramaccioli
& Filippini,1984).

References

Ammeter,J.H., Bürgi,H.B., Gamp,E.,Meyer-Sandrin,V.,
    Jensen,V.P. (1979) Inorganic Chemistry 18,733.
Brown,G.M. and Levy,H.A. (1979) Acta Crystallogra-
    phica B35, 785-8.
Dorner,B.,Bokhenkov,E.L.,Chaplot,S.L.,Kalus,J.,Nat-
    kaniec,I.,Pawley,G.S.,Schmelzer,U.,and Sheka,E.
    F. (1982) Journal of Physics C15,2353-65.
Eastman,E.D. and McGavock,W.C. (1937)Journal Ameri-
    can Chemical Society 59,145.
Filippini,G. and Gramaccioli,C.M. (1986) Acta Crys-
    tallographica B42, in the press.

Filippini,G.,Gramaccioli,C.M.,Simonetta,M. and Suf-
    fritti,G.B.(1976) Acta Crystallographica A32,
    259-64.

Filippini,G.,Simonetta,M. and Gramaccioli,C.M.
    (1984) Molecular Physics 51,445-59.

Gramaccioli,C.M. and Filippini,G. (1983) Acta Crys-
    tallographica A39, 784-91.

Gramaccioli,C.M. and Filippini,G. (1984) Chemical
    Physics Letters 104, 50-3.

Gramaccioli,C.M. and Filippini,G. (1985) Acta Crys-
    tallographica A41, 361-5.

Guthrie,G.B., Scott,D.W. and Waddington,G. (1954)
    Journal American Chemical Society 76,1488.

Pawley,G.S. (1967) Physica Status Solidi 20,347-60.

Pawley,G.S. (1968) Acta Crystallographica B24,485-6.

Pawley,G.S.,Mackenzie,G.A.,Bokhenkov,E.L.,Sheka,E.F.,
    Dorner,B.,Kalus,J.,Schmelzer,U. and Natkaniec,I.
    (1980) Molecular Physics 39, 251-60.

Rinaldi,R. and Pawley,G.S. (1975) Journal of Physics
    8, 599-616.

Robbins,A.,Jeffrey,G.A.,Chesick,J.P.,Donohue,H.,Cot-
    ton,F.A.,Frenz,B.A. and Murillo,C.A. (1975) Acta
    Crystallographica B31, 2395-99.

Schomaker,V. and Trueblood,K.N. (1968) Acta Crystal-
    lographica B24, 63-76.

58. The early days of molecular graphics

Robert Langridge

"We may say most aptly that the Analytical Engine weaves alge-
braical patterns just as the Jacquard-loom weaves flowers and
leaves" (Lovelace, 1871).

As foreseen by Ada Lovelace, the analytical engines of
today do indeed weave, not only algebraic patterns, but also
insights limited mainly by the imagination of the user. Not
only does this "enchanted loom" weave patterns to illustrate
scientific concepts but one can reach into the loom itself and
manipulate the warp and weft of the concepts and visualize the
consequences.

This powerful tool is now an integral part of almost all
research, and nowhere does it have a greater impact than in
studies of the structure of molecules and their interactions.

The fist direct production of molecular structure illustration on the CRT of a computer was by Busing (1951). In the
field of large molecules, Bennett and Kendrew (1952) made the
first application of a digital computer (EDSAC) to protein
structure followed by the use of an IBM 650 for calculations
of the Fourier transform of DNA (Langridge 1957). But is was
not until 1964 that the first major developments in molecular
graphics occurred, with the development of ORTEP for illustration graphics by Johnson (1964) and the first three dimensional interactive graphics built at Project MAC MIT (Stotz and

Ward, 1965) applied to protein and nucleic acid structural research by Levinthal and Langridge (Langridge and MacEwan, 1965, Levinthal, 1965, 1966).

The use of a pen plotter in ORTEP became a standard for illustration graphics for many years and is still actively used.   Raster graphics is now so commonplace that it is hard to remember that it was a relative latecomer - because of the expense of the necessary memory. The Brookhaven Raster Display (BRAD), refreshed from a drum system, was the first to be applied to molecular structure (Meyer, 1971), and allowed real-time red/green stereo display.   Among the pioneers of raster-based displays were Max (1979) and Feldman (Feldman et al, 1976).   The hidden-surface algorithm developed by he and Porter (1979) has been extensively used.

Interactive three dimensional graphics, although beginning in 1964, did not come into general use until considerably later, partly due, of course, to the cost of the necessary vector systems, but partly also to lack of direct experience with hands on use of a truly powerful three dimensional interactive system.   Developments continued on the relatively small systems (Levinthal et al, 1968, Barry at al, 1969, Barry and North, 1971 and Katz and Levinthal, 1972).

One of the important milestones in the field was the first Evans & Sutherland display, the LDS-1, which for the first time provided in high speed digital hardware 4X4 matrix multiplication and concatenation, three dimensional clipping and perspective.   We installed the first production model of the LDS-1 at the Computer Graphics Laboratory in Princeton in 1970.   The host computer was an early DEC PDP-10. This was funded as Research Resource by the National Institutes of Health. To quote from the text of my original grant proposal:

> "...to develop methods for constructing and refining
> molecular structures as large as for example proteins
> ... and for studying their interaction with each other

and also with smaller molecules such as enzyme substrates and drugs... Interaction with real models at
this level of flexibility is very difficult and recov-
ery of models at various stages a nightmare... the use
of a dynamic computer display is likely to be far
superior to real models and allow the investigator to
pursue ideas which, although conceptually clear, cannot be easily written down as three dimensional coordinates at each stage of analysis" (Langridge, proposal
"Special Research Resource for Biomolecular Graphics"
from Princeton University to US National Institutes of
Health, 1969)

The interactive display program which we developed at
Princeton (Bond, 1972, Lesk, 1972, Langridge, 1974) provided
many of the tools now required for protein engineering such as
amino acid replacement, bond manipulation, distance monitoring
and of course time sliced stereoscopic viewing, but the high
cost of this system prevented its general use. It was not until much less expensive systems became available that the use
spread.

Since the greatest need for graphics systems was in studies of large molecules, particularly proteins, and because of
the sheer size and unwieldiness of the models, the problem of
electron density fitting was addressed first by Avrin (1967)
and particularly later by Diamond (1981) and Jones (1978,
1982) The system FRODO, written by Jones, has come into wide
general use for electron density fitting in protein crystallography.

The use of interactive graphics in drug design came more
slowly, partly due to cost, and partly due to an initial (and
lingering) skepticism of its value as anything more than a pro-
ducer of pretty pictures. The first pharmaceutical company to
devote major resources to molecular graphics was Merck, Sharp
& Dohme led by Peter Gund who began his work on graphics dur-

ing the early days of our Princeton LDS 1/PDP-10 system. Rather than invoke my obvious prejudices on the value of interactive graphics in drug design, I will instead quote from a Merck chemist who has used the system developed by Gund and his colleagues (Gund et al, 1980), Joshua Boger, at a meeting of the Royal Society of Chemistry in 1985:

> "The role of computer-assisted molecular modelling in drug design is difficult to demonstrate. The medicinal chemist meets little skepticism when claiming that a high pressure liquid chromatograph or a nuclear magnetic resonance spectrometer was useful in solving a medicinal chemistry puzzle. The information gained by these instruments is obviously distinct in kind from that otherwise observable, and the role of this information in problem resolution can be made unambiguous, logical, and convincing. In contrast, the role of computer-assisted molecular modelling often is in the realm of ideation, for which there are few other instrumental paradigms. Unlike other instrumental techniques, the outputs of which are reducible, usually without information loss, to a few numerical values, the output of molecular modelling is often a picture. It is in the interaction of that picture with a chemist's hypothesis that modelling characteristically functions. This interface of hypothesis and picture is difficult to describe." (Boger, 1985).

Molecular graphics can only increase in importance over the next few years, as the cost of hardware decreases and software becomes more generally useful. The cost of the original Project MAC system used in our first three dimensional interactive molecular graphics in 1964 was of the order of $2 million. Our E&S LDS-1/PDP-10 installation at Princeton cost $700,000. In 1986 workstations more powerful than either of these are sold for less than $50,000. Extrapolating to the millenium I expect to see, by year 2000, workstations on each scientist's desk with the local power of a present day super-computer and full three dimensional graphics, networked to large scale data bases and with software to enable the investigator to weave his or her ideas into the scientific research of the 21st century.

Ada Lovelace would be pleased.

ARVIN, D.E. (1967). Computer display of protein electron density functions. Ph.D. thesis, MIT.

BARRY, C.D., ELLIS, R.A., GRAESSER, S.M. and MARSHALL, G.R. (1969). In _Pertinent concepts in computer graphics_ (ed. Faiman and Nievergelt) P. 104, University of Illinois Press.

BARRY, C.D. and NORTH, A.C.T. (1971). _Cold Spring Harbor Symposium on Quantitative Biology_ 36, 577.

BENNETT, J.M. and KENDREW, J.C. (1952). _Acta Crystallographica_ 5, 109.

BOGER, J. (1985) in Proceedings of the 3rd SCI-RSC Medicinal Chemistry Symposium, Royal Society of Chemistry.

BOND, P.J. (1972). _Computer Graphics ACM SIGGRAPH_ 6, 13.

BUSING, W. (1951). Abstracts, American Crystallographic Association Meeting.

DIAMOND, R. (1981). In _Twelth international congress of crystallography_ Abstract 18 5-20.

FELDMAN, R.J., HELLER, S.R. and BACON, C.R.T. (1974). _Journal of Chemical Doccumentation_ 12, 234.

GUND, P. ANDOSE, J.D., RHODES, J.B., and SMITH, G.M. (1980). _Science_ 208, 1425.

JOHNSON, C.K. (1965) ORTEP, ORNL Technical Report 3794, Oak Ridge National Laboratory, TN.

JONES, T.A. (1978). _Journal of Applied Crystallography_ 11, 268.

JONES, T.A. (1982) FRODO, In _Computational crystallography_ (ed. D. Sayre) p 303. Oxford University Press.

KATZ, L. and LEVINTHAL, C. (1972). Annual Reviews of Biophysics and Bioengineering 1, 465.

LANGRIDGE, R. (1957). Federation Proceedings 33, 2332.

LANGRIDGE, R. and MACEWAN, A.W. (1965). In IBM scientific computing symposium on computer aided experimentation. p. 305, IBM, Yorktown Heights, NY.

LESK, A. (1972). Software 2, 259.

LEVINTHAL, C. (1965). In IBM scientific computing symposium on computer aided experimentation. p. 315, IBM, Yorktown Heights, NY.

LEVINTHAL, C. (1966). Scientific American 214, 42.

LEVINTHAL, C., BARRY, C.D., WARD, S.A. and ZWICK, M. (1968). In Emerging concepts in computer graphics (eds Secrest and Nievergelt) p. 231. W.A. Benjamin, Inc.

LOVELACE, A.A. (1871). In Taylor's Scientific Memoirs pp. 666-731. Editorial notes on the memoir "On the Mathematical Principles of the Analytical Engine" by Menabrea.

MAX, N.L. (1979). Computational Graphics 13, 165.

MEYER, E.F. (1970). Journal of Applied Crystallography 3, 392.

PORTER, T. (1979). Journal of Computer Graphics 13, 234.

STOTZ, R.H. and WARD, J.F. (1965). Operating Manual for the ESL Display Console. Project MAC Internal Memorandum MAC-M-217, MIT.

59. Penicillin

Dorothy Hodgkin

When I was first invited to present this lecture by Dr. Duax, I thought, yes, I should like to describe our old determination of the crystal and chemical structure of penicillin and also to add to this some new structure analysis of the acid stable penicillin by accurate modern methods, which would bear on the relation between structure and biological activity. The early work is buried in the large book published by Princeton University Press and was carried out under conditions of wartime secrecy - it is still available, but the new work, sad to say, has not proceeded as desired.

You will all probably know the beginning of the story of penicillin itself, the story of the discovery in 1929 of a natural substance of high antibiotic activity produced by the mould, _penicillin nonatum_. The early work of Fleming and others in London on its isolation and biological activity ran slowly down owing to its instability until Florey and Chain in Oxford, in a general search for antibiotics, took it up again just before the war in 1938. They found that it was indeed very unstable in a number of conditions, acid and alkaline solutions, but that they could prepare a rather stable barium salt, solid, not crystalline, with which they did a number, first of all, of animal experiments on mice, and then on human beings.

I was walking down South Parks Road in the usual Oxford way when I met Ernst Chain in a state of great excitement, the day after the night during which they had carried out one of their critical experiments on mice, in which they had injected four mice with <u>streptococcus aureus</u> and four mice with <u>streptococcus aureus</u> plus their penicillin extract. And the four mice who were unprotected died and the other four lived, Chain told me all. It was Florey who added later "you don't need any statistics to interpret that experiment." They set out confidently to add other workers to their group and to organise a small "factory" in the Pathology Laboratory basement with milk extracting machinery to extract quantities sufficient to try further experiments. Chain said hopefully, "Some day we will have crystals for you."

I first tried to crystallise the barium salt and never succeeded. But they were convinced it was fairly pure so they started degradation experiments on it before it was fully purified and crystalline. In a note they published early in the war in 1942, they described their material as $C_{24}H_{32}O_{10}N_2Ba$, rather far from the correct formula.

Meanwhile, the clinical work was going ahead under Florey's direction at the Radcliffe Infirmary, Oxford. Their first patient, a postman suffering from acute blood poisoning, appeared to be recovering after three day's treatment; then their supplies ran out and he died. So they had to go back and make much more of the material. And the next trials they made were all on children who shouldn't need so much of the antibiotic for a cure. Following the work with children, Professor Florey and Professor Hugh Cairns took their penicillin preparation to the war front in North Africa, and tried their material on war wounds; they found it very effective.

At this point, the Ministry of Supply took over the organisation, in the national interest, of people working on penicillin. We all were asked to write reports on our results,

Scheme 1

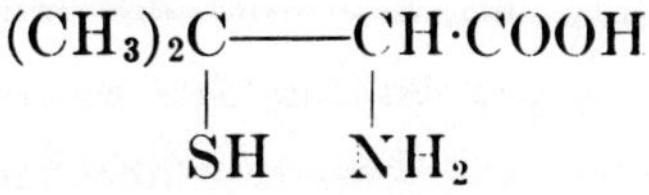

$$(CH_3)_2C\!\!-\!\!-\!\!CH\cdot COOH \qquad R\cdot CONH\cdot CHCHO$$
$$\qquad\ \ SH\quad NH_2 \qquad\qquad\qquad\qquad COOH$$
$$\mathbf{I} \qquad\qquad\qquad\qquad\qquad \mathbf{II}$$

$$R\cdot CONH\cdot CH_2CHO + CO_2$$
$$\mathbf{III}$$

Degradation products for penicillins

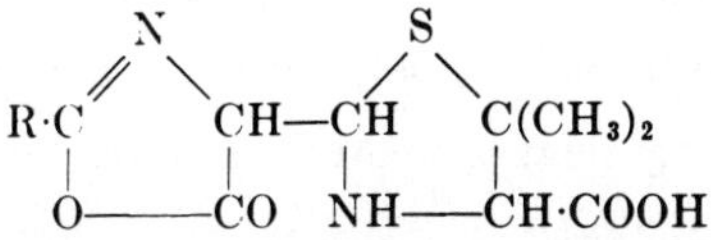

IV

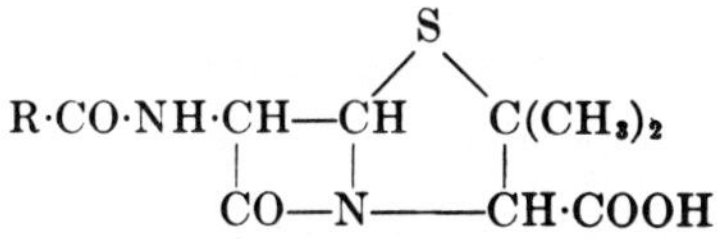

V

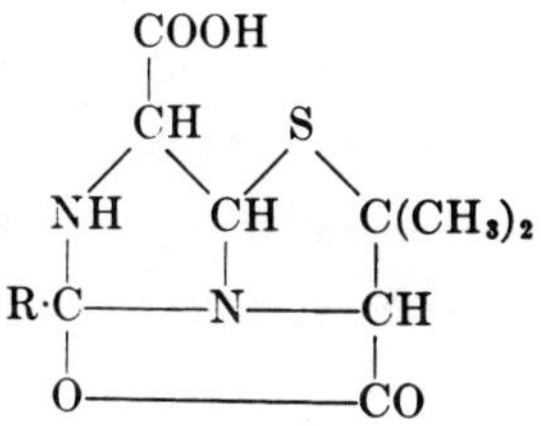

VIa

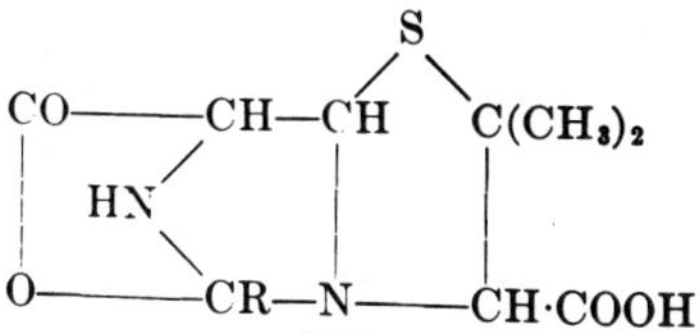

VIb

Potential chemical structures for penicillin

which were circulated as pen reports amongst us, instead of publishing in the open literature. Since it was difficult to obtain enough supplies to use in the war in this country, Florey and Norman Heatley went to the United States and saw both government research institutes and the big chemical and pharmaceutical firms, to bring them into an organisation for the production of penicillin.

By this time the first work on the degradation of penicillin had proved successful at Oxford. The two most important degradation products: I and II, were penicillamine a small amino acid, and penicilloic acid, also isolated as the barium salt, of which the group R varied according to the source of the penicillin. The structure determination and chemical synthesis of penicillamine was largely carried out by J. H. Cornforth as a young post doctoral fellow; much later he gained the Nobel prize for very different research, on sterol biosynthesis.

The structure of the degradation products suggested a number of possible structures for the penicillin nucleus IV, the thiazolidine-oxazolone structure was first written and strongly favored by Sir Robert Robinson, V, the β-lactam structure was probably first written by E. P. Abraham - Robinson commented anyone could write this structure, the more complicated tricyclic formulae were supported by Heilbron's group in London. By the time the Americans were joining the penicillin group all that seemed necessary was to synthesise the different proposed structures and see which was right; the new coordinating committee was set up, to which all reports of progress in penicillin research were sent via the Committee for Penicillin Synthesis, CPS.

It was from the United States we first heard that the sodium salt of penicillin had been crystallised in the Squibb laboratories. I was rather shocked with myself to have let this happen, and rang up Edward Abraham to make some sodium

Table I. Unit cell dimensions of the salts of penicillin (**P**).

Salt	a	b	c	β	Space Group	n
Na 2-pentenylP	37.08	6.0	18.4	106°	C2	8
Na benzylP	8.48	6.33	15.63	94.2°	$P2_1$	2
K benzylP	9.36	6.37	30.35		$P2_12_12_1$	4
Rb benzylP	9.45	6.44	30.2		$P2_12_12_1$	4

salt. And he said "Oh we have lots of it, we keep it in a dessicator." He brought some over from the Pathology laboratory, and we put some on a microscope slide. And as were were talking to one another, I looked at the slide - on it was a mass of little crystals, formed through picking up water of crystallisation from the atmosphere. They were small and fragile, and proved on X-ray examination, to have a rather complicated unit cell, 2 molecules in the asymmetric unit, Table I. I asked for preparations of the potassium, rubidium and other salts, hoping all the same for a detailed analysis.

You will realise from the structures of the degradation products it was now known that the formula given for the old barium salt was wrong, and that penicillin contained sulfur. In Oxford this was found from the study of the degradation products helped by X-ray measurements - the formula indicated for penicillamic acid, an oxidation product of penicillamine, had too many oxygens for the available bonds. When I reported this to Wilson Baker, he said "I have suspected this for some time" and tested for sulfur, lost at an earlier stage of the analysis but now clearly there. In the U.S. the correct analysis was reported on the crystalline sodium salt.

From the sodium salt analysis and from a small sample Sir Robert Robinson brought over in the autumn of 1943, it became clear that this differed from the Oxford penicillin; in the Oxford penicillin named F, or I, the R group was aliphatic, 2-pentenyl; in the American specimen, G or II, it was benzyl.

Sir Henry Dale, who knew the research director of the Merck laboratories, wrote asking for a purified sample of sodium benzyl penicillin for crystallographic analysis. The sample, 10 mgr, was brought over early in February 1944 (I suppose in a military aircraft), to Sir Henry Dale in the Royal Institution. He sent it by the hand of Kathleen Lonsdale to Oxford. It was accompanied by a telegram on how to crystallise it, saying "Dissolve in minimum, **emphasise minimum,** absolute methyl alcohol; add slowly dropwise ethyl acetate." So with Kathleen Lonsdale watching me I weighed out 3 mgr. sodium benzyl penicillin into a small tube approximately 2 cm. long and 1/2 cm. across, and added 2 drops of methyl alcohol and then about 10 drops ethyl acetate followed later by a further 15 drops. The crystals slowly began to grow so we left them for a time. When growth had apparently ceased I withdrew all solvent, allowed them to dry, and then securely corked the tube. I withdrew one crystal with a needle and took an X-ray photograph, and this showed straight away the crystal was monoclinic, $P2_1$, two molecules in the unit cell, very nice for X-ray analysis - and the reflections were good too, sharp and clear though the crystal was small (Table I). We decided to go right ahead with a full analysis. So we checked on a few of the best looking crystals and started straight away to collect three dimensional data on that crop of crystals. I made one horrible error. I thought I might get larger crystals if I brought them out of solution more slowly. I dissolved in methyl alcohol and added again ethyl acetate. But the penicillin - unstable if too long in solution - decomposed. However, the whole structure was solved from the crystals from the first 3 mg.

While Barbara Low, my research student was collecting the sodium salt data (she had learnt the subject on the penicillin degradation products) we asked for potassium and rubidium salts of benzyl penicillin, and we were sent the potassium salt from

Squibb and the rubidium salt from I.C.I. in England. They arrived at the end of March and beginning of April, 1944. We grew crystals again and went straight ahead collecting three dimensional data on both of them.

There was no difficulty in recording the reflections on Weissenberg photographs and measuring the intensities by eye estimation. We could then carry out the first calculations, Patterson projections and Harker sections for the sodium salts with assistance of Beevers and Lipson strips, and an adding machine recommended to us by J. M. Robertson. We had more difficulty calculating structure factors. So we used at first structure factor graphs as suggested by Bragg and Lipson. For the monoclinic b projection these were very simple to draw for each h0l reflection - bands across the unit cell of positive and negative density.

We had a feeling we should go into three dimensions, and this presented us with further computing difficulties. And when we began to work on the potassium and rubidium salts, we found that they were not isomorphous with the sodium salt, which as educated crystallographers we should have expected. The unit cell of each of these were clearly related to that of the sodium salt but orthorhombic, with approximately doubled c axis, 30 Å, rather large to handle with 6° Beevers-Lipson strips (Table I).

Just at that time there was a meeting in Oxford of our newly formed X-ray analysis group. The first meeting had taken place at Cambridge the year before. At the Oxford one, in April, 1944, we first heard the full story, from Paul Ewald himself of the discovery of X-ray diffraction in a very wonderful evening lecture, and there was discussion between him and W.L. Bragg and others on trying to form an International Union of Crystallography. And, as for me, everybody was very anxious to know what we had found out about penicillin, and extremely anxious to help in its X-ray analysis. So, Beevers set out to

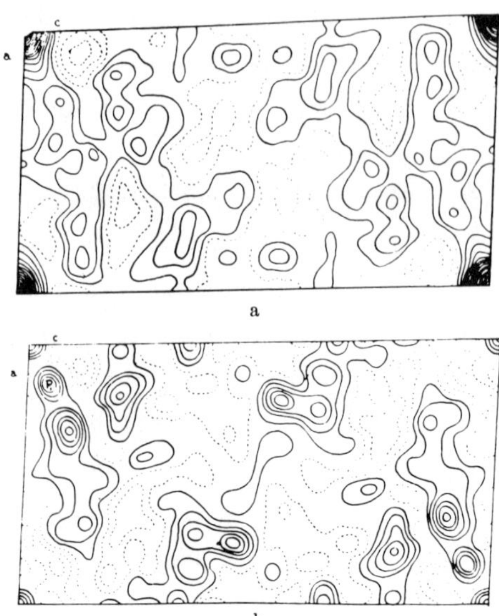

Figure 1. Sodium benzylpenicillin a) Patterson projection Pxz
and b) Patterson section PxOz.

extend his strips to 3° intervals. And W. H. Bragg had devised
an X-ray microscope" as a means of determining the structure
factors for a set of atoms seen in projection by means of optical diffraction.   Charles Bunn had developed Bragg's idea and
set up a very good X-ray microscope in his laboratory at Northwich.   Both offered the use of their instruments.   And Comrie
discussed the making of punched cards to help us to do three
dimensional calculations in full.   We accepted all help.
Actually, the structure was solved from projections before
either Beevers' extended strips or Comrie's cards were ready.
But we did use the X-ray microscopes.

Figure 1 shows the Harker section we had calculated for
the sodium salt.   It is a beautiful map with sharp discrete
peaks in spite of limitations in the X-ray data.   It suggests
that the whole structure could have come out of the full three
dimensional Patterson if we had computed it.   But since by now

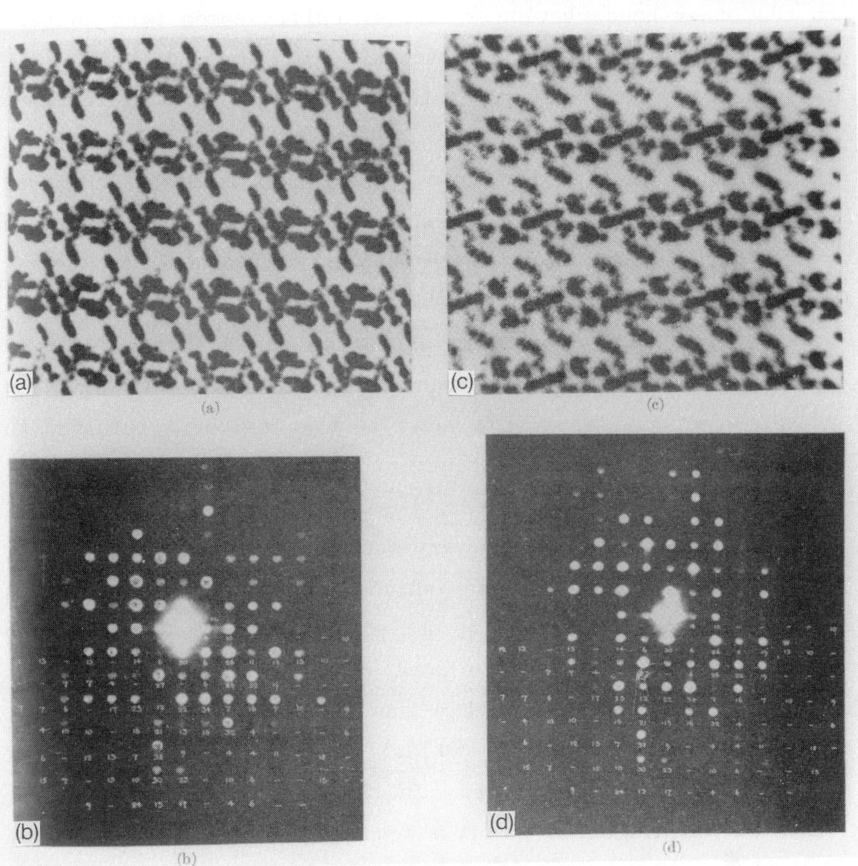

Plate 1.   Optical diffraction effects.

we had the isomorphous pair of potassium and rubidium salts
from which we could obviously obtain direct phase information,
we thought it best to concentrate on these. The sodium salt,
on the other hand, having a smaller cell, seemed best to test
out the use of the X-ray microscope.

The first experiments in the use of the X-ray microscope
were carried out with Kathleen Lonsdale on the instrument at
the Royal Institution in London. But it was wartime; it seemed
a good plan to take the children to the country in vacation to
a farm between Northwich, where Charles Bunn worked, and Stoke-
on-Trent where my husband worked. Charles Bunn and I started
working together, after hours as far as he was concerned. We
used to look at night at Manchester burning in one direction
and Stoke in the other from heavy bombing. Soon I returned to
Oxford, leaving Charles to continue experiments on his own.

At the end of the summer, I got a letter from him saying
"I have not been able to do very much" and later asking for details of the crystal optics - which I sent him. The situation
changed, he was encouraged to use working time on the structure
solving and soon he had placed a penicillin model of the
thiazolidine-oxazolone within the confines of the unit cell,
in such a way that it gave optical diffraction effects similar
to the observed intensities. In this apparatus the structure
drawn out is distorted to fit a square lattice and photographed
through a square array of pinholes to provide the experimental
diffraction grating. Plate 1a shows the first stage of promising agreement, at which he decided to calculate structure factors and phases and hence an electron density map, as in Figure
2 - which he sent to us in Oxford, in return for a projection
we had obtained for the rubidium benzylpenicillin.

When I looked at Charles' map I was very worried. It
seemed to me that atoms placed as he had placed them would not
look in an electron density projection in the way that his appeared. Also, the atomic arrangement he showed did not look

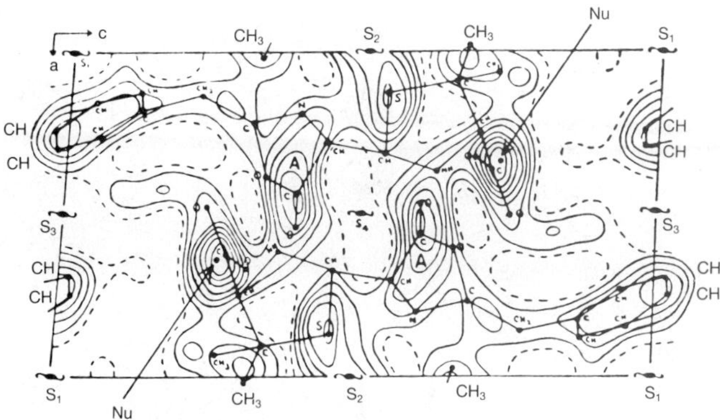

Figure 2. Sodium benzylpenicillin trial structure.

anything like the peaks in our map. Our map was imperfect.
The rubidium and potassium ions could be placed with certainty,
but their positions were such that they did not contribute to
half the reflections. The first projections had false symmetry, which we had broken by adding terms based on the assumption that the second heavy peak present was sulfur. But apart
from the rubidium and sulfur positions, the interpretation was
obscure. And our rubidium sites did not fit with Charles' suggested sodium sites. I then decided that there must be something right about each map - both should show the penicillin
molecule viewed in the same direction but differently packed.
If we compared the maps, putting one over the other, we could
easily pick out the individual penicillin unit, curled up instead of extended as we had somehow expected (Figure 3).

The positions of a number of atoms could be easily recog-
nised, once the unit had been identified, particularly the
position of the benzene ring in our map, the only part of
Charles' map which was correct. We immediately calculated
phases for extra terms. Figure 4 shows the maps with 5 terms
extra, then with a further 4 terms which improved it still further. At first we thought an oxazolone ring could be placed
between the thiazolidine and phenyl groups. But gradually the
contours shifted to show a β-lactam.

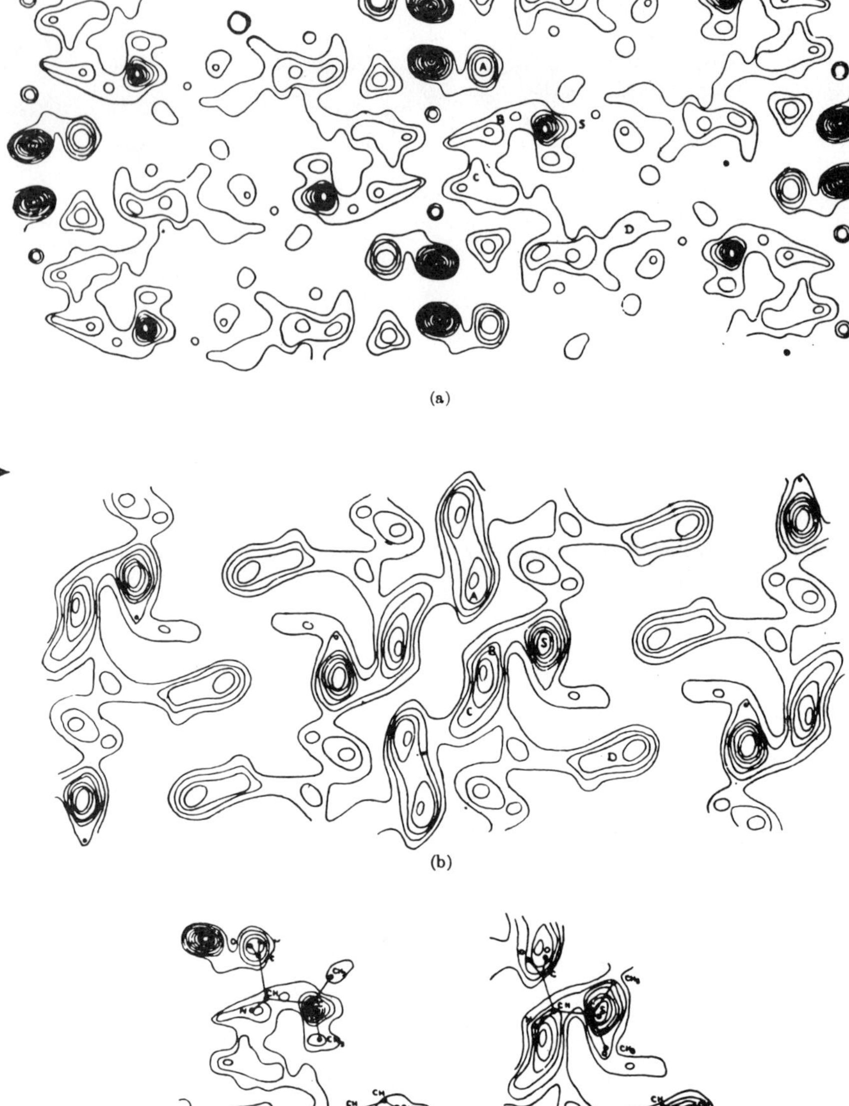

Figure 3. Appropriate electron density projections compared a) rubidium benzylpenicillin, b) sodium benzylpenicillin and c) the units.

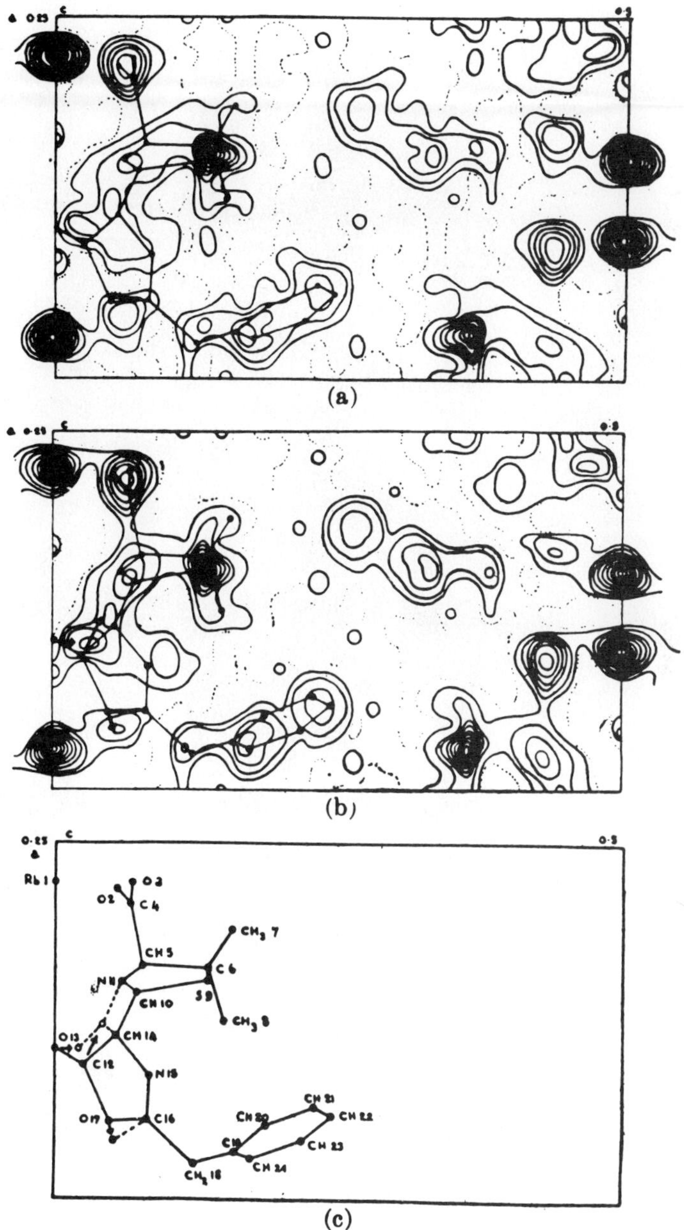

Figure 4. Rubidium benzylpenicillin electron density projections with added terms

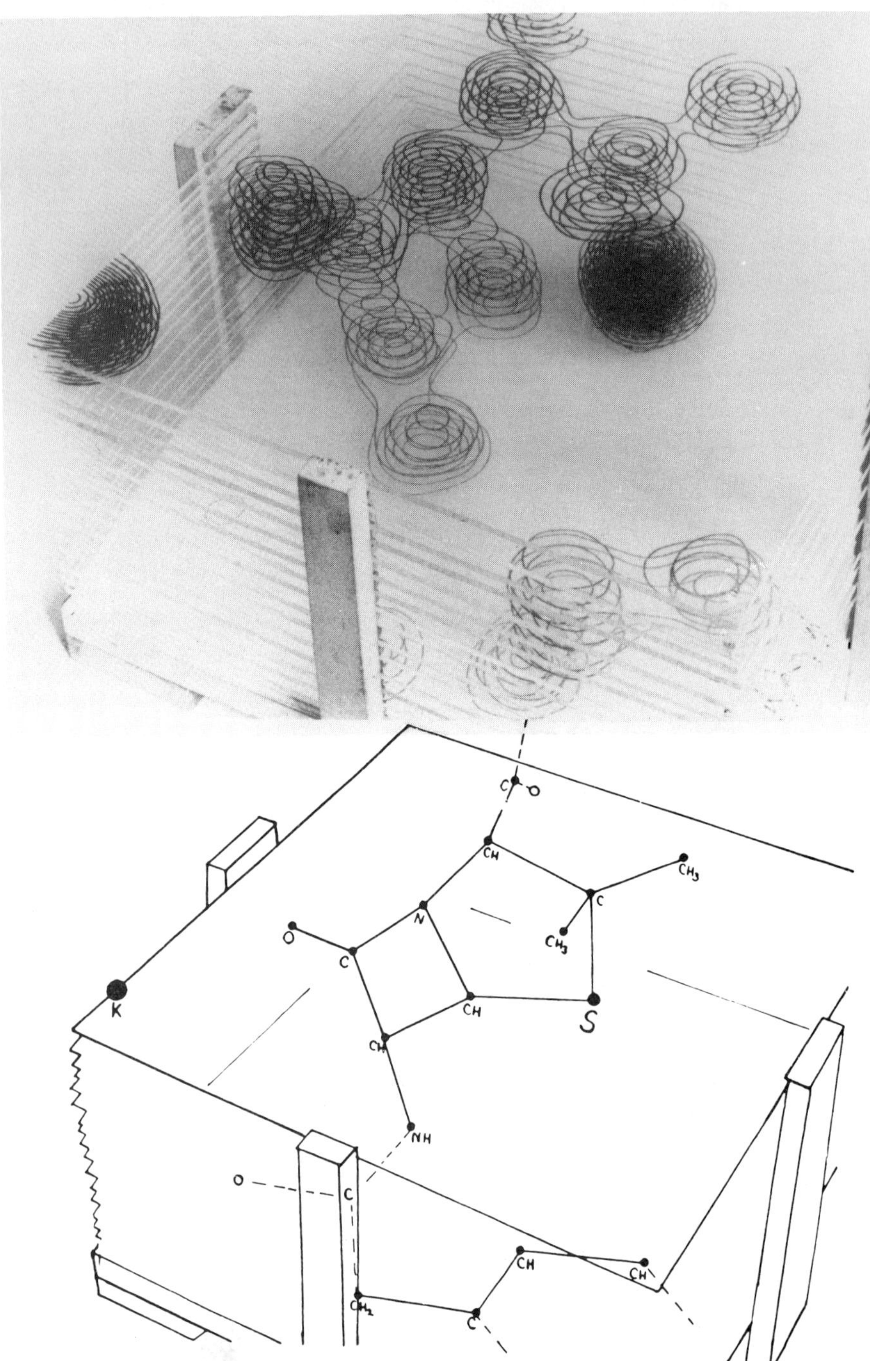

Plate 2.   Potassium benzylpenicillin - part of the three dimensional electron density map plotted on plastic sheets. The lower image presents an interpretation of the map in terms of the structure of penicillin.

I had three letters dated 14th February, 1945. The first
was from Charles, somewhat worried about the drastic rearrangement of his structure, the second written a few hours later,
also from Charles, with more time to think, enthusiastic, and
also favouring the β-lactam. The third was from Dr. W. T.
Astbury, saying that he had had a letter from Maureen (his
daughter at Somerville) "saying how excited and happy you are
about your penicillin results", "I feel you must have got the
thing out".

It was true. I did think we had essentially solved the
structure, but I thought we should not say definitely the β-
lactam structure was correct until we had refined the three
projections and checked the three dimensional arrangement of
all the atoms in the molecule. By the end of April these operations were completed; we were all agreed, and we thought we
should report it. When I went over early in May to report the
structure to Sir Robert Robinson, he was still unconvinced.
"You must be wrong" he said, "the molecule must have been inac-
tivated by X-rays in the crystal." So I said we could easily
check that - which we did with the help of Dr. Heatley. And I
thought to myself "If there had not been so much controversy, I
could have said I knew the structure in February."

We did go on to use the punched cards Comrie designed to
calculate the electron density in three dimensions for both
sodium and potassium benzylpenicillins, and we checked the determination of the electron density if wrongly placed atoms
were used in structure factor calculations. We made a model
for the first time on plastic sheets to show the three dimensional electron density (Plate 2). I used to enjoy very much
building up the sheets over the atoms - now no one shows such
models of simple structures, but they still are used in proteins.

Figures 5 and 6 show the final electron density projections and the positions of the atoms in the sodium and potas-

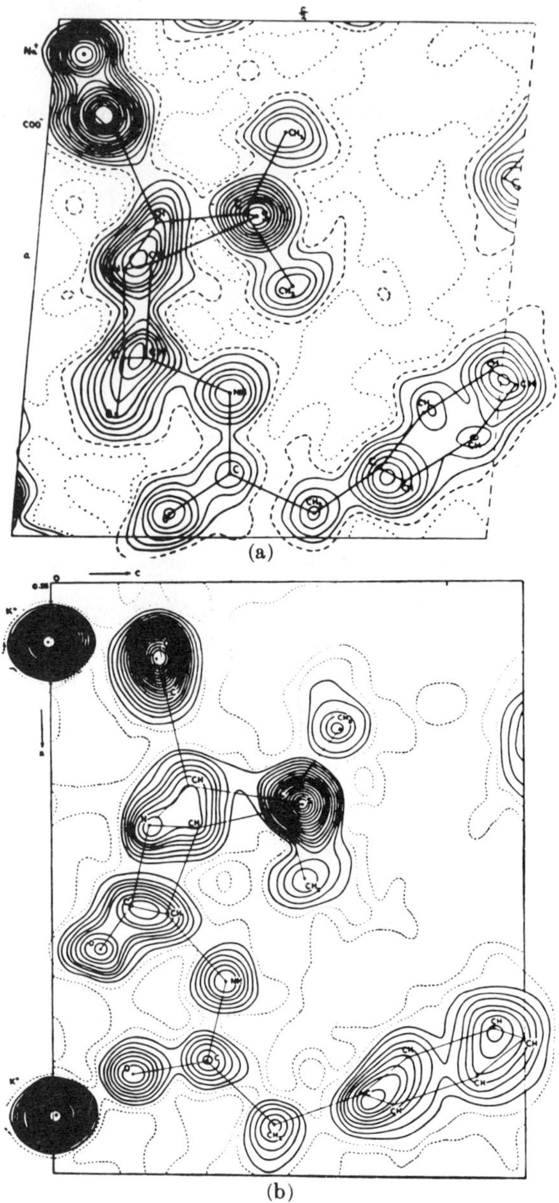

Figure 5. Refined electronn density projections for a) sodium benzylpenicillin and b) potassium benzylpenicillin.

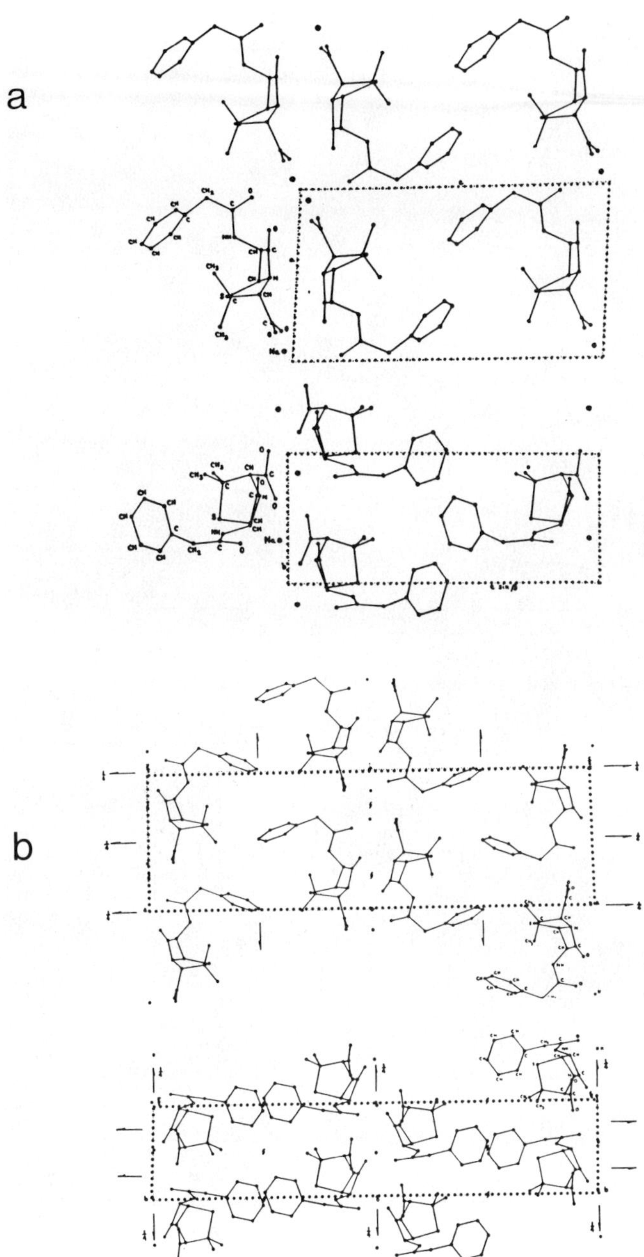

Figure 6 Atomic positions displayed in a and c projections for
a) sodium benzyl penicillins and b) potassium benzylpenicillins

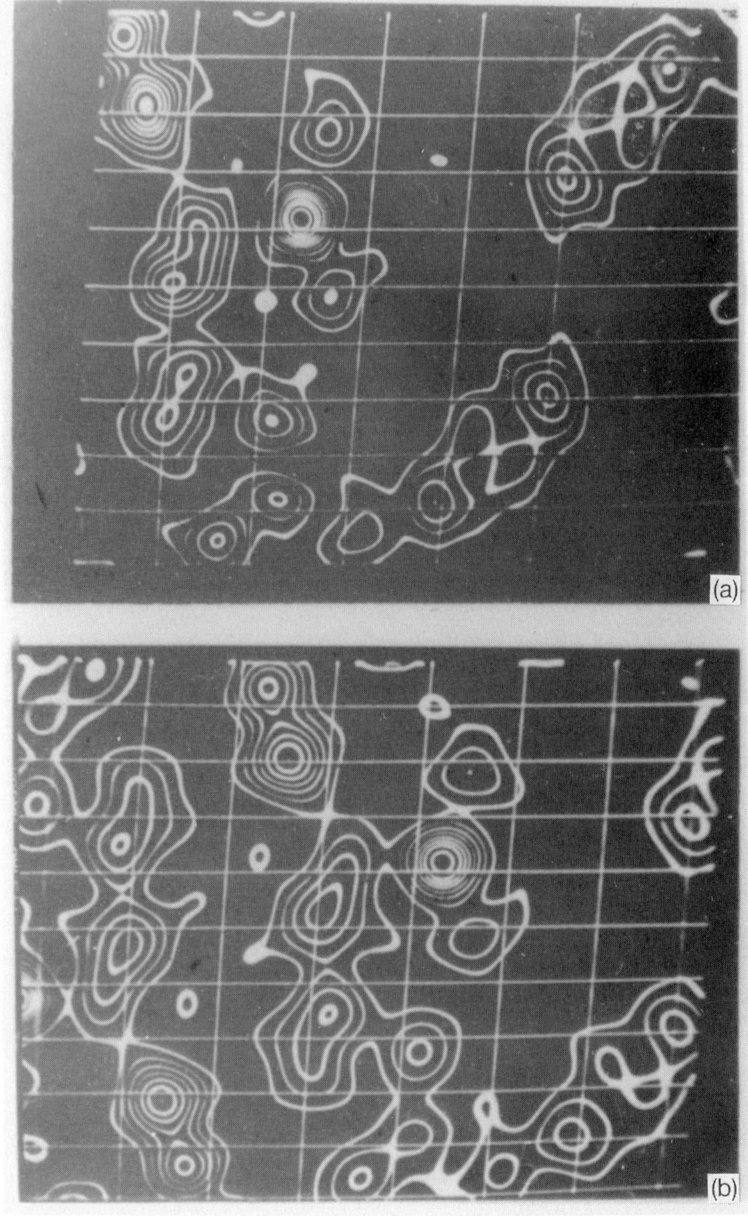

Plate 3. Sodium benzyl pinicillin - Electron density maps computed on XRAC. Map a) was calculated with the original phase set, b) with the extended phase set.

sium salts seen in the main a and b projections; these illustrate the character of the crystal structures, with alternate packing in layers of the ionic groups and the aromatic rings. The atomic positions are not very accurately defined, however, owing to the limitation of the original data collection, particularly of the sodium salt. In projections the β-lactam oxygen is poorly defined (Figure 5).

When first I came to the United States in 1947, I visited the different laboratories which had worked on penicillin, and I found in Peoria, Dr. Schiedtz who had grown a large crystal of sodium benzyl penicillin and taken a very good Weissenberg photograph of the h0l reflections. These extended to the limit of the copper K_α radiation, but were characteristically very weak beyond the limit of about 1 Å spacing to which the first measurements extended. I borrowed the photograph and estimated the new intensities, scaling them together with the old one. At Auburn, Alabama, I found Ray Pepinsky with his X-ray machine working, and we put my data on it to calculate a new projection. First we assigned the inner reflections their calculated signs, and then added in, one by one, the new terms giving them each time the sign that would reduce negative background density. The resulting map (Plate 3) was greatly improved and the β-lactam oxygen clearly defined as if every new term had contributed positive density to that one site. We later found a large potassium benzylpenicillin crystal and obtained better data on this structure - but still not such accurate data as would be measured today.

With new and more accurate data we ought to be able to find whether the bond lengths in different penicillins vary with their activity and stability. Activity probably depends on the interaction of the penicillins with the growing polypeptide system in the bacterial cell wall, or blocking the enzyme systems controlling cell growth, so that its understanding involves more than the structure of penicillin alone. Stability seemed possibly more likely to be correlated with precise bond lengths and molecular geometry.

So we began a study of the acid stable penicillins, particularly penicillin V, phenoxymethylpenicillin. E. C. Maslen and Sixten Abrahamsson solved the crystal structure of the free acid and showed it had an extended conformation and rather different apparent bond lengths within the β-lactam ring. But then we had qualms - salts should be compared perhaps with salts.

So the extension on which I hoped to embark at the beginning of this paper was the solution of the crystal structures of potassium and sodium phenoxymethylpenicillin, on which we had done a little preliminary work at Oxford. Dr. David Smith took up the challenge to find the structures in time for the Chinese meeting. He had no difficulty in obtaining crystals of the potassium salt and measuring the X-ray data by a diffractometer. The structure was complicated, four molecules in a triclinic cell. In Oxford Tom Blundell and Francisca Borras had found this complexity and made measurements from which they calculated a three dimensional Patterson. The general character of this showed that, in spite of triclinic symmetry, the packing was not unlike that in the original potassium benzylpenicillin - David Smith solved the structure from his data by direct methods, but has had great difficulty with refinement. The differences in the four molecules depend on different packing of the aromatic rings, which had not been sorted out by our meeting. As for the sodium salt, no crystal good enough for diffractometry had been grown by September, 1986.

So I found myself at the meeting still with my old problem unsolved. How lucky it was that crystals from our first 3 mgr of sodium benzylpenicillin grew so nicely. How lucky we were that potassium benzylpenicillin had a beautiful symmetric orthorhombic crystal structure and not the complex triclinic structure of potassium phenoxymethylpenicillin. There are still problems to be solved for the future.

Reference

CROWFOOT, D., BUNN, C.W., ROGERS-LOW, B.W. and TURNER-JONES, A., in "The Chemistry of Penicillin" (Clark, H.T., Johnson, J.H. and Robinson R. Eds) Princeton University Press, 1949.

Index